Khan Towhid Osman

Forest Soils

Properties and Management

Springer

Khan Towhid Osman
Department of Soil Science
University of Chittagong
Chittagong
Bangladesh

ISBN 978-3-319-02540-7 ISBN 978-3-319-02541-4 (eBook)
DOI 10.1007/978-3-319-02541-4
Springer Cham Heidelberg New York Dordrecht London

Library of Congress Control Number: 2013951530

Printed on acid-free paper

Springer is part of Springer Science+Business Media (www.springer.com)

Preface

In an account of Forest Soils (Chap. 147, Encyclopedia of Soil Science by Ratan Lal, published by CRC Press in 2005), Nicholas B. Comerford asked "Why study forest soils? If there is a subdiscipline called forest soils, why not study corn or tomato soils?" He replied himself by saying that studying forest soil science is justified because "a long-lived forest cover imposes a unique set of characteristics in which it grows". He further argued that "soil characteristics and processes important to the forest soil scientists are not of great concern to cropland soil scientists for whom the biological and economic time frames of food production are relatively short". Forest soil characteristics are not only unique but their interpretations are also different. A dense and thick vegetative cover, a continuous forest floor, very deep root systems, long duration (centuries for natural forests; decades for man-made forests) of occupancy of site, and natural succession of vegetation and soil as a coupled system make forest soils a distinctly different entity. For example, we do not get a good crop harvest from a poorly fertile soil unless we add adequate fertilizers to an agricultural soil, but very productive tropical rainforests often can occur in poorly fertile Oxisols and Ultisols. Consideration of both native soil fertility and nutrient recycling may better give a projection of the forest site quality and productivity.

Forests cover almost one-third land area of the world. Scientific investigations of forest soil properties have also been considerable in Europe, America, and the former USSR. But I could find only a few books on forest soils when I had to deliver lectures on the subject to my second semester B.Sc. (Hons.) students in the Institute of Forestry and Environmental Sciences as a guest faculty during 1990s. The following books were very good books of soil science for foresters: *Forest Soils and Forest Growth*, S. A. Wilde (1950) Chronica Botanica Co.; *Forest Soils: Properties and Processes*, K. A. Armson (1977) University of Toronto Press; *Properties and Management of Forest Soils*, William L. Pritchett and Richard F. Fisher (1987) Academic Press (latter edition *Ecology and Management of Forest Soils*, Richard F. Fisher and Dan Binkley 2000, 3rd Edition, Wiley), but I felt the necessity of a good book on forest soil science that would meet the requirements of general soil, forest soil, and environmental scientists.

Several dozens of good books are available on agricultural soils probably because agricultural soils are related to our food security, agricultural crop lands are intensively managed systems and farmers often find problems in their management. Now-a-days, forest plantations, particularly those for biomass energy production, are also being intensively managed with weeding, thinning, irrigation, and fertilizer addition. Good management without being well acquainted with the resource base is not possible. Forests are also useful in the context of global climate change adaptation and mitigation. These views inspired me to write this book *Forest Soils: Properties and Management* suited to students of soil science, forestry, and environmental sciences, and professionals of related disciplines. It has ten chapters. Chapter 1 gives background information on pedology. It will familiarize readers, may be even non-soil people, with the origin and development of soils. Chapters 2, 3 and 5 describe physical, chemical, and biological properties of forest soils and their impacts on forest vegetation. Chapter 4 narrates forest soil organic matter. Nutrient dynamics in forest ecosystems are explained and elaborated in Chap. 6. An account of soil classification systems of the world and a description of the types of soils in different forest types of the world are given in Chap. 7. Forest soil degradation due to natural and human induced forest disturbances is discussed in Chap. 8. Chapter 9 presents the relationship between forest soil quality and climate change. Forest management and silvicultural treatments related to seedling development, site preparation, planting, tending, weeding, fertilization, and irrigation and their effects on soil and environment have been discussed in considerable details in Chap. 10.

For last 20 years, I taught forest soils to undergraduate students of Soil Science and Forestry and Environmental Sciences. The general outline of the preset book has been tested on these students during this period. I am grateful to my students. Their reactions and responses inspired me to shape this book. I tried to incorporate recent information and I hope this book will be helpful to foresters, forest managers, environmental experts, and students.

Khan Towhid Osman

Contents

List of Abbreviations

AEC	Anion Exchange Capacity
BA	Basal Area
BNF	Biological Nitrogen Fixation
BSP	Base Saturation Percentage
CEC	Cation Exchange Capacity
DOM	Dissolved Organic Matter
DON	Dissolved Organic Nitrogen
ECe	Electrical Conductivity of Saturation Extract
ECEC	Effective Cation Exchange Capacity
ET	Evapotranspiration
FAO	Food and Agriculture Organization of the United Nations
GWT	Ground Water Table
IAA	Indole Acetic Acid
IPCC	Intergovernmental Panel on Climate Change
ISSS	International Soil Science Society
NPR	Nodule Producing Rhizobacteria
NRCS	Natural Resource Conservation Service
PGPR	Plant Growth Promoting Rhizobacteria
PHPR	Plant Health Promoting Rhizobacteria
SAR	Sodium Adsorption Ratio
SOC	Soil Organic Carbon
SOM	Soil Organic Matter
TEB	Total Exchangeable Bases
USDA	United States Department of America
VAM	Vesicular Arbuscular Mycorrhiza
WRB	World Reference Base (for Soil Resources)

1 Rocks, Minerals, and Soils

Rocks, minerals, and soils are components of the lithosphere (upper part of the earth's crust). Rocks and minerals provide raw materials (parent materials) for development of most of the soils of the world. The processes of disintegration (fragmentation, splitting, and detachment) and decomposition (formation of simpler substances from complex substances) are responsible for the production of loose and unconsolidated parent materials from rocks and minerals. These processes are collectively called weathering. Only a few soils develop from organic parent materials which are formed by deposition of residues of past vegetation, usually accumulated under wet conditions.

The unconsolidated products of weathering may remain in place of their origin (residual parent materials) or carried away to other places by various agents (transported parent materials). Soil formation may occur in both places by the action of climate, organism, and topography on parent materials through a long period of time. Natural processes of addition, removal, transformation, and translocation continuously operate for soil formation. These processes convert an apparently homogeneous and static parent material into a complex heterogeneous dynamic soil, differentiated into several layers or horizons, the collection of which is called the soil profile. Ultimately, the soil appears as a three-dimensional body on the surface of the earth.

1.1 Rocks

Rocks are defined as the naturally occurring hard and consolidated inorganic materials of the earth. Rocks are aggregates of minerals. A rock can be composed of one mineral or of a large number of different minerals. There are three major types of rocks on the basis of origin—igneous (formed by the solidification of magma and lava), sedimentary (primarily formed by consolidation and lithification of sediments), and metamorphic (formed by change of preexisted rocks in mineralogical composition or structure due to high temperature and pressure). Igneous and metamorphic rocks are weathered and form sediments which are deposited and lithified later into sedimentary rocks. On the other hand, flow of magma through preexisting igneous or sedimentary rocks gives rise to metamorphic rocks. Figure 1.1 represents the rock cycle indicating the processes of formation of different rocks.

1.1.1 Igneous Rocks

Igneous rocks are formed by the solidification of magma and lava. Magma is the hot, molten, viscous material within the earth's interior. If magma erupts through lines of weaknesses, fractures, fissures, joints, etc. on the earth's surface, it is called lava. Magma generally originates in the asthenosphere, at a depth between 100 and 200 km but possibly extending up to 700 km, and rises toward the surface of the earth. Magma, on its way to the surface through the crust, exerts tremendous pressure and heat, and modifies (or metamorphoses) the adjacent rocks. However, magma may solidify at very great depths forming intrusive or plutonic rocks. Granite, gabbro, and diorite are examples of intrusive rocks. When solidification occurs near the surface of the earth, the resulting rock is called hypabyssal rock. Magma solidifies in strands and pockets within the existing rocks, or between layers of sedimentary rocks of the crust. This gives rise to different rock structures. There are concordant and discordant intrusive rocks. Concordant rocks lie parallel to the layers of existing rocks. Discordant rocks are intrusive rocks which are not parallel to existing rock layers. Laccolith accumulates within layers of sedimentary rocks and up-arches the overlying sediments; it is concordant. Sills are thin concordant intrusive rocks between sedimentary rock strata. Batholith is a large discordant intrusive rock; discordant dike occupies cracks and parallel walls. Thus, igneous rocks are sometimes found to overlap sedimentary and metamorphic rocks.

Magma cools very slowly within the earth's interior because rocks are very poor conductor of heat. Coarse-grained

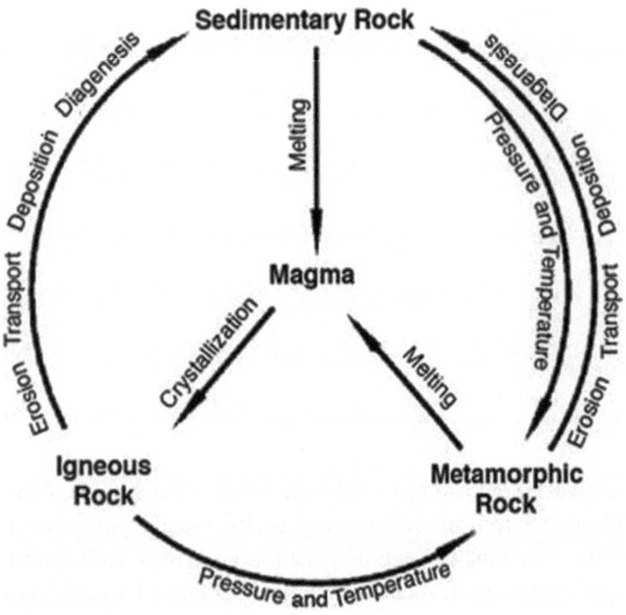

Fig. 1.1 Main types of rocks and the rock cycle

minerals can develop due to this slow cooling and solidification during formation of intrusive rocks. In entrapped magma with dissolved water, very large crystals (greater than 1 cm) of minerals are formed. This is known as pegmatitic texture of igneous rocks. On the earth's surface, lava cools relatively rapidly. Generally, fine-grained minerals are developed due to the rapid cooling of lava. Sometimes lava may cool so rapidly that crystallization process does not take place. The resulting rock has a glassy appearance. Obsidian is such a type of extrusive rock. Examples of extrusive rocks are basalt, rhyolite, and andesite; they are all fine-grained rocks. However, many igneous rocks are mixtures of coarse and fine crystals, and this texture is called porphyritic. This mixed texture probably results from two-step cooling.

Igneous rocks are also categorized into basic/mafic and acidic/silicic/felsic on the basis of percentage of silica (SiO_2) content (Table 1.1). Rocks with low silica content are called basic rocks. Basic rocks contain high ferromagnesian minerals. Therefore, basic rocks are also known as mafic rocks.

1.1.2 Sedimentary Rocks

The principal process of sedimentary rock formation is the lithification of sediments containing fragments of pre-existing rocks. The fragments may be rock pieces, such as pebbles of granite or shale and individual mineral grains, such as sand-sized quartz and feldspar crystals. Sedimentary rocks formed by the consolidation of rock fragments are called clastic sedimentary rocks. Clastic rocks may be classified as coarse-grained, medium-grained, and fine-grained sedimentary rocks. Coarse-grained sedimentary rocks include conglomerate formed by cementation of rounded gravels and breccia formed by cementation of angular fragments. Medium-grained sedimentary rocks include sandstone formed by cementation of sand grains and siltstone formed by cementation of silt-sized particles. There are some sandstones in which more than 90 % of grains are quartz; they are called quartz sandstone. Sandstones with more than 25 % of feldspar are called arkose. Fine-grained sedimentary rocks include shale formed by cementation of clay-sized particles having fissility and lamination, claystone formed by cementation of clay-sized particles but lacking fissility and lamination, and mudstone formed by cementation of silt- and clay-sized particles together.

Table 1.1 Categories of igneous rocks on the basis of silica content

Categories	Silica content	Rock examples
Ultrabasic/ultramafic	<45 %	Serpentine, periotite
Basic/mafic	45–55 %	Basalt, gabbro, dolerite
Intermediate	55–65 %	Amphibolite, andesite
Acidic/silicic/felsic	65–85 % or higher	Granite, rhyolite

Some sedimentary rocks are formed by precipitation from solution; they are called chemically formed sedimentary rocks. Chemically formed sedimentary rocks include carbonate-containing rocks, such as limestone, which is composed of the mineral calcite ($CaCO_3$), dolostone or dolomite composed of the mineral dolomite ($CaCO_3$. $MgCO_3$), chert composed of hard re-precipitated silica, rocksalt composed of halite (NaCl), etc. The third process of sedimentary rock formation is the consolidation of remnants of animals and plants. They are organically formed sedimentary rocks. Many limestones are formed by consolidation of $CaCO_3$-rich shells of marine organisms. Other organically formed sedimentary rocks include coal and peat. Plants, animals, and their remnants are sometimes preserved and silicified beneath earth's surface. These are called fossils.

1.1.3 Metamorphic Rocks

Metamorphic rocks are formed from preexisting rocks (igneous, sedimentary, or other metamorphic rocks) by the change in solid state either in texture or in mineralogical composition through the action of heat, pressure, chemically active solution, etc. Thermal or contact metamorphic rocks are formed by the action of heat during intrusion of magma and found along margins of intrusive bodies, while regional metamorphic rocks form by tremendous pressure due to deep burial and collision of tectonic plates. They occupy large areas and reveal much uplift and erosion after their formation. Some examples of metamorphic rocks are gneiss, slate, marble, schist, and quartzite.

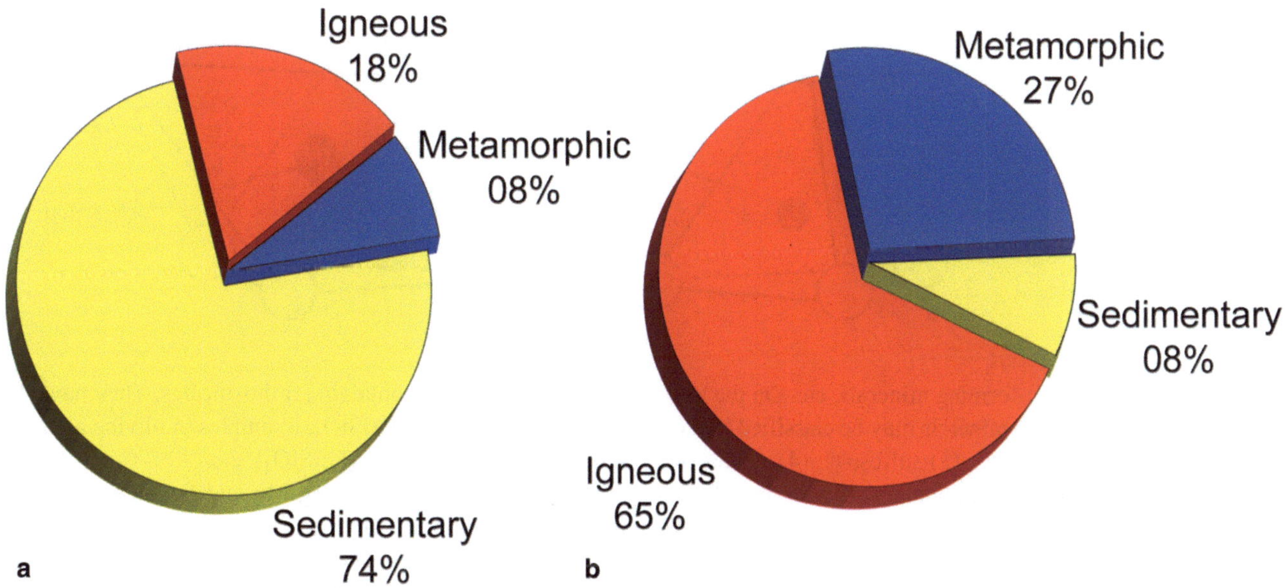

Fig. 1.2 Abundance of rocks in (**a**) earth surface and (**b**) earth crust

Examples of metamorphic rock include slate, schist, gneiss, marble, quartzite, etc. Slate is a fine-grained metamorphic rock created by minor metamorphism of shale or mudstone. This rock is characterized by the foliation of its mineral grains which causes it to have cleavage that is parallel. Schist is a medium- to coarse-grained foliated rock. Schists result from more intense metamorphism than slate. Gneiss is a coarse-grained metamorphic rock formed by metamorphosis of igneous rock. Foliation of quartz, feldspars, micas, and amphiboles into alternating light- and dark-colored bands is found in gneiss. Marble is a non-foliated metamorphic rock formed from limestone or dolomite. Quartzite forms from the recrystallization of silica found in sandstone.

Three-fourth of the earth's surface is composed of sedimentary rocks. Proportion of igneous rocks increases with the increase in depth beneath the surface. When the entire crust is considered, igneous rocks constitute about 65 %. Figure 1.2a shows relative abundance of different rocks within 5 km of the earth's surface. Sedimentary rocks constitute 74 %, including 52 % in shale, 15 % in sandstone, and 7 % in limestone. Igneous rocks contribute 18 %, including 15 % in granite and 3 % in basalt (Foster 1988). Figure 1.2b presents the abundance of rocks in the earth's crust.

1.2 Minerals

A mineral is defined as a naturally occurring, inorganic solid that has a definite chemical composition (or a definite compositional range) and a crystalline structure. If a naturally occurring inorganic solid, such as opal, is not crystalline, it is not considered as a mineral; it is a mineraloid. Minerals may be composed of only one or many elements. For example, diamond and graphite are composed of carbon alone. Many elements, including copper, gold, lead, mercury, nickel, platinum, silver, etc., may occur almost in the pure state in the earth's crust. They are called native elements. Some minerals may consist of even more than ten elements. Some minerals have the same chemical formula but differ in crystalline structure. For example, all of anatase, brookite, and rutile have the same the formula TiO_2, but they are different minerals because their crystalline structures are different. Most minerals of the earth's crust are compounds of silicon and oxygen in addition to other elements. The simplest form is the oxide of silicon, SiO_2. The most common mineral with this formula is quartz. However, the most abundant minerals in rocks of the earth's crust are the silicates. Since the third abundant element in the earth's crust is aluminum, most minerals are aluminosilicates of one or more of the remaining elements.

1.2.1 Classification of Minerals

Based on the mode of origin, minerals may be primary or secondary. Primary minerals are formed by crystallization during the solidification of magma and lava. In other words, minerals present in igneous rocks are primary minerals. There are two types of primary minerals—essential and accessory minerals. Essential minerals are those minerals which determine the character of the rocks, while the presence or absence of accessory minerals does not alter the character of the rocks. Secondary minerals are formed by alteration of the primary minerals and by resynthesis and recrystallization during weathering. Minerals abundant in rocks are rock-forming minerals; minerals concentrated somewhere in the crust that can be mined are ore-forming minerals; minerals

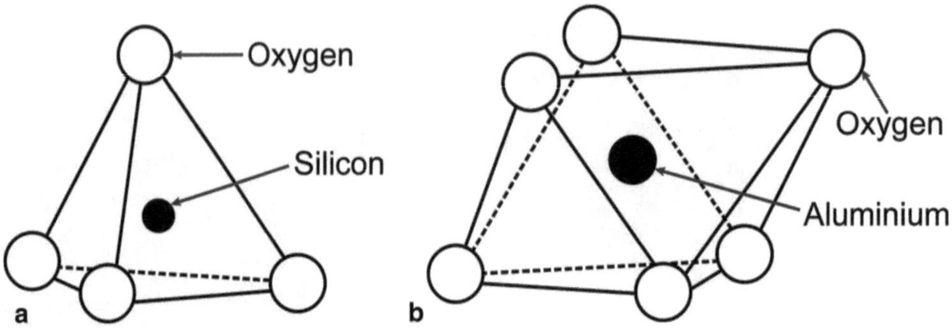

Fig. 1.3 Silica tetrahedron (**a**) and alumina octahedron (**b**)

abundant in soil are soil-forming minerals, etc. On the basis of chemical composition, minerals may be classified into (1) native elements, (2) sulfides, (3) sulfosalts, (4) oxides and hydroxides, (5) halides, (6) carbonates, (7) nitrates, (8) borates, (9) sulfates, (10) phosphates, and (11) silicates. However, minerals are classified in many different ways.

1.2.1.1 Silicate Minerals

The most prominent rock-forming silicate minerals are quartz, feldspars, pyroxenes, amphiboles, mica, and olivine. Quartz (SiO_2) is dominant in all types of rocks. Chalcedony has also the formula SiO_2 but differs in crystalline structure. It is dominant in chert and flint. Feldspars include framework silicates, which in addition to silicon, aluminum, and oxygen contains potassium, sodium, or calcium. Those that contain potassium are called potassium feldspars. Examples of potassium feldspars are orthoclase, microcline, and sanidine (All have the formula $KAlSi_3O_8$, but differ in crystalline structure. The most dominant rock-forming K- feldspar is orthoclase.) If sodium or calcium or both are incorporated into the feldspar crystal, the mineral is plagioclase feldspar. Plagioclase feldspars contain albite [$NaAlSi_3O_8$] and anorthite [$Ca_2Al_2Si_2O_8$], and a series of integrated minerals containing both sodium and calcium at various proportions. Pyroxenes have the general formula (Na, Ca, Mg, Fe, $Al)_2Si_2O_6$. They represent a group of important rock-forming silicate minerals of variable composition. Among these, calcium-, magnesium-, and iron-rich varieties predominate. Examples of pyroxenes are augite, enstatite, hepersthene, diopside, rhodonite, etc. The main rock-forming mineral of the family is augite [(Ca, Na)(Mg, Fe^{2+},Fe^{3+},Al)(Si, $Al)_2O_6$]. Amphibole is the family name of another very large and complex group of minerals. Examples of amphiboles are actinolite, aenigmatite, anthrophyllite, tremolite, hornblende, etc. Hornblende is the main rock-forming mineral of the family. Hornblende has the general formula [Ca_2,Na(Mg, $Fe^{2+})_4$(Al, Fe.Ti). (Al, $Si)_8O_{22}(O,–OH)_2$]. It has four end-members—common hornblende, tschermakite, edenite, and pargasite. Mica includes minerals with a sheet silicate structure. The two most common micas are biotite or black mica [$K(Mg, Fe)_3(AlSi_3O_{10})(OH)_2$] and muscovite or white mica [$KAl_2(AlSi_3O_{10})(OH)_2$]. Olivines are magnesium and magnesium-ferrous-containing orthosilicates. They have the general formula (Mg, Fe)SiO_4. Examples of olivine are forsterite (Mg_2SiO_4), fayalite (Fe_2SiO_4), etc.

Silicate minerals containing Mg and Fe such as biotite, pyroxene, amphibole, olivine, etc. are called ferromagnesian or mafic minerals. They are usually dark colored (dark greenish to black). They are also called dark minerals. Feldspars and muscovite are non-ferromagnesian minerals.

1.2.1.2 Non-silicate Minerals

Important non-silicate minerals include carbonate minerals such as calcite ($CaCO_3$), magnesite ($MgCO_3$), siderite ($FeCO_3$), dolomite ($CaCO_3.MgCO_3$), etc.; halide minerals such as halite (NaCl), synvite (KCl), fluorite (CaF_2), etc.; hydroxide minerals such as gibbsite [$Al(OH)_3$], boehmite [AlO(OH)], goethite [FeO(OH)], etc.; oxide minerals such as hematite (Fe_2O_3), magnetite (Fe_3O_4), rutile, (TiO_2), ilmenite ($FeTiO_3$), pyrolusite (MnO_2), zircon (ZnO_2SiO_2), etc.; phosphate minerals such as apatite [$Ca_5(PO_4)_3$.(F, Cl,OH)], vivianite [$Fe_3(PO4)_2.8H_2O$], etc.; sulfate minerals such as gypsum ($CaSO_4$. $2H_2O$), barite ($BaSO_4$) etc.; and sulfide minerals such as chalcopyrite ($CuFeS_2$), galena (PbS), pyrite (FeS_2), etc. The oxide, carbonate, and sulfate minerals mentioned earlier may be primary or secondary. They are found in abundance in some rocks. For example, limestone is dominated by calcite and the rock phosphate is mainly composed of apatites. The most important secondary minerals are the clay minerals such as kaolinite, smectite, vermiculite, illite, chlorite, etc. The secondary minerals are abundant in sedimentary rocks and soils.

1.2.1.3 Clay Minerals

Clay minerals are formed during biogeochemical weathering of silicate minerals. Clay minerals are phyllosilicates (from Greek *phyllon* meaning leaf), so called because they are foliated. They are constituted by the organization of silica tetrahedral and alumina octahedral sheets. The basic unit of the silica tetrahedral sheet is the silica tetrahedron composed of one silicon atom surrounded by four oxygen atoms (Fig. 1.3a). A tetrahedral sheet is built by sharing basal oxygen of one tetrahedron by its neighboring tetrahedron. On the other hand, the basic unit of the octahedral sheet is the alumina octahedron composed of one aluminum atom surrounded by six oxygen/hydroxyls giving an eight-sided geometric shape (Fig. 1.3b).

In the tetrahedral sheet, some Si may be substituted by Al, and in the octahedral sheet Al may be substituted by Mg or Fe. In some clay minerals, most Al atoms in the octahedral sheets are substituted by Mg. The process is known as isomorphous substitution because the substituting ion has similar ionic radii to the ion that is being substituted. Al octahedral sheets are known as dioctahedral and Mg octahedral sheets are known as trioctahedral sheets. Clay mineral crystals are formed by either one tetrahedral sheet bound to one octahedral sheet (1:1 type clay minerals) or two tetrahedral sheets sandwiching one octahedral sheet (2:1 type clay minerals). In some clays (such as chlorite), an additional magnesium hydroxide [$Mg_6(OH)_{12}$] sheet is bound to the 2:1 structure. It is, therefore, regarded as a 2:1:1-type clay mineral.

1:1-Type Clay Minerals A layer of 1:1 clay crystal consists of one tetrahedral and one octahedral sheet held tightly together by oxygen atoms mutually shared by silicon and aluminum cations in their respective sheets. These layers, in turn, are held together by hydrogen bonding. Consequently, the structure is fixed and no expansion ordinarily occurs between layers when the clay is wetted. Such a clay mineral is, therefore, known as non-expanding clays. The 1:1-type clay includes the kandite group [$Al_4Si_4O_{10}(OH)_8$], the chief members of which are kaolinite, dickite, nacrite, and halloysite; it is formed mainly by the decomposition of orthoclase feldspars.

2:1-Type Clay Minerals The 2:1 clay crystal units are characterized by an octahedral sheet sandwiched between two tetrahedral sheets. Three general groups have this basic crystal structure: smectite, vermiculite, and illite. Smectites and vermiculites are expanding type of clay, while illite is non-expanding. The smectite group has the general formula $(Na, Ca)_{0.33}(Al, Mg)_2Si_4O_{10}(OH)_2.nH_2O$. It includes montmorillonite, bentonite, nontronite, hectorite, saponite, and sauconite; it is formed by the alteration of mafic igneous rocks rich in Ca and Mg. These minerals have high swelling/shrinking potential. Wide cracks commonly form when smectite-dominated soils (e.g., Vertisols) are dried.

Vermiculite group has the general formula $(Mg, Fe,Al)_3(Al, Si)_4O_{10}(OH)_2.4H_2O$. In the tetrahedral sheet of most vermiculite, considerable substitution of aluminum for silicon has taken place. For this reason, a high proportion of net negative charge develops in vermiculite. There are many strongly adsorbed cations including Ca^{2+} and Mg^{2+} and water molecules in the interlayer space of vermiculites. They bind the units tightly together. The degree of swelling and shrinking is, therefore, considerably less in vermiculites than in smectites. Vermiculites are considered limited-expanding type of clay minerals, expanding more than kaolinite but much less than the smectites.

Illite has the general formula $(KH_3O)(Al, Mg,Fe)_2(Si, Al)_4O_{10}(OH)_2.H_2O$. It includes hydrous micas, phengite, brammalite, celadonite, and glauconite; it is formed mainly by the decomposition of some micas and feldspars. They are predominant in marine clays and shales. The K^+, Ca^{2+}, and Mg^{2+} cations in interlayers of illite prevent the entrance of H_2O into the structure. So, the illite clays are non-expanding clays.

2:1:1-Type Clay Minerals This silicate group is represented by chlorites having the general formula $(Mg, Fe)_3(Si, Al)_4O_{10}(OH)_2{\cdot}(Mg, Fe)_3(OH)_6$. In chlorite clay crystals, the 2:1 layers alternate with a magnesium-dominated trioctahedral sheet. Magnesium replaces a considerable amount of aluminum in the octahedral sheet of the 2:1 layer. Thus, the crystal unit contains two silica tetrahedral sheets and two magnesium-dominated trioctahedral sheets giving rise to the term 2:1:1-type structure. Important members of the group are clinochlore, chamosite, nimite, pennantite, and cookeite.

1.2.1.4 Other Clays

Sesquioxide clays Sesquioxides form clay-sized precipitates predominantly in tropical soils. Sesquioxides are mixtures of aluminum hydroxide [$Al(OH)_3$] and iron oxide (Fe_2O_3), or iron hydroxide [$Fe(OH)_3$]. Examples of iron and aluminum oxides common in soils are gibbsite ($Al_2O_3.3H_2O$) and goethite ($Fe_2O_3.H_2O$). These substances form amorphous to crystalline clay-sized materials.

Allophanes Allophanes are amorphous (non-crystalline) clays common in soils forming from volcanic ash. They have high cation and anion exchange capacity. Almost all of their charge is from accessible hydroxyl ions (OH–), which can attract a positive ion or cause the loss of the attached H^+. These clays have a variable charge that depends on H^+ in solution. Types and contents of clay minerals vary widely in different climates and soils. Some examples are given here.

1.2.1.5 Abundance of Minerals in the Earth's Crust

About 3,000 minerals have so far been identified, but most of them are rare. Only about 100 minerals constitute the major mineral components of rocks. Most of these minerals are also relatively uncommon. The bulk of the earth's crust is made up of a few silicate minerals. Figure 1.4 shows the relative abundance of minerals in the earth's crust. Feldspars constitute more than 50 %, including 39 % as plagioclase and 12 % as orthoclase. Quartz and pyroxene also contribute similar proportions to orthoclase. Olivine constitutes only 3 %. This could be the reason behind the high proportions of quartz and feldspars found in soils.

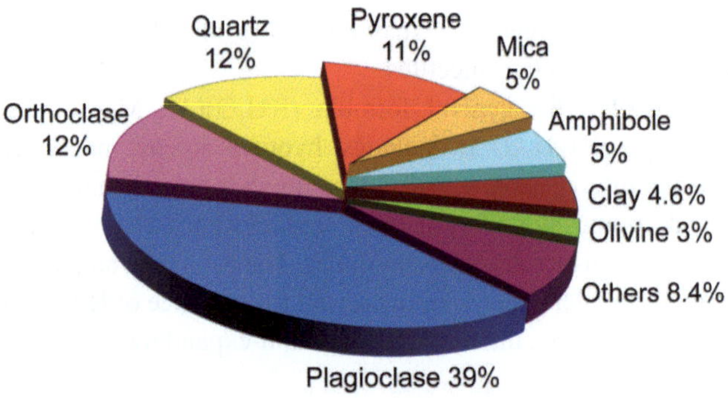

Fig. 1.4 Relative abundance of minerals in the earth's crust

1.3 Weathering

The processes of disintegration and decomposition of rocks and minerals are collectively called weathering. Rocks and minerals are weathered by the action of natural forces like heat, water, wind, glacier, organisms, chemical reactants, etc. Because the changes in rocks and minerals are either physical or chemical, there exist two types of weathering, physical weathering and chemical weathering. A large rock mass is fragmented into smaller pieces due to physical weathering. By chemical weathering, the chemical and mineralogical composition of rocks is altered, or a mineral is decomposed into soluble and insoluble products or a new mineral is formed. An additional type of weathering, called biological weathering, is sometimes recognized, but this is actually physical or chemical weathering caused by biota.

The principal agents of physical weathering are heat, water, wind, glaciers, and organisms. The most noticeable physical weathering is caused by temperature. So, physical weathering may be divided into thermal weathering and mechanical weathering. Chemical weathering occurs by reactions of minerals with acids, bases, salts, ions, etc. released by decomposition of other minerals, or even with water. Organisms also produce substances such as carbonic acid and other organic acids which actively participate in chemical alteration of rocks. Chemical weathering is, therefore, sometimes divided into geochemical weathering and biochemical weathering. Chemical weathering may also be termed as biogeochemical weathering. Weathering is the process of forming parent materials of mineral soils. However, weathering and soil genesis may happen simultaneously.

1.3.1 Physical Weathering

Thermal Weathering Thermal weathering is caused by expansion and contraction of rocks due to variation in temperature in the environment. All rocks and minerals expand when heated by solar radiation and contract when cooled. Rocks expand during the day and contract at night. More expansion occurs during summer than in winter. However, not all parts of the rock body are equally exposed to heat. So, different parts of rock expand and contract at different rates. Different minerals expand and contract at different rates as well. Since a rock is usually composed of a variety of minerals, and they expand and contract differently, internal pressures are created when rocks expand or contract. Due to this internal pressure, a rock on the earth's surface is ultimately fragmented after prolonged exposure to fluctuations of temperature. Another process of thermal weathering is the exfoliation which occurs in layered rocks, rocks with approximately homogeneous composition at different zones, or rocks expanding almost at similar rates up to a depth. The outer part of a boulder is expanded and contracted usually at higher rates than the inner parts and may be detached in thin layers. In this process, a thin layer of rock is detached from the rock body. This detachment resembles the peeling up of onions.

Mechanical Weathering Mechanical weathering is the fragmentation of rocks by the action of water, wind, glacier, gravity, etc. Water may cause physical weathering in different ways. First, rocks may absorb water and swell when wetted and shrink when dried. Expansion and contraction in volume due to alternate wetting and drying create stresses within rocks. As a result, rocks are fragmented. Second, water may accumulate in fractures and joints of rocks. Ice is formed when the temperature is lowered below the freezing point. Ice has a greater volume and it exerts tremendous lateral pressure on surrounding rocks. Because of this pressure, already-fractured rocks are fragmented into several pieces. This process is called frost wedging. Another type of mechanical weathering by freezing water is known as frost heaving, by means of which rock fragments are lifted upwards. Third, flowing water has a considerable cutting action. Presence of suspended materials and rock fragments in water highly increases the capacity of this cutting action. By the water flow, the suspended materials roll and disintegrate themselves as well as the walls and beds of rocks through which they travel. Large bodies of ice slowly moving along slopes are called glaciers. Glaciers forming on already-weathered surfaces contain variously sized rock fragments, including boulders. When glaciers flow down the slope, ice begins to melt and the rock fragments shatter one another. The flowing glaciers tend to or do smash whatever rock they find obstructing their way. Wind alone is a weak weathering agent. Presence of suspended particles increases the abrasive effect of wind. Abrasion by wind over a long time may considerably weather rock surfaces. Large rock masses at higher elevation are pulled and detached by gravity. Such materials deposited on piedmont and foothill are called colluvium. Plants may extend their roots in rock fractures and crevices. When they grow, the roots exert lateral pressure. Many burrowing animals ingest sand and gravel and as this material passes through their gut, it gets physically comminuted (Suzuki et al. 2003).

1.3.2 Chemical Weathering

The principal processes of chemical weathering are solution, hydration, hydrolysis, oxidation, reduction, carbonation, and other acid reactions.

Solution Minerals are not ordinarily soluble in water. Natural water contains some dissolved acids which enhance the dissolving power of water. Minerals on prolonged wetting in acidic water may considerably be dissolved. Gypsum on wetting for a long period of time may be dissolved with the production of soluble calcium and sulfate ions.

$$\underset{\text{(Gypsum - insoluble)}}{CaSO_4.2H_2O} \xrightarrow{Water} \underset{\text{(Soluble calcium and sulfate ions)}}{Ca^{2+} + SO_4^{\;2-} + 4H_2O}$$

Hydration Hydration is the addition of water to the mineral molecule. Iron and aluminum oxides are readily hydrated in soil.

$$\underset{\text{(Hematite - red)}}{Fe_2O_3 + 3H_2O} \xrightarrow{Hydration} \underset{\text{(Limonite - yellow)}}{Fe_2O_3.3H_2O}$$

Hydrolysis Hydrolysis is probably the most effective chemical weathering process by which minerals are decomposed. During hydrolysis, water molecule splits into hydrogen and hydroxyl ions and the hydrogen replaces a cation from the mineral structure. Due to this replacement, the mineral becomes susceptible to decomposition by further hydrolysis or other chemical reactions.

$$\underset{\text{(Orthoclase - insoluble)}}{KAlSi_3O_8 + H_2O} \xrightarrow{Hydrolysis} \underset{\text{(Hydrogen aluminium silicate - insoluble)(soluble)}}{HAlSi_3O_8 + KOH}$$

On further hydrolysis,

$$\underset{\text{(Hydrated Al oxide - insoluble)}}{HAlSi_3O_8 + H_2O} \xrightarrow{Hydrolysis} \underset{\text{(Silicic acid - soluble)}}{Al_2O_3.3H_2SiO_4}$$

Oxidation and reduction Some minerals are oxidized; some others are reduced depending on the conditions.

$$\underset{\text{(Pyrite - insoluble)}}{FeS_2 + O_2} \xrightarrow{Oxidation} \underset{\text{(Ferrous sulfate - soluble)}}{FeSO_4}$$

$$\underset{\text{(Goethite - insoluble)}}{FeOOH} \xrightarrow{Reduction} \underset{\text{(Ferrous oxide)}}{FeO + O + H_2O}$$

Carbonation Carbonation is a very active and common chemical weathering process. Most carbonation processes are biochemical in nature. Carbonic acid is naturally produced by dissolution of CO_2 in water. Carbon dioxide is available from the atmosphere and as products of microbial and root respiration. Carbonic acid dissolves many minerals.

$$\underset{\text{(Calcite - insoluble)}}{CaCO_3 + H_2CO_3} \xrightarrow{Carbonation} \underset{\text{(Soluble)}}{Ca^{2+} + HCO_3^-}$$

The processes of chemical weathering may operate simultaneously or one process may follow another. For example, hydration and carbonation may occur at the same time, or oxidation may follow hydrolysis as— (Olivine) (Serpentine) (Hematite)

$$\underset{\text{(Olivine)}}{MgFeSiO_4 + H_2O} \xrightarrow{Hydrolysis} \underset{\text{(Serpentine)}}{H_4Mg_3Si_2O_9 + SiO_2 + FeO} \xrightarrow{Oxidation} \underset{\text{(Hematite)}}{Fe_2O_3}$$

1.3.3 Biochemical Weathering

Plants and their associated microbiota directly impact chemical weathering by generating chelating compounds, by modifying pH through production of CO_2 and organic acids, by altering the exposed surface areas of minerals, via nitrification and by affecting the residence time of water (Drever 1994). Of these, the production of organic acids may be the most important contribution of biochemical weathering in soils. These acids can either be released directly from the organism or be derived as by-products of organic decomposition.

1.3.4 Weatherabilty of Minerals and Weathering Products

Under similar conditions different minerals weather at different rates. Generally, the more complex the mineralogical composition is, the more easily the mineral weathers. Plagioclase weathers more rapidly than orthoclase. Hornblende weathers fairly rapidly. Quartz is very resistant to weathering.

Products of weathering depend on the minerals themselves and the conditions of weathering (Ollier 1975). Minerals differ in the kinds of reactions they undergo. In most instances, calcite dissolves completely and the products are all soluble. Silicates are comparatively less susceptible to chemical weathering. When they weather, they produce some soluble substances, including bases and some other minerals. Quartz is the most resistant to weathering and tends to accumulate as weathering proceeds. Feldspars weather more easily and form clay minerals. Ferromagnesian silicate minerals produce insoluble iron oxides and hydroxides and some clays. Most other chemical components are dissolved away. Intensity of weathering determines to a great extent the products of weathering. Most of the chemical weathering processes occur in the presence of water. The intensity of chemical weathering increases with increasing rainfall and temperature. Chemical weathering is very intense in warm moist climate. Therefore, starting with the same rock, different weathering products will be obtained under different climatic conditions. Commonly quartz, feldspars, and mica dominate among the primary minerals in soils under most situations. Presence of some feldspars, mica, and ferromagnesian minerals in soils indicate its inherent capacity to supply nutrients through weathering process. Soil clays are either produced by alteration of primary minerals or synthesized during weathering and soil genesis. They include clay minerals and other clay-sized materials, such as allophanes (amorphous clays), hydrous oxides of Fe and Al, etc.

1.4 Soils

Soils are natural unconsolidated materials on the surface of the earth. Soils have evolved through natural processes as a dynamic and a three-dimensional body. Soils have three phases—solid, liquid, and gas. The solid phase contains organic and inorganic materials intimately mixed together by natural processes. The liquid phase consists of water with dissolved salts, acids, bases, and ions. The gaseous phase consists of soil air which renews from time to time from the atmosphere. Water and air occupy, respectively, small and large pores of the soil left by the close packing of soil solids. So, a soil has four major components—mineral matter, organic matter, water, and air. Soils support plant growth by providing anchorage, nutrients, water, air, and warmth and protecting from toxins. In addition, soils have many crucial ecological functions: it is a transformer of energy; it is a recycler of materials; it is a purifier of water; and above all it functions as an ecosystem component.

Kellogg (1960) defined soil as, "Soil is a collection of natural bodies occupying a portion of the earth surface that supports plant growth and that has properties due to the integrated effect of climate and vegetation acting upon parent material, as conditioned by relief, over a period of time." More recently, USDA and NRCS (2003) defined soil as, "Soil is a natural body comprised of solids (minerals and organic matter), liquid, and gases that occurs on the land surface, occupies space, and is characterized by one or both of the following: horizons, or layers, that are distinguishable from the initial material as a result of additions, losses, transfers, and transformations of energy and matter or the ability to support rooted plants in a natural environment."

1.4.1 Pedon

A soil has three dimensions—length, breadth, and depth. There are variations in soil properties in all these directions. After some distance in the lateral direction, a different soil having different properties may occur. A soil has a lower limit too. The upper part of the regolith (unconsolidated and loose materials above bedrock) is soil. Therefore, a soil occupies a volume. The smallest volume of soil having similar characteristics throughout the lateral dimensions is called a pedon. Simonson (1978) defined pedon as the smallest effective unit of soil. A pedon is actually an arbitrary unit chosen for examination of soil in field and sampling for study. A pedon may have an area of 1–10 square meters and it is usually hexagonal in shape. It extends in depth into the parent material from which the soil has developed. A group of similar pedons that occur together in a landscape is called a polypedon. A polypedon represents a soil individual. Different soils in a landscape have polypedons of different pedon characteristics.

1.4.2 Soil Profile

The pedon characteristics are exhibited by the exposed sides of the pedon. An exposed side of the pedon actually represents the soil profile. Soil profile is defined as a vertical section of soil from surface down to the parent material. The horizontal layers in a soil profile roughly parallel to one another and formed by pedogenic processes are known as horizons. Horizons are, however, different from layers. A soil layer is any stratum which is not formed by pedogenic processes. A layer is generally formed by geological processes or by human activity. There are five master horizons denoted by O, A, E, B, and C. O is an organic horizon found in some temperate forest soils; most tropical soils lack it. In forestry literature, the O horizon is designated as 'forest floor'. A mineral soil may contain all these horizons or may lack one or more of them. Below the C horizon, there is usually the bedrock which is denoted by R. It is not a soil horizon. Rocks are weathered to parent materials which give rise to the soil on further pedogenic transformation. Figure 1.5 shows an arbitrary soil profile with different master horizons.

Characteristics of master horizons In Fig. 1.5, each master horizon is shown in the relative position in which it occurs in a soil profile. The horizon boundaries in the field soil are not very regular, smooth, and distinct as shown in Fig. 1.5. Generally, the upper horizon diffuses into the lower horizon. Sometimes, gradual changes from one master horizon to another give rise to transitional horizons. These are named with two letters, for example, AB, BA, and BC. Subordinate divisions of master horizons are named by adding lowercase letters, for example, Ap, Bt, and so on. Thick horizons may

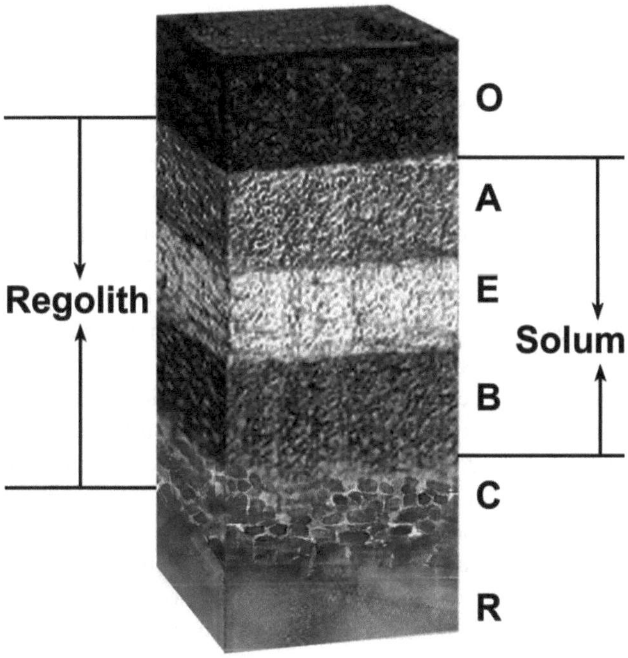

Fig. 1.5 An arbitrary soil profile

be subdivided using Arabic numerals, such as A1 and A2, or Bw1, Bw2, and Bw3. Principal characteristics of the master horizons are summarized further.

O Horizon O horizon consists of dead organic residues including leaves, twigs, barks, branches, flowers, fruits, etc. in various stages of decomposition. Generally, a thick O horizon develops under coniferous forests of the temperate region. The O horizon may be divided into three subhorizons, Oi (the uppermost layer of fresh organic residues), Oe (the middle partially decomposed layer), and Oa (completely decomposed organic layer). In forestry literature, they are correspondingly called L (litter), F (fermented), and H (humus) layers (also see Sect. 4.5).

A Horizon The A horizon is the surface horizon of most mineral soils. It is characterized by the intimate mixing of organic matter with mineral matter and considerable eluviation. The A horizon is also typified by a granular or fine blocky structure and friable consistency. A horizon is the most fertile portion of the soil body. It provides a favorable environment for root growth and biological activity.

E Horizon The E horizon is the mineral horizons in which the main feature is loss of silicate clay, iron, aluminum, or some combination of these, leaving a concentration of sand and silt particles of quartz or other resistant materials.

B horizon B horizons form below an A, E, or O horizon and are dominated by obliteration of all or much of the original rock structure and show one or more of the following: illuvial concentration of silicate clay, iron, aluminum, humus, carbonates, gypsum, or silica, alone or in combination; evidence of removal of carbonates; and residual concentration of sesquioxides.

C Horizon The C horizon is weathered, unconsolidated geologic material below the A or B horizon. Anything that one can dig with a spade, which has not been changed very much by the soil-forming processes, is considered the C horizon.

Some subordinate designations are made by specific small letters followed by the master horizon capital letters (Soil Survey Staff 1993, 1994) to indicate the main feature of the horizon. Thus, Ap is used to denote A horizon modified by ploughing, Bk is a B horizon having accumulation of alkaline earth carbonates, and so on.

Small letters for subordinate designation of horizons

a Highly decomposed organic matter
b Buried genetic horizon
c Presence of concretions or nodules
d Physical root restriction
e Intermediately decomposed organic matter
f Frozen soil
ff Dry permafrost
g Strong gleying
h Accumulation of illuvial organic matter
i Slightly decomposed organic matter
k Accumulation of carbonates
n Accumulation of sodium
o Residual accumulation of sesquioxides
p Disturbance due to plowing
q Accumulation of silica
s Accumulation of illuvial sesquioxides and organic matter
t Accumulation of silicate clay
y Accumulation of gypsum

1.5 Soil Formation

1.5.1 Soil-Forming Factors

Materials and conditions that determine the course of soil development and affect the characteristics of soils are called soil-forming factors. Jenny (1941) identified five soil-forming factors—climate, organism, relief (topography), parent material, and time—and proposed an equation to show relationships of a soil property with these soil-forming factors. It is known as fundamental soil-forming equation. In this equation, a soil property is regarded as a dependent variable, which depends on the soil-forming factors—the independent variables. The soil-forming equation may be written as—where S = any soil property, Cl = climate, o = organism, r = relief, or topography, p = parent material, t = time, and *f* = function.

Climate and organisms are said to be active factors because their effects are direct and more noticeable, and some soils are associated with a particular climatic and vegetational zones. Relief, parent material, and time are passive factors because their effects are less apparent. The fundamental equation of soil formation gives the impression that any soil property can be estimated from the variables—climate, organism, parent material, relief, and time. But some of these factors cannot be quantified. So, the above equation remains to be descriptive rather than determinative.

The soil-forming factors act individually and jointly, and the property of a soil is determined by the resultant of the interactions of these factors. The soil-forming equation expresses that similar soils will develop under similar sets of the soil-forming factors. Jenny (1980) states that for a given combination of Cl, o, r, p, and t, the state of the soil system is fixed; only one type of soil exists under these conditions. Any particular combination of these factors will give rise to a certain soil-forming process, a set of physical, chemical, and biological processes that create a particular soil (van Breemen and Buurman 2002). The equation shows that under similar sets of organism, parent material, relief, and time, differences in properties between two soils would be brought about by climate alone. It also indicates that similar soils may develop under dissimilar sets of conditions at two separate places provided that the resultant effects of the interacting factors are the same at both places. So, if resultant effects of the interactions of the factors are regarded as causal, dividing soil-forming factors into active and passive appears to be impractical. Buol et al. (1997) regarded climate, parent material, topography, and land use as the driving variables in soil formation. On the other hand, vegetation, animals, microbes, and physical and chemical properties of soil constitute the state variable. Driving variables interact with the state variables over time so that soil and vegetation development proceed together.

1.5.1.1 Climate as a Soil-Forming Factor

Conditions of rainfall, temperature, humidity, solar radiation, day-length, etc. at a place over a long period is called climate. Climatic conditions at a place depend on geographic location, elevation, velocity and direction of prevalent wind, position of land in relation to sea, ocean, and mountains, characteristics of oceanic current, etc. The two most important elements of climate that affect soil formation are rainfall and temperature. Rainfall and temperature influence the kind and degree of weathering, nature of weathering products retained or lost, production of biomass and decomposition of organic matter, biogeochemical processes operating in the parent materials, circulation of materials within the soil, etc.

Precipitation effects Water acts as a reactant and a medium of translocation. As rainfall increases, rate of weathering, depth of weathering, diversity of weathering products, formation of clay, release of soluble substances, and organic matter production increase. However, leaching of materials in solution and suspension also increase when precipitation exceeds evapotranspiration. In very dry conditions, physical weathering is relatively intense and the resulting parent material is sandy. Due to scarcity of water there is little chemical weathering and inadequate leaching. In moister regions where there is some rainfall but the potential evaporation is higher than precipitation, there is some chemical weathering and formation of soluble salts but due to insufficient water to leach them downward (and by capillary rise of groundwater), a salt crust may develop on the surface soil. Sometimes, $CaCO_3$ may also accumulate on the surface (lime layer). In further moister areas, lime and salt layers are present at higher depth below the surface or may be lost. In very moist conditions, such as in humid tropics, both formation and leaching of clay, soluble salts, humus, etc. are high. Thus, contents of clay, organic matter, cation-exchange capacity (CEC), and available nutrients increase as rainfall increases up to a limit and decrease afterwards. Under intense weathering in humid conditions, feldspars and other easily weatherable minerals are removed causing enrichment of quartz. Some soluble Fe/Al may reprecipitate as their hydrous oxides.

Soil Moisture Regimes Recognized by Soil Survey Staff 1999

- Aquic: saturated with reducing conditions most of the year
- Udic: soil moisture control section is dry for <90 cumulative days per year
- Ustic: soils that remain dry for >90 cumulative days per year
- Aridic: soils that remain dry in all parts for > half of the year

Temperature effects Temperature regulates the rate of chemical reactions. For every 10 °C rise in temperature, the rate of chemical reactions is almost doubled. Increased weathering and clay formation occur with an average increase in soil temperature at optimum soil moisture level. Both organic matter production and decomposition increase with the increase in soil temperature. The organic matter content of soil is the balance between organic matter production and mineralization. Both production and mineralization of organic matter are low in boreal forest soils; they are usually rich in organic matter because the microbial decomposition rate is slower. Tropical soils are low in organic matter, although tropical rain forests produce more organic matter.

This is because of a relatively higher rate of decomposition at elevated temperature. Temperate region forests have a long dormant winter period, resulting in less plant growth and decomposition of organic matter than that in tropical rain forests. Therefore, the content of soil organic matter tends to be similar in the temperate rain forests and tropical rain forests.

Soil Temperature Regimes Recognized by Soil Survey Staff 1999

- Cryic—mean annual temperature < 8 °C
- Frigid—mean annual temperature <8 °C; difference between mean summer and mean winter temperature is > 6 °C
- Mesic—mean annual temperature > 8 °C and < 15 °C; difference between mean summer and mean winter temperature is >6 °C
- Thermic—mean annual temperature >15 °C and <22 °C; difference between mean summer and mean winter temperature is >6 °C
- Hyperthermic—mean annual temperature >22 °C; difference between mean summer and mean winter temperature is >6 °C

Climate and distribution of Soil Orders A variety of soil orders (USDA Soil Taxonomy; Soil Survey Staff 1999) may be encountered in any climatic zone of the world. For example, all of the following soil orders Alfisols (fine-textured soils with base status >35 % BSP), Andisols (soils characterized by andic materials, i.e., originating from volcanic ash, pumice, cinders, etc.), Entisols (soils that have little or no evidence of horizon differentiation), Histosols (soils dominated by organic soil materials having >20 % organic carbon), Inceptisols (soils that exhibit only the beginning of soil-profile development), Mollisols (very dark-colored, base-rich, mineral soils of the grasslands), Oxisols (soils having the upper boundary of an oxic horizon; sandy loam or finer texture with <10 % weatherable minerals), Ultisols (highly weathered soils that have low base saturation; <35 % BSP), and Vertisols (clayey soils that have deep, wide cracks for a considerable time of the year) are found in the humid tropical region. Again, many soil orders such as Andisols, Entisols, Histosols, Inceptisols, and Vertisols occur in almost all climatic regions. However, earlier pedologists noticed that some particular soil types are associated with particular climatic and vegetational zones of the world. These soils were termed as zonal soils. Major examples of zonal soils are laterites, lateritic soils (Oxisols, Ultisols, and Alfisols), podzols, podzolic soils (Spodosols and Alfisols; Spodosols are soils that contain spodic horizon, in which amorphous mixtures of organic matter and aluminum, with or without iron, have accumulated), etc. The zonality concept of soil has been abandoned and the earlier soil nomenclature no longer exists. Still, there are some soils typical of a climatic zone. For example, Oxisols and Ultisols develop only in humid tropical climate, Alfisols in humid tropical and humid temperate climates, Spodosols mainly in humid temperate climate, Aridisols (soils that have limited availability of soil moisture for sustained plant growth) in arid climate, and Gelisols (soils with gelic materials underlain by permafrost) develop in the cold climate.

Pedoclimate

Texture of parent material and soil greatly influences the proportion of rainfall that would be effective in soil formation. A coarse textured parent material absorbs water rapidly but it drains out quickly. The effective soil moisture is low but percolating water leaches most suspended and soluble materials out of the soil profile. On the other hand, a fine textured soil retains more water in the profile which favors higher chemical weathering. On a sloping land a part of the rain-water moves down the slope and reaches nearby plains. So, soil on a sloping land, has less effective moisture for soil development. On the other hand, effective soil moisture in a depression is higher than total amount of rainfall. Moreover, a forest vegetation creates a microclimate through interception, stemflow, and transpiration different from a cultivated land. Soil temperature is greatly influenced by vegetation cover, humus and moisture contents. Therefore, pedoclimate (moisture and temperature condition of the soil) may differ significantly within the same climatic zone for differences in parent material, relief and vegetation. It appears that pedoclimate may have a higher impact on soil formation than atmospheric climate.

1.5.1.2 Organism as a Soil-Forming Factor

Organisms that influence soil formation include plants (natural vegetation or crops), microorganisms (algae, bacteria, actinomycetes, and fungi), animals, including vertebrates (burrowing animals like moles), and mesofauna (earthworms, ants, termites). These organisms live in and on the soil or exploit it. Among the changes they cause are addition and decomposition of organic matter, additions and losses of plant nutrients, biogeochemical cycling, soil mixing, and changes in soil structure and porosity.

Plants extend roots to greater depths of soil and create channels which facilitate movement of water within the soil.

Plants absorb nutrients from deeper zones of the soil and return them in litter on the surface. Thus, plants redistribute materials within the soil. The amount and proportion of nutrients absorbed from the soil differs with plant species. Some plants absorb and return more bases than others. Litter from such vegetation tends to prevent soils against becoming more acidic. Plants shed leaves and other litter, which on decomposition form humus. Humus becomes admixed with mineral matter in soil. Plants absorb water and nutrients and secrete metabolites through their roots. Studies in Tasmanian native forests have shown that soils formed on the same rock substrate but under different native forest communities often have contrasting profile morphology and chemical properties. Differences in soil properties are generally most pronounced in soils formed under dry (dominated by *Eucalyptus obliqua* and *Eucalyptus amygdalina,* with *Eucalyptus. viminalis*) or wet eucalypt forests (characterized by *Eucalyptus obliqua* and/or *Eucalyptus regnans,* with *Eucalyptus delegatensis*) (Laffan et al. 1998). Examples of the effects of some native vegetation on soil formation are given in the following.

Soils under Different Native Vegetations

Grasslands High organic matter content to a considerable depth below the surface; continuous root production, high interception of rain. Typical soil order—Mollisols.

Temperate coniferous forests Low retention of bases (Ca, Mg, K) in vegetation, low recycling of nutrients, highly leached, acidic soils. Typical soil order—Spodosols.

Temperate deciduous forests High retention of bases in biomass, high recycling of nutrients, slightly to moderately acidic soils. Typical soil order—lfisols.

Tropical broadleaf forests Intensive weathering and leaching, enrichment of oxides and hydroxides of Fe and Al, low organic matter accumulation, moderately acidic soils. Typical soil order—Oxisols.

Soil microorganisms are important transformers. They mineralize organic matter, transform nitrogen, phosphorus, sulfur, iron, etc., and produce organic substances, which form complexes and chelates with metals. Microorganisms humify bulk of the organic residues into humus which is intimately mixed with mineral particles. Soil insects and animals add organic matter and mix soil materials laterally and vertically. Some insects consume soil particles and transform them biochemically. Termites and ants contribute to the decay of dead plant materials and, thereby, nutrient circulation; they also play an important role in soil formation (Lobry de Bruyn 1999). In the savannah region, termites build huge mounds; some of these mounds are several meters high. In some areas these mounds dominate the landscape. Earthworms are well-documented and dominant member of the soil macrofauna influencing the soil formation processes in the temperate zone. Earthworms ingest, transform, and excrete huge quantity of soil materials increasing the strength of soil aggregates and leaving channels that improve permeability and aeration.

Human effects on soil formation Human effects on soil formation are either direct—through plowing, cropping, liming, manuring—or indirect—through changing the natural soil-forming factors, changing the vegetation by deforestation, changing the relief by leveling and terracing, changing the moisture regime through irrigation and drainage, changing the parent material through transport, dumping, or the exploitation of peat and rocks, and erosion and sedimentation (Dudal 2004). Corresponding soil changes vary greatly in kind, intensity, time, and spatial scale, and significance for soil and environmental quality. In land uses such as agriculture, surface horizons have usually been more altered than subsurface horizons, while engineering activities often change the entire sequence of soil horizons. Engineering activities lead to some of the most intensive impacts by removing soils entirely and replacing them with different earth materials, as well as by reshaping landforms or constructing new ones.

Anthropogenic Soils Anthropogenic soils are soils that bear the signs of long-term human impacts in their characteristics. They are human-altered and human-transported soils (USDA and NRCS 2007). These soils are classified as Anthrepts (Inceptisols having an anthropic or a plaggen epipedon), and in other categories having the formative elements Anthraltic, Anthraquic, Anthropic Humic, Anthropic, Anthroportic, Plaggic, and Plagghaplic in United States Department of Agriculture (USDA) Soil Taxonomy. In WRB 2006 (FAO 2006), two reference soil groups of anthropogenic soils were recognized—Anthrosols (soils that have been formed or profoundly modified through long-term human activities, such as addition of organic materials or household wastes, irrigation, or cultivation) and Technosols (soils dominated by artifacts and having properties substantially different from those of natural soil materials, and recognizable as such in the soil).

1.5.1.3 Parent Material as a Soil-Forming Factor

Parent materials are the main raw materials from which a soil is ultimately formed. So, many of the properties of parent materials may inherit to the resulting soils. Soil texture is mainly inherited from the parent materials. Coarse-textured parent materials generally produce coarse-textured soils. Mineralogical and chemical compositions of parent materials are also very important in soil formation. For example, acid parent materials tend to produce acid soils. Lime-rich parent materials produce calcareous soils. Mineralogical composition and fertility of soil also depend on the mineralogical composition of the parent materials. However, the effects of parent materials are dominant on young and immature soils. Under warm and humid climate, the effects of parent materials largely disappear as the soils grow older. Actually, many different parent materials may produce similar soils depending on other soil-forming factors, particularly climate.

Types of parent materials Organic soils develop from organic parent materials formed by the deposition of residues of many generations of vegetation, usually under wetland conditions, such as swamps and marshes. There are two types of organic deposition—peat (if the residues are sufficiently intact to permit the fibers to be identified) and muck (if most of the material has sufficiently decomposed so that little fiber remains). Mineral parent materials form by the weathering of rocks and minerals. When the unconsolidated materials remain in place of origin of the rock and soil is formed there, they are called residual parent materials. In many instances, weathered materials are carried away from the place of origin to other places where soil formation takes place. These materials are called transported parent materials.

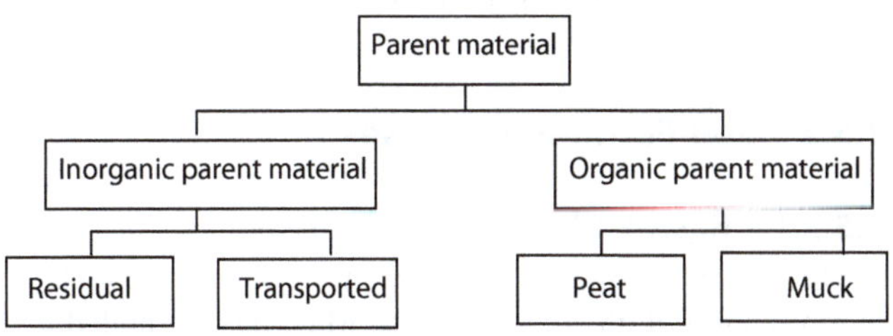

Transported parent materials are sediments transported by water, wind, glacier, or gravity. Water transported materials deposited in riverbeds, lakes, and oceans are respectively called alluvial, lacustrine, and marine parent materials. Windblown transported parent materials are eolian parent materials; sediments carried by glaciers are glacial parent materials, and those transported by gravity are colluvial parent materials. Materials directly deposited by ice are glacial till. These parent materials contain rock debris of mixed sizes ranging from boulders to clays. Glacial till may be deposited in irregular ridges known as moraines. Eolian parent materials include (1) dune sand, (2) loess, and (3) aerosolic dust. Dune sand is formed by piling-up of windblown medium- and fine-grained sand into hills. These are found along beaches of large lakes, seas, oceans, and in barren deserts. Loess is formed by deposition of windblown materials largely silt sized with some fine sand and coarse clay. Aerosolic dust is composed of very fine particles (approximately 1–10 μm) and deposited generally with rainwater. Colluvial deposits are found on valleys around mountains and foothills.

Characteristics of soils formed from different rocks

Volcanic ash: Fine-textured soils with plenty of easily weatherable minerals. They contain high organic matter and available nutrients.

Granite: Coarse-textured soils with medium fertility. Most soils are friable and permeable but acidic and low in base status.

Basalt: Fine-textured soils with sufficient easily weatherable minerals. In temperate regions where water is available, these soils are very productive.

Sandstone: Poorly fertile coarse-textured soils.

Limestone: Fine-textured soils with variable fertility

1.5.1.4 Topography as a Soil-Forming Factor

Topography refers to the configuration (surface feature) of the land. Topography has a strong influence on soil development. Soils on the side of hills tend to be shallow; surface materials are continually removed by erosion. Soils on the tops of hills tend to be deep. Soils in the valleys are deeper and contain more horizons. Soil materials are continually deposited from upper slopes in foothills. Topography influences soil formation through its effect on drainage, infiltration, runoff, erosion, and exposure to sunlight and wind. It causes localized changes in moisture and temperature. Soils on hilltops are drier than soils at the bottom of hillslopes. When rain falls on a hillslope, for example, water runs down from the top of the hill. Excess water collects at the bottom of the hill. The drier soils at the top are quite different from the wetter soils at the bottom, even if both soils form under the same overall conditions of climate, organisms, parent material, and time.

Topography was found to be a major factor controlling both hydrological and soil processes at the landscape scale in boreal forest region. Seibert et al. (2007) found several significant correlations between topographic indices and soil properties. For example, the thickness of the organic layer increased with topographical wetness index and the thickness of the leached E horizon increased with upslope area.

Important components of topography are altitude (height above mean sea level), degree of slope (inclination against the horizontal land axis), and aspect (face of the sloping surface). The degree of slope affects erosion, natural or accelerated, so that soil materials are continually removed from sloping surfaces and deposited in valleys. The steeper the slope (higher degree), the greater is the loss of soil materials.

Generally shallow soils above bedrocks develop on sloping surfaces. Usually a thin A horizon (surface horizon) develops over the thin C horizon (parent materials). Such thin soils were earlier known as lithosols. On gentler slopes, well-developed B horizon may be found. The depth of soil, thus, depends on the degree of slope. Sometimes erosion removes the entire regolith (unconsolidated materials over the bedrock) exposing the bedrock. Another important component of topography is aspect. North- and east-facing slopes in the Southern Hemisphere receives more sunlight and soils are warmer and drier. The reverse slopes are affected similarly in the Northern Hemisphere. Therefore, soils and natural distribution of vegetation are different in opposing aspects.

1.5.1.5 Time as a Soil-Forming Factor

The soil originates, grows, and matures with time. The time required to develop a soil varies from some centuries to some millions of years depending on other soil-forming factors. Generally profile development and horizon differentiation are faster under warm, humid, and forested conditions because of higher weathering and leaching of bases, colloids, and other materials. On an average, it takes 200–400 years to form 1 cm soil in environments characterized by mild climate. In tropical areas soil formation is faster; it takes 200 years. Soil formation is also fast in some parent materials, including alluviums, loess, and volcanic ash. Mohr and van Baren (1954) reported that 1 cm soil on volcanic ash parent material developed within only 1.3 years. On the other hand, 1 cm soil formation in an Oxisol in Africa needed about 750 years (Aubert 1960).

There is considerable confusion on how to estimate the age of a soil. Since parent material is the initial state of the soil system, the time of soil development is the difference from the time of parent material formation to the time of studying the soil. However, it is difficult to determine when parent material has completely formed and when pedogenesis has started. Weathering and pedogenesis occur simultaneously in nature; so the age of a soil is sometimes measured from the time the rock was first exposed to the surface or from the time the alluvial material was deposited. However, soil incepts, grows, and matures with time through several stages of development. Figure 1.6 shows the succession of horizons in a soil over time.

Over time, vegetation at a space develops through a series of stages of succession. As a coupled system component, soil development keeps pace with the vegetation development. In other words, succession of soil and vegetation occurs simultaneously up to a steady state. Both soil and vegetation tend to retain this steady state, unless or until abrupt changes in the state factors or driving factors take place. When the steady state is reached, further changes may occur if abrupt environmental changes occur. Otherwise, the state remains fairly stable. However, natural changes in climatic conditions often take place over time with accompanying changes in the soil-vegetation steady state.

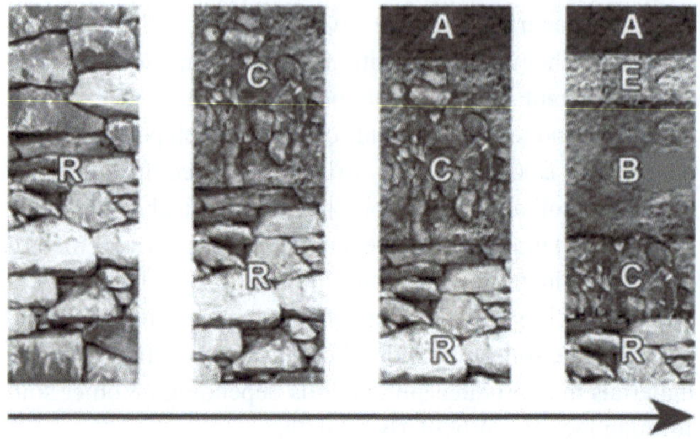

Fig. 1.6 Succession of soil horizons over time

> **Soil is a non-renewable resource**
> Forming soil of a furrow slice (15 cm depth) and accumulating enough materials to make the soil fertile needs about 3000 years. Degradation of soil fertility may occur within some years. Soil has limited resilience capacity, but it takes thousands of years to recover if its fertility is lost. For these reasons, the soil is considered as a non-renewable resource: once it has been destroyed, it is lost forever.

1.5.2 Soil-Forming Processes

1.5.2.1 Basic Soil-Forming Processes

There are four basic soil-forming processes that operate in the formation of all soils in nature. Materials are added and lost, transformed and circulated within and outside the soil by these processes, so that the parent material is differentiated into several horizons. The basic soil-forming processes are addition, removal, transformation, and translocation. These processes are often simultaneous, and occur usually in conjunction. Some transformations aid in translocation and removal, some others may cause chemical precipitation and addition. Addition of rainwater may cause transformation, translocation, and removal.

Addition Addition of materials and energy in soil naturally take place from the atmosphere, organisms, rocks and minerals, and groundwater. Humans also add some materials to the soil. Inputs from the atmosphere are solar energy, moisture, gases, dusts, and sometimes pollutants. The amount of solar energy depends on geographical locations, topography, mineralogical composition, degree of moisture saturation, etc. Addition of water through rainfall depends on climatic conditions. The amount of water intake by a soil may, on the other hand, depend on soil, vegetation, and slope characteristics. Addition of air is needed for the aeration and redox systems. Important gases added from atmosphere to soil to participate in various functions are O_2, CO_2, and N_2. Aerosolic dusts include fine mineral particles carried by wind or particulates emitted by industries. Organisms provide the organic inputs—plant and animal litter on the surface, and decaying roots, dead burrowing animals, soil insects, microorganisms, etc. within the soil. Rocks and minerals provide the raw materials to form most soils. Moisture and soluble substances may also be added through capillary rise of groundwater. Materials added to soil by humans in managed soil systems include manures, fertilizers, soil conditioners, lime, and other agrochemicals. Different inorganic and organic materials are added to soil in waste disposal sites. However, maximum organic and atmospheric additions take place on the surface of the soil. Addition of organic matter and its transformation to humus make the surface soil darker than the subsoil. This is the A horizon.

Removal Materials are removed from soil by volatilization and evaporation to the atmosphere, in solution and suspension in water downward, and outside the soil by biomass harvest. Gaseous losses take place due to evapotranspiration, volatilization, diffusion, etc. Moisture is continuously lost from soil through evapotranspiration and percolation. Gases produced by biochemical or geochemical transformations—CO_2, NH_3, N_2, NO_x, SO_2, H_2S, CH_4, etc.—are always lost to variable extents depending on soil and other environmental conditions. Erosion is the loss of soil material with the running water. Under natural conditions, both erosion and soil formation proceeds simultaneously and erosion losses are compensated by gains through soil formation. At the steady state, there is a fairly stable equilibrium between losses and gains through several processes involving the interactive systems. However, accelerated erosion causes loss of the fertile upper soil and finer soil particles, and sometimes, if the slope is steep, the A horizon or even the entire soil body is lost. Loss of soluble substances such as ions, bases, acids, salts, inorganic and organic colloids, complexes, etc. accompanies the percolating water to the groundwater. It is called leaching which is the complete removal of soluble materials from the soil profile (Schaetzl and Anderson 2005).

Transformations Complex transformations occur in soil all the time. These transformations include physical, physicochemical, chemical, and biological processes. These transformations may be reversible and irreversible. The most important physical transformation that occurs in soil is perhaps the aggregation of soil particles and formation of peds. Aggregation is one of the criteria that distinguish soils from other natural loose materials on the surface of the earth. Aggregation is effected by colloids—clays and humus; cementing agents—lime, Ca^{2+} ions, microbial products; and binding agents—plant roots and fungal hyphae, etc. Chemical transformations include solution, hydration, hydrolysis, oxidation, reduction, carbonation, and other acid reactions. Complexation, precipitation, chelation, ion exchange, dispersion, coagulation, etc. are other important chemical and physicochemical transformations related to soil formation. Decomposition of organic matter and humification are important biological transformations related with soil formation. Organic substances form complexes with mineral matter and circulate them throughout the soil profile.

Translocation Translocation of materials occurs in all directions of the soil—lateral, upward, and downward. This may take place in solution and suspension, or as mass movement along cracks and channels. When soil materials move downward with water and are lost from the soil body, it is called leaching. If movement takes place in solution or suspension from an upper horizon to a lower horizon, it is known as eluviation. Eluviation occurs usually from the surface horizons (A horizon). Under prolonged weathering in humid conditions, eluviation of bases, humus, clay minerals, and sesquioxides may create a bleached gray horizon enriched with silica (E horizon) below the A horizon. Translocation of clay-sized particles in suspension is called lessivage or argilluviation (L. *argilla*, white clay, *luv*, washed). Most lessivage occurs from the upper profile to the lower (from A and E horizons to B horizon). Eluvial materials are deposited in the B horizon (either by coagulation, precipitation, or mechanical impedance). This process of deposition is known as illuviation. Illuviation of humus, clay, iron oxides or hydroxides, lime, etc. are common occurrences. In arid climates, soluble materials move upward with capillary rise of water. Water is evaporated leaving the salts in the surface soil. Soil water with dissolved materials may move within a lateral continuum, particularly from a relatively moister zone to a drier zone. A considerable amount of soil materials is translocated vertically and laterally by such soil animals as rodents, earthworms, and insects.

1.5.2.2 Specific Soil-Forming Processes

Soils evolve in natural conditions by the action of natural forces. Conditions over the globe are varied and forces are diverse. So, there are many different types of soil in nature. These soils must have formed by different processes. In other words, a set of natural processes should be responsible for the formation of a specific soil. These processes are called specific soil-forming processes. Some of these processes are laterization, latosolization, ferralitization, podzolization, salinization, and calcification.

Laterization, latosolization, and ferralitization Laterization, latosolization, and ferralitization are processes of

enrichment of iron, manganese, and aluminum oxides and hydroxides in the soil profile. When iron and aluminum come in solution to the soil profile from outside, such as saprolite below or from upslope, and precipitated, the process is called laterization (Fanning and Fanning 1989). It is a dominant soil-forming process of the humid tropics, usually beneath broadleaf evergreen forests. Long-term intensive weathering is followed by intensive leaching. Potential decomposition is higher than biomass production leading to low accumulation of organic matter. Weathering dissolves most minerals and releases bases which are lost by runoff and leaching.
In latosolization, residual sesquioxides accumulate; bases and silica are leached out from the profile. The residual material is rich in oxides of Fe and Al, Ti, and Mn and oxyhydroxides and hydroxide minerals. Clay minerals such as kaolinite neoform from these residual ions. The Fe and Al compounds in these soils are typically highly crystalline. That is, they have neoformed into minerals with discrete ordered structures that are not easily weatherable. An oxic (Bo) horizon develops in this way. However, Oxisols may contain both residual and migrated iron and aluminum (Schaetzl and Anderson 2005). Van Breemen and Buurman (2002) used the term 'ferralitization' for the processes associated with strong weathering that lead to the formation of Oxisols (Soil Taxonomy; Soil Survey Staff 1999) or Ferralsols (World Reference Base for Soil Resources; FAO 2006). These soils have an oxic or ferralic B horizon affected by extreme weathering of primary and secondary minerals. Almost all weatherable minerals are removed from the sand and silt fractions. In the clay fraction, weathering causes a dominance of kaolinite, gibbsite, and iron minerals, resulting in a low CEC (<16 $cmol_c\ kg^{-1}$ at pH 7).

Podzolization Podzolization is the dominant soil-forming process of the humid temperate climate under forest vegetation. Podzolization occurs under vegetation that produces acidic litter, such as coniferous forest and heather, and on coarse-textured parent materials (Schaetzl and Anderson 2005). In such situation, plenty of organic matter is produced but the decomposition of organic matter is relatively slow. Organic residues accumulate in the surface and their decomposition produces appreciable organic acids. The pH of the surface soil is lowered below even 4 and at this strong acidity most minerals are weathered. Some of these weathered minerals are lost from the profile, but the great majority are redistributed, in suspension, to the lower profile. In podzolization, humus, iron, and aluminum, in some combination, are translocated from the A and E horizons to the B horizon. A spodic horizon develops by podzolization (Lundstrom et al. 2000). The resulting soils are generally coarse textured with few primary minerals and silicate clays.

Calcification Calcification is the process of precipitation and accumulation of secondary calcium carbonates in soils. The accumulation of $CaCO_3$ may result in the development of a calcic horizon. It occurs mainly in the arid regions where the availability of water for leaching is low. Solubility of $CaCO_3$ is also low and it accumulates in the profile (Southard 2000). Calcium carbonate may precipitate from soluble calcium released by weathering of minerals. Calcification is also common in subhumid grassland soils and in humid climates where soil horizons are clay rich and slowly permeable (Schaetzl et al. 1996).

Salinization Salinization is the accumulation of soluble salts in soil-producing saline soils. Natural salts often accumulate in surface of arid land soils as a white crust. The salts are generally chlorides and sulfates of sodium, potassium, calcium, and magnesium. In regions where the potential evaporation is greater than precipitation, some soluble salts are liberated by weathering but due to scarcity of water, these salts cannot be leached to any considerable extent. Moreover, capillary rise of water brings some salts on surface. Water is evaporated leaving the salts in soil. A salic horizon may develop this way. When sodium carbonate and bicarbonate predominate among salts, exchangeable Na may be high ($>15\%$ of the cation exchange capacity). The resulting soil is then called a sodic soil. The process is alkalization. Salinization may also occur in poorly drained soils of humid areas by the capillary rise of salts from ground water.

Study Questions

1. Mention the characteristics of different types of rocks. What types of soil can form from granite rocks in warm humid climate?
2. Give names of five potassium-containing minerals. Mention the chemical weathering products of these minerals.
3. Explain that any soil property is the function of the resultant of the interactions of five soil-forming factors. Discuss human as a soil-forming sub-factor.
4. Why the soil is considered as a non-renewable natural resource? Explain the factors that should determine the time required for soil development.
5. What is a soil profile? How does it form? Why O horizons are generally lacking in tropical forest soils?

References

Aubert G (1960) Influences de la vegetation sur le sol en zone tropicale humide et semi-humide. Rapp Du Sol et de la Vege Colloq Soc Bot Fr 1959:11–13
Buol SW, Hole FD, McCracken RJ, Southard RJ (1997) Soil genesis and classification, 4th edn. Iowa State University Press, Ames
Drever JI (1994) The effect of land plants on weathering rates of silicate minerals. Geochem. Cosmochem. Acta 58:2325–2332

Dudal R (2004) The sixth factor of soil formation. Presented at the International Conference on Soil Classification 2004. Petrozavodsk, Russia (3–5 August, 2004)

Fanning DS, Fanning MCB (1989) Soil Morphology, Genesis and Classification. John Wiley, New York

FAO (2006) World reference base for soil resources 2006. A framework for international classification, correlation and communication. FAO–UNESCO–ISRIC.Food and Agriculture Organization of the United Nations, Rome

Foster RJ (1988) General geology, 5th edn. Merrill Publishing Co., New York

Jenny H (1941) Factors of soil formation a system of quantitative pedology. McGraw- Hill Book Company Inc., New York

Jenny H (1980) The soil resource: origin and behavior. Springer-Verlag, New York

Kellogg CE (1960) Foreword. Soil Classification, a comprehensive System 7th Approximation. J Soil Sci 11:172–185

Laffan MD, Grant JC, Hill RB (1998) Some properties of soils on sandstone, granite and dolerite in relation to dry and wet eucalypt forest types in northern Tasmania. Tasforests 10:49–58

Lobry de Bruyn LA (1999) Ants as bio indicators of soil function in Rural Environments. Agricul Ecosys Environ 74:425–441

Lundstrom US, van Breemen N, Bain DC et al (2000) Advances in understanding the podzolization process resulting from a multidisciplinary study of three coniferous forest soils in the Nordic Countries. Geoderma 94:335–353

Mohr ECJ, van Baren FA (1954) Tropical soils: a critical study of soil genesis as related to climate, rock and vegetation. R. Trop. Inst., Amsterdam. Inter-science, New York

Ollier CD (1975) Weathering. Longman, London

Schaetzl RJ, Anderson S (2005) Soils: genesis and geomorphology. Cambridge University Press, Cambridge

Schaetzl RJ, Frederick WE, Tornes L (1996) Secondary carbonates in three fine and fine-loamy Alfisols in Michigan. Soil Sci Soc Am J 60:1862–1870

Seibert J, Stendahl J, Sørensen R (2007) Topographical influences on soil properties in boreal forest. Geoderma 141:139–148

Simonson RW (1978) A multiple-process model of soil genesis. In: Mahaney WC (ed) Quaternary Soils. Norwich, UK

Soil Survey Staff (1993) Soil Survey Manual. Handbook 18. U.S. Govt Printing Office, Washington DC

Soil Survey Staff (1994) Keys to Soil Taxonomy. USDA Soil Conservation Service, 6th edn. Washinton DC

Soil Survey Staff (1999) Soil Taxonomy, US Department of Agriculture Handbook No. 436. US Government Printing Office, Washington DC

Southard RJ (2000) Aridisols. In: Sumner ME (ed) Handbook of Soil Science. Boca Raton, FL, CRC Press, London

Suzuki Y, Matsubara T, Hoshino M (2003) Breakdown of mineral grains by earthworms and beetle larvae. Geoderma 112:131–142

USDA NRCS (2003) Keys to Soil Taxonomy, 9th edn. United States Department of Agriculture. Govt. Printing Office, Washington DC

USDA NRCS (2007) Anthropogenic Soils. ICOMANTH Report No. 2—Version 2.0, August, 2007

van Breemen N, Buurman P (2002) Soil Formation, 2nd edn. Kluwer Academic Publishers, New York

Physical Properties of Forest Soils

2

Physical properties of forest soils develop under natural conditions by the influence of permanent vegetation over a long period of time. Physical properties of forest soils may be almost permanent properties unless modified by harvesting operations, shifting cultivation, and forest fires. Important physical properties of forest soils include texture, structure, porosity, density, aeration, temperature, water retention, and movement. The physical properties of forest soils affect every aspect of soil fertility and productivity. Soil physical properties determine the ease of root penetration, the availability of water and the ease of water absorption by plants, the amount of oxygen and other gases in the soil, and the degree to which water moves both laterally and vertically through the soil. Soil physical properties also influence the natural distribution of forest tree species, growth, and forest biomass production. However, soil physical properties are largely controlled by the size, distribution, and arrangement of soil particles.

2.1 Soil Particles

Soils are composed of variously sized particles. There are two types of particles—primary particles and secondary particles. Individual discrete particles are called primary particles and their aggregates are known as secondary particles. Particles greater than 2 mm diameter are known as gravels which include pebbles (2–7.5 cm), cobbles (7.5–25 cm), stones (25–60 cm), and boulders (>60 cm). Although particles larger than 2 mm are less common and they hardly affect soil fertility and productivity, many productive forests have developed on gravelly or stony soils. Primary particles with the maximum "effective diameter" of 2 mm are classified into three categories—sand, silt, and clay. Effective diameter, which is the diameter of a sphere that has a velocity of fall in a liquid medium equal to the particle in question, has been considered because soil particles are not all spherical. The dimensions of the different categories of soil particles differ between the United States Department of Agriculture (USDA) and the International Soil Science Society (ISSS) classification systems (Table 2.1).

Table 2.1 Dimensions of sand, silt, and clay particles

ISSS System		USDA System	
Soil particles	Diameter (mm)	Soil particles	Diameter (mm)
Coarse sand	2.00–0.20	Very coarse sand	2.00–1.00
Fine sand	0.20–0.02	Coarse sand	1.00–0.50
Silt	0.02–0.002	Medium sand	0.50–0.25
Clay	<0.002	Fine sand	0.25–0.10
		Very fine sand	0.10–0.05
		Silt	0.05–0.002
		Clay	<0.002

Soil separates do not differ in size alone; there are important differences in their physical behavior and mineralogical makeup.

Characteristics of Soil Particles

Sand: Sand particles are mainly fragments of quartz and some feldspars and mica. They have little surface area exposed (0.1 m^2 g^{-1} specific area). Sand particles are visible to the naked eye, gritty in feeling, having little or no capacity to hold water or nutrients, and bind other particles. They are loose when wet, very loose when dry. Sand does not absorb water and does not exhibit swelling and shrinkages, stickiness and plasticity.

Silt: Silt particles are also fragments of primary minerals. Most silt particles are not visible to the naked eye, but can be seen through an ordinary microscope. They feel smooth when wet and like talcum powder when dry. They have low to medium capacity to attract water, nutrients and other particles. Because of adhering film of clay, they exhibit some plasticity, cohesion, adhesion and absorption and can hold more amount of water than sand but less than clay.

Clay: Clay particles are mainly secondary minerals such as kaolinite, smectite, vermiculite, illite, chlorite, hydrated oxides of Fe and Al, etc. Clay particles can be seen by an electron microscope and have large surface area (10–1000 $m^2\ g^{-1}$). They have electrical charges, both negative and positive, on their surfaces. Because of these properties, clays have high water and nutrient holding capacity and they participate in chemical reactions in the soil.

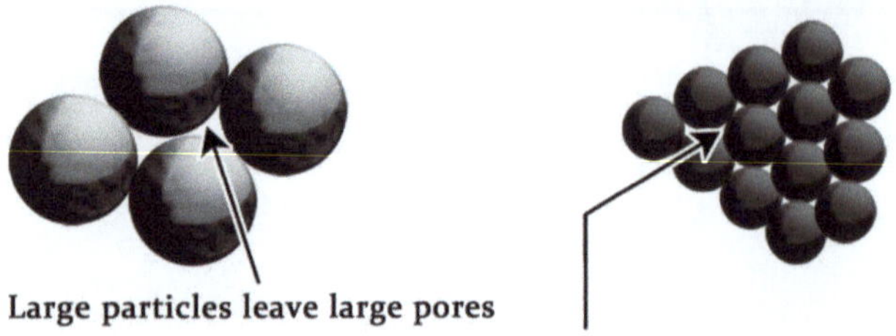

Fig. 2.1 Pores left by soil particles of different sizes

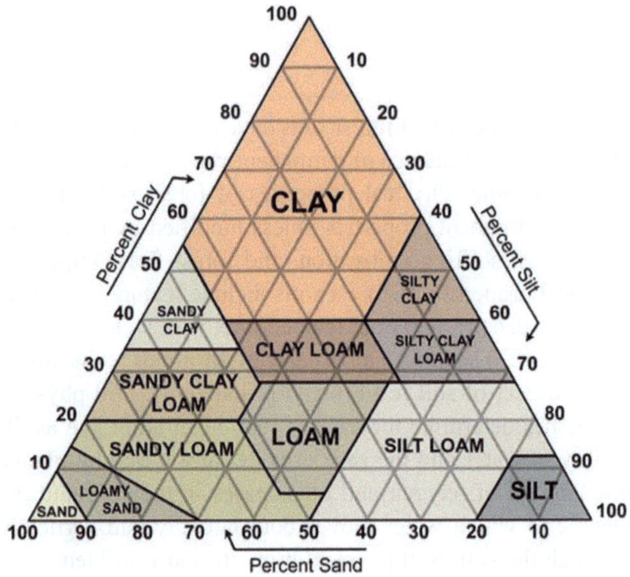

Fig. 2.2 Soil textural triangle devised by the USDA

2.2 Soil Texture

Soil texture refers to the degree of fineness or coarseness created by the close packing of variously sized particles together in a soil. It is determined by the relative proportion of sand, silt, and clay in a soil. Soils are rarely composed of a single size class of particles; they are mixtures of different size classes. However, one or two size classes usually dominate the physical behavior of the soil. The soil is then named after the name of that soil separate. Thus, a sandy soil displays the properties of sand particles. When a soil equally exhibits the properties of sand, silt, and clay, then it is called loam (approximately 40% sand, 40% silt, and 20% clay). A higher proportion of sand in loam produces sandy loam. In this way, 12 textural classes have so far been identified. They are (from coarse to fine) sand, loamy sand, sandy loam, loam, silt loam, silt, sandy clay loam, clay loam, silty clay loam, sandy clay, silty clay, and clay. Soil texture is not usually changed by management practices. Soil texture is inherited from the parent materials and it originates through weathering and pedogenic processes, including recrystallization, eluviation, and illuviation. It may, however, be altered by erosion, deposition, truncation, landfill, etc.

Sand particle is coarse, clay is fine, and silt is medium. When sand particles are packed together (i.e., if the soil is sandy), they leave larger gaps or "pores" between them (Fig. 2.1). On the other hand, gaps between clay particles are small. Larger pores, the macropores, generally accommodate air and smaller pores, whereas the micropores contain water. Thus, the coarseness or fineness of soil determines its air–water relationships. Texture is one of the most important physical properties of soil that affects its fertility and productivity. Actually, the whole soil environment is regulated by soil texture.

Soil texture is determined in the laboratory by a technique based on the velocity of fall of a particle in a liquid medium, which is proportional directly to the square of the radius of the particle and inversely to the viscosity (a fluid's internal resistance to flow) of the liquid (Stokes' Law, Hillel 1980). Percentages of sand, silt, and clay are determined by either "pipette method" or "hydrometer method" (Day 1965). Soil textural class names can be obtained from the "USDA textural triangle" (Fig. 2.2), if the percentages of any two size fractions are known. For example, lines for 40% sand and 20% clay intercept in the area demarcated as loam. Experienced foresters and soil survey personnel have learned the technique of determining soil texture in the field fairly accurately by feeling a lump of soil between their fingers. Readers are referred to Thien (1979) who has provided a flowchart for the determination of texture by the feel method. (For this section, the reader may further consult Foth 1990 and Brady and Weil 2002.)

2.2.1 Suitable Forest Tree Species for Different Soil Textures

A tree species may occur on a variety of soil textures in their natural habitat, but their best growth occurs in the most suitable sites. For example, black cherry (*Prunus serotina* Ehrh.) grows well on a wide variety of soils throughout its range in eastern North America, if summer growing conditions are cool and moist. On the Allegheny Plateau, black cherry develops well on all soils except for the very wettest and very driest (Hough 1965). Site quality does not differ between soils developed from glacial till and from residual parent material. It tolerates a wide range of soil drainage and grows about the same on well-drained sites as on somewhat poorly drained sites but shows rapid loss in productivity with increasingly wetter conditions (Becker et al. 1977; Cerutti et al. 1971). Cottonwood (*Populus deltoides*) does well in a heavy cold damp soil (Chittendon 1956; Duke 1983) but thrives best on moist well-drained, fine sandy loams or silts close to streams (Duke 1983). On the other hand, European larch (*Larix decidua*) grows better in fertile soil consisting of loam, sandy loam, or silty loam. Shade and poorly drained conditions are not well tolerated (IUCN 2006). Hackberry (*Celtis occidentalis*) grows in many soils, and although principally a bottomland tree, it is frequently found on limestone outcrops or limestone soils (usually fine textured). In western Nebraska, hackberry grows on the north side of sand dunes and in river valleys (Eyre 1980).

Sandy soils are coarse textured soils being loose and friable; they absorb water rapidly and drain it quickly, and can be worked easily in both moist and dry conditions. They are called "light textured soils." They are usually poor in fertility and suffer from water scarcity. Sandy soils are widely distributed in the tropics occupying most of arid and semiarid areas. For instance, the total estimated extent of Arenosols is 900 million hectares, mainly in Western Australia, South America, South Africa, Sahel, and Arabia (FAO 2006). These soils are characterized by a low soil organic carbon, a low cation exchange capacity, a high risk of nutrient leaching, a low structural stability, and a high sensitivity to erosion and to crusting (Pieri 1992; Sanchez and Logan 1992). Some tree species requiring low water and nutrients can grow there; some even prefer sandy soils but growth of most tree species in sandy soil is stunted. The following is a list of tree species that can be found naturally growing well on sandy soils, or can be grown successfully on sandy soils.

List of trees suitable for sandy soils

Common name	Scientific name
Amur Maple	*Acer ginnala*[a]
Black cherry	*Prunus serotina*[a]
Black locust	*Robinia pseudoacacia*[a]
Box Elder	*Acer negundo*[a]
Canadian spruce	*Picea alba*[a]
Cedar wattle	*Acacia elata*[b]
Heath-leaved banksia	*Banksia ericifolia*[b]
Chinese juniper	*Juniperus chinensis*[a]
Coastal she-oak	*Casuarina equisetifolia*[b]
European white birch	*Betula alba*[a]
Gray birch	*Betula populifolia*[a]
Hedge maple	Acer campestre[a]
Hop tree	*Ptelea trifoliate*[a]
Large tooth aspen	*Populus grandidentata*[a]
Monarch Birch	*Betula maximowicziana*[a]
Norway spruce	*Picea excels*[a]
Pignut	*Carya glabra*[a]
Pitch pine	*Pinus rigida*[a]
Quaking aspen	*Populus tremuloides*[a]
Red cedar	*Juniperus virginiana*[a]
Red swamp banksia	*Banksia occidentalis*[b]
Scarlet oak	*Quercus coccinea*[a]
Scots pine	*Pinus sylvestris*[a]
Sour cherry	*Prunus cerasus*[a]
Swiss mountain pine	*Pinus Montana*[a]
Tatarian Maple	*Acer tataricum*[a]
Tree of Heaven	*Ailanthus glandulosa*[a]
White pine	*Pinus strobes*[a]
White poplar	*Populus alba*[a]

[a] http://www.backyardgardener.com/tree/indexlist5.html
[b] Evans (2010) Native Australian trees suited to sandy soils

Loam soils are medium textured soils having generally more nutrients and humus than sandy soils, and have better infiltration and drainage than clay soils. Loams are desirably porous and retain sufficient water, nutrients, and air. Unless adapted in extreme textures, most tree species do well in sandy loam to clay loam soils for adequate water, air, and nutrients. To mention a few—Balsam fir (*Abies balsamea*), Basswood (*Tilia americana*), and Hackberry (*Celtis occidentalis*) (USDA NRCS 1995); Bitternut hickory (*Carya cordiformis*) (Voss 1985), European larch (*Larix decidua*) (Matras and Paques 2008), Gamar (*Gmelina arborea*) (Onyekwelu et al. 2006), Mahagoni (*Swietenia mahagoni*), and Garjan (*Dipterocarpus turbinatus*) (Hossain 2012); and Red maple (*Acer rubrum*) (Abrams 1998), Red oak (*Quercus rubra*) (Celine et al. 1996), Red pine (*Pinus resinosa*) (Leaf et al. 1971), Teak (*Tectona grandis* (Zanin 2005) White ash (*Fraxinus americana*) and White oak (*Quercus alba*) (USDA NRCS 1995).

Clay soils are generally compact and stiff; sticky when wet and hard when dry and require much energy to work in both wet and dry conditions. They are called "heavy textured soils." Clay soils retain large amount of water but drain very slowly. They are often waterlogged. They are fertile but plants usually suffer from oxygen stress in clay soils. Suitable tree species for clay soils are listed as follows:

List of trees suitable for clay soils

Common name	Scientific name
Amur corktree	*Phellodendron amurense*[c]
Bitternut hickory	*Carya cordiformis*[c]
Black ash	*Fraxinus nigra*[b]
Butternut	*Juglans cinerea*[c]
Common hackberry	*Celtis occidentalis*[b]
Common honey locust	*Gleditsia triacanthos*[c]
Cottonwood	*Populus deltoides*[c]
Crabapple	'Prairie Fire' Malus spp[a]
European alder	*Alnus glutinosa*[c]
European larch	*Larix deciduas*[c]

(continued)

Common name	Scientific name
Golden Black Spruce	Picea mariana[a]
Green ash	*Fraxinus pennsylvanica*[b]
Hawthorn	*Crataegus species*[c]
Japanese Plume Cedar	Cryptomeria japonica[a]
Kentucky coffee tree	*Gymnocladus dioicus*[c]
Norway maple	*Acer platanoides*[c]
River Birch	Betula nigra[b]
Shagbark hickory	*Carya ovate*[c]
Silver maple	*Acer saccharinum*[b]
Southern Magnolia	*Magnolia grandiflora*[a]
Tamarack	*Larix laricina*[c]

[a] http://voices.yahoo.com/trees-heavy-clay-soil-4924142.html. Accessed 7 June 2013

[b] Ware (1980) Selecting trees for clay soils. Metro Tree Impr Alliance (METRIA) Proc. 3:102–106

[c] http://www.extension.umn.edu/yardandgarden/ygbriefs/h-claytrees.html. Accessed 7 June 2013

2.2.2 Soil Texture and Species Distribution

According to Ramade (1981), soil texture governs most of the properties of the soil, its permeability, its capacity to retain water, its degree of aeration, its ability to make the nutrients stored in the clay–humus complex available to plants, its ability to withstand mechanical working of the top soil, and, finally, its ability to support a permanent plant cover. In the Lapland of Finland, Sepponen et al. (1979) demonstrated that pine stands are found on more coarse-grained soils than spruce stands. Jha and Singh (1990) suggested that soil texture is an important factor in the constitution and distribution of dry tropical forest communities. Soil texture was found to be largely responsible for the distribution of hardwood species within an old growth forest in the Sandhills region of southeastern USA. The higher clay content (and presumably higher moisture and fertility) of the upland soils have allowed for the development of a forest type that closely resembles the oak-hickory forests (Gilliam et al. 1993). Levula et al. (2003) found an obvious relation between soil particle-size distribution and tree species composition in a forest stand of Finland. The basal area (BA) of pine and the pine–spruce BA ratio correlated positively, and the BA of spruce negatively, with the mean particle size of the B1 horizon. Also, the coarse sand and gravel fraction in the B1 horizon correlated positively with the BA of pine and negatively with the BA of spruce. Scull and Harman (2004) studied species distribution and site quality in forests of southern lower Michigan, USA. They observed that poorly drained sites (depending on whether they were sandy or loamy) supported either pine communities (on the coarser sandy sites) or broad-leaved communities [ash, elm (*Ulmus* spp.), red maple, sugar maple, and yellow birch] mixed with the needle-leaved species. Well-drained sandy sites supported several different vegetation mixes of pine with northern pin oak (*Quercus ellipsoidalis*) and black oak (*Q. velutina*) depending on whether the sand was aeolian, outwash, or morainic. Well-drained loamy soils were broken down into two groups. Sugar maple and basswood were most distinctive of silt loams, whereas clay loams supported populations of maple, elm, yellow birch, and occasionally, hemlock (*Tsuga canadensis*), and balsam fir. Four soil types have been identified in a 52-ha forest dynamics plot in Bornean mixed dipterocarp forest (ranked by increasing fertility and moisture: sandy loam, loam, fine loam, and clay). The distributions of 73 % of tree species in the plot are significantly aggregated on one of these soil types (Russo et al. 2005).

2.3 Soil Structure

Soil structure is the arrangement of soil particles into units of different sizes and shapes. These units are called peds or aggregates and the processes of formation of peds are collectively called aggregation. According to Lal (1991), soil structure refers to the size, shape, and arrangement of solids and voids, continuity of pores and voids, their capacity to retain and to transmit fluids and organic and inorganic substances, and ability to support vigorous root growth and development. Peds differ from "clods" and "concretions"; clods or chunks are artificially formed (such as by plowing) hard soil mass. Concretions are hard lumps produced by the precipitation of dissolved substances (usually iron and manganese oxides). In some soils and sediments, the particles are not aggregated but remain separated; such soils are called single grained, such as some sandy soils. Most soils are structured soils. In some soils such as heavy clays, all the particles adhere together. Structure of these soils is called massive.

Soil structure determines pore-size distribution, which affects water flow and erosion potential (White 1985), microbial and faunal behavior (Edwards and Bremner 1967; Elliott et al. 1980; van Veen et al. 1984), and organic matter dynamics (Campbell and Souster 1982; Tisdall and Oades 1982; Schimel et al. 1985; Schimel 1986). Aggregation may affect nutrient turnover by controlling microbial predation (van Veen et al. 1984) and by protecting organic matter from microbial degradation (Young and Spycher 1979; McGill et al. 1981; Van Veen and Paul 1981; Voroney et al. 1981; Tisdall and Oades 1982). Aggregates influence microbial community structure (Hattori 1988), limit oxygen diffusion (Sexstone et al. 1985), regulate water flow, determine nutrient adsorption and desorption (Linquist et al. 1997; Wang et al. 2001), and reduce run-off and erosion (Barthes and Roose 2002). All of these processes have profound effects on soil organic matter dynamics and nutrient cycling (Six et al. 2004).

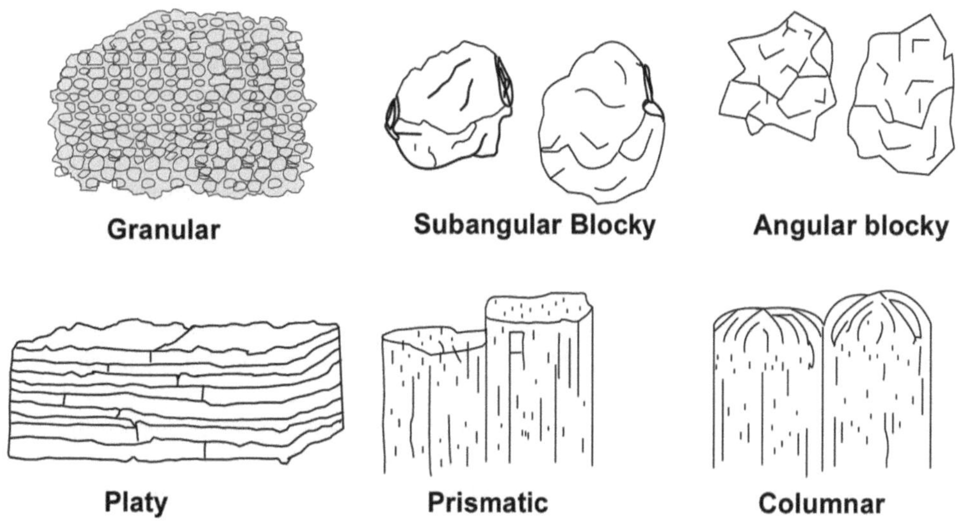

Fig. 2.3 Approximate shapes of different soil structures

Soil structure is classified based on shapes of peds into types, on size of peds into classes, and on distinctness and stability of aggregates into grades. There are four types of soil structure—spheroidal (granular and crumb), block-like (angular blocky and sub-angular blocky), plate-like, and prism-like (prismatic and columnar) (Fig. 2.3). On the basis of size of aggregates, soil structure is divided into five classes: very fine, fine, medium, coarse, and very coarse. Grades of soil structure are identified on the basis of the visibility in a horizon and stability of aggregates. Stability of aggregates refers to their resistance to destruction by water. Three structural grades are identified: weak (poorly formed, indistinct peds which are not durable), moderate (moderately well-developed peds which are fairly durable), and strong (very well-formed peds which are quite durable and distinct) (Hillel 1980).

2.3.1 Soil Structure Formation

Aggregation results from complex interactions of many factors including the environment, soil management, plant influences, and soil properties, such as mineral composition, texture, soil organic matter, pedogenic processes, microbial activities, exchangeable ions, nutrient reserves, and moisture availability (Kay 1998). There are several mechanisms of aggregation. Aggregates are formed in stages, with different bonding mechanisms dominating at each stage (Tisdall and Oades 1982). According to Duiker et al. (2003), aggregation results from the rearrangement of particles, flocculation, and cementation. Clay particles are electrically charged particles. They carry both negative and positive charges, but there are, generally, higher number of negative charges on their surfaces. They attract cations. When clay particles adsorb and get satisfied largely with monovalent cations like Na^+ and K^+, they remain dispersed in a liquid medium. Clay particles may come closer in a suspension only when they are flocculated. Flocculation (a process wherein colloids come out of suspension in the form of flock or flakes by the addition of a clarifying agent) is effected by replacing monovalent cations on colloidal surfaces by polyvalent cations like Ca^{2+}, Mg^{2+}, Fe^{2+}, Fe^{3+}, Al^{3+}, etc. Flocculation of clay particles forms clay flocks which are cemented together to form microaggregates. Microaggregates are bound together to produce macroaggregates. However, flocculation alone is not enough for aggregation; it involves a combination of different processes such as hydration, pressure, dehydration, etc. and requires cementation of flocculated particles. Well-known natural organic cements are the polysaccharides—levans and dextrans, polyuronides, etc. They stabilize the aggregates when they are dehydrated. Dehydrated oxides of Fe and Al bind soil particles firmly together. Colloidal clays are strong cementing agents; their cementation depends on their mineralogical composition, specific surface, and the state of deflocculation and dehydration.

Sand and silt particles are inert materials. They can come closer but cannot hold themselves together because they do not possess the power of adhesion and cohesion. Clays form coatings on them and cement several sand and silt particles into a larger unit. Soil particles are bound mechanically by plant roots and fungal hyphae. There are also various chemical compounds which act as cements and gums in soils. These materials include flocculated clays, hydrated oxides of Fe and Al, organic compounds produced by root exudation, microbial synthesis and excretion and organic matter decomposition, polyvalent cations, etc. (Osman 2013). Thus, soil particles aggregate as a result of physical binding by roots and hyphae, gluing by bacterially produced polysaccharides, and physical–chemical interactions between silicate clay surfaces and functional groups of partially decomposed organic matter (Aspiras et al. 1971; Aringhieri and Sequi 1978; Tisdall and Oades 1979; Sorensen 1981).

2.3.2 Soil Structure and Tree Growth

Soil structure, which modifies soil texture, exerts important influences on the edaphic conditions and the environment (Bronick and Lal 2005). Soil structure regulates pore size, number of pores, and distribution of pores and total porosity of the soil. Thus, retention and movement of soil water including infiltration, permeability, percolation, drainage, leaching, etc. all depend on soil structure. Soil structure affects tree growth by influencing root distribution and the ability to take up water and nutrients (Pardo et al. 2000). Soil structure facilitates oxygen and water infiltration and can improve water storage (Franzluebbers 2002). Microbial processes like organic matter decomposition, mineralization, stabilization, nitrification, and nitrogen fixation are influenced by soil structural conditions. Passioura (1991) reviewed literature on the relationship of soil structure and plant growth and suggested that the root–soil contact, which influences water and nutrient uptake, depends on soil structure. Aggregation may affect nutrient turnover by controlling microbial predation (Van Veen et al. 1984) and by protecting organic matter from microbial degradation (Voroney et al. 1981).

2.4 Density, Porosity, and Compaction

Density is defined as the mass per unit volume. There are two kinds of soil density—bulk density and particle density. Bulk density is estimated by dividing the mass (dry weight, usually after oven drying at 105 °C until constant weight) by total volume of soil (volume of solid + volume of pores) and particle density is estimated by dividing mass by the volume of the solids.

If a core of soil has a total volume V_T, which is the sum of the volume of solids, V_S and volume of pores, V_P and if the dry weight of the soil is W_S, then

$$\text{Bulk density, } D_b = \frac{W_s}{V_T}, \text{ and}$$

$$\text{Particle density, } D_p = \frac{W_s}{V_T}$$

Bulk density can be determined by the core method. Standard stainless steel cores with sharp edges may be used. The cores are pushed into the soil with the help of a wooden hammer at soft strokes so that the natural structure is not disturbed. After collection, the core with the soil is dried in an oven at 105 °C for 24 h and the dry weight of the soil is divided by the volume of the core to obtain bulk density. Particle density may be measured by a pycnometer (Pall and Mohsenen 1980).

Bulk density in surface soils in oak and pine forests of Indian Central Himalaya was found to range between 1.24 and 1.94 Mg m^{-3} (Jina et al. 2011). Bulk density of soil varies due to organic matter content, texture, compaction, and porosity of the soil. Histosols have very low bulk density (about 0.70 Mg m^{-3}; Brady and Weil 2002) and particle density (<1.0 Mg m^{-3}). Loose soils have lower bulk densities and compact soils have higher bulk densities. Coarse textured or sandy soils tend to have a higher bulk density. Average bulk density of a loam-textured mineral soil is taken to be 1.3 Mg m^{-3}. Forest soils are more porous soils and have lower bulk density than cultivated soils. Bulk density is an important physical property. It is needed in order to convert soil chemical data, routinely expressed as mass concentrations, into amounts per unit area or unit volume (Viro 1951; Tamminen 1991). As it is an index of soil compaction, bulk density affects site productivity (Strong and La Roi 1985; Gale and Grigal 1987) and tree growth (Hamilton and Krause 1985; Froehlich et al. 1986).

Problem

The internal diameter of a soil core is 4.2 cm and the length of the core is 6 cm. You have collected a full core field moist soil, weighed and dried in an oven at 105 °C for 24 h. Weight of the empty core (W_1) is 160 g, weight of core + field moist soil (W_2) is 320 g, weight of core + dry soil (W_3) is 275 g. Calculate field moisture content and bulk density.

Solution

Volume of core, $V = \pi r^2 L = 3.14 \times 2.1^2 \times 6 = 83.08\ cm^{-3}$
Weight of moist soil, $W_4 = W_2 - W_1 = 320 - 160 = 160$ g
Weight of dry soil, $W_5 = W_3 - W_1 = 275 - 160 = 115$ g
Weight of moisture $W_6 = W_2 - W_3 = 160 - 115 = 45$ g
% field moisture content $= (W_6 \div W_5) \times 100 = 39\,\%$, and
Bulk density $= W_5 \div V = 115 \div 83.08 = 1.38\ g\ cm^{-3} = 1.38\ Mg\ m^{-3}$.

Bulk density may differ in different depths of soil. In forested mineral soils of central and northern Finland, Tamminen and Starr (1994) observed that bulk density rapidly increased with depth in the surface but remained uniform at depths >20 cm. This was related with distribution of organic matter and compaction. Bulk density tends to increase with depth primarily due to the lack of organic matter and aggregation.

Density of dominant soil forming silicate minerals ranges from 2.60 to 2.75 Mg m^{-3}. Therefore, particle density of mineral soils usually remains within this range. Mineral soils are considered to have an average particle density of 2.65 Mg m^{-3}. Bulk density and particle density values are used to estimate the porosity (the percentage of soil volume that is void) according to the following relationship:

$$\text{Porosity} = \left\{1 - \frac{\text{Bulk density}}{\text{Particle density}}\right\} \times 100$$

Spaces in soil that are not occupied by solids are called pores and the total volume of pores is the porosity. Soil pores are filled with either air or water, or both depending on the soil moisture content, and condition of rainfall and irrigation. Roots and soil organisms, both macro and microflora and fauna, occupy these pores. There are small or large pores depending on the texture and structure. Large pore spaces allow fast infiltration and percolation of water through a soil. On the other hand, small pores have strong attractive forces to hold water in the pore. Large pores, also called macropores (>0.75 mm), accommodate air and conduct water into the soil where they fill the micropores (<0.75 mm). Clay soils have numerous micropores and hold large quantities of water, but since they have few macropores, they produce very slow infiltration rates. The pores in the clays may be so small and hold water so tenaciously that the water is not available to plants. Sandy soils have numerous macropores, but few micropores. Total porosity of soil is of little practical significance. The number and proportion of macropores and micropores and distribution of different size classes of the pores seem to be more important. However, porosity of a soil is an index of its state of compaction.

Problem

A soil has a bulk density of 1.40 Mg m−3. Calculate the weight of soil of 1 ha area and 20 cm depth. The soil has 0.23 % total N (w/w), calculate nitrogen content per hectare basis. Taking particle density as 2.65 Mg m−3, calculate its porosity.

Solution

Volume of 1 ha-20 cm soil is $= 10{,}000 \times 20/100 = 2000\,m^3$

Bulk density is $= 1.40$ Mg m^{-3}

Weight of 1 ha-20 cm soil is $= 2000 \times 1.4 = 2800$ Mg (2,800,000 kg)

Total N $= 0.23\,\% = 0.23 \times 2800 \div 100 = 6.44$ Mg/ha-20 cm.

Particle density is $= 2.65$ Mg m^{-3}

So, porosity is $= [(1 - 1.40\} \div 2.65] \times 100 = 47\,\%$

Soil compaction is the physical consolidation of the soil by an applied force that destroys soil structure, compresses soil volume, increases bulk density, reduces porosity, and limits water and air movement (Osman 2013). Stunted growth of trees, badly formed tree roots, standing water, and physically dense soil are some signs of forest soil compaction. Texture, structure, organic matter, and water are important soil factors that determine susceptibility of a soil to compaction. Soils made up of particles of about the same size compact less than soil with a variety of particle sizes. Hard, dense, low-organic matter soils suffer more from compaction than loose, friable, high-organic matter soils. A dry soil is not easily compacted; soil compaction increases with the increase in soil water content until a moist condition is reached. Wet soils are also less susceptible to compaction.

When soil is compacted, soil strength is increased and total porosity is reduced at the expense of the large pores. Thus, volumetric water content and field capacity are increased, while air content, water infiltration rate, and saturated hydraulic conductivity are decreased, if a soil is compacted. Compacted soils offer physical impedance to root extension, which affects water and nutrient absorption. Soil compaction caused by harvesting operations can affect future regeneration and growth of trees (Sakai et al. 2008). Available soil water content is reduced when a soil is compacted (Rockich et al. 2001). When this occurs, nutrients and water may become limited because the plant's demands exceed the ability of the root system to access these resources (Froehlich and McNabb 1984). Nutrients have limited mobility in compacted soil (Unger and Kaspar 1994); so nutrient deficiencies may seriously limit tree growth there.

Forest soils may be compacted by grazing animals and by expanding roots of the trees themselves, but more noticeably by vehicles used for a range of mechanized forest operations. Harvesting machinery may be very heavy and, combined with the pushing and pulling and lifting of logs, may exert large pressures on the soil. Some harvesting operations may greatly disturb the soil (Greacen and Sands 1980). Compaction of forest soils may be caused by:

1. trafficking by heavy equipment during felling, forwarding, skidding, and site preparation operations;
2. dragging action of logs as they are moved from the stump to the landing; and
3. slash disposal and the creation of planting or seeding sites during site preparation.

The main forces causing compaction of forest soils are mostly heavy machinery used during timber harvesting and mechanical site preparation (Greacen and Sands 1980; Froehlich and McNabb 1984). Soil compaction is a major factor affecting forest production (Mckee et al. 1985; Firth and Murphy 1989). Froehlich (1973) concluded that 50 % of the harvested area is disturbed and 25 % can be considered as compacted in western USA forest under tractor logging. In British Columbia, over 20 % of the harvested area has been compacted (Utzig and Walmsley 1988). Nowadays, forestry machineries are becoming heavier and more damaging on forest soil compaction. Deep rutting strip off the top layers of soil from the ground and push it beside skid trail, thus resulting in heavy soil erosion (Zeleznik et al. 2009). Mechanization of some forest operations has greatly intensified over the last decade, particularly in plantations, and forest managers are currently expressing concern about the compaction of forest soils and its consequences. Approximate contact pressures of a range of machines used in the logging of forests are mentioned below (Greacen and Sands 1980):

Vehicle	Pressure (kPa)
Flexible tracked skidder	30–40
Crawler tractor	50–60
Rubber-tyred skidder	55–85
Forwarder with rear bogie	85–100
Forwarder with single rear axle	105–125

Recovery of compacted forest soils is variable depending on the severity of the compaction and local conditions. Most studies, however, indicate that recovery from compaction and its effects are a long-term affair. Severely compacted soils may require up to 40 years or more to recover naturally, according to Hatchell and Ralston (1971). Froehlich and McNabb (1984) state that the effects of soil compaction should be assumed to persist for several decades on forest sites.

2.4.1 Effect of Soil Compaction on Forest Tree Growth

Soil compaction affects ecosystem stability and site productivity (Froehlich 1979; Wronski and Murphy 1994; Kuan et al. 2007). Soil compaction is particularly severe on temporary access areas such as forest landings and skid trails. These areas may be unproductive unless soil rehabilitation is carried out. Tree species growing on compacted soil show reduced root elongation rate (Whalley et al. 1995) and reduced height growth (Greacen and Sands 1980; Ares et al. 2007; Bulmer et al. 2007).

Soil compaction affects seed germination and seedling emergence of forest trees (Taylor 1971) and natural regeneration in forests (Sokolovskaya et al. 1977). Complex interactions occur among soil strength, water, nutrient availability, and aeration due to compaction. Roots may fail to penetrate smaller pores than their diameter. Because compaction both increases soil strength and decreases the number of macropores, the rate of root elongation and therefore root length is reduced. Zyuz (1968) reported abundant pine roots in soils of strength less than 1,700 kPa. Root penetration was restricted above 2,500 kPa. The penetration of roots of Radiata Pine (*Pinus radiata*) into sandy soil in South Australia was found to be greatly restricted at soil strengths greater than 3,000 kPa (Sands et al. 1979). Soil compaction decreases growth of forest trees; however, this is not always true and not of all tree species. Growth reduction has been reported in economically significant forest tree species such as *Pinus radiata* (Potter and Lamb 1974), *Pinus elliottii* (Haines et al. 1975), *Pinus taeda* (Duffy and McClurkin 1974), *Pinus ponderosa* (Trujillo 1976), *Pinus nigra* var. *mararima* (Berben 1973), *Picea abies* (Sokolovskaya et al. 1977), and *Pseudotsuga menziesii* (Berben 1973). Lodgepole pine and white spruce seedlings were grown on four compacted forest soils of Alberta, Canada. Significant reduction in seedling growth (maximum root depth, maximum root depth in soil core, total weight, shoot weight, root weight, stem diameter, shoot height, seedling survival, and shoot weight: root weight ratio) was observed with increased bulk density (Corns 1988).

2.5 Soil Air

Well-drained forest soils are generally porous, and the large pores contain air if they are not saturated with water. Approximately 25 % volume of a loam-textured surface soil is filled with air under field condition. In dry soils, water from many micropores is evaporated and these micropores may also be filled with air for some time. On the other hand, air in the pores may be expelled from the soil by water immediately after rainfall or irrigation. Then the macropores are also filled, although temporarily, with water. After cessation of rain or irrigation, say some hours or days later, gravity pulls this water into the groundwater table, leaving the spaces for air. Therefore, soil air is a dynamic component of the soil.

Soils get air from the atmosphere, for which soil air has a composition closely related to the atmospheric air. The atmosphere consists of nitrogen (78.084 % by volume) and oxygen (20.9476 %) with smaller quantities of water vapor (variable), carbon dioxide (CO_2) (0.0314 %), inert gases such as argon (0.934 %), and some other rare gases. The concentration of oxygen in soil is less because of utilization by living organisms, roots, and chemical reactants. The concentration of CO_2 in soil air is higher because of biological respiration. The soil CO_2level in active soil reaches 0.15 % at 15 cm, but can reach 5 % and affect the pedosphere chemistry (Adl 2003). O_2 content in well-aerated soils may be about 20 %. It is reduced as the water content increases and the rate of exchange of gas between soil and atmosphere decreases. In some cases, oxygen can be as low as 5–10 %. If the soil remains saturated with water or waterlogged for a long period, say several days, O_2 content may completely be depleted. However, the composition of soil air is highly variable. It depends on the ease of exchange of air between the atmosphere and the soil. The more readily the exchange occurs, the closer is the composition of soil and atmospheric air. Renewal of air in soil is not so rapid because of the discontinuity of soil pores and the obstruction offered by soil water.

Soil texture, structure, and porosity affect the amount and composition of soil air and its renewal. Light textured soil or sandy soil contains much higher air than heavy soil. Diffusion of air occurs slowly in a waterlogged soil. Under prolonged waterlogging, the O_2 in soil air is diminished and the CO_2 concentration highly increases. In waterlogged soils, some volatile gases like H_2S, CH_4, and C_2H_4 may accumulate and exert toxic effects on plant roots. Soil microorganisms utilize O_2 for respiration and decomposition of organic matter and produce CO_2. Hence, soils rich in organic matter contain higher percentage of CO_2 in soil air.

2.5.1 Aeration

Aeration refers to the process of exchange of air and its constituent gases between soil and the atmosphere. Aeration takes place by two physical processes: mass flow and diffusion.

Mass Flow Mass flow occurs when the entire mass of air moves from one place to another, particularly from the soil to the atmosphere and from the atmosphere to the soil under conditions of fluctuating air pressure. Air is driven away from or into the soil pores by this pressure. Movement occurs due to atmospheric pressure changes, soil temperature changes, entry of water following rain and irrigation, evaporation of water from pores, and driving force of wind near soil surface. However, mass flow is much less important than diffusion, except perhaps in layers at or very near the soil surface.

Diffusion Diffusion is the predominant process of soil aeration. In diffusion, individual gas constituent moves separately under the influence of partial pressure gradient. When partial pressure of CO_2 in soil air increases due to root and microbial activity, molecules of CO_2 diffuse from soil to the atmosphere. When O_2 in soil air is consumed for respiration in soil, its partial pressure is reduced and a partial pressure gradient is created between the soil and the atmosphere; then O_2 diffuses into the soil. Generally, there is net diffusion of CO_2 into soil and net diffusion of O_2into the atmosphere. Through diffusion, each gas moves in a direction determined by its own partial pressure. The diffusion of gases in soil is regulated by numbers of air-filled pores and their continuity, both decreasing with increasing soil water content, and the tortuosity of air-filled pores which increase with soil water content.

2.5.2 Effect of Soil Air on Forest Tree Growth

Root respiration is an essential physiological function of a tree related to its healthy growth. The energy released in root respiration is necessary for mineral uptake, synthesis of new protoplasm, and maintenance of cell membranes. In the absence of an adequate supply of free O_2, root respiration is very inefficient and does not release enough energy to maintain essential root functions, particularly mineral uptake (Rosen and Carlson 1984). Richards and Cockroft (1974) observed that root growth was inhibited when air space in soil was reduced to less than 15 %. Root growth was negligible when air space dropped to 2 %. Growth of trees in a number of investigations, including *Pinus sylvestris*, *Picea abies*(Youngberg 1970), *Pinus contorta*(Lees 1972), and citrus (Patt et al. 1966), was found to be severely reduced when the O_2 content dropped below 10 %.

Tree roots generally function well at oxygen levels above 10 % (Kozlowski 1985). At lower concentrations, roots will cease growth and lose their selective permeability (Costello et al. 1991). Metabolic processes are altered and cellular metabolites may accumulate to toxic levels. By-products of anaerobic respiration (lactate, pyruvate, and ethanol) accumulate to toxic levels (Levitt 1980). Also, some root pathogens become active under anaerobic conditions. For example, Oak root fungus, *Armillaria mellea*, may colonize root tissues killed by oxygen stress, and from that food base, invade healthier portions of the root system (Wargo 1981). Likewise, accumulation of ethanol in roots may stimulate attack by *Armillaria* (Wargo and Montgomery 1983). Low oxygen in soil reduces height growth, leaf growth, cambial growth, and reproductive growth of trees that grow on poorly aerated soils, although the amount of growth reduction varies widely among species. Some tree species can tolerate low O_2 supply by (1) producing hypertrophied lenticels which assist in aeration of the stem and release of toxic compounds and (2) growing new roots to replace loss of original roots under anaerobic conditions (Kozlowski 1986). Not only do flooded or compacted forest soils have limited oxygen in the root zone, but vast areas of heavy textured soils may also be poorly aerated (Kramer and Kozlowski 1979). It has been estimated that the soil air to a depth of 1 m does not contain enough O_2 to maintain respiration of plants and microorganisms for more than a few days (Drew 1983).

2.6 Soil Temperature

Soil temperature is an important physical property that regulates most of the physical, chemical, and biological processes of the soil, and the physiological processes of soil organisms and forest plants. Soil temperature has tremendous ecological impacts through evaporation, transpiration, organic matter decomposition, CO_2 emission due to soil respiration, and permafrost thawing. In forest ecosystems, soil temperature regulates microbial transformations of nitrogen (Plymale et al. 1987) and sulfur (Strickland and Fitzgerald 1984), and other nutrients. It controls decomposition of organic matter and formation of humus (Waide et al. 1988), as well as root growth, activity, and respiration (Marshall and Waring 1985).

2.6.1 Factors Affecting Forest Soil Temperature

Temperature of forest soil is influenced by a number of factors, namely meteorological conditions (i.e., solar radiation and air temperature), site topography, soil water, organic matter, texture, and the area of surface covered by litter and canopies of plants (Paul et al. 2004). Since soils obtain heat mainly from the solar radiation, and the amount and intensity of solar radiation depend on geographic, climatic, edaphic, and topographic conditions, soil temperature varies widely among forest types. For example, tropical forest soils are warmer than temperate and tundra soils.

Plant cover, which reduces albedo, can be extremely important to the energy balance of high-latitude ecosystems. For example, Hall et al. (1996) found that jack pine-spruce forests in Saskatchewan and Manitoba had an average albedo of 0.08 compared to a value of 0.8 for bare snow. It has also been shown that the tussock growth form of *Eriophorum vaginatum* improves the soil thermal regime in Alaskan tundra (Chapin et al. 1979). The insulating effect of vegetation arises partly from development of a layer of organic detritus on the soil surface, which reduces daily and seasonal temperature fluctuations. Breshears et al. (1998) reported that winter soil temperatures under patches of woody plants (*Pinus edulis* and *Juniperus monosperma*) in semiarid woodland were warmer than interpatch areas due to the buildup of a litter layer. The insulating properties of litter result from the relatively low thermal conductivity of organic matter compared to mineral soil and the large proportion of air spaces.

The hydrologic cycle has an important role in controlling soil temperature. In a bare soil, color and moisture content mainly determine the albedo (ratio of incident radiation to reflection from surface) (Hanks and Ashcroft 1980). Dark soils absorb more energy than lighter ones, and wetting the soil effectively darkens it. Snow cover has a dominant effect on albedo; the albedo of snow- and ice-covered surfaces ranges from 0.45 to 0.90 (Strahler and Strahler 1983). Snow also acts as an effective insulator which retards the loss of heat from the soil. Bertrand et al. (1994) and Stadler et al. (1996) observed an inverse relationship between depth of snowpack and depth of soil freezing at mid-latitudes. Soil freezing can be prevented entirely under very deep snow (Isard and Schaetzl 1995).

Under field condition, soil temperature can be estimated from air temperature with enough precision (Soil Survey Staff 1975). To get mean annual soil temperature (MAST) in tropical regions, 2.5 °C is added to the air temperature in the soil–atmosphere interface. Since the canopy and the forest vegetation and the litter layers have insulating and buffering effects, this relation may not hold good for forest soils. Experimental studies have shown that vegetation canopies can lower soil temperature during growing season significantly and reduce MAST(Qashu and Zinke 1964; Cermak et al. 1992). Qashu and Zinke (1964) concluded that lower soil temperatures under oak or pine canopies may result from a lower rate of soil warming during springtime. Soils are not good conductors of heat; and, therefore, subsoils are cooler than the surface soil. For the same amount of solar radiation, the soil is heated more and faster than the air in the soil–atmosphere interface. Therefore, surface soil temperature is generally 1–5 °C higher than the air in the immediate soil surface. The surface soil temperature fluctuates with seasons more than that of the subsoil. There may be a depth in the soil where soil temperature fluctuates very little with seasons.

Soil temperature depends on soil cover. Generally, forested sites are cooler than field sites in the summer and warmer in the winter. Forest canopy intercepts solar radiation and acts as an insulator. Clear cutting forests, that is, removing the vegetative cover, increases soil temperature. Spittlehouse and Stathers (1990) observed that maximum daily soil temperature (1-cm depth) in a clear-cut exceeded 50 °C compared to 16 °C in an adjacent mature, western hemlock-fir forest in coastal British Columbia. Forests regulate soil temperature by influencing (1) air temperature, (2) amount and intensity of precipitation reaching the soil, (3) depth and duration of snow cover, (4) wind velocity, (5) shade of living cover, (6) thickness of the forest floor, (7) water content of the soil, (8) humus content of the soil, (9) color of the soil, and (10) structure of the soil. Denuded sites and fields showed greater temperature variation in the winter than forests. Jeffrey (1963) observed that canopy covers of different forest tree species differ in trapping and absorbing solar radiation. Mueller (2002) studied three forest types and observed that a mountain White Pine–Hemlock–Hardwood forest, with boreal components in a valley flat at Ramsey's Draft in Virginia, had growing season soil temperatures lower than those of Oak- Hickory Ridge and Mesic Slope Forests in the Shenandoah Valley. Some early season temperature differences may be as great as 4 °C. Balisky and Burton (1995) suggested that percent canopy cover was the primary control on soil temperature in high-elevation Engelmann Spruce (*Picea engelmannii* Parry ex. Engelm.)-subalpine fir forests.

Leaf litters on the forest floor act as insulators, keeping the soils warmer in the winter by trapping radiant heat. This blanketing effect also decreases the diurnal range in soil temperature especially in the spring. During the summer, forest cover lowers the maximum soil temperature when compared to the fields. The forest floor is a highly efficient insulating layer maintaining low soil temperatures and often a shallow permafrost table (Dyrness 1982). In several experiments, removal of the forest-floor layer was found to exert profound changes in the thermal regime of the soil. For example, forest-floor removal resulted in an active layer 132 cm thick (compared to 47 cm in an undisturbed area) after 3 years (Viereck and Dyrness 1979). Brown et al. (1969) reported similar results after tractor removal of surface organic matter; they stress that melting permafrost often leads to serious problems with erosion. By far, the greatest increases in depth to permafrost were measured on the plots that had the entire forest-floor layer mechanically removed. In these plots, the average increase in active-layer thickness was almost sevenfold (from about 20–137 cm) after 4 years.

Modifying the scheme of Stathers and Spittlehouse (1990), the factors that affect forest soil temperature may be summarized as:

Category	Factor
Weather	Solar radiation
	Air temperature
	Precipitation
	Wind speed
Site	Latitude
	Elevation
	Slope
	Aspect
Surface	Vegetation cover
	Forest floor
	Snow cover
	Surface roughness
Soil	Moisture
	Soil color
	Organic matter
	Texture
	Compaction

2.6.2 Effect of Soil Temperature on Forest Tree Growth and Physiology

It has been shown that root growth of conifer seedlings depends on several environmental factors, including soil temperature and it is a limiting factor for root growth under boreal conditions (Tryon and Chapin 1983; Andersen et al. 1986; Kozlowski and Pallardy 1997). Root growth of several conifer species begins when soil temperature exceeds 5 °C, accelerates rapidly at soil temperatures above 10 °C, reaches a maximum rate at 20 °C, and decreases thereafter (Lopushinsky and Max 1990). In Norway spruce seedlings, new root tips appear after the soil temperature has increased to 5 °C, root growth is limited in the 5–8 °C range (Vapaavuori et al. 1992), and reaches a maximum at around 26 °C (Lyr and Hoffman 1967). Thus, responses to soil temperature are species specific and may vary depending on genotype, local race, or ecotype (Lahti et al. 2005).

Soil temperature influences forest tree growth through its effects on soil processes and tree physiology. Several experimental findings in boreal ecosystems have provided support for this fact (Van Cleve et al. 1990; Stromgren and Linder 2002; Jyske et al. 2011). It is widely accepted that low temperature is the major limiting factor to the growth of tree species in subalpine and boreal forests. However, forest tree species differ in their response to soil temperature. Greenhouse-grown white spruce seedlings were found to be more sensitive to soil temperatures than black spruce seedlings. Root elongation of Alaskan tree seedlings, such as aspen and poplar, increased with increasing temperature under controlled condition. However, the results could be different in field conditions due to modifying effects of soil moisture and other properties. Poor rooting due to low soil temperatures has been suggested as a major cause for failure of reforestation efforts in some unfavorable environments. Soil temperature has been shown to be a major determinant of the length of the growing season in cold ecosystems (Chapin et al. 1979), and it influences community composition and structure (Van Cleve et al. 1981; Farnsworth et al. 1995; Thompson and Naeem 1996).

Soil temperature affects many physiological processes of trees including stomatal conductance, transpiration, nutrient translocation and retranslocation, carbon dioxide fixation (Lawrence and Oechel 1983; DeLucia 1987; Day et al. 1991; Harper and Camm 1993; DeLucia et al. 1991; Landhaausser et al. 1996), carboxylation and electron transport during photosynthesis (Cai and Dang 2002), leaf resistance to heat (Chaisompongpan et al. 1990), carbon/biomass allocation (Pereira 1990; DeLucia et al. 1992), and growth (Heninger and White 1974; Lopushinsky and Max 1990). Low soil temperatures decrease metabolic activity and the turgor of root cells due to reduced water uptake. Nutrient and water uptake are reduced by low soil temperatures. In an experiment, needles of black spruce growing on heated soils contained greater amounts of nitrogen, phosphorus, and potassium and showed 20% higher photosynthetic rates (Van Cleve et al. 1990). Temperature alters lipid and protein contents in root cell plasma membranes (Clarkson et al. 1988; Yoshida and Uemura 1989), affecting their transport properties (Iswari and Palta 1989). The resulting decrease in water uptake rate can cause reduction in photosynthesis by inducing partial stomatal closure (Farquhar et al. 1989) and a consequent decrease in intercellular carbon dioxide concentration in leaves, decreasing the availability of CO_2 for assimilation. Soil temperature can affect nutrient uptake by changing water viscosity and root permeability or through effects on the structure and function of cell membranes (Marschner 1990; Kozlowski and Pallardy 1997). These effects of temperature are, however, short term because both physiological and morphological adjustments in roots can match ion uptake to shoot demand in the long run (Clarkson et al. 1988). At low soil temperatures, decreased nutrient demand of the shoot can retard nutrient uptake (Lambers et al. 1998).

Soil temperature has been shown to influence the rate of symbiotic nitrogen fixation by forest trees. Optimum soil temperature for N fixation by tropical forest tree legumes appears to be between 25 and 30 °C (Reddell et al. 1985). Nitrogen fixation may be limited in higher soil temperatures. High soil temperatures may affect population of rhizobium in tropical soils (Bowen and Kennedy 1959). *Casuarina cunninghamiana*plants showed no nitrogen fixation at 15 °C soil temperature (Reddell et al. 1985). Nodulation and nitrogen fixation occurred at 20 °C and above. The optimum soil temperature was 25 °C for the symbiotic plant and 20 °C for nitrogen-fertilized plants. Reddell et al. (1985) suggested that such temperature requirements for significant nodulation and nitrogen fixation affect the natural distribution of *Casuarina* species. Habish (1970) recorded nodulation and nitrogen fixation by *Acacia mellifera* up to 35 °C soil temperature.

2.6.3 Effect of Soil Temperature on Forest Soil Processes

Soil chemical and biological processes usually have an optimal temperature range where the reaction rate is maximal, and higher and lower temperatures beyond this range reduce the rate of reaction. The temperature that allows for the most rapid growth of microorganisms during a short period, 12 or 24 h, is known as "optimum growth temperature." On the basis of optimum growth temperature, microorganisms are divided into the following categories: psychrophiles (organisms that have optimum growth temperature of 15 °C or lower); mesophiles (organisms that grow best within a temperature range of approximately 25–40 °C); and thermophiles (organisms that grow best at temperature above 45 °C). Since most soil microbes are mesophiles, microbial transformations occur more rapidly in the range of 25–40 °C soil temperature. The optimum temperature for most biological processes in soil lies between 20 and 40 °C; these processes are strongly limited by temperatures below 10 °C and above 40 °C (Stathers and Spittlehouse 1990). Forest soil temperatures in boreal region are well below the optimum values. In many temperate forests, soil temperature is much lower in the winter. Biological and chemical reactions involved in the decomposition of organic matter and evolution of CO_2, are, therefore, much higher in the tropical forests than temperate forests. Many experiments have been conducted by soil temperature manipulations to observe their effects on organic matter decomposition, mineralization, and soil respiration in the climate change. The results may convincingly show the impact of soil temperature on soil processes. More rapid decomposition of litter and greater efflux of CO_2 from soil occur at higher soil temperatures. For example, mass of American beech leaf litter remaining after 2 years of decomposition in a northern hardwood forest at ambient, ambient +5 °C, and ambient +7.5 °C soil temperature treatments was approximately 60, 50, and 42 %, respectively (McHale et al. 1998). The decomposition of sugar maple litter was, on the other hand, unaffected by the soil temperature treatments. In a mixed deciduous forest, Peterjohn et al. (1994) observed that heating soil 5 °C doubled daily rates of net N mineralization in both mineral soil and forest floor, which persisted for the length of the growing season. Heating soil 5 °C increased the estimated annual soil C efflux from 712–1,250 g m^{-2}. The temperature dependence of net N mineralization has been reported for a range of ecosystems, including longleaf pine–wiregrass (Wilson et al. 1999), alpine tundra (Fisk and Schmidt 1995), oak woodland–annual grassland (Leirós et al. 1999), arctic dwarf-shrub heath (Schmidt et al. 1999), and a range of communities along an altitudinal gradient (Powers 1990). Microbial respiration and the mineralization of N and S increase over the range of temperatures most commonly experienced in natural soils. The relationship between soil temperature and CO_2 efflux was found to be strongly exponential at lower temperature ranges but it does not persist at higher temperatures. Microbial sensitivity to temperature diminishes at higher temperatures, indicating the exponential model may not always be appropriate. Kirschbaum (1995) determined the Q_{10} for organic matter decomposition to be ~8.0 at 0 °C, 4.5 at 10 °C, 2.5 at 20 °C, and ~1.0 at 35 °C. Optimum nitrification in forest soils occurs at a temperature between 20 and 35 °C. The decline at higher temperatures may be partially due to increased biological O2 consumption (Parton et al. 2001; Avrahami et al. 2003). The response of denitrification to temperature is similar to that of nitrification, but can have a higher temperature maximum (Strong and Fillery 2002; Simek and Cooper 2002).

Soil respiration refers to the release of CO_2 from soils due to production of CO_2 by roots and soil organisms and, to a lesser extent, chemical oxidation of carbon compounds. Soil temperature strongly affects microbial activity (Macdonald et al. 1995), and hence microbial respiration in soil (Hendrickson 1985). Soil respiration is also sensitive to moisture, and under intermediate moisture conditions, microbial respiration has been found to increase with increasing soil temperature (Schlentner and Van Cleve 1985). Raich and Schlesinger (1992) showed that temperature was the single best predictor of annual soil respiration rates. To explain the relationship between soil temperature and soil respiration, both linear regression (Froment 1972) and Q_{10} relationship (exponential dependence) (Raich and Schlesinger 1992) were used. However, Lloyd and Taylor (1994) evaluated the goodness of fit of various soil temperature (R) and soil temperature (T) relationships and concluded that none of the exponential (Q_{10}) and the simple linear relationships hold good for explaining soil respiration under every situation. They presented an empirical equation which yields an unbiased estimator of respiration rates over a wide range of temperature. In the soil, temperature, substrate availability, and matric potential strongly control microbial metabolism. The temperature dependence of microbial respiration and net N mineralization appears to involve the access to or metabolism of larger substrate pools as soil temperature increases (Macdonald et al. 1995). There may be a shift in microbial community composition with temperatures. Members of the microbial community using the substrates at higher temperature might be different from the microbial community at lower temperatures.

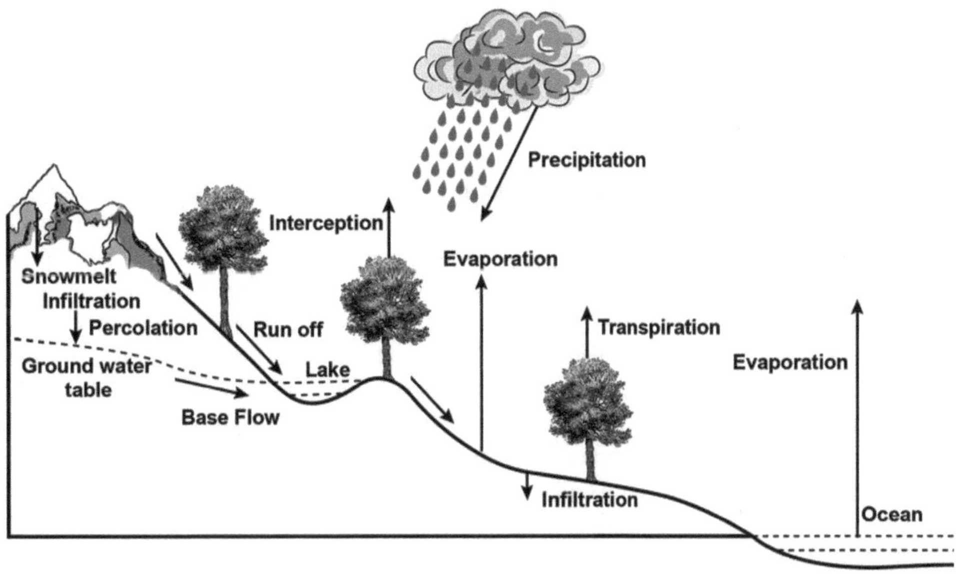

Fig. 2.4 The hydrologic cycle

2.6.4 Control of Forest Soil Temperature

Sometimes forest soil temperature needs to be controlled for tree seed germination, seedling establishment, regeneration, and growth. Different techniques are used for this purpose. One is the regulation of soil moisture status. Since water has higher specific heat than soil particles and needs more energy to be warm, higher moisture content would decrease temperature of warm soils. On the other hand, soil particles get more easily heated than water, so draining the land would improve temperature of cold soils. Covering the soil with plant residues and other organic mulches, or by black plastic cover, may increase or decrease soil temperature depending on their effects on absorption and reflection of solar radiation, infiltration, and evaporation of water. Mulches will tend to lower soil temperature because more solar radiation will be reflected and less absorbed, and the water content of the soil will tend to be greater because more water will infiltrate. Mulches will, again, tend to increase soil temperature because they reduce loss of heat by radiation, and reduce water evaporation from the soil surface. The net effect is to reduce soil temperature in the warm season and increase temperature in the cool season.

Forest clearance, that is, removing the vegetative cover, increases soil temperature. Spittlehouse and Stathers (1990) observed that maximum daily soil temperature (1-cm depth) in a clear-cut exceeded 50 °C compared to 16 °C in an adjacent mature, western hemlock-fir forest in coastal British Columbia. Opening the canopy by thinning is, sometimes, done to increase soil temperature. Mounding and trenching, which are forms of site preparation widely practiced in cold, wet soils of northern forests, have been shown to increase soil temperature, significantly increasing tree seedling survival and growth (Weber et al. 1995; Paterson and Mason 1999).

2.7 Hydrological Properties of Forest Soils

The major reservoirs of water on the earth are oceans (1350×10^6 km^3), polar ice and glaciers (29×10^6 km^3), aquifers (8×10^6 km^3), lakes (0.1×10^6 km^3), soil moisture (0.1×10^6 km^3), atmosphere (0.013×10^6 km^3), rivers (0.002×10^6 km^3), and biosphere (0.001×10^6 km^3). Thus, although water covers 70 % of the Earth's surface, freshwater is scarce. Water in oceans and seas comprises about 97.2 % of the total volume of water. Freshwater is only the remaining 2.8 % including 0.6 % in groundwater and less than 0.1 % in soil water. Three fourths of the freshwater are in the form of ice and permanent snow cover, and approximately one fourth is in groundwater. Only 0.3 % of the total amount of freshwater is stored in lakes, reservoirs, and rivers (Lal and Shukla 2004).

2.7.1 Hydrologic Cycle

Water continuously changes forms and pools among the earth's biosphere, atmosphere, lithosphere, and hydrosphere. This movement of water in a cycle within the pools is known as the hydrological cycle. The pools include various reservoirs such as atmosphere, oceans, lakes, rivers, soils, glaciers, snowfields, and groundwater. The chief hydrologic processes are evaporation, transpiration, condensation, precipitation, runoff, infiltration, percolation, groundwater flow, sublimation, and melting (Fig. 2.4). Forests have an important role in determining the nature and intensity of these hydrologic processes. The generation of precipitation is commonly considered as the beginning of the terrestrial hydrological cycle. Precipitation may be in the form of rainfall or snow. The rain or melt water may be intercepted by vegetation cover or detained by land surface depressions, may infiltrate into the soil, or may run over land surface into streams. The infiltrated water may get stored in the soil as soil moisture or may percolate to deeper layers and ultimately to groundwater.

Circulation of water in the soil–vegetation atmosphere continuum is vital to the energy, carbon, and solute balances of the system. In this cycle, water falls as rainfall and returns to the atmosphere as evapotranspiration (ET). It may be intercepted by vegetation and evaporated directly to the atmosphere. A part of the water that reaches the ground may infiltrate into the soil, may be evaporated from the soil surface, or may be transpired by vegetation. Water may move as surface runoff and it may percolate through the soil to groundwater as recharge. Forests play an important role in the hydrological cycle through the exchange of energy, water, carbon, and other substances.

Soil hydrology is an environmental component playing a strong role in shaping forest structure and composition. For example, inundation of soil with water can lead to death of roots of maladapted species (Vartapetian and Jackson 1997). Impeded drainage, due to either topographic position (Vormisto et al. 2004) or soil type (Pelissier et al. 2001), has been shown to affect plant species distribution. Certain palm species grow preferentially in low-lying areas affected by poor drainage (Svenning 2001; Souza and Martins 2004) or seasonal inundation (Frangi and Lugo 1998). On the other hand, limited water availability under drought conditions can lead to reduced plant growth (Baker et al. 2003), net biomass loss via increased tree mortality (Rolim et al. 2005), and seedling death (Engelbrecht and Kursar 2003; Engelbrecht et al. 2005). Hydrologic factors, such as flooding regime (Salo et al. 1986) and precipitation gradients (Pyke et al. 2001), have been found to covary with patterns in vegetation distribution.

2.7.1.1 Precipitation

Precipitation is the main source of the earth's water supply. It comes from condensation of small water vapor droplets around dust particles, or ice crystals forming clouds. Collision and coalescence make water droplets increase in size and when they attain about 2 mm in diameter, they begin to descend to the earth's surface as rain. By collision and sticking together, ice crystals may form snowflakes which can reach the ground in the form of snow or rain, depending on the temperature of the lower atmosphere.

Precipitation is one of the most basic processes in hydrological cycle and also the basic factor in water balance (Zhang and Xu 2009). The amount of precipitation in a region depends on many factors, including latitude, longitude, solar radiation, temperature, direction of wind, position of oceans, mountains, forests, etc. There has been much discussion on whether forests increase precipitation. Data in support of this proposition have been provided by many climate models. Forests may affect precipitation by modifying temperature, humidity, evaporation, transpiration, and condensation. Large-scale tropical deforestation results in warmer and drier climate. Deforestation of Montane forests can seriously impact the hydrologic cycle of surrounding landscape. Forests increase amount of precipitation, not only in the forest areas themselves, but also in the country surrounding them. In Amazon, 25–30 % of rainfall comes from precipitation recycling. Some studies using general circulation models (GCMs) suggest that on a continental scale forests may affect precipitation (Rowntree 1988; Institute of Hydrology 1994; Xue 1997). Researchers of Duke University analyzed multiple years of data using the National Aeronautics and Space Administration (NASA) Goddard Institute for Space Studies General Circulation Computer Model and Global Precipitation Climatology Project to produce several climate simulations. They found that deforestation in different areas of the globe affects rainfall patterns over a considerable region. Deforestation in the Amazon region of South America (Amazonia) influences rainfall from Mexico to Texas and in the Gulf of Mexico. Similarly, deforesting lands in Central Africa affects precipitation in the upper and lower US Midwest, while deforestation in Southeast Asia was found to alter rainfall in China and the Balkan Peninsula. Tropical deforestation results in decrease in moisture convergence and precipitation because of reduction of ET.

2.7.1.2 Interception

Rainfall interception is an important hydrological process in forested catchments. A part of the precipitation, rain or snow, may be intercepted by vegetation and other surface covers before reaching to the land surface. Some of the rainfall evaporates from the vegetative cover, say canopy of forests, and some may flow down along vegetation stems and through the canopy. Precipitation that is not intercepted will fall as throughfall or stemflow on the forest floor. Rainfall interception varies with species composition, age, and density of vegetation cover. Twenty-five to thirty-five percent rains may be intercepted by a densely growing coniferous forest. Rainfall interception by deciduous hardwood forests may be about 15 % for the period with leaves and about 7 % for the period without leaves. Net interception in the tropical rainforest is approximately 10 % of rainfall. Annual interception of 22–49 % was measured in New Zealand forests (Fahey 1964; Pearce and Rowe 1979), and losses of 20–45 % were found in mature coastal forests of the Pacific Northwest (McMinn 1960; Spittlehouse 1998). Lewis (2003) found 22.3 % canopy interception in a second growth redwood forest at the Caspar Creek Experimental Watersheds. Reid and Lewis (2007) estimated that interception accounted for about 70 % of wet-season ET before logging in a coastal redwood forest. Interception losses of 2–5 % have been measured for litter in deciduous forests of eastern North America (Helvey and Patric 1965). The proportion of rainfall intercepted by forest canopy varies considerably among species. On an average, pine forests intercept 28 % of rainfall compared to 14 % for eucalyptus forests based on Table 2.1. Short grass and crops intercept 20–48 % of rainfall during the growing season. However, on an annual basis the percentage is much smaller compared to forests (Zang et al. 1999). Interception of rainfall measured in aspen, pine and spruce stands in the boreal forest ranges from 9 to 55 % of rainfall, increasing with canopy density. Up to 70 % of intercepted rainfall evaporates directly from the canopy. As a result, direct intercepted rain evaporation from the canopy increases with canopy density and amounts to approximately 25 % of total evaporation for mature evergreens, declining to negligible amounts for clearings (Elliott et al. 1998).

2.7.1.3 Evapotranspiration

Evaporation is the change of state of water from liquid to gas and its transfer from any surface to the atmosphere. Land, lakes, rivers, and oceans send up a steady stream of water vapor. Transpiration is another process of vaporization of water by a biological process where water vapor is released to the atmosphere from vegetation largely through their stomata. The combined loss of water to the atmosphere via the processes of evaporation and transpiration is called ET. At any given temperature, the air can hold only a given amount of moisture known as the saturation humidity. The relative humidity (expressed as percentage) is the ratio of the measured humidity (g water m^{-3} air) to the saturation humidity. Evaporation and transpiration continue until 100 % relative humidity is reached.

ET is a major component of the hydrologic balances in terrestrial ecosystems. It represents 60–75 % of precipitation inputs at the global (Vorosmarty et al. 1998), continental (Sun et al. 2002), and regional scales (Lu et al. 2005). It influences regional water availability and use (Zhang et al. 2004; Sun et al. 2006). In southern China, ET from well-forested eucalyptus plantations can be as high as 90 % of annual precipitation (Zhou et al. 2002). Thus, some tree species may have a drying or moisture-depleting effect. The most direct effect of climate and land use changes on hydrology is alteration of the magnitude and distribution of ET (Dow and DeWalle 2000). Further, ET is also an indicator of ecosystem productivity and biodiversity (Currie 1991; Law et al. 2002), and in fact, it is the only variable that links hydrology and biological processes in most ecosystem models.

ET is difficult to measure on an ecosystem or watershed scale. Methods have been developed to measure ET at the leaf level, the tree level, and the stand level. At the stand level, instruments mounted on a tower above the canopy are routinely used to measure humidity and wind velocities at high frequency, with water fluxes out of the forest canopy calculated by the eddy covariance method (Fisher et al. 2005). Thus, when forests are disturbed on a small scale, the effects on water are relatively well known. Full or partial removal of a small forest, through natural causes or harvest, ET is immediately reduced with a net result of more runoff. ET increases when forests regrow with ET and runoff usually returning to pre-disturbance levels in a few years. In a forest basin, ET is mainly governed by forest structure and climate. Bosch and Hewlett (1982) reviewed 94 catchment experiments and showed statistically that annual streamflow increment following the treatments is proportional to the reduction in forest cover. ET decreases with the reduction in forest cover.

Evaporation in the boreal forest is influenced by interception, soil moisture storage, the albedo (reflectance) of forest canopies, the roughness of canopies, and the ability of roots to access water (Granger and Pomeroy 1997). Aspen canopies reflect twice as much solar radiation as evergreen canopies do, thus remain cooler, with lower evaporation over the summer. Forests are rougher and have better-developed root systems than do the shrubs, grasses, and barren sites in clearings, and so evaporation rates are one-third less in clearings than in mature boreal forests.

2.7.1.4 Infiltration

In the beginning of a rainfall event, water is rapidly soaked into the soil. Water enters the soil and the wetting front advances gradually below the surface. This process by which water (rain, irrigation, snowmelt, wastewater, etc.) on the ground surface enters the soil is called infiltration. The rate of water entry in soil surface is called the infiltration rate, which measures the meters of water entering the soil per unit time. Infiltration rate is the highest at the beginning of rains and it gradually decreases as the rainfall continues and the soil is moistened. Infiltration rate is the minimum when the soil becomes saturated with water. If precipitation exceeds infiltration, water will first be ponded in depressions and then start running off as overland flow or streamflow. Soil properties that affect the infiltration rate include soil texture, structure, porosity, organic matter content, water content, compaction, frozen surface, etc. A sandy soil usually has a higher infiltration rate than a clayey soil. A crust on the soil surface can seal the pores and restrict the entry of water and a compacted zone or an impervious layer close to the surface restricts the entry of water into the soil. The content of water in the soil affects the infiltration rate of the soil. The infiltration rate is generally higher when the soil is initially dry; it decreases as the soil becomes wet. Factors that affect the porosity, especially macroporosity, also affect infiltration rate.

From results of 153 soils, Pitt et al. (1999) observed infiltration values for clays and silts to be 170 mm h^{-1} for non-compacted and 10 mm h^{-1} for compacted soils. Values for sands were 380 mm h^{-1} for non-compacted and 46 mm h^{-1} for compacted soils. Simcock (2006) and Tian et al. (2006) measured infiltration rates for several forest and agricultural soils. They observed higher infiltration rates (121–1,207 mm h^{-1}) in pine forest soils than pasture

(220–330 mm h^{-1}) and agricultural soils (between 3 and 99 mm h^{-1}). In situ infiltration measurements showed that infiltration under grazed pasture was an order of magnitude less than that under pine forest (Taylor et al. 2009). Simcock (2006) showed that infiltration rates of pumice soils were decreased by 5–10 times when vegetation is changed from native shrub land to pasture, and decreased again when vegetation is changed from pasture to crops. Measured macroporosities for forest soils were five times higher than pasture. The high average infiltration capacities measured for soils under forest suggest that very high rainfall intensities are required to generate surface runoff.

Duan et al. (2011) studied infiltration characteristics of forest soils in Yuanyang terrace under six typical plantations (*Alnus cremastogyne*, *Alnus cremastogyne*, *Cunninghamia lanceolata*, *Dillenia indica*, *Cunninghamia lanceolata*+*Quercus cerris*, and *Castanopsis orthacantha*. Results showed that infiltration was closely related with soil physicochemical properties and litter water-holding abilities, etc. The infiltration rate was related positively with soil macroporosity, total porosity, saturated water content, soil organic matter content, etc, and negatively with soil bulk density and capillary porosity. There were great differences among forest types for initial, steady-state, and mean infiltration rates. Burning forests adversely affects the forest floor and thereby reduces infiltration (Robichaud et al. 1993; Robichaud and Waldrop 1994; Robichaud 1996). Fires may result in the development of water repellent soil conditions (DeBano 1981).

In respect to the fact that nearly 60% of the land surface of the Northern Hemisphere is periodically frozen, the infiltration of melt water is an important hydro- and cryological process (Iwata et al. 2007). Snowmelt infiltration into frozen soils has been quantified by percolation tests using ring infiltrometers (van der Kamp et al. 2003), monitoring of the total soil water content (Gray et al. 2001), and monitoring of runoff (Bayard et al. 2005). Soil hydraulic conductivity is generally lower under frozen conditions than under unfrozen conditions. Soil freezing reduces snowmelt infiltration rates and generates runoff (Bayard et al. 2005). When the soil has a relatively high intrinsic permeability and the amount of snowmelt is relatively small, however, snowmelt water may infiltrate into frozen soil completely (Stahli et al. 1999). Snowmelt infiltration processes are affected by many factors including soil temperature, frost depth, pre-winter water content, snowpack thickness, and their complex interactions (Stahli 2005).

2.7.1.5 Percolation

Percolation is the downward movement of water through the soil profile. Water that is infiltrated into soil moves down to the water table by percolation. A part of the water that infiltrates gets evaporated, some is absorbed and transpired by plants, some is retained in pores and on particles, and the remaining water percolates. The factors that affect infiltration also affect percolation. In addition, the depth of the groundwater table (GWT) and presence of compacted layer also influence the percolation. The deeper the GWT, the higher is the rate of percolation. Percolation affects hydraulic conductivity, soil permeability, and drainage.

2.7.1.6 Runoff

Runoff is the movement of water across the surface of the land. Rainfall or snowmelt, in excess of infiltration and ponding, runs off the ground as overland flow and stream flow. There is a general relation—the greater the infiltration, the smaller is the runoff. Infiltration can be increased and runoff can be reduced by soil management practices. Excess water in saturated soils flows into channels, streams and rivers, to the ocean, terminal lakes, or swamps. Groundwater can meet (base flow) stream flow in rivers and lakes if the water table is near the surface. Runoff is influenced by soil physical properties such as texture, structure, porosity, compaction and chemical properties such as organic matter content and exchangeable bases, etc. and slope characteristics.

$R = P + S - I - ET$, where
R = runoff
P = precipitation
S = storage
I = infiltration
ET = evapotranspiration

The runoff response of a catchment is controlled predominantly by rainfall regime, topography, vegetation, and soil hydraulic properties (Dunne 1978). Bonell and Balek (1993) highlighted the differences in these driving variables between temperate landscapes and the humid tropics, where, for example, predominantly clay-rich soils, and high precipitation intensities can generate extensive saturation overland flow where an impeding layer exists near the surface. This combination also suggests that disturbance arising from forest conversion could extend the impeding layer to the surface, favoring the occurrence of infiltration-excess runoff (de Moraes et al. 2006). Degradation of the vegetative cover leads to increasing surface runoff and soil erosion. Raindrops detach soil particles (Rose 1960) which clog soil pores forming a soil crust (Bryan et al. 1984). Raindrops on bare soil may seal the soil surface which results in low infiltration rates and consequently high run off.

The major annual runoff events in boreal forests are usually linked to snowmelt because of saturated or frozen soils. Soils are normally frozen at the time of snowmelt, which restricts the infiltration capacity. Relatively high ice-content soils in clearings (because of low evaporation the previous summer) produce the greatest runoff during spring snowmelt (Pomeroy and Granger 1997). The effect is magnified because of high snow accumulation and rapid snowmelt in clear-cut areas. In summer, the effect of relatively high soil moisture reserves and minimal intercepted rain evaporation in clearings is to promote runoff generation from rainfall. In contrast, jack pine stands over sandy soils produce minimal runoff and the other mature forest sites produce little runoff; however, up to one-third of rainfall becomes runoff from clearings.

2.7.2 Effect of Water Stress on Forest Trees

As soil–water content in soil declines, trees become stressed and begin to react to resource availability changes. A point is reached when water is so inadequately available that tree tissues and processes are damaged. Lack of water eventually leads to catastrophic biological failures and death. Water deficits have both long- and short-term effects on plants. Processes such as stomatal opening and closure can be affected within minutes, whereas processes such as leaf expansion may be affected over a period of months (Myers 1988).

Water stress is the most common limitation on tree growth in forests. Zahner (1969) reported that 80 % of the year-to-year variation in diameter growth of forest trees in humid areas can be related to differences in rainfall and the related variations in water stress. The effects of water stress on tree growth may be classified as directly resulting from loss of turgor and indirectly resulting from disturbance of physiological processes that may or may not be related to loss of turgor. The most important direct effect is decrease in cell enlargement, which results in decreased leaf area, decreased width of annual rings, and decreased size of trees. The first visible effect of water stress is wilting, followed by leaf scorching and defoliation. Among the important indirect effects are those resulting from closure of stomata, inhibition of the photosynthetic machinery, disturbance of enzyme-mediated steps in metabolic processes such as carbohydrate and nitrogen metabolism, and the functioning of growth regulators (Kramer 1987). A major drought effect is the reduction of photosynthesis. This is caused by a decline in leaf expansion, reduction of photosynthetic machinery, premature leaf senescence, and associated reduction in food production.

Effects of water stress on phenology, growth, stomatal activity, and water status were assessed in 2-year-old seedlings of *Quercus frainetto* Ten. (*Quercus conferta* Kit.), *Quercus pubescens* Willd., *Quercus macrolepis* Kotschy (*Quercus aegilops* auct.), and *Quercus ilex* L. growing in containers. When irrigated well, these oak plants tended to develop greater individual leaf area, number of leaves per plant, total plant leaf area, height and root: shoot ratios than water-stressed plants, but the difference between treatments was not significant for any parameter in any species (Fotelli et al. 2000). The variety of ecosystem types in the Mediterranean region reflects the adaptability of native woody plants to drought, because water availability is the main factor regulating the productivity of these ecosystems (Dafis 1982).

Premature senescence and shedding of leaves can be induced by drought. There may be some difference among tree species in this regard. For example, yellow poplar (Liriodendron) is notorious for shedding many leaves during summer droughts, sycamore (Platanus) sheds some leaves, and buckeye (Aesculus) may shed all of its leaves as drought continues. On the other hand, leaves of dogwood (Cornus) usually wilt and die rather than abscise. Growth of vegetative and reproductive tissues is constrained at water stress by cell initiation shortages, cell enlargement problems, and inefficient food supplies. Cell enlargement depends upon hydraulic pressure for expansion and is especially sensitive to water stress.

2.7.3 Effect of Waterlogging on Forest Trees

Plants become affected in a number of ways when soils are waterlogged. When soil is flooded, the flow of oxygen through the soil becomes restricted. Pezeshki (1994) stated that once the soil becomes anaerobic, adverse effects on plants occur such as chlorosis, reduced growth rate, disruption of cell membranes, reduced mineral uptake, altered growth regulator relationships, stomatal closure, leaf wilting and epinasty, reduced photosynthesis and respiration, altered carbohydrate partitioning, and potentially death (Hamid et al. 2009). Soil water saturation over prolonged periods of time produces a negative impact on many terrestrial plants, due to slow diffusion rates of gases in water (Armstrong 2002), which hampers oxygen supply to roots and reduces respiration and photosynthesis rates (Voesenek et al. 2006). Waterlogging by periodic to continuous flooding of bottomland is the major factor affecting tree regeneration in many forested wetlands (Kevin and Brooks 2003; Sakio 2005; Trowbridge et al. 2005; Battaglia and Sharitz 2006). Mediterranean ecosystems receive almost all of their yearly rainfall during the autumn–spring cold period, in contrast with the prolonged, warm dry summer. Precipitation can be very abundant in autumn–spring, with frequent events of soil waterlogging. Tree species can show important differences in tolerance to waterlogging during early stages of regeneration, even within the same genus, depending on the environmental conditions of their habitats. Several authors have reported that seeds of tree species occurring in potentially flooded sites are not affected by waterlogging (Guo et al. 1998) or they are able to keep under water during long periods of time without significant loss of viability (DuBarry 1963). On the other hand, excess water affects negatively the capacity of seed germination in many forest tree species that are less adapted to flooding (Walls et al. 2005). Inter-specific differences in tolerance to waterlogging may determine the structure and composition of the community along a gradient of soil humidity (Perez-Ramos and Maranon 2009). Natural forests in floodplains have evolved to handle periodic flooding and, therefore, many trees in the flooded areas will indeed survive. Flood stressed trees exhibit a range of symptoms including leaf chlorosis (yellowing), defoliation, reduced leaf size and shoot growth, sprouting, and crown dieback. Early fall coloration and leaf drop often occur. There have been numerous investigations on the effects of waterlogging on woody species but most on seedlings for convenience of flooding treatments (Coutts 1982). An experiment was conducted to test the tolerance of recently planted *Corymbia maculata* (spotted gum, syn. *Eucalyptus maculata*), *Lophostemon confertus* (brush box), *Platanus orientalis* (oriental plane), and *Platanus acerifolia* (London plane) to waterlogging. The species were found to vary considerably in their ability to tolerate and recover from waterlogging. Waterlogging suppressed root and shoot growth in all experimental species (Smith et al. 2001). Tolerance of seedlings of three oak species (*Quercus robur, Q. rubra, Q. palustris*) to waterlogging for a period of 7 weeks was tested by Colin-Belgrand et al. (1991) under greenhouse conditions. A permanent water table was maintained at 6 cm below the soil surface. Waterlogging had strong effects on root development; flooded roots decayed, while hypertrophied lenticels and subsequently adventitious roots appeared on the taproot. Leaf nutrient contents decreased markedly as a result of waterlogging. A simulated field experiment was conducted to study the response in growth and energy-metabolic enzyme activities of 1-year-old seedlings of *Taxodium distichum*, *Carya illinoensis*, and *Sapium sebiferum* to waterlogging and flooding for a period of 60 days. Under waterlogging and flooding, the relative growth of test tree species was in the order of *T. distichum* > *C. illinoensis* > *S. sebiferum*, indicating that *T. distichum* had the strongest tolerance against waterlogging and flooding, while *S. sebiferum* had the weakest one (Wang et al. 2010).

2.8 Soil Water

The amount of water present in soil is determined by the loss in weight of soil after drying in an oven at 105 °C for a period needed to reach a constant weight. It can be expressed as volume per unit volume ($m^3\ m^{-3}$) or mass per unit mass ($g\ kg^{-1}$) or their percentages. The volume of water per unit volume of soil is called the volumetric water content, denoted by θ_v. The mass of water per unit mass of soil (dry weight) is called the mass water content which is denoted by θ_m. The two are related as:

$\theta_v = \theta_v \frac{p_b}{p_w}$, where ρ_w is the density of water (1000 kg m^{-3} at 20 °C) and ρ_b is the bulk density.

Water is present in soils in different forms, including hygroscopic water, capillary water, and non-capillary or gravitational water. Whatever the form of water may be, soil water is retained by two different forces, adhesion and cohesion. Adhesion is the attraction between a particle and a water molecule. It is the retention of water vapor on soil colloids. Water molecules are arranged in the form of a thin film around the particles by adhesion. The force of attraction in the inner side of the film is greater than that in the outer side. Some water may also be retained on pore walls by the force of adhesion. Water on particle and pore surfaces is known as hygroscopic water. Hygroscopic water is so strongly held in soil that it cannot be withdrawn and absorbed by plant roots. The force of attraction between water molecules is called cohesion. Cohesion is a much weaker force than adhesion. Most water in capillary pores is retained by cohesion. This is known as capillary water, a considerable part of which can be absorbed by plant roots. After a rainfall event or irrigation, the large soil pores, the non-capillary pores also become filled with water. This water is easily pulled downward by the gravity. This water is very temporarily held in soil, and some hours after the cessation of rain, it drains away. This water is called gravitational or superfluous water.

When an oven dry soil is exposed to the humid air, water vapor is adsorbed on the soil particles. During soil wetting, water saturates the particles first, entering the capillary pores next. The amount of water retained in soil when particle surfaces and all the capillary pores are filled with water is called the field capacity. Since water retained above field capacity, the gravitational water, is not of particular benefit for plants because of its immediate removal, field capacity is taken to be the upper limit of plant's available (absorbable) water. Suppose a plant is growing in a soil at field capacity. If further water is not applied, water will be reduced with time in the absence of percolation by evaporation and transpiration. At a point of water loss, water will be held in soil so strongly that plant roots will not be able to absorb it. The leaves of many plants will lose turgidity and plants will wilt (some plants may avoid wilting by closing stomata, shedding leaves, stunting growth, etc.). If the soil is watered again, plants may recover from wilting. But if further water loss proceeds, a point will be reached when plants will not be able to recover; they will wilt permanently. The amount of water retained in soil in this condition is called wilting point; the percentage of water retained is called permanent wilting percentage. In any way, the difference between field capacity and wilting point is theoretically the available water. The field capacity varies widely among soils depending on texture, structure, bulk density, porosity, and organic matter. The wilting point also varies enormously depending on these soil properties and the plant characteristics, including leaf area, stomatal density, leaf resistance, rooting density, rooting depth, active roots, etc. Thus, available water is a variable entity. All plants do not wilt at the same water content. The same plant may wilt in different soils at different water contents. Estimating available water content in soil is, therefore, difficult. It is more difficult in forest soils because roots are not distributed uniformly throughout the whole profile, and wilting point is not reached in different horizons all at the same time. The amount of water that could be retained when all the pores, large or small, are filled with water is called maximum water holding capacity. It is not a good index of water status of arable soils. It may be important in saturated soil.

Water is held in soil by variable suction (negative pressure). It is customary to express soil water relations in terms of water potential instead of pressure. Soil water potential is the amount of free energy of soil water relative to the free energy content of pure free water at 20 °C under one atmospheric pressure. Water potential is the capacity of the water to move within the soil plant system. Water moves in soil from one point to another because there is difference in soil water potential between the two points. Water moves from the point of higher potential to a point of lower potential. In other words, water moves in the soil–plant system along a potential gradient. Soil water potential is the sum of two potentials, matric potential (Ψ_m; energy of water attached to soil matrix) and solute or osmotic potential (Ψ_s; energy of orientation of water molecules around solute molecules).

Soil water potential, $\psi_{soil} = \psi_m + \psi_s$

Since matric potential and solute potential are both negative entities, soil water potential (Ψ_{soil}) is also negative. When the soil is at the maximum water holding capacity, soil water has a potential nearer to pure free water (0 kPa); at field capacity, average Ψ_{soil} is around −0.3 MPa and at wilting point it is −15 MPa. Therefore, the amount of water that can be held in soil between −15 MPa and −0.3 MPa soil water potential is taken to be the available water.

Let there be two points in soil: point A and point B. If soil water potential is higher at point A than at B, then water will move from point A to point B.

$$\Psi_A > \Psi_B$$

Direction of water movement

$$\text{Point A} \longrightarrow \text{Point B}$$

The rate of movement of water will increase as the magnitude of difference in potentials between two points increases. That is, the rate of water movement in soil (and soil plant system) is proportional to the potential gradient, Ψ.

$R \alpha \Delta\psi$, or $R = K\Delta\psi$ where R is the rate of flow and K is the proportionality constant, popularly known as the hydraulic conductivity. Hydraulic conductivity is the ease of transmittance of water through soil. Hydraulic conductivity ($K\theta$) is defined as "the metres per day of water seeping into the soil under the pull of gravity or under a unit hydraulic gradient" (Kirkham 2005). Hydraulic conductivity depends on porosity (pore size, pore volume, continuity of pores, etc.) and soil water content. As long as the physical properties of a soil and water content do not change, the hydraulic conductivity is constant. Water moves in soil under saturated and unsaturated conditions. Rate of flow under these two conditions at a particular hydraulic gradient is usually different, and consequently there is saturated soil hydraulic conductivity (K_s) and unsaturated soil hydraulic conductivity (K_θ).

Compared with agricultural soils, forest soils and those under native vegetation generally feature low bulk density and high saturated hydraulic conductivity, total porosity, and macroporosity, as a result of ample litter cover, organic inputs, root growth and decay, and abundant burrowing fauna (Lee and Foster 1991). Replacement of natural vegetation with managed land cover is generally associated with decreased rooting networks and faunal activity, thereby reducing the potential for well-developed macropore networks (Reiners et al. 1994; Schwartz et al. 2003). The rooting systems of woody vegetation such as forest and shrubland demonstrate substantially greater depth, diameter, dispersion, and biomass than rooting systems of herbaceous plants or cultivated crops (Lee and Lauenroth 1994; Jackson et al. 1996; Messing et al. 1997). Forest cover has been associated with lower bulk density and greater saturated hydraulic conductivity than pasture in different climates and parent materials throughout the world (Reiners et al. 1994; Godsey and Elsenbeer 2002; Jimenez et al. 2006; Li and Shao 2006; Abbasi et al. 2007).

Several studies have shown significant negative correlations between saturated hydraulic conductivity and clay content (Talsma and Flint 1958; Bonsu and Lal 1982) and positive correlations with sand content. Bonsu and Masopeh (1996) investigated saturated hydraulic conductivity of some forest soils of Ghana and observed it to vary from 0.24×10^4 ms^{-1}–8.4×10^4 ms^{-1}. Saturated hydraulic conductivity may be employed in modeling infiltration, which is used in predicting erosion and runoff of a watershed (Mein and Larson 1973). Saturated hydraulic conductivity may also be used in the study of spatial variability of soils (Ahuja and Nielson 1984; Bonsu and Laryea 1989).

Sparling et al. (2000) observed a wide range of variation in unsaturated hydraulic conductivity of forest soils. In his study, the highest unsaturated hydraulic conductivity was found under the beech forest remnant (31 mm h^{-1}). It was also high in the topsoil under podocarp forest (31 mm h^{-1}). Unsaturated hydraulic conductivity (K_θ) is the single most important hydraulic parameter for flow and transport of water and related phenomena in soil. Olorunfemi and Fasinmirin (2011) studied the effects of soil moisture contents, bulk density, total porosity, water holding capacity, organic matter content, and cation exchange capacity on unsaturated hydraulic conductivity of various sized aggregates of the A horizon of the tropical rain forest zone of Nigeria. At 2 cm s^{-1} suction rate, the mean value of K_θ ranged from

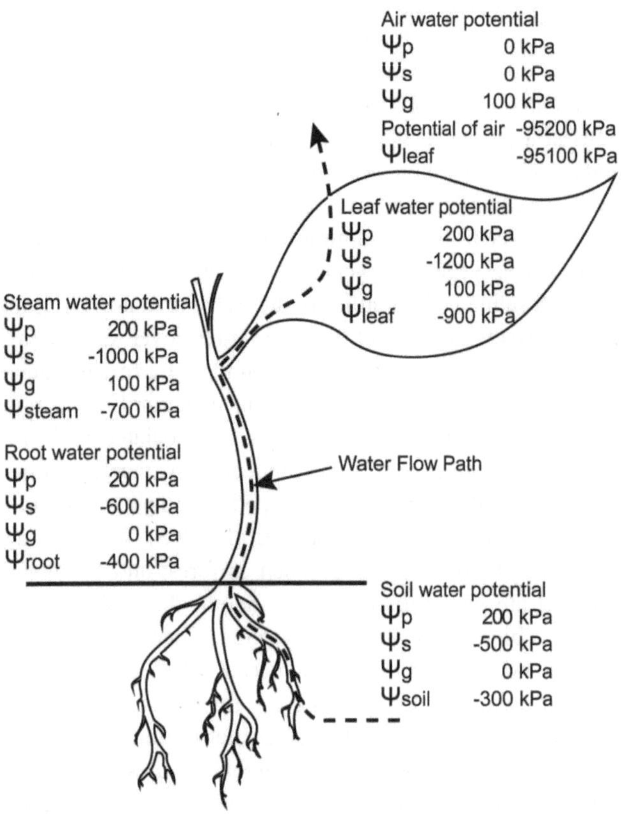

Fig. 2.5 Water potential gradient along soil–plant–atmosphere system

0.0022 ± 0.001 cm s^{-1}–0.00071 ± 0.0004 cm s^{-1}. Unsaturated hydraulic conductivity was positively and significantly correlated with total porosity and water holding capacity. Abrisqueta et al. (2006) compared unsaturated hydraulic conductivity for disturbed and undisturbed forest soil. For soil moisture values close to saturation, K_θ values were >0.392 and >0.019 cm h^{-1}, for undisturbed and disturbed soils, respectively.

Plant water potential is the sum of three potentials—osmotic potential (ψ_s), pressure potential (ψ_p), and gravitational potential (ψ_g). Matric potential (ψ_m) is negligible in plant tissue because little suction is applied by cell walls on water molecules. However, turgidity exerts positive pressure on water increasing its tendency to move from cell to cell; so the pressure potential is positive. Pressure potential may be negative in highly flaccid cells but this is a rare case. Osmotic potential is created by the dissolved ions in cell sap. It is always negative.

Plant water potential, $\Psi plant = \psi s + \psi p + \psi g$

ψ_s is negative, ψ_p and ψ_g are positive; and thus, theoretically, Ψ_{plant} may be negative or positive depending on the magnitude of the pressure and gravitational potential. In practicality, ψ_s dominates over $\psi_p + \psi_g$ in plant tissues and cell water potential is usually negative. Soil water is absorbed by plant roots, transmitted through the stem to the leaves and is transpired through the leaves. Along this soil–plant–atmosphere continuum, water moves along a potential gradient (Fig. 2.5).

Study Questions

1. What are the different types of mineral particles in soil? Discuss the significance of different assemblages and arrangements of these particles in a forest soil. Make a list of forest trees suitable for different soil textures.
2. What do you mean by density of soil? How does it influence porosity, water movement, and aeration of soil? Why is the bulk density of forest soil usually lower than that of agricultural soils?
3. Discuss forest soil temperature with regard to variation and impact on forest tree growth.
4. Discuss important hydrological processes of forest soils in respect of water storage and movement.
5. What is water potential? What is the relation between water potential and hydraulic conductivity? Discuss the role of water potential in the movement of water within the soil–plant–atmosphere continuum.

References

Abbasi MK, Zafar M, Khan SR (2007) Influence of different land-cover types on the changes of selected soil properties in the mountain region of Rawalakot Azad Jammu and Kashmir. Nutr Cycl Agroecosyst 78:97–110

Abrams MD (1998) The red maple paradox. BioScience 48(5):355

Abrisqueta JM, Plana V, Ruiz-Canales A, Ruiz-Sánchez MC (2006) Unsaturated hydraulic conductivity of disturbed and undisturbed loam soil. Span J Agric Res 4(1):91–96

Adl SM (2003) The ecology of soil decomposition. CAB International, Oxon

Ahuja LR, Nelson JW (1984) Scaling soil water properties and infiltration modeling. Soil Sci Soc Am J 48:699–702

Andersen CP, Sucoff EI, Dixon RK (1986) Effects of root zone temperature on root initiation and elongation in red pine seedlings. Can J For Res 16:696–700

Ares A, Terry T, Harrington C, Devine W, Peter D, Bailey J (2007) Biomass removal, soil compaction, and vegetation control effects on five-year growth of Douglas-fir in coastal Washington. For Sci 53:600–610

Aringhieri R, Sequi P (1978) The arrangement of organic matter in a soil crumb. In: Emerson WW et al. (eds) Modification of soil structure. Wiley, New York

Armstrong WDM (2002) Root growth and metabolism under oxygen deficiency. In: Waisel Y, Eshel A, Kafkafi U (ed) Plant roots: the hidden half. Marcel Dekker, New York

Aspiras RB, Allen ON, Chesters G, Harris RF (1971) Chemical and physical stability of microbially stabilized aggregates. Soil Sci Soc Am Proc 35:283–286

Avrahami S, Liesack W, Conrad R (2003) Effects of temperature and fertilizer on activity and community structure of soil ammonia oxidizers. Environ Microbiol 5:691–705

Baker IT, Denning AS, Hanan N, Prihodko L, Vidale PL, Davis K, Bakwin P (2003) Simulated and observed fluxes of sensible and latent heat and CO_2 at the WLEF-TV Tower using SiB2.5. Global Change Biol 9:1262–1277

Balisky AC, Burton PJ (1995) Root-zone soil temperature variation associated with microsite characteristics in high-elevation forest openings in the interior of British Columbia. Agr Forest Meteorol 77:31–54

Barthes B, Roose E (2002) Aggregate stability as an indicator of soil susceptibility to runoff and erosion validation at several levels. Catena 47:133–149

Battaglia LL, Sharitz RR (2006) Responses of floodplain forest species to spatially condensed gradients: a test of the flood-shade tolerance trade-off hypothesis. Oecologia 147:108–118

Bayard D, Stahli M, Parriaux A, Fluhler H (2005) The influence of seasonally frozen soil on snowmelt runoff at two Alpine sites in southern Switzerland. J Hydrol 209:66–84

Becker DD, Haagen JT, Knight WR (1977) Interim soil survey report for Cameron and Elk Counties, Pennsylvania. US Department of Agriculture, Soil Conservation Service, Washington, DC

Berben JC (1973) Effect of soil density and precipitation on the growth and root development of some forest tree species. Bull Soc R For Beig 80:377–401

Bertrand A, Robitaille G, Nadeau P, Boutin R (1994) Effects of soil freezing and drought stress on abscisic acid content of sugar maple sap and leaves. Tree Physiol 14:413–425

Bonell M, Balek J (1993) Recent scientific developments and research needs in hydrological processes of the humid tropics. In: Bonell M, Hufschmidt MM, Gladwell JS (eds) Hydrology and water management in the humid tropics. Hydrological research and strategies for water management. Cambridge University Press, UNESCO

Bonsu M, Lal R (1982) Hydraulic properties of some Alfisols of Western Nigeria: a comparison of field and laboratory methods. Niger J Soil Sci 35:198–206

Bonsu M, Laryea KB (1989) Scaling the saturated hydraulic conductivity of an Alfisol. J Soil Sci 40:731–742

Bonsu M, Masopeh BA (1996) Saturated hydraulic conductivity values of some forest soils of Ghana determined by a simple method. Ghana J Agric Sci 29:75–80

Bosch JM, Hewlett JD (1982) A review of catchment experiments to determine the effect of vegetation changes on water yield and evapotranspiration. J Hydrol 55:3–23

Bowen GD, Kennedy M (1959) Effect of high soil temperatures on *Rhizobium* spp. Queensland Journal of Agricultural Science 16:177–197

Brady NC, Weil RR (2002) The nature and properties of Soils, 13th edn. Pearson Education, Inc., New Delhi

Breshears D, Nyhan JW, Heil CE, Wilcox BP (1998) Effects of woody plants on microclimate in a semiarid woodland: soil temperature and evaporation in canopy and intercanopy patches. Int J Plant Sci 159:101–1017

Bronick CJ, Lal R (2005) Soil structure and management: a review. Geoderma 124:3–22

Brown J, Warren R, Donald V (1969) The effect of disturbance on permafrost terrain. Spec Rep 138. (Hanover)

Bryan RB, Imeson AC, Campbell IA (1984) Solute release and sediment entrainment on micro-catchments in the Dinosaur Park Badlands Alberta, Canada. Hydrology 71:79–106

Bulmer C, Venner K, Prescott C (2007) Forest soil rehabilitation with tillage and wood waste enhances seedling establishment but not height after 8 years. Can J Forest Res 37:1894–1906

Cai T, Dang QL (2002) Effects of soil temperature on parameters of a coupled photosynthesis-stomatal conductance model. Tree Physiol 22:819–827

Campbell CA, Souster W (1982) Loss of organic matter and potentially mineralizable nitrogen from Saskatchewan soils due to cropping. Can J Soil Sci 62:651–656

Celine M, Dominique A, Michel G, Daniel A (1996) Short range soil variability and its effect on red oak growth (*Quercus rubra* L.). Soil Sci 161(1):29–38

Cermak V, Bodri L, Safanda J (1992) Underground temperature fields and changing climate: evident from Cuba. Global Planet Change, 5(4):325–337

Cerutti JR, James R, Backer AD (1971) Warren County Pennsylvania interim soil survey report. U.S. Department of Agriculture, Soil Conservation Service, Pennsylvania State University, and Pennsylvania Department of Environmental Resources. University Park, PA

Chaisompongpan N, Li PH, Davis DW, Markhart III AH (1990) Root cooling protects plants from heat injury. In Davies WJ, Jeffcoat B (ed) Importance of root to shoot communication in the responses to environmental stress. Bri Soc Plant Growth Reg Mon No 21

Chapin FSI, Van Cleve K, Chapin MC (1979) Soil temperature and nutrient cycling in the tussock growth form of *Eriophorum vaginatum*. J Ecol 67:169–189

Chittendon F (1956) RHS Dictionary of Plants. Oxford University Press, Oxford

Clarkson DT, Earnshaw MJ, White PJ, Cooper HD (1988) Temperature dependent factors influencing nutrient uptake: an analysis of responses at different levels of organization. In Long SP, Woodward FI (ed) Plants and temperature. Symp 42, Society for Experimental Biology, Cambridge

Colin-Belgrand M, Dreyer E, Biron P (1991) Sensitivity of seedlings from different oak species to waterlogging: effects on root growth and mineral nutrition. Ann For Sci 48:193–204

Corns IGW (1988) Compaction by forestry equipment and effects on coniferous seedling growth on four soils in the Alberta foothills. Can J For Res 18:75–84

Costello LR, MacDonald JD, Jacobs KA (1991) Soil aeration and tree health: correlating soil oxygen measurements with the decline of established oaks. USDA Forest Service Gen. Tech. Rep. PSW–126

Coutts MP (1982) The tolerance of tree roots to waterlogging v. growth of woody roots of sitka spruce and lodgepole pine in waterlogged soil. New Phytol 90:467–476

Currie DJ (1991) Energy and Large-Scale Patterns of Animal and Plant-Species Richness. The American Naturalist 137:27–49

Dafis S (1982) The present state of the forests in our country. Scientific Bull Geotechnica 18–22

Day PR (1965) Particle fractionation and particle-size analysis. p. 545–567. In CA Black et al (ed) Methods of soil analysis, Part I. Agronomy 9:545–567

Day TA, Heckathorn SA, Delucia EH (1991) Limitations of photosynthesis in *Pinus taeda* L. (loblolly pine) at low soil temperatures. Plant Physiol 96:1246–1254

de Moraes JM, Schuler AE, Dunne T, Figueiredo RO, Victoria RL (2006) Water storage and runoff processes in plinthic soils under forest and pasture in Eastern Amazonia. Hydrol Process 20:2509–2526

DeBano LF (1981) Water repellent soils: a state-of-the-art. USDA Forest Service General Technical Report PSW-46, USDA

DeLucia EH (1987) The effect of freezing nights on photosynthesis, stomatal conductance in seedlings of Engelmann spruce. Plant Cell Environ 10:333–338

DeLucia EH, Day TA, Oquist G (1991) The potential for photoinhibition of *Pinus sylvestris* L. seedlings exposed to high light and low soil temperature. J Experiment Bot 42:611–617

DeLucia EH, Heckathorn SA, Day TA (1992) Effects of soil temperature on growth, biomass allocation and resource acquisition of *Andropogon gerardii* Vitman. New Phytol 120:543–549

Dow CL, DeWalle DR (2000) Trends in evaporation and Bowen ratio on urbanizing watersheds in eastern United States. Water Resour Res 36:1835–1843

Drew MC (1983) Plant injury and adaptation to oxygen deficiency in the root environment. A review. Plant Soil 75:179–199

Duan XF, Song WF, Li J, Wang L (2011) Soil infiltration characteristics in water conservation forest of terrace in Yuanyang County of Yunnan Province. Bulletin of Soil and Water Conservation 2011(4):47–52

DuBarry AP (1963) Germination of bottomland tree seed while immersed in water. J For 61:225–226

Duffy PD, McClurkin DC (1974) Difficult eroded planting sites in North Mississippi evaluated by discriminant analysis. Soil Sci Soc Am Proc 38:676–378

Duiker SW, Rhoton FE, Torrent J, Smeck NE, Lal R (2003) Iron (hydr) oxide crystallinity effects on soil aggregation. Soil Sci Soc Am J 67:606–611

Duke JA (1983) Handbook of energy crops (unpublished). Published only on the Internet, http://www.hort.purdue.edu/newcrop/duke_energy/refa-f.html. (Accessed on 12 Aug, 2012)

Dunne T (1978) Field studies of hillslope flow processes. In: Kirkby MJ (ed) Hillslope hydrology. Wiley, Chichester

Dyrness CT (1982) Control of depth to permafrost and soil temperature by the forest floor in black spruce/feather moss communities. Research Note, PNW-396, USDA

Edwards AP, Bremner JM (1967) Microaggregates in soils. Eur J Soil Science 18(1):64–73

Elliott ET, Anderson RV, Coleman DC, Cole CV (1980) Habitable pore space and microbial trophic interactions. Oikos 35:327–335

Elliott JA, Toth BM, Granger RJ, Pomeroy JW (1998) Soil moisture storage in mature and replanted sub-humid boreal forest stands. Can J Soil Sci 78:17–27

Engelbrecht BM, Kursar TA (2003) Comparative drought-resistance of seedlings of 28 species of co-occurring tropical woody plants. Oecologia 136:383–393

Engelbrecht BMJ, Kursar TA, Tyree MT (2005) Drought effects on seedling survival in a tropical moist forest. Trees 19:312–321

Evans J (2010) Native Australian trees suited to sandy soils. http://www.helium.com/items/1881299-native-australian-trees-suited-to-sandy-soils. (Accessed on 14 Aug, 2012)

Eyre FH (1980) Forest cover types of the United States and Canada. Society of American Foresters, Washington, DC

Fahey BD (1964) Throughfall and interception of rainfall in a stand of radiata pine. J Hydrol 3:17–26

FAO (2006) World Reference Base for soil resources 2006. FAO, Rome

Farnsworth EJ, Nunez-Farfan J, Careaga SA, Bazzaz FA (1995) Phenology and growth of three temperate forest life forms in response to artificial soil warming. J Ecol 83:967–977

Farquhar GD, Wong SC, Evans JR, Hubick KT (1989) Photosynthesis and gas exchange. In: Jones HG, Flowers TJ, Jones MB (ed) Plants under stress. Cambridge University Press, Cambridge

Firth I, Murphy G (1989) Skidtrails and their effect on the growth and management of young Pinus rudiata. New Zealand J Forest Sci 19:22–28

Fisher JB, DeBiase TA, Ye Q, Xu M, Goldstein AH (2005) Evapotranspiration models compared on a Sierra Nevada forest ecosystem. Environmental Modeling Software 20:783–796

Fisk MC, Schmidt SK (1995) Nitrogen mineralization and microbial biomass nitrogen dynamics in three alpine tundra communities. Soil Sci Soc Am J 59:1036–1043

Fotelli MN, Radoglou KM, Constantinidou HIA (2000) Water stress responses of seedlings of four Mediterranean oak species. Tree Physiol 20:1065–1075

Foth HD (1990) Fundamentals of soil science, 8th edn. Wiley. New York

Frangi JL, Lugo AE (1998) A floodplain palm forest in the Luquillo Mountains of Puerto Rico five years after hurricane Hugo. Biotropica 30:339–348

Franzluebbers AJ (2002) Water infiltration and soil structure related to organic matter and its stratification with depth. Soil Tillage Res 66:197–205

Froelich HA (1973) The impact of even-aged forest management on physical properties of soils. In: Herman RK, Lavender DP (eds) Even-age management. School of Forestry, Oreg, State Univ., Corvallis, pp 199–220

Froehlich HA (1979) Soil compaction from logging equipment: effects on growth of ponderosa pine. J Soil Water Conserv 34:276–278

Froehlich HA, McNabb DH (1984) Minimizing soil compaction in Pacific Northwest forests. In Stone EL (ed) Forest soils and treatment impacts. Department of Forestry, University of Tennessee: Knoxville, pp 159–192

Froelich HA, Miles DWR, Robbins RW (1986) Growth of young Pinus ponderosa and Pinus contorta on compacted soil in Central Washington. Forest Ecol Manage 15:285–294

Froment A (1972) Soil respiration in a mixed oak forest. Oikos 23:273–277

Gale MR, Grigal DF (1987) Vertical root distribution of northern tree species in relation to successional status. Can J For Res 17:829–834

Gilliam FS, Yurish BM, Goodwin LM (1993) Community composition of an old growth Longleaf pine forest: relationship of soil texture. Bull Torrey Bot Club 120:287–294

Godsey S, Elsenbeer H (2002) The soil hydrologic response to forest regrowth: a case study from southwestern Amazonia. Hydrol Process 16:1519–1522

Granger RJ, Pomeroy JW (1997) Sustainability of the western Canadian boreal forest under changing hydrological conditions—2—summer energy and water use. In Rosjberg D, Boutayeb N, Gustard A, Kundzewicz Z Rasmussen P (ed) Sustainability of water resources under increasing uncertainty. IAHS Publ No. 240. IAHS, Wallingford

Gray DM, Toth B, Zhao L, Pomeroy JW, Granger RJ (2001) Estimating areal snowmelt infiltration into frozen soils. Hydrol Process 15:3095–3111

Greacen EL, Sands R (1980) Compaction of forest soils: a review. Aust J Soil Res 18:163–189

Guo Y, Shelton MG, Lockhart BR (1998) Effects of flood duration and season on germination of black, cherrybark, northern red, and water oak acorns. New For 15:69–76

Habish HA (1970) Effect of certain soil conditions on nodulation of *Acacia* spp. Plant Soil 33:1–6

Haines LW, Maki TE, Sanderford SC (1975) The effect of mechanical site preparation treatments on soil, productivity and tree growth. For Soils and For Land Management. Les Presses de I'Universite Laval: Quebec, pp 379–395

Hall FG, Sellers PJ, Williams DI (1996) Initial results from the Boreal ecosystem-atmosphere experiment, BOREAS. Silva Fenn 30:109–212

Hamid HA, Shukor NAA, Mat S, Senin AL (2009) Effects of Waterlogging on Growth and Physiology of *Hopea odorata* Roxb. Intl J Biol 1(2):87–93

Hamilton WN, Krause HH (1985) Relationship between jack pine growth and site variables in New Brunswick plantations. Can J Forest Res 15:922–926

Hanks RJ, Ashcroft GL (1980) Applied soil physics. Springer, Berlin

Harper GJ, Camm EL (1993) Effects of frozen storage duration and soil temperature on the stomatal conductance and net photosynthesis of *Picea glauca* seedlings. Can J Forest Res 23:2459–2466

Hatchell GE, Ralston CW (1971) Natural recovery of surface soils disturbed in logging. US For Serv Tree Plant Notes 22:5–9

Hattori T (1988) Soil aggregates in microhabitats of microorganisms. Rep Inst Agric Res Tohoku Univ 37:23–36

Helvey JD, Patric JH (1965) Canopy and litter interception by hardwoods. Water Resour Res 1:193–206

Hendrickson OQ (1985) Variation in the C/N ratio of substrate mineralized during forest humus decomposition. Soil Biol Biochem 17:435–440

Heninger RL, White DP (1974) Tree seedling growth at different soil temperatures. Forest Sci 20:363–367

Hillel D (1980) Fundamentals of soil physics. Academic, New York

Hossain MM (2012) Germination and initial growth of tree seedlings on deforested and natural forest soil at Dulhazara, Bangladesh. Rev biol trop 60(4):

Hough AF (1965) Black cherry (*Prunus serotina* Ehrh.). In: Fowells HA (ed) Silvics of forest trees of the United States. U.S. Department of Agriculture, Agriculture Handbook 271, Washington, DC

Institute of Hydrology (1994) Amazonia: forest, pasture and climate—results from ABRACOS. Wallingford, UK

Isard SA, Schaetzl RJ (1995) Estimating soil temperatures and frost in the lake effect snowbelt region, Michigan, USA. Cold Reg Sci Tech 23:317–332

Iswari S, Palta JP (1989) Plasma membrane H+-ATPase as a site of functional alteration during cold acclimation and freezing injury. In: Li PH (ed) Low temperature stress physiology in crops. CRC Press, Boca Raton

IUCN (2006) IUCN red list of threatened species. www.iucnredlist.org. Accessed 12 May 2006

Iwata Y, Hayashi M, Hirota T (2007) Comparison of Snowmelt Infiltration under different soil-freezing conditions influenced by snow cover. Vadose Zone J 7:79–86

Jackson RB, Canadell J, Ehlerlinger JR, Mooney HA, Sala OE, Schulze ED (1996) A global analysis of root distributions for terrestrial biomes. Oeceologica 108(3):389–411

Jeffrey WW (1963) Soil temperature measurements in forests of northwestern Canada. Ecology 44:151–153

Jha CS, Singh JS (1990) Composition and dynamics of dry tropical forest in relation to soil texture. J Veg Sci 1:609–614

Jimenez CC, Tejedor M, Morillas G, Neris J (2006) Infiltration rate in Andisols: effect of changes in vegetation cover (Tenerife, Spain). J Soil Water Conserv 61(3):153–158

Jina BS, Bohra CS, Lodhiyal LS, Sah P (2011) Soil characteristics in oak and pine forests of Indian Central Himalaya. E-Int Scientific Res J 3(1):19–22

Jyske T, Manner M, Mäkinen H, Nojd P, Peltola H, Repo T (2011) The effects of artificial soil frost on cambial activity and xylem formation in Norway spruce. Trees doi:10.1007/s00468-011-0601-7

Kay BD (1998) Soil structure and organic carbon: a review. In: Lal R, Kimble JM, Follett RF, Stewart BA (ed) Soil processes and the carbon cycle. CRC Press, Boca Raton

Kevin AE, Brooks JR (2003) Prolonged flooding decreased stem density, tree size and shifted composition towards clonal species in a central Florida hardwood swamp. Forest Ecol Manage 173:261–279

Kirkham MB (2005) Principles of soil and plant water relations. Elsevier Academic, Burlington

Kirschbaum MU (1995) The temperature dependence of soil organic matter decomposition, and the effect of global warming on soil organic C storage. Soil Biol Biochem 27:753–760

Kozlowski TT (1985) Soil aeration, flooding and tree growth. J Arboric 11:85–96

Kozlowski TT (1986) Soil aeration and growth of forest trees (Review article). Scand J For Res 1:113–123

Kozlowski TT, Pallardy SG (1997) Physiology of woody plants, 2nd edn. Academic, San Diego

Kramer PJ (1987) The role of water stress in tree growth. J Arboric 13(2):33–38

Kramer PJ, Kozlowski TT (1979) Physiology of woody plants. Academic, New York

Kuan HL, Hallett PD, Griffiths BS, Gregory AS, Watts CW, Whitmore AP (2007) The biological and physical stability and resilience of a selection of Scottish soils to stresses. Eur J Soil Sci 58:811–821

Lahti M, Aphalo PJ, Finer L, Ryyppo A, Lehto T, Mannerkoski H (2005) Effects of soil temperature on shoot and root growth and nutrient uptake of 5-year-old Norway spruce seedlings. Tree Physiol 25:115–122

Lal R (1991) Soil structure and sustainability. J Sustain Agric 1:67–92

Lal R, Shukla A (2004) Principles of soil physics. CRC, Amstardam

Lambers H, Chapin FS III, Pons TL (1998) Plant physiological ecology. Springer-Verlag, New York

Landhaausser SM, Wein RW, Lange P (1996) Gas exchange and growth of three arctic tree-line tree species under different soil temperature and drought preconditioning regime. Can J Bot 74(1):686–693

Law BE, Falge E, Gu L, Baldocchi DD, Bakwin P, Berbigier P, Davis K, Dolman AJ, Falk M, Fuentes JD, Goldstein A, Granier A, Grelle A, Hollinger D, Janssens IA, Jarvis P, Jensen NO, Katul G, Mahli Y, Matteucci G, Meyers T, Monson R, Munger W, Oechel W, Olson R, Pilegaard K, Paw KT, Thorgeirsson UH, Valentini R, Verma S, Vesala T, Wilson K, Wofsy S (2002) Environmental control over carbon dioxide and water vapor exchange of terrestrial vegetation. Agr Forest Meteor 113:97–120

Lawrence WT, Oechel WC (1983) Effects of soil temperature on the carbon exchange of taiga seedlings. II. Photosynthesis, respiration and conductance. Can J Forest Res 13:850–859

Leaf AL, Leonard RE, Berglund JV (1971) Root distribution of a plantation-grown red pine in an outwash soil. Ecology 52(1):153–158

Lee KE, Foster RC (1991) Soil fauna and soil structure. Aust J Soil Res 29(6):745–775

Lee CA, Lauenroth WK (1994) Spatial distributions of grass and shrub root systems in the short grass steppe. The Am Midland Naturalist 132(1):117–123

Lees JC (1972) Soil aeration response to drainage intensity in basin peat. Forestry 45:135–143

Leiros MC, Trasar-Cepeda C, Seoane S, Gil-Sotres F (1999) Dependence of mineralization of soil organic matter on temperature and moisture. Soil Biol Biochem 31:327–335

Levitt J (1980) Responses of plants to environmental stresses. Vol. II. Water, radiation, salt, and other stresses. Academic, New York

Levula J, Ilvesniemi H, Westman CJ (2003) Relation between soil properties and tree species composition in a Scots pine-Norway spruce stand in southern Finland. Silva Fennica 37(2):205–218

Lewis J (2003) Stemflow estimation in a redwood forest using model-based stratified random sampling. Environmetrics 14(6):559–571

Li YY, Shao MA (2006) Change of soil physical properties under long-term natural vegetation restoration in the Loess Plateau of China. J Arid Environ 64:77–96

Linquist BA, Singleton PW, Yost RS, Cassman KG (1997) Aggregate size effects on the sorption and release of phosphorus in an Ultisol. Soil Sci Soc Am J 61:160–166

Lloyd J, Taylor JA (1994) On the temperature dependence of soil respiration. Funct Ecol 8:315–323

Lopushinsky W, Max TA (1990) Effect of soil temperature on root and shoot growth and on budburst timing in conifer seedling transplant. New For 4:107–124

Lu J, Sun G, McNulty SG, Amatya DM (2005) A comparison of six potential evapotranspiration methods for regional use in the southeastern United States. J Am Water Resour Assoc 41:621–623

Lyr H, Hoffman G (1967) Growth rates and periodicity of tree roots. In: Romberger JA, Mikola P (eds) International review of forest research Vol 2. Academic, New York

MacDonald NW, Zak DR, Pregitzer KS (1995) Temperature effects on the kinetics of microbial respiration and net nitrogen and sulfur mineralization. Soil Sci Soc Am J 59:233

Marschner H (1990) Mineral nutrition of higher plants. Academic, New York

Marshall JD, Waring RH (1985) Predicting fine root production and turnover by monitoring root starch and soil temperature. Can J For Res 15:791–800

Matras J, Pâques L (2008) EUFORGEN Technical Guidelines for genetic conservation and use for European Larch (*Larix decidua*). Bioversity International, Rome

McGill WB, Hunt HW, Woodmansee RG, Reuss JO (1981) Phoenix. A model of the dynamics of carbon and nitrogen in grassland soils. Ecol Bull 33:49–115

McHale PJ, Mitchell MJ, Bowles FP (1998) Soil warming in a northern hardwood forest: trace gas fluxes and leaf litter decomposition. Can J For Res 28:1365–1372

McKee WH Jr, Hatchell GE, Tiarks AE (1985) Managing site damage from logging. A loblolly pine management guide. General technical Report, SE-32, USDA For. Serv., pp 1–21
McMinn RG (1960) Water relations and forest distribution in the Douglas-fir region on Vancouver Island. Publication 1091. Canada Department of Agriculture, Forest Biology Division, Ottawa, Canada
Mein RC, Larson CL (1973) Modelling infiltration during a steady rain. Wat Resour Res 9:384–397
Messing I, Alrikkson A, Johansson W (1997) Soil physical properties of afforested and arable land. Soil Use Manag 13:209–217
Mueller RF (2002) Soil Temperature and Forest Type. Forests of the Central Appalachians Project. Virginians for Wilderness Web Site. http://www.asecular.com/forests/soiltemperatures/soil_2002.html. (Accessed on 20 Aug, 2012)
Myers BJ (1988) Water stress integral-a link between short-term stress and long-term growth. Tree Physiol 4:315–323
Olorunfemi IE, Fasinmirin JT (2011) Unsaturated hydraulic conductivity and infiltration of soils of tropical rain forest climate of Nigeria. Proceedings of the Environmental Management Conference, Federal University of Agriculture, Abeokuta, Nigeria
Onyekwelu JC, Mosandl R, Stimm B (2006) Productivity, site evaluation and state of nutrition of *Gmelina arborea* plantations in Oluwa and Omo forest reserves, Nigeria. For Ecol Manag 229(1-3):214–227
Osman KT (2013) Soils: principles, properties and management. Springer, Dordrecht
Pall R, Mohsenin NN (1980) A soil air pycnometer for determination of porosity and particle density. American Society of Agricultural and Biological Engineers, St. Joseph, Michigan
Pardo A, Amato M, Chiaranda FQ (2000) Relationships between soil structure, root distribution and water uptake of chickpea (*Cicer arietinum* L.). Plant growth and water distribution. Eur J Agron 13:39–45
Parton WJ, Holland EA, Del Grosso SJ, Hartman MD, Martin RE, Mosier AR, Ojima DS, Schimel DS (2001) Generalized model for NO_x and N_2O emissions from soils. J Geophys Res 106:17403–17419
Passioura JB (1991) Soil structure and plant growth. Aust J Soil Res 29(6):717–728
Paterson DB, Mason WL (1999) Cultivation of soils for forestry. Forestry Commission Bull 119, Purely Print, Blandford Forum, Dorset
Patt J, Carmeli D, Zafrir T (1966) Influence of soil physical condition on root development and on production of citrus trees. Soil Sci 102:82–84
Paul KI, Polglase PJ, Smethurst PJ, O'Connell AM, Carlyle CJ, Khanna PK (2004) Soil temperature under forests: a simple model for predicting soil temperature under a range of forest types. Agri Forest Meteorol 121:167–182
Pearce AJ, Rowe LK (1979) Forest management effects of interception, evaporation, and water yield. J Hydrol 18:73–87
Pelissier R, Dray S, Sabatier D (2001) Within-plot relationships between tree species occurrences and hydrological soil constraints: an example in French Guiana investigated through canonical correlation analysis. Plant Ecol 162:143–156
Pereira JS (1990) Whole-plant regulation and productivity in forest trees. In Davies WJ, Jeffcoat B (ed) Importance of root to shoot communication in the responses to environmental stress. Bri Soc Plant Growth Reg Mon No 21
Perez-Ramos IM, Maranon T (2009) Effects of waterlogging on seed germination of three Mediterranean oak species: ecological implications. Acta Oecologica 35:422–428
Peterjohn WT, Melillo JM, Steudler PA, Newkirk KM, Bowles FP, Aber JD (1994) Responses of trace gas fluxes and N availability to experimentally elevated soil temperatures. Ecol Appl 4:617–625
Pezeshki SR (1994) Plant responses to flooding. In Wilkinson RE (ed) Plant environment interactions. Marcel-Dekker, New York
Pieri C (1992) Fertility of soils: a future for farming in the West African Savannah. Springer Series in Physical Environment. Springer-Verlag, Berlin
Pitt R, Lantrip J, Harrison R, Henry CL, Xue D, O'Connor TP (1999) Infiltration through disturbed urban soils and compost-amended soil effects on runoff quality and quantity. Report EPA/600/R-00/016. Edison, NJ
Plymale AE, Boerner REJ, Logan TJ (1987) Relative nitrogen mineralization and nitrification in soils of two contrasting hardwood forests: effects of site microclimate and initial chemistry. Forest Ecol Manage 21:21–36
Pomeroy JW, Granger RJ (1997) Sustainability of the Western Canadian boreal forest under changing hydrological conditions. I. Snow accumulation and ablation. In: Rosbjerg D., Boutayeb N-E, Gustard A, Kundzewicz ZW, Rasmussen PF (eds) Sustainability of water resources under increasing uncertainty (Proceedings of the Rabat Symposium S1, April 1997). IAHS Publ No 240
Potter MK, Lamb KM (1974) Root development of radiata pine in the ground soils of Eyrewell Forest, Canterbury. NZ J For 19:264–275
Powers RF (1990) Nitrogen mineralization along an altitudinal gradient: Interactions of soil temperature, moisture, and substrate quality. Forest Ecol Manage 30:19–29
Pyke CR, Condit R, Aguilar S, Lao S (2001) Floristic composition across a climatic gradient in a neotropical lowland forest. J Veg Sci 12:553–566
Qashu HK, Zinke PJ (1964) The influence of vegetation on soil thermal regime at the San Dimas Lysimeters. Soil Sci Soc Am Proc 28:703–706
Raich JW, Schlesinger H (1992) The global carbon dioxide flux in soil respiration and its relationship to vegetation and climate. Tellus 44B:81–99
Ramade F (1981) Ecology of natural resources. Wiley, Chichester
Reddell P, Bowen GD, Robson AD (1985) The effects of soil temperature on plant growth, nodulation and nitrogen fixation in *Casuarina cunninghamiana*. New Phytol 101:441–44450
Reid LM, Lewis J (2007) Rates and implications of rainfall interception in a coastal redwood forest. In: Standiford RB, Giusti GA, Valachovic Y, Zielinski WJ, Furniss MJ, (eds) Proceedings of the redwood region forest science symposium: What does the future hold? Gen Tech Rep PSW-GTR-194. Pacific Southwest Research Station, Forest Service, USDA, Albany, CA
Reiners WA, Bouwman AF, Parsons WFJ, Keller M (1994) Tropical rain forest conversion to pasture: changes in vegetation and soil properties. Ecol Appl 4(2):363–377
Richards D, Cockroft B (1974) Soil physical properties and root concentrations in an irrigated apple orchard. Aust J Exptl Agric Animal Husb 14:103–107
Robichaud PR (1996) Spatially-varied erosion potential from harvested hillslopes after prescribed fire in the Interior Northwest. PhD Diss., University of Idaho, Moscow
Robichaud PR, Waldrop TA (1994) A comparison of surface runoff and sediment yields from low- and high-severity site preparation burns. Water Resour Bull 30:27–34
Robichaud PR, Luce CH, Brown RE (1993) Variation among different surface conditions in timber harvest sites in the Southern Appalachians. International workshop on soil erosion: Proc.Moscow, Russia. Purdue University, West Lafayette, IN, pp 231–241
Rockich DP, Meney KA, Dixon KW, Sivasithamparam K (2001) The impact of soil disturbance on root development in woodland communities in Western Australia. Aust J Bot 49:169–183
Rolim SG, Jesus RM, Nascimento HEM, do Couto HTZ, Chambers JQ (2005) Biomass change in an Atlantic tropical moist forest: the ENSO effect in permanent sample plots over a 22-year period. Oecologia 142:238–246
Rose CW (1960) Soil detachment caused by rainfall. Soil Sci 89:28–35
Rosen CJ, Carlson RM (1984) Influence of root zone oxygen stress on potassium and ammonium absorption by Myrobalan plum rootstock. Plant Soil 80:345–353
Rowntree PR (1988) Review of general circulation models as a basis for predicting the effects of vegetation changes on climate. In Reynolds

ERC, Thompson FB (ed) Forests, climate and hydrology: regional impacts. Kefford, UK
Russo SE, Davies SJ, King DA, Tan S (2005) Soil-related performance variation and distributions of tree species in a Bornean rain forest. J Ecol 93:879–889
Sakai H, Nordfjell T, Saudicani K, Talbot B, Bollehuus E (2008) Soil compaction on forest soils from different kinds of tyres and trucks and possibility of accurate estimate. Croatian J Forest Engg 29(1):15–27
Sakio H (2005) Effects of flooding on growth of seedlings of woody riparian species. J Forest Res 10:341–346
Salo J, Kalliola R, Hakkinen I, Makinen Y, Niemela P, Puhakka M, Coley PD (1986) River dynamics and the diversity of Amazon lowland forest. Nature 322:254–258
Sanchez PA, Logan TJ (1992) Myths and science about the chemistry and fertility of soils in the tropics. In: Lal R, Sanchez PA (eds) Myths and science of soils of the tropics, SSSA, Madison, Wisconsin, SSSA Special Publication 29:35–46
Sands R, Greaten EL, Gerard CJ (1979) Compaction of sandy soils in radiata pine forests. I. A penetrometer study. Aust J Soil Res 17:101–113
Schimel DS (1986) Carbon and nitrogen turnover in adjacent grassland and cropland ecosystems. Biogeochem 2:345–357
Schimel DS, Coleman DC, Horton KA (1985) Soil organic matter dynamics in paired rangeland and cropland toposequences in North Dakota. Geoderma 36:201–214
Schlentner RE, Van Cleve K (1985) Relationships between CO_2 evolution from soil, substrate temperature, and substrate moisture in four mature forest types in interior Alaska. Can J Forest Res 15:97–106
Schmidt IK, Jonasson S, Michelsen A (1999) Mineralization and microbial immobilization of N and P in arctic soils in relation to season, temperature and nutrient amendment. Appl Soil Ecol 11:147–160
Schwartz RC, Evett SR, Unger PW (2003) Soil hydraulic properties of cropland compared with reestablished and native grassland. Geoderma 116:47–60
Scull PR, Harman JR (2004) Forest distribution and site quality in southern Lower Michigan, USA. J Biogeogr 31:1503–1514
Sepponen P, Lähde E, Roiko-Jokela P (1979) On the relationship between the forest vegetation and the soil physical properties in Finnish Lapland. Folia Forestalia. 402. P 31 (In Finnish with English summary)
Sexstone AJ, Parkin TP, Tiedje JM (1985) Temporal response of soil denitrification to rainfall and irrigation. Soil Sci Soc Am J 49:99–103
Simcock R (2006) Stormwater and erosion in the urbanising Acacia Heights Catchment, Taupo. Landcare Research Contract Report: LC0607/034. Taupo District Council, Taupo
Simek M, Cooper JE (2002) The influence of soil pH on denitrification: Progress towards the understanding of this interaction over the last 50 years. Eur J Soil Sci 53:345–354
Six J, Bossuyt H, Degryze S, Denef K (2004) A history of research on the link between (micro) aggregates, soil biota, and soil organic matter dynamics. Soil Tillage Research 79:7–31
Sokolovskaya NA, Revut IB, Markov IA, Shevlyakov IR (1977) The role of soil density in forest regeneration. Lesovedenie 2:44–51
Smith KD, May PB, Moore GM (2001) The influence of waterlogging on the establishment of four Australian landscape trees. J Arboricul 27(2):49
Soil Survey Staff (1975) Soil Taxonomy. USDA Handbook 436. U.S. Govt. Printing Office, Washington DC
Sorensen LH (1981) Carbon-nitrogen relationships during thehumification of cellulose in soils containing different amounts ofclay. Soil Biol Biochem 13:313–321
Souza AF, Martins FR (2004) Microsite specialization and spatial distribution of *Geonoma brevispatha*, a clonal palm in south-eastern Brazil. Ecol Res 19:521–532
Sparling GP, Shepherd TG, Schipper LA (2000) Topsoil characteristics of three contrasting New Zealand soils under four long-term land uses. New Zealand Journal of Agricultural Research 43:569–583
Spittlehouse DL (1998) Rainfall interception in young and mature coastal conifer forests. In: Alila Y (ed) Mountains to sea: human interaction with the hydrologic cycle, 51st Annual Conference Proceedings; 10-12 June 1998, Canadian Water Resources Association, Victoria BC, Canada
Spittlehouse DL, Stathers RJ (1990) Seedling microclimate. British Columbia Ministry of Forests, Victoria, Land Management Rep no 65
Stadler D, Wunderli H, Auckenthaler A, Flühler H (1996) Measurement of frost-induced snowmelt runoff in a forest soil. Hydrol Proc 10:1293 1304
Stahli M (2005) Freezing and thawing phenomena in soils. In: Anderson MG (ed) Encyclopedia of hydrological sciences, Vol. 2. Wiley, New York, pp 1069–1076
Stähli M, Jansson PE, Lundin LC (1999) Soil moisture redistribution and infiltration in frozen sandy soils. Water Resour Res 35:95–103
Stathers RJ, Spittlehouse DL (1990) Forest Soil Temperature Manual. FRDA Rep 130, BC Min For Victoria, BC
Strahler AN, Strahler AH (1983) Modern physical geography, 2nd edn. Wiley, New York
Strickland TC, Fitzgerald JW (1984) Formation and mineralization of organic sulfur in forest soils. Biogeochem 1:79–95
Stromgren M, Linder S (2002) Effects of nutrition and soil warming on stemwood production in a boreal Norway spruce stand. Glob Change Biol 8:1195–1204
Strong DT, Fillery IRP (2002) Denitrification response to nitrate concentrations in sandy soils. Soil Biol Biochem 34:945–954
Strong WL, La Roi GH (1985) Root density-soil relationship in selected boreal forests of central Alberta. Canada. Forest Ecol Manage 12(3/4):233–251
Sun G, McNulty SG, Amatya DM, Skaggs RW, Swift LW, Shepard JP, Riekerk H (2002) A comparison of the watershed hydrology of coastal forested wetlands and the mountainous uplands in the southern US. J Hydrol 263:92–104
Sun G, Zhou G, Zhang Z, Wei X, McNulty SG, Vose JM (2006) Potential water yield reduction due to reforestation across China. J Hydrol 328:548–558
Svenning JC (2001) On the role of microenvironmental heterogeneity in the ecology and diversification of neotropical rain-forest palms (Arecaceae). Bot Rev 67:1–53
Talsma T, Flint SE (1958) Some factors determining the hydraulic conductivity of subsoils with special reference to tile drainage problems. Soil Sci 35:198–206
Taminnen P, Starr M (1994) Bulk density of forested mineral soils. Silva Fennica 28(1):53–60
Tamminen P (1991) Kangasmaan ravinnetunnusten ilmaiseminen ja viljavuuden alueellinen vaihtelu Etela-Suomessa. Summary: expression of soil nutrient status and regional variation in soil fertility of forested sites in southern Finland. Folia Forestalia 777:40
Taylor HM (1971) Effects of soil strength on seedling emergence, root growth and crop yield. In: Barnes KK, Carleton WM, Taylor HM, Throckmorton RI, Vanden berg GE (eds) Compaction of agricultural soils. St Joseph. (ASAE. Monogr.)
Taylor M, Mulholland M, Thornburrow D (2009) Infiltration characteristics of soils under forestry and agriculture in the upper Waikato catchment environment Waikato Technical Report 2009/18
Thien SJ (1979) A flow diagram for teaching texture by feel analysis. J Agron Educ 8:54–55
Thompson LJ, Naeem S (1996) The effects of soil warming on plant recruitment. Plant Soil 182:339–343
Tian YQ, McDowell R, Yu Q, Sheath GW, Carlson WT, Gong P (2006) Modelling to analyse the impacts of animal treading effects on soil infiltration. Hydrol Process 21:1106–1114
Tisdall JM, Oades JM (1979) Stabilization of soil aggregates by the root systems of ryegrass. Aust J Soil Res 17:429–441
Tisdall JM, Oades JM (1982) Organic matter and water-stable aggregates in soils. J Soil Sci 33:141–163

Trowbridge WB, Kalmanovitz S, Schwartz MW (2005) Growth of valley oak (Quercus Lobata Nee) in four floodplain environments in the Central Valley of California. Plant Ecol 176:157–164

Trujillo DP (1976) Container-grown ponderosa pine seedlings respond to fertilization. USDA. For Serv Res Note RM 319

Tryon PR, Chapin FS III (1983) Temperature control over root growth and root biomass in taiga forest trees. Can J For Res 13:827–833

Unger PW, Kaspar TC (1994) Soil compaction and root growth: a review. Agronomy J 86:759–766

USDA NRCS (1995) Northeast Wetland Flora, USDA NRCS Plants Database, USDA NRCS National Plant Data Center & the Biota of North America Program, Washington, DC

Utzig GF, Walmsley ME 1988. Evaluation of soil degradation as a factor affecting forest productivity in British Columbia: a problem analysis Phase I. For. Can. and B.C. Min. For., Victoria, B.C. FRDA Rep. No. 025

Van Cleve K, Barney R, Schlentner R (1981) Evidence of temperature control of production and nutrient cycling in two interior Alaska black spruce ecosystems. Can J For Res 11:258–273

Van Cleve K, Oechel WC, Hom JL (1990) Response of black spruce (*Picea mariana*) ecosystems to soil temperature modification in interior Alaska. Can J Forest Res 20:1530–1535

van der Kamp G, Hayashi M, Gallen D (2003) Comparing the hydrology of grassed and cultivated catchments in the semi-arid Canadian prairies. Hydrol Process 17:559–575

van Veen JA, Paul EA (1981) Organic carbon dynamics in grassland soils. 1. Background information and computer simulation. Can J Soil Sci 61:185–201

van Veen JA, Ladd JN, Frissel MJ (1984) Modelling C and N turnover through the microbial biomass in soil. Plant Soil 76:257–274

Vapaavuori EM, Rikala R, Ryyppö A (1992) Effects of root temperature on growth and photosynthesis in conifer seedlings during shoot elongation. Tree Physiol 10:217–230

Vartapetian BB, Jackson M (1997) Plant adaptations to anaerobic stress. Ann Bot 79:3–20

Viereck LA, Dyrness CT (1979) Ecological effects of Wickersham Dome Fire near Fairbanks, Alaska. Gen Tech Rep PNW-90. US Department of Agriculture, Forest Service, Pacific Northwest Forest and Range Experiment Station, Portland

Viro PJ (1951) Nutrient status and fertility of forest soil. 1. Pine stands. Selostus: Metsamaan ravinnesuhtte ja viljavuus. 1. Mannikot. Communicaciones Instituti Forestalis Fenniae 39(4):54

Voesenek LACJ, Colmer TD, Pierik R, Millenaar FF, Peeters AJM (2006) How plants cope with complete submergence. New Phytol 170:213–226

Vormisto J, Tuomisto H, Oksanen J (2004) Palm distribution patterns in Amazonian rainforests: what is the role of topographic variation? J Veg Sci 15:485–494

Voroney RP, Van Veen JA, Paul EA (1981) Organic C dynamics in grassland soils. 2. Model validation and simulation of the long-term effects of cultivation and rainfall erosion. Can J Soil Sci 61:211–224

Vorosmarty CJ, Federer CA, Schloss AL (1998) Potential Evaporation Functions Compared on US Watersheds: Possible Implications for Global-Scale Water Balance and Terrestrial Ecosystem Modeling. J Hydrol 207:147–169

Voss EG (1985) Michigan flora. Part II. Dicots (Saururaceae-Cornaceae. Bull 59. Cranbrook Institute of Science; Ann Arbor, University of Michigan]

Waide JB, Caskey WH, Todd RL et al (1988) Changes in soil nitrogen pools and transformations following clearcutting. In Swank WT, Crossley DA Jr, (ed) Forest hydrology and ecology at Coweeta. Ecol Studies 66:221–232

Walls RL, Wardrop DH, Brooks RP (2005) The impact of experimental sedimentation and flooding on the growth and germination of floodplain trees. Plant Ecol 176:203–213

Wang X, Yost RS, Linquist BA (2001) Soil aggregate size affects phosphorus desorption from highly weathered soils and plant growth. Soil Sci Soc Am J 65:139–146

Wang GB, Cao FL, Zhang XY, Zhang WX (2010) Effects of waterlogging on the growth and energy-metabolic enzyme activities of different tree species (Article in Chinese). Ying Yong Sheng Tai Xue Bao 21(3):590–596

Ware GH (1980) Selecting trees for clay soils. Metro. Tree Impr Alliance (METRIA) Proc. 3:102–106

Wargo PM (1981) Defoliation and secondary-action organism attack: With emphasis on *Armillaria mellea*. J Arbor 7:64–69

Wargo PM, Montgomery ME (1983) Colonization by *Armillaria mellea* and *Agrilus bilineatus* of oaks injected with ethanol. Forest Sci 29:848–857

Weber MG, McAlpine RS, Wotton BM, Donnelly JG, Hobbs MW (1995) Prescribed burning and disk trenching effects on early plantation performance in eastern Ontario, Canada. Forest Ecol Manage 78:159–171

Whalley WR, Dumitru E, Dexter AR (1995) Biological effects of soil compaction. Soil Tillage Res 35:53–68

White RE (1985) The influence of macropores on the transport of dissolved and suspended matter through soil. Adv Soil Sci 3:95–120

Wilson CA, Mitchell RJ, Hendricks JJ, Boring LR (1999) Patterns and controls of ecosystem function in longleaf pine-wiregrass savannas. II. Nitrogen dynamics. Can J Forest Res 29:752–760

Wronski EB, Murphy G (1994) Responses of forest crops to soil compaction. In: Soane BD, van Ouwerkerk C (eds) Soil compaction in crop production. Elsevier, Amsterdam

Xue Y (1997) Biosphere feedback on regional climate in tropical north Africa. Q J Roy Meteor Soc 123:1483–1515

Yoshida S, Uemura M (1989) Alterations of plasma membranes related to cold acclimation of plants. In: Li PH (ed) Low temperature stress physiology in crops. CRC Press, Boca Raton

Young JL, Spycher G (1979) Water-dispersible soil organic-mineral particles: I. carbon and nitrogen distribution. Soil Sci Soc Am J 43:324–328

Youngberg CT (1970) Soils and tree growth requirements. In: Berg AB (ed) Management of young-growth Douglas-fir and Western Hemlock. Ore State Univ Sen Forestry 34–37

Zahner R (1969) Water deficits and growth of trees. In: Kozlowski TT (ed.) Water deficits and plant growth. Academic, New York

Zanin DK (2005) Feasibility of teak plantation for smallholders in Eastern Panama. Michigan Technological University, Michigan

Zeleznik P, Mali B, Robek R, Kraigher H (2009) Alteration of root growth caused by heavy forest machinery on Osankarica. International Symposium "Root Research and Applications" RootRAP, 2–4 September 2009, Boku—Vienna, Austria

Zhang Z, Xu Z (2009) Rough set method to identify key factors affecting precipitation in Lhasa. Stoch Environ Res Risk Assess 23:1181–1186

Zhang L, Dawes WR, Walker GR (1999) Predicting the effect of vegetation changes on catchment average water balance. Technical Report 99/12 Cooperative Research Centre for Catchment Hydrology

Zhang L, Hickel K, Dawes WR (2004) A rational function approach for estimating mean annual evapotranspiration. Water Resour Res 40:1–14

Zhou G, Morris JD, Yan JH, Yu Z, Peng SL (2002) Hydrological impacts of reforestation with eucalyptus and indigenous species: a case study in southern China. Forest Ecol Manage 167:209–222

Zyuz NS (1968) Bulk density and hardness of the Hillocky Sands of the Middle Don. Sol Sod Sci 13:1769–1776

3 Chemical Properties of Forest Soils

In soil, there are inorganic and organic solids, solutes, liquids, and gases. There are larger and smaller particles, including sand, silt, and clay, and colloids—fine crystalline minerals and amorphous humus. Fine silicate clays and oxides and hydroxides of iron and aluminum, lime, gypsum, and phosphates are there along with hundreds of many other compounds, and nutrient ions. These materials are variably active and reactive; some are almost inert such as the sand grains, and some undergo continuous dynamic reactions such as the colloidal and charged clay particles. Insoluble materials are made soluble, and soluble materials are insolubilized by diverse chemical and biochemical reactions. Important indices of the chemical behavior of all soils, including forest soils, are pH, cation-exchange capacity (CEC), anion-exchange capacity (AEC), base saturation (BS) percentage, exchangeable sodium percentage (ESP), electrical conductivity, and redox potential. These indices characterize the forest soils and affect the growth and distribution of forest tree species.

3.1 Chemical Composition of Soils

There are more than 100 chemical elements in soil, but its greater mass is made up of oxygen (49.0 %), silicon (33.0 %), aluminum (7.1 %), iron (4.0 %), calcium (5.0 %), magnesium (0.5 %), and potassium (1.4 %). Plant nutrient elements like Ca, Mg, K, Na, P, and S are present in the minerals, as sorbed complexes and in the soil solution. O, Si, and Al occur as constituents of minerals and as oxides. Iron occurs mainly in the form of oxides and hydroxides. Most elements are not generally found in pure elemental forms; they are bound in the minerals (Chap. 1). There are a variety of primary minerals mostly associated with sand and silt fractions (quartz, feldspars, micas, olivines, pyroxenes, amphiboles, etc.) and secondary minerals associated mostly with the clay fraction (kaolinite, montmorillonite, illite, chlorite, vermiculite, halloysite, allophanes, etc.). Important Ca-containing minerals are calcite, gypsum, apatite, and dolomite and common Mg-containing minerals are dolomite and hornblende. Potassium occurs mainly in feldspar and mica. The most important phosphate mineral is the apatite; P also occurs as aluminum phosphate and calcium phosphate and as organic compounds such as phospholipids, inositol, nucleic acids, etc. Nitrogen occurs in soil mainly in the organic form as proteins, peptides, and amino acids. All micronutrients like Mo, Fe, Mn, Zn, Cu, and B occur in the inorganic form. There is little difference in the chemical composition of forest soils from that of the cultivated soils, except in the nature of organic matter.

Soils are composed of both inorganic and organic substances, but the proportion of organic matter in mineral soils is relatively low. Mineral soils contain less than 5 % of their dry mass as organic matter. On the other hand, organic soils contain more than 20 % of their mass in organic matter. Interesting differences occur in organic matter content and organic matter distribution between forest soils and agricultural soils. Generally, forest soils have higher organic matter content than adjacent cultivated soils. Also, organic matter content decreases gradually with the depth of forest soils. But most agricultural soils have lower organic matter and its content abruptly decreases with depth of soil. Forest soils have plenty of organic residues on the surface, and most temperate and boreal forest soils have an organic O horizon known as the forest floor on the surface of the mineral soils. The nature of organic matter in forest soils is discussed in Chap. 4.

3.2 Soil Solution

The term soil solution refers to the aqueous liquid phase of the soil with its dissolved solutes. The solutes include ions, ion pairs, ligands, and complexes. The principal cations and anions in soil solution respectively are Ca^{2+}, Mg^{2+}, Na^+, K^+, NH_4^+, Al^{3+}, Fe^{3+}, Mn^{2+} (other trace metal ions) and $H_2PO_4^-$, SO_4^{2-}, PO_4^{3-}, CO_3^{2-}, HCO_3^-, NO_3^-, Cl^-, etc. The most prevalent metal cation in the soil solution is the Ca^{2+} ion, typically for most soils. The prevalent anions are nitrate, chloride, and sulfate in acidic soils and carbonate and bicar-

K. T. Osman, *Forest Soils*, DOI 10.1007/978-3-319-02541-4_3, © Springer International Publishing Switzerland 2013

bonate in alkaline soils. There are many other cations and anions in smaller proportions in soil solution. Plants and soil organisms take up the nutrients and elements they need from the soil solution. Thus, ions are continuously removed from soil solution by plant roots. The soil solution is a dynamic entity. Plant roots take up ions from soil solution and release exudates into it. Dissolved ions can be sorbed on organic and inorganic components of the soil. Some ions can be desorbed into the soil solution from the sorption complexes. Some substances in soil solution can precipitate. Weathering of minerals releases fresh soluble ions to the soil solution. Ions are transported through the soil into groundwater or removed through surface runoff processes. Microorganisms can remove ions from the soil solution, and when they die and are decomposed, ions are released to the soil solution (Sparks 2003). The composition of soil solution is, thus, highly variable. Soils have limited capacity to replenish the solutes in soil solution. However, the actual concentration of the ion species in the soil solution changes with changes in soil moisture content (Wolt 1994).

The composition of the soil solution is of great importance to biological activity because it represents the easily available fraction of the pools of chemical constituents in the soil. The solution chemistry may reflect and regulate soil processes and soil chemical changes, the availability of nutrient ions to plant roots, and the potential phytotoxicity of, e.g., H and Al ions. Six beech forest sites, of *Galium odoratum* and *Stellaria nemorum* in South Sweden, were sampled during a year using percolation lysimeters installed immediately under the root zone by Falkengren-Grerup (1994). Chemical analyses were made of dissolved organic carbon (DOC), Al, Ca, Cl, Fe, K, Mg, Mn, Na, NH_4, NO_3, PO_4, SO_4, and Zn. The concentrations in the percolation water varied greatly between sampling events, this variation being especially pronounced for NH_4, NO_3, and phosphate, but tenfold differences in concentrations were also found for Al, Cl, Fe, K, and Mn. Soil solution chemistry at three northwestern European *Calluna vulgaris* heathland sites in Denmark (DK), the Netherlands (NL), and Wales (UK) were studied by Schmidt et al. (2004). Soil solution chemistry measured below the organic soil layer and below the rooting zone and water fluxes estimated with hydrological models were combined to calculate element budgets. Remarkably high N leaching was observed at the NL heath with 18 and 6.4 kg N ha^{-1} per year of NO_3-N and NH_4-N leached from the control plots, respectively, indicating that this site is nitrogen saturated.

Nutrient ion concentrations in the soil solution of three deciduous forests in the Sierra de Gata Mountains of central-western Spain were determined and the variation with soil moisture and solution pH and exchange capacity (EC) were examined by Menendez et al. (1995). Ion concentration and pH both decreased with depth of soil. There was a strong correlation between soil solution composition and soil moisture content. Skyllberg et al. (2001) determined soil solution chemistry of two soil profiles developed under a mixed spruce (*Picea abies* and *Picea sitchensis*) stand and of one soil profile developed under an oak (*Quercus robur*) forest stand. Soils under spruce were classified as Spodosols and soils under oak were classified as Inceptisols. All profiles developed from the same parent material, a Saahlian sandy till, containing less than 2% clay. Soil solution pH was substantially lower in the upper 35–40 cm of the two Spodosols, as compared to the Inceptisol. The total amount of soil-adsorbed base cations did not differ among the three profiles on an area basis down to 1 m soil depth. The results confirmed that organic acids reacting with Al to form Al-OM complexes is a major pH-buffering process in acidic forest soils. Soluble organic acids, especially those that complex with Al or Fe, may perform several critical functions in the biogeochemistry of the rhizosphere and the productivity of the aboveground ecosystem, especially by enhancing dissolution or inhibiting precipitation of soil minerals containing PO_4 or SO_4 (Fox et al. 1990). Such acids have been identified from forest soils in several regions (Fox and Comerford 1990). The soluble oxalate anion is an effective weathering agent in mineral soils (Pohlman and McColl 1986) and has been implicated in release of K from a hydroxy-interlayered vermiculite (Comerford et al. 1990). Organic acids (e.g., oxalic, citric) also can form soluble complexes with Al and other metals, ameliorating the toxicity of the free metal cations to higher plants and, thereby, increasing biological productivity (Hue et al. 1986). Concentrations of the principal inorganic and organic solutes in the soil root zone were measured in six contrasting lodgepole pine (*Pinus contorta* ssp. *latifolia*) forest ecosystems for 5 years by Fahey and Yavitt (1988). Consistent temporal changes in the principal inorganic solutes (Ca, Mg, K, Na, SO_4, and Cl) were observed in all the forest stands and years of study: High concentrations at the initiation of snowmelt in the spring were followed by rapid declines to rather constant values in the mid to late stages of snowmelt. Except for K, concentrations of these solutes differed significantly between sites and between years. Sulfate was the principal mobile anion in the root-zone soil solutions, but contributions of bicarbonate and organic anions were also important. The pH of root-zone solutions was relatively high (6.0), did not change significantly as snowmelt proceeded, and was significantly lower in high-clay soils. Little seasonal or spatial variation of inorganic N and P concentrations was observed in root-zone solutions, probably as a result of high biotic demand for these limiting nutrients. The flux of N and P in these ecosystems was predominately via organic forms so that losses of these nutrients were strongly linked to the mobility of DOC.

Soil solutions contain varying amounts of dissolved organic matter (DOM), which originate from plant litter, soil humus, microbial biomass, or from root exudates. Only a small proportion of DOM, mostly low molecular substances such as organic acids, sugars, and amino acids, can be identified chemically (Herbert and Bertsch 1995). Most of the DOM in soils is complex molecules of high molecular weight, namely humic substances. Soil solutions contain some dissolved organic nitrogen (DON) which is often the dominant nitrogen form leached from the forest floor of both deciduous and coniferous forests (Sollins and McCorison 1981, Michalzik et al. 2001). Similarly, nitrogen export in stream waters from many forested watersheds is often dominated by DON (Campbell et al. 2000). DON is thought to be a major constituent of terrestrial and marine N cycles (Neff and Hooper 2002). DON consists of proteins and amino acids (Yu et al. 2002). Only a small proportion (less than 5 %) of total amino acids in DON is in the free form (Westerhoff and Mash 2002). Other compounds include amides (20 %) and heterocyclic compounds such as pyroles, pyrrolidines, pyridines, pyrines, and pyrazoles (25 %) (Westerhoff and Mash 2002). The chemical composition of DON in forest soils can vary with tree species and degree of decomposition. Litter from different plant species differs not only in the composition but also in concentrations of water-extractable compounds (Don and Kalbitz 2005). Montagnini et al. (1991) studied soil-solution chemistry over a 15-month period in three forest stands of contrasting nitrogen mineralization and nitrification rates in the southern Appalachians of North Carolina, USA, using porous-cup lysimeters. In a black-locust-dominated stand, soil solution NO_3–N was 3.73 and 5.04 mg l^{-1} at 30- and 60-cm depth, respectively, and DON was 0.718 and 0.582 mg l^{-1}, respectively.

3.3 Soil Reaction

The extent of acidity and alkalinity in a soil is known as soil reaction, and it is denoted by soil pH. The pH is defined as the negative logarithm of hydrogen ion activity which is equal to the hydrogen ion concentration (moles l^{-1}) in a very dilute solution. The pH scale of a solution runs from 0 to 14; 7.0 is the neutral point. A pH value less than 7 indicates an acidic solution, while a pH greater than 7 indicates an alkaline solution. For a very heterogeneous media like soil, pH values between 6.5 and 7.5 may be taken as fairly neutral. For soil pH measurement, a soil–water suspension, usually at a soil to water ratio of 1:2.5, is prepared by stirring air-dry soil in distilled water for half an hour and the pH is recorded in an electronic pH meter. Other ratios such as 1:1 and 1:2 could be used, and there would be variable readings if the ratio is changed. So, the ratio of the measurement is reported along with the pH values.

Acidic soils with pH < 6.5 are formed due to the acidifying properties of parent material, organic matter, aluminum, carbon dioxide (CO_2), and the presence of very low quantities of clay minerals (Tisdale et al. 1993). Coarse-textured soils, highly weathered soils, and intensely leached soils are usually acidic. Soils may be acidified by the decomposition of organic matter and production of acids. The degree of acidification by soil organic matter (SOM) or humus will vary with the type of vegetation. For example, coniferous forests have generally a lower pH than deciduous forests. Acidification is controlled by the carboxylic and phenolic reactive groups, within SOM and humus, behaving as weak acids and dissociating, releasing hydrogen ions. Soluble salts are derived from organic matter decomposition or mineral weathering. In acid soils, the cations of these salts will displace Al^{3+} and decrease the pH further. The high concentration of H^+ ions is not the limiting factor to plant growth, but rather the toxicity of other nutrients at low pH values and/or the deficiency in essential nutrients that occurs as a result of soil acidity. Large quantities of aluminum, iron, and manganese ions are liberated at low pH values and are toxic to the majority of plant species (Larcher 1980). The nutrients Ca^{2+}, Mg^{2+}, K^+, PO_4^{3-}, and MoO^{2-} are depleted or made unavailable when a soil becomes acidic. In some instances, low levels of sulfur, potassium, molybdenum, copper, and zinc may also occur. Since many of the macro- and micronutrients are found at low concentrations in acid soils, soil fertility is considered low (Marschner 1986). Alkaline soils have a pH greater than 7.0, due to the presence of calcium carbonate from a calcareous parent material or due to a high proportion of exchangeable Na^+ in exchange complexes. When pH is greater than 8, soils are sodic or alkaline and have a sodium adsorption ratio (SAR) of 15 or more. Sodic soils have nutrient limitations and are deficient in zinc, iron, phosphorus, and occasionally calcium, potassium, and magnesium. However, nitrogen is the most limiting nutrient when soils are alkaline. In sodic soils, boron may be at phytotoxic concentrations. High bicarbonate concentrations in the soil are mainly responsible for iron deficiency that is further increased by high soil moisture or anaerobic conditions.

The pH of a soil influences several soil characteristics: weathering processes, soil structure, humification and mineral transformation, microbial population and activity, mobilization of nutrients, and ion exchange. Soil pH changes seasonally with the distribution of precipitation, and an accurate characterization of pH must be done over a full year in the zone penetrated by roots (Larcher 1980). Most forest soils have acidic soil reaction, with pH values lying within 3.0–7.0. Some forest soils have pH values greater than 7.0. Forest soils are generally more acidic (have less pH) than grassland and agricultural soils due to the continuous production of organic acids and CO_2 from decomposing litters. Additionally, trees may acidify the soil by taking up calcium, magnesium, and other bases from soil and storing them in woody tissues. Dryland forest soils have pH values between 7 and 8, humid temperate forests may have 3–5, tropical forests may have

Table 3.1 Ranges of soil pH and prevalent ions in soil solution in different types of soil

Type of soil	pH range	Prevalent ions in soil solution
Sodic forest soils	8.5–10.0	Na^+, Ca^{2+}, Mg^{2+}, Cl^-, HCO_3^-
Calcareous forest soils	7.5–8.0	Ca^{2+}, Mg^{2+}, K^+, Na^+, CO_3^{2-}, NO_3^-
Tropical forest soils	4.0–6.0	Ca^{2+}, Mg^{2+}, K^+, $Al(OH)_3^+$, H^+, CO_3^{2-}, NO_3^-, SO_4^{2-}
Mangrove forest soils	7.0–8.0	Ca^{2+}, Mg^{2+}, K^+, Na^+, Cl^-, HCO_3^-, CO_3^{2-}, NO_3^-, SO_4^{2-}
Temperate forest soils	3.0–5.0	Ca^{2+}, Mg^{2+}, K^+, $Al(OH)_3^+$, H^+, CO_3^2, NO_3, SO_4^{2-}
Peat soils	3.5–4.5	H^+, Fe^{2+}, Al^{3+}, CO_3^{2-}
Coastal acid sulfate forest soils	2.0–4.0	H^+, Fe^{2+}, Fe^{3+}, Al^{3+}, SO_4^{2-}

4–6, coastal acid sulfate forest soils may have pH<3, and mangrove forest soils have pH values around 7.5. Table 3.1 shows pH ranges of different types of forest soils. Soil pH values at the extremes (<4.0 and >8.5) can make some nutrients toxic and others unavailable to plants.

An interesting affinity exists between forest tree species and soil reaction (Ewald 2003). It appears well documented that certain species are calcicole with a preference to soils rich in calcium carbonate (*Allium ursinum, Fragaria vesca*), while others are so-called calcifuge species (*Vaccinium myrtillus, Deschampsia flexuosa*) that grow on soils with little or no calcium carbonate and with an acidic reaction (Rorison 1960; Klnzel 1983). Nevertheless, typical calcicole species (*Mercurialis perennis, Arum maculatum*) may occur in abundance on moderately acidic soils, with very little or no calcium carbonate, while typical calcifuge species (*Pteridium aquilinum*) are also found on soils with relatively high contents of calcium carbonate (De Silva 1934). The coexistence of typical calcicole and calcifuge species has also been reported (Cooper 1997).

Most foresters believe that pines grow best on acidic soils while hardwoods prefer slightly acidic to neutral soils. Actually, most tree species will grow well over a broad range of pH values (Williston and LaFayette 1978). Some may have a narrow desirable range. For example, the Caribbean pine (*Pinus caribaea*) grows best in frost-free areas in loam or sandy loam soils of pH between 5.0 and 5.5. European larch (*Larix decidua*) prefers mountainous regions with well-drained, slightly acidic to neutral soil conditions (pH from 6 to 7). Radiata pine (*Pinus radiata*) occurs naturally on acidic soils. Sitka spruce (*Picea sitchensis*) develops best on fresh to very moist sites with pH typically ranging from 4.0 to 5.5. A list of tropical, temperate, and boreal forest trees with their suitable pH ranges is given in Table 3.2.

External sources as well as internal processes may bring about changes in soil chemistry. Internal H^+ sources include soil respiration (formation of carbonic acid), plant and microbial uptake of base cations, and NH_4^+ with an equivalent release of H^+, nitrification of NH_4^+, oxidation of sulfides, and other reduced sulfur compounds, and the production of organic acids. The natural acidification of forest soils due to a number of processes including humus disintegration and nitrification is accelerated by the chronic atmospheric deposition of wet or dry acidic, or acidifying substances (Schutt and Cowling 1985). Acids drastically alter the chemical structure of soils and plants. In ecosystems subjected to a constant "acid bath," nutrients essential for growth such as calcium, magnesium, and potassium are eventually leached from the soil by acidified particles, while aluminum is liberated from minerals and becomes toxic (Hinrichsen 1986).

External causes of soil acidification include the deposition of acidic or potentially acidifying N and S compounds and the use of acid or potentially acidifying fertilizers (Nilsson and Tyler 1995). The present bulk deposition in southern Finland is about 500 mg m^{-2} a^{-1} for SO4–S, 250 mg m^2 a^1 for NO3-N, and 250 mg m^2 a^{-1} for NH4-N with higher values close to the southern coast of Finland (Leinonen and Juntto 1992; Lövblad et al. 1992). The average pH of the bulk deposition is around 4.4 in southern Finland. The present bulk deposition in Kevo in Northern Lapland is about 150 mg m^2 a^1 for SO_4-S, 45 mg m^2 a^1 for NO_3-N, and 25 mg m^2 a^1 for NH4-N, and the pH of the bulk deposition is about 4.8. Rodhe et al. (1995) refer to studies, in which the bulk deposition of SO_4 in remote areas of Northern America has been less than 200 mg S m^2 a^1.

The acidification of forest soils is currently an important ecological issue. It is generally attributed to acid inputs in forests by rain precipitation or dry deposition which leads to an exchange between H^+ ions in the throughfall and Ca^{2+}, Mg^{2+}, or K^+ ions on the exchange complexes (Bonneau 2005). The dominant cause for the acidification of forest soils is the atmospheric deposition of acidifying compounds, like sulfuric and nitric acids and ammonia in Central Europe. Matzner (1992) reported that tree removal of bases is the dominant cause of forest soil acidification in northern Scandinavia and parts of North America. Forest soil acidification and depletion of nutrient cations have been reported for several forested regions in North America, predominantly in the Eastern USA, including the northeast and the central Appalachians, and also in parts of southeastern Canada and the Southern USA due to continuing regional inputs of nitrogen and sulfur causing leaching of base cations and increased availability of Al in soil (Fenn et al 2006). Lawrence et al. (1997) suggested that the depletion of base cations from the soil is the process that leads to Al mobilization. The accumulation and ultimate transmission of acidity from forest soils to streams is of concern, as is the increase in Al concentrations in surface waters. The loss of Ca is an early indicator of the soil response to

Table 3.2 A list of important tropical, temperate, and boreal forest trees and their most suitable pH ranges

Name of tree	Suitable pH range	Name of tree	Suitable pH range
Acacia (Acacia mangium)	4.5–6.5	Oak, Live (*Quercus virginiana*)	6.0–7.5
African Wattle (Peltophorum africanum)	6.5–7.5	Oak, Pin (*Quercus palustris*)	5.0–6.5
Alder (*Alnus rubra*)	6.0–8.0	Oak, Shumard (*Quercus shumardii*)	4.4–6.2
American Beech (*Fagus grandifolia*)	5–7.5	Oak, Southern Red (*Quercus falcata*)	5.0–7.0
Ash, Green (*Fraximus pennsylvanica*)	3.6–7.5	Oak, Water (*Quercus nigra*)	3.6–6.3
Ash, White (*Fraxinus Americana*)	6.0–7.5	Oak, White (*Quercus alba*)	4.5–6.2
Aspen, American (*Populus tremuloides*)	3.8–5.5	Oak, Willow (Quercus phellos)	3.6–6.3
Avocado (*Persea Americana*)	6.1–6.5	Pecan (*Carya illinoiensis*)	4.8–7.5
Birch, Silver (*Betula pendula*)	4.5–6.0	Pine, Jack (*Pinus banksiana*)	5.5–6.5
Blackgum (*Nyssa sylvatica*)	4.6–7.0	Pine, Loblolly (*Pinus taeda*)	4.5–7.0
Blackcypress (*Taxodium distichum*)	4.6–7.5	Pine, Longleaf (*Pinus palustris*)	4.5–7.0
Chestnut (*Castanea sativa*)	4.5–6.5	Pine, Slash (*Pinus elliotti*)	4.5–7.0
Cottonwood (*Populus deltoids*)	3.6–7.5	Pine, White (*Pinus strobus*)	4.5–6.0
Dogwood (*Cornus spp.*)	5.0–8.0	Poplar (*Populus balsamifera*)	4.5–7.5
Eucalyptus (*Eucalyptus camaldulensis*)	5.5–8.0	Rainbow Eucalyptus (*Eucalyptus deglupta*)	5.5–6.5
Fir, Balsam (*Abies balsamea*)	5.0–6.0	Redcedar, Eastern (*Juniperus virginiana*)	6.0–7.5
Fir, Douglas (*Pseudotsuga*spp.)	6.0–7.0	River Birch (*Betula nigra*)	4.6–6.0
Gamar (*Gmelina arborea*)	5.0–7.0	Rubber (*Hevea brasiliensis*)	4.0–6.5
Garjan (*Dipterocarpus turbinatus*)	5.0–6.0	Sal (*Shorea robusta*)	5.0–6.0
Hackberry (*Celtis accidentalis*)	5.0–7.5	Spruce, Black (Picea mariana)	4.7–6.5
Hemlock (*Conium spp.*)	5.0–6.0	Spruce, Colorado (Picea pungens)	6.0–7.0
Magnolia, Southern (*Magnolia grandiflora*)	6.5–7.5	Sweetgum (*Liquidambar styraciflua*)	3.6–7.5
Mahagoni (*Swietenia mahagoni*)	6.1–7.4	Sycamore (*Plantanus occidentalis*)	4.4–7.5
Maple, Red (*Acer rubrum*)	4.4–7.5	Tamarind (*Tamarindus indica*)	4.5–7.0
Maple, Sugar (*Acer saccharum*)	6.0–7.5	Walnut, Black (*Juglans nigra*)	5.0–7.5
Neem (*Azadirachta indica*)	5.0–8.0		
Oak, Black (*Quercus velutina*)	6.0–7.0		

acid deposition for a wide range of forest soils in the USA (Lawrence et al. 1999; Huntington et al. 2000).

Massive forest damage has been observed in certain mountainous regions affected by soil acidification due to acid rain. Acid deposition, commonly known as acid rain, occurs when emissions from the combustion of fossil fuels and other industrial processes undergo complex chemical reactions in the atmosphere and fall to the earth as wet deposition (rain, snow, cloud, and fog). The main chemical precursors leading to acidic conditions are sulfur dioxide (SO_2) and nitrogen oxides (NO_x). When these two compounds react with water, oxygen, carbon dioxide, and sunlight in the atmosphere, the result is sulfuric (H_2SO_4) and nitric acids (HNO_3), the primary agents of acid deposition. Rain in equilibrium with carbon dioxide in the air has a pH of about 5.6. Rain at pH 4.6 is ten times more acidic (most acid rain ranges from pH 4.3 to 5.0). So, acid rain is strongly acidic in nature. An annual rainfall of 1 m per hectare of pH 5.6 can dissolve 400–500 kg of $CaCO_3$ per hectare. Dissolved calcium is eventually lost from the watersheds. In addition, cation-exchange reactions occur rapidly and hydrogen ions in acid rain could acidify soil by depleting exchangeable base nutrients (Krug and Charles 1983).

Large-scale death and decline of trees is ascribed to direct effect of SO_2 and also increasing concentration of mobile forms of Al in soils. About 1,000 km^2 of spruce stands died during the last 3 decades in the Czech Republic (Mladkova et al. 2005). High concentrations of acidificants in the atmosphere have led to this damage of forest and soils. Approximately 38% of the upland forest area in Canada receives acid deposition in exceedance of the critical load. In some cases, exceedances occur in areas where the critical load is low and the acid loadings are high. In other cases, although the critical loads are comparatively high, high levels of acid deposition exceed the natural buffering capacity of the soils (Carou et al. 2008).

The acidification of soils leads not only to decreased pH but also to leaching of base cations and increasing Al^{3+} concentrations (Pierzynski et al. 2000). Aluminum forms play a decisive role in its potential bioavailability and toxicity. Toxicity to plants decreases in the order: Al (not in a form of phosphates or silicates), Al^{3+}, $Al(OH)^{2+}$, and $Al(OH)_2{}^{+}$. Aluminum bound in fluoride or organic complexes and $Al(OH)_3$ are supposed to be nontoxic (Sposito 1996; Gauthier 2002). The distribution of toxic Al^{3+} is strongly affected by content of humic and fulvic acids in surface horizons. Aluminum complexes with both types of acids decrease the toxic impact of Al^{3+} through decreasing its bioavailability and activity in soil solution (Adams et al. 2000).

3.4 Liming

Extensive forest areas were ameliorated by large-scale liming in recent years in order to prevent proceeding acidification and degradation of forest soils. Liming favorably affected the sorption properties of soils and acidity, and the effect was usually manifested in the layer of forest floor humus and in mineral soil within 10 years. Liming reduces acidity and stimulates development of microflora and soil edaphon along with releasing nutrients from SOM. Mineralization of nitrogen is stimulated at nitrogen-rich sites with $C/N<30$. A positive effect of liming on nutrition with bases is generally accompanied by an adverse influence on N dynamics in acidic soils under spruce monocultures (Formanek and Vranova 2002).

Abrahamsen (1994) reported a decrease in active and exchangeable acidity in mineral soil after liming. The response is usually delayed (Bakker et al. 1999) and this beneficial effect persisted for decades (Meiwes 1995). The mineral soil pH was not affected immediately after liming (Mares 1992) and aerial application of limestone without soil pretreatment (Perina and Podrazsky 1988). Lochman and Sebkova (1998) observed a positive influence of liming on soil to a depth of 40 cm under birch but the mineral soil pH dropped under spruce despite a positive effect on the layer of forest floor humus. The value of CEC increased after liming (Podrazsky 1992; Kreutzer 1995). Liming increased the abundance of earthworms in mineral soil (Makeschin 1991), and the contents of available potassium (Podrazsky 1992) and phosphorus.

Historically, Central Pennsylvania has received and currently receives some of the highest levels of acidic deposition in North America (Lynch 1998). Most soils in this area including the Mosquito Creek watershed in Central Pennsylvania occur as naturally acidic and poorly buffered soils (Pabian and Bittingham 2007). High levels of acidic deposition for a long time have acidified drainage water and extirpated fish populations from many headwater areas (Hoover and Rightnour 2002). For restoration of stream water quality to support historical levels of fish populations and to revitalize the forest, a large-scale liming project was conducted in several headwater terrestrial areas (Sharpe et al. 2006). The objectives were to investigate the effects of terrestrial liming on water quality, soil quality, vegetation, aquatic and terrestrial macroinvertebrates, and terrestrial vertebrates, such as birds. Birds have high calcium demands for reproduction. Pabian and Brittingham (2007) selected sites in a contiguous forest on the Mosquito Creek watershed. They quantified the bird community and reproduction and determined whether bird abundance was correlated with soil calcium availability and calcium-rich food (snail) abundance, and they tested whether changing the soil parameters with liming would result in changes in the bird community and bird reproductive parameters. They found that the availability of calcium-rich food items (snails) was a potential link between high levels of acid deposition and low bird abundances in North America.

Studies on sugar maple (*Acer saccharum*) dieback in northeastern North American forests suggested that Ca deficiency due to acid deposition could be a cause of tree growth reduction or decline (Kolb and McCormick 1993; Heisey 1995; Ouimet and Camiré 1995; Sharpe and Sunderland 1995; Wilmot et al. 1995; Long et al. 1997; Moore et al. 2000; Bailey et al. 2004). A positive growth and vigor response of sugar maple to liming in base-poor northern hardwood stands demonstrated that Ca deficiency was linked to sugar maple decline in these ecosystems (Wilmot et al. 1996; Long et al. 1997; Moore et al. 2000).

3.5 Buffering Capacity

Buffering capacity of a soil can be defined as its ability to resist changes in pH. A soil possesses this capacity due to the presence of colloids and electrolytes such as clay, humus, and other colloidal materials, weak organic and inorganic acids and bases, etc. Fine-textured forest soils with high organic matter contents have high buffering capacity. The phenomenon appears to be simple, but the mechanism of buffering involves complex physicochemical, chemical, and biochemical reactions. Buffering capacity can be expressed in units of meq $H^+(\text{kg soil})^{-1}$ $(\text{pH})^{-1}$. Many different factors may influence the pH values and their changes in forest soils. Bloom et al. (1979) proposed exchange processes between H^+ and Al cations to be an important regulator in acidic tropical forest soils. Many studies have verified the effect of SOM on Al solubility in other acidic soils (Simonsson and Berggren 1998). As pointed out by Skyllberg (1999), chemical models based on a pH-dependent Al-mineral solubility combined with cation-exchange reactions between dissolved Al ions (mainly Al^{3+}) and base cations (Reuss et al. 1990) are restricted to a situation when soil chemical processes give rise to a positive relationship between pH and BS. Skyllberg and Borggaard (1998) showed that acid–base titration data for the O, E, Bh, and Bs horizons of the Spodosol profile could be described by a simple weak organic acid analogue, when the effect on pH caused by organically complexed Al was taken into consideration. This might suggest that weak acidic groups of SOM are the main pH buffers throughout acid Spodosol profiles and that pH effects caused by changes in the quality of SOM play a relatively minor role as compared to the composition of organically adsorbed major cations.

Federer and Hornbeck (1985) measured buffer capacity for major horizons of forest soils from four locations in New England by titration of field-moist samples with either HCl or NaOH. Titration curves for O horizons were nearly linear over a wide pH range, that is, buffer capacity was independent of pH. Titration curves for mineral horizons were S shaped with ambient pH roughly in the middle of the least-buffered part of the curve. We also measured exchangeable acid cations and NH_4^+ in unbuffered KCl extractions and exchangeable bases in NH_4OAc extraction at pH 7.0. The Ca^{2+} and Mg^{2+} ions in KCl extractions at ambient pH were only slightly less than in NH_4OAc extractions at pH 7.0, implying that exchangeable bases did not depend much on the extraction pH. The O horizons were generally highly base saturated at ambient pH even though their pH was low; mineral soils had lower BS. Buffer capacity measured over the first 0.5 pH unit to the acid side depended strongly on organic matter fraction in the sample. They estimated that acid rain at pH 4.0 in New England would take at least several decades to lower pH of the soil profile by a whole pH unit. James and Riha (1985) equilibrated samples of organic surface horizons (Oi, Oe, and Oa) from New York State forest soils with 0–20 cmol HNO_3 kg^1 soil in the laboratory by a batch technique designed to simulate reactions of acid precipitation with forest floors. Each organic horizon retained a constant percentage of added H^+ (67–96 %) regardless of quantity added, and buffer capacities of the samples ranged from 18 to 36 cmol (H^+) kg^1 (pH unit)^1, an order of magnitude higher than those measured in underlying mineral horizons. Calcium was the dominant cation in unacidified equilibrium solutions, and its concentration changed the most in response to acidification of the soils. Aluminum contributed < 1 % to total cationic charge in solution and Fe concentration decreased upon addition of up to at least 5 cmol (H^+) kg^1. From 38 to 79 % of H^+ removed by the soils was not balanced by increases in soluble Ca, Mg, K, Mn, Fe, Zn, Al, and Cu, and buffering by soluble organic and inorganic constituents accounted for 8–58 % of H^+ removal. The protonation of anions of weak organic acids in solution and solid phases probably accounted for this H^+ removal without increasing cation concentrations.

3.6 Ion Exchange

Very fine soil particles (or colloids) are electrically charged and have high surface area; and they can attract and bind ions from soil solution at their surfaces. The bond is unspecified and oppositely charged ions, cations against negative charges and anions against positive charges, can take part in the bond. These adsorbed ions are exchangeable and can change places with other positively charged ions depending on their charges, concentration, and activity (Fig. 3.1). The capacity of a soil to bind cations at particle surfaces is denoted by CEC which is defined as the number of centimoles of charges per kilogram of soil. For anions, the corresponding expression is AEC. Most soils have more negative charges (higher CEC) than positive charges (lower AEC). Most common exchangeable cations are Ca^{2+}, Mg^{2+}, K^+, NH_4^+, H^+, and Al^{3+}. Common anions are SO_4^2, NO_3, Cl, and PO_4^3. Cation exchange is the property of clay minerals and humus in soil. During ion exchange processes, an equilibrium is maintained in the soil solution. If the composition of soil solution is changed,

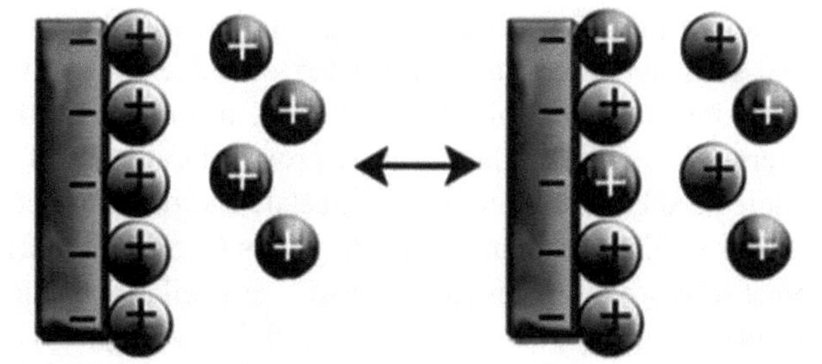

Fig. 3.1 Exchange of cations between soil colloidal phase and soil solution phase

Table 3.3 Average cation capacities of different colloids

Type of colloids	CEC, $cmol_c$ kg^1
Kaolinite	1–10
Montmorillonite	80–120
Vermiculite	120–150
Chlorite	20–40
Gibbsite, Goethite	1–10
Humus	100–300

there is a concomitant change in the composition of adsorbed cations. Cations may be introduced to soil by weathering, organic matter decomposition, atmospheric deposition, and fertilizers. Soil colloids differ in their number of negative charges; so, there is an enormous variation in CEC of different colloids (Table 3.3). Charges on soil colloids are of two types: (i) permanent charges or charges independent of pH and (ii) pH-dependent or variable charges. The permanent charges develop on layer silicate clays (except kaolinite) by isomorphous substitution. This charge arises by atomic proxying during crystallization in the structure of clay minerals where a cation of lower charge substitutes for either Si^{4+} or Al^{3+} in octahedral or tetrahedral positions. This substitution results in a charge deficit (fewer cationic charges compared to anionic charges) and negative charge on the clay which is directly dependent on the amount of substitution. On the other hand, pH-dependent charges develop on all types of colloids, on layer silicate colloids from broken crystal edges, and on amorphous inorganic colloids, oxides and hydroxide colloids, and humus by dissociation of active groups and association of hydrogen ions or protonation. Since this pH-dependent charge is a protonation or deprotonation reaction, it can be described by standard methods used in acid–base chemistry. There may be both negative and positive pH-dependent and -independent charges. For a brief description of the processes of charge development on crystalline and amorphous soil colloids, readers may consult Osman (2013).

A good number of methods have been employed for determining CEC of soils, and the most common method uses 1M ammonium acetate at a pH of 7.0 (Chapman 1965). This method was developed for agricultural soils; an extraction solution of pH 7.0 is suitable for agricultural soils because these soils are often limed to pH 6.5 or higher. Negative correlations between soil pH and CEC in soils of the northeastern USA and Scandinavia have provided evidences for criticizing the neutral ammonium acetate saturation procedure for CEC measurement of forest soils (Johnson 2002). Forest soils are rarely limed, and most forest soils are acidic. The pH of forest soils is commonly well below 7 and may be as low as a pH of 3. Using a solution of pH 7.0 for forest soils has, therefore, been questioned (Skinner et al 2001). Extracting exchangeable cations with extracting solutions of pH 7.0 may result in the overestimation of CEC in soils with pH-dependent charges (Rhoades 1982). Gillman et al. (1983) compared methods of CEC analysis and found that CEC determined at a pH appreciably higher than soil pH will overestimate the CEC of soils. The CEC determined by an unbuffered extracting solution is sometimes termed as effective cation-exchange capacity (ECEC). Recently, many forest soils researchers have used unbuffered NH4Cl to measure CEC of forest soils (Johnson et al. 1982; Lindberg et al. 1989; Richter et al. 1992; Johnson et al. 1997; Trettin et al. 1999). Johnson and Todd (1998) used NH4Cl to determine how harvesting affects CEC in a mixed oak forest soil. In a study of CEC of forest floor materials from pine, hardwood, and mixed stands, Wells and Davey (1966) observed that CEC was highly dependent upon the saturating cation used in the determination and somewhat less dependent upon the replacing cation. They found barium saturation and calcium replacement as the best cations for CEC determination of these materials. Buffering properties of the forest floor organic materials were strongest primarily at slightly acid to alkaline reactions. Kalisz and Stone (1980) measured CECs of acid forest humus layers and mineral soils at the pH of air-dried samples using a neutral salt procedure. The values obtained were 0.25–0.5 as measured using NH_4OAc at pH 7.0. Organic matter was the chief source of EC at all localities. CEC (field pH) of mors (raw humus formed in acid conditions where the plant litter is low in bases; bogs, heaths, and pine-forest soils have thick layers of mor humus) from the Adirondack Mt. area and mulls (mild humus formed in slightly acid to neutral soils usually under broadleaf deciduous forests, crumby and richer in calcium) from central New York averaged from 52 to 123 cmole$_c$ kg^1 organic matter, respectively, as compared with 183 and 236 cmole$_c$ kg^1 when determined at pH 7.0. Tropical forest soils are usually devoid of high activity clays like montmorillonite and vermiculite. They contain a high proportion of kaolinite and sesquioxide clays. They are poor in organic matter too. So, tropical forest soils have lower CEC than temperate forest soils. For example, the exchangeable cations and CEC across the pedons of tropical teak forests studied by Egbuchua and Bosah (2011) were low with mean values of 3.74 cmolkg^1 for exchangeable Ca, 1.89 cmolkg^1 for Mg, 0.12 cmolkg^{-1} for K, 0.17 cmolkg^1 for Na, and 9.55 cmolkg^{-1} for CEC. The ECEC ranged from 6.46 to 10.60 cmolkg^{-1} with a mean of 8.41 cmolkg^1. The critical value of CEC in tropical soils is 4 cmolkg^1 as established by Agboola and Ayodele (1987) and Sanchez (1976). The low values could be attributed to the low colloidal content of clays which are the essential materials for cation exchange and adsorption in the soil. Sukardjo (1994) reported chemical and nutritional properties of some mangrove forest soils of Indonesia. Dominant trees were *Avicennia officinalis* and *Ceriops tagal*. The soils had pH values (4.35–5.29) unusually lower for mangrove forests. Probably, sulfidic materials have accumulated in these mangrove areas. The average CEC of *A. officinalis* soil was 23.72 cmol$_c$ kg^1 and in *C. tagal* soil was 14.15 cmol$_c$ kg^1. The CEC of forest soils derived from volcanic ash is regulated more by variable or pH-dependent charges associated with the colloidal fraction. The CEC of andic (volcanic ash) soils decreases with decreasing pH, and may, in fact, be quite low at the acidic pH values that characterize many forest soils. Moreover, some andic soils may have substantial AEC at acidic field pH values (Kimble et al. 2000), influencing the retention of ions such as chloride and nitrate. Although PO_4^{3-} and SO_4^{2-} ions are sorbed through anion exchange, fixation of these ions (described later) is much greater than this exchange reaction. McDaniel and Wilson (2007) observed average ECEC, CEC (pH 7.0), and CEC (pH 8.2) of volcanic ash soils of USA to be 6.5, 19.1, and 26.4 cmol$_c$ kg^1, respectively.

Ross et al. (2008) have questioned traditional views developed primarily in an agricultural context that use CEC, base cation saturation and exchangeable cation concentrations as indicators of base cation availability in forest soils. They argue that for acid forest soils (1) new measures of exchangeable Al and H are needed, (2) BS should be abandoned as a measure of base cation availability, (3) the paradigm that higher pH accompanies higher CEC does not hold, and (4) CEC should not be used to indirectly infer base cation availability. Still CEC is an important determinant of forest soil chemical quality, particularly the retention of major nutrient cations Ca, Mg, and K, and immobilization of potentially toxic cations Al and Mn; these properties can, thus, be useful indicators of soil health, informing of a soil's capacity to absorb nutrients, as well as pesticides and chemicals.

3.7 Base Saturation Percentage

BS refers to the proportion of the CEC in soil which is occupied by base cations (calcium, magnesium, sodium, and potassium). It is estimated by dividing the total exchangeable bases (TEB) with CEC. Total TEB of a soil can be obtained by subtracting exchangeable hydrogen and exchangeable aluminum from CEC, or approximately by summation of exchangeable sodium, potassium, calcium, and magnesium. BS percentage can be expressed as

$$BSP = \frac{TEB}{CEC} \times 100.$$

There is a positive relation between pH and BS percentage. Acidic soils have low and alkaline soils have high BS percentage. BS percentage in forest soils vary within a wide range, from <5% to >95% depending on the pH of the soils. Roskams (1997) studied pH and BS in ten podzol forest soil plots. BS values lower than 20% in at least two mineral depth layers were found in seven plots. Very low BS values (<5%) were found in three plots. Clark (1965) reported very high BS percentage of Gray Brown Podzolic, Gray Wooded, Brown Wooded, and many of the Acid Brown Wooded soils with values more than about 90%. Only soils with well-developed Podzol features and a Gray Wooded soil developed from acid shale had BS values below about 50%. The upper horizons of some Brown Podzolic and some weakly developed Podzols had intermediate BS values. The BS in the 0–5-cm layer of Polish forest soils ranged from 6.76 to 97.9%. It is accepted that soils with BS<25% are particularly susceptible to acidification. BS values for forest soils of less than 10% are problematic with respect to aluminum availability (Cronan et al. 1989). Sharpe and Sunderland (1995) studied the chemical and nutritional properties in some forest soils severely damaged by acidity. Both the declining and non-declining stands had BS values of less than 10% although the non-declining mean was 9.39%. Estimates of exchangeable Ca, Mg, and K indicate significant differences between declining and non-declining stands. Richter et al. (2011) showed from the results of a soil monolith study that BS affected root growth of European beech seedlings.

3.8 Salinity and Sodicity

Soils are said to be saline when they contain excess soluble salts. These salts include chlorides and sulfates of sodium, potassium, calcium, and magnesium. Sometimes, there may be some nitrates, carbonates, bicarbonates, and borates. The concentration of salts in saturation extracts of soil is proportional to its electrical conductivity. Therefore, electrical conductivity of the saturation extract of the soil (ECe) is taken to be the basis of classifying soils into saline and nonsaline categories. The saturation extract of soil is obtained by adding water to soil up to the maximum water-holding capacity level, stirred and equilibrated, and the extract is taken under suction. The ECe value of the extract is measured with a conductivity meter. Agricultural soils with ECe values greater than 4.0 dS m^1 (decisiemens per meter) are called saline soils according to United States Department of Agriculture (USDA) Salinity Laboratory Staff (Richards 1954). Along with the presence of high concentration of soluble salts, saline soils have pH around 7.0–7.5. If the soils are not sodic at the same time, physical conditions of saline soils are satisfactory. But, if the soil is sodic (ESP>15; SAR>13), its structural quality is degraded. On the basis of salinity and sodicity, there are four classes of forest soils. However, there must be some difference between salt tolerance of crop plants and forest trees, and taking 4.0 dS m^1 as the basis of salinity in forest soils may be debated.

Categories	Characteristics
Nonsaline non-sodic soil (normal soil)	ECe<4.0 dS m^1; SAR<13, ESP<15
Saline non-sodic soil	ECe>4.0 dS m^1; SAR<13, ESP<15
Nonsaline sodic soil	ECe<4.0 dS m^1; SAR>13, ESP>15
Saline sodic soil	ECe>4.0 dS m^1; SAR>13, ESP>15

Soil salinity develops under three environmental settings: arid and semiarid regions, coastal tidal areas, and humid areas with high water table. There are saltbush, shrubs, and salt-tolerant trees in arid saline soils, mangrove forests in coastal tidal saline soils, and halophytes in waterlogged saline soils. Soil salinity induces injury, inhibits seed germination and vegetative and reproductive growth, alters plant morphology and anatomy, and often kills nonhalophytes (Kozlowski 1997). Plants suffer from soil salinity mainly for two reasons: (i) reduced water absorption due to reduction in osmotic potential and (ii) specific ion effects. Water does not move into the plant and could even move osmotically from the cells to the soil with elevated salt content. Increased salt contents tend to draw water toward the salt solution. The predominance of some ions in soil, for example, sodium, may hamper the absorption of other ions, say potassium. Excessive sodium in soils causes soil aggregates to break down, resulting in poor aeration and slow water permeability. The resulting soil lacks good drainage and proper oxygen concentrations and leads to reduced moisture uptake by roots. On the other hand, the accumulation of specific salts within plants may exert toxic effects. Salts absorbed by the plant can desiccate leaf cells, causing browning and leaf abscission. Plant growth is directly affected by high levels of sodium chloride and other salts. Excess sodium and more importantly chloride has the potential to affect plant enzymes and cause cell swelling, resulting in reduced energy production and other physiological changes. In woody species, salt damage includes late, stunted buds, small leaves, and necroses in buds, roots, leaf margins, and shoot tips. Before the growing season ends, leaves become yellow and dry and root and shoot sectors die (Larcher 1980).

Khan et al. (2009) studied the effects of salinity on growth of some forest tree species at their seedling stage in nursery during 2007. Four forest tree species, i.e., *Acacia ampliceps*,

Acacia nilotica, *Eucalyptus camaldulensis*, and *Azadirachta indica* were tested against 3.61, 6.0, 12.0, and 18.0 dS m^{-1} ECe levels. The length of root and shoot and root-shoot ratio of these forest tree species were significantly affected by salinity. Moreover, growth parameters of each forest tree species decreased as salinity increased. *E. camaldulensis* was found as the most susceptible to salinity. Another study was carried out to compare tree response to soil salinity of Casuarina cunninghamiana and E. camaldulensis grown on a saline site in tropical north Australia by Sun and Dickinson (1995). Tree response was examined with regard to survival, height, diameter at 1.4 m height, root growth, transpiration rate, and plant chemical components at 24–36 months. Both species had a similarly high survival rate. C. cunninghamiana appeared to use less water than E. camaldulensis in both summer and winter. The effect of salinity on seed germination characteristics of five arid zone tree species, *Acacia cyanophylla, A. seyal, A. tortilis, A. tumida*, and *Parkinsonia aculeata*, was determined under ambient laboratory conditions by Nour et al. (2006). Seeds were sown on filter papers soaked with salt solutions of 0, 1.0, 2.0, 5.0, 10.0, and 20.0 dS m^1 salinity levels. Considerable interspecific variation in the response of seed germination to salinity was observed. Salinity levels of 5.0 and 10.0°dS°m^1 significantly (P>0.05) delayed the initiation of seed germination of all the species except *A. cyanophylla*, which appeared less sensitive to salinity stress than the other species. The high salinity level (10 dS m^1) significantly induced (P>0.05) germination speed of seeds of *A. seyal*and *A. tumida*, while it had no significant effect on the germination speeds of the other species. Salt concentration of 20 dS m^1 completely inhibited seed germination. Germination percentage of *A. seyal*was significantly (P>0.05) retarded by salinity level as low as 2 dS m^1.

Mangrove is an ecological term referring to an association of woody species which forms the dominant vegetation in tidal, saline wetlands of tropical and subtropical coastal environments (Hutchings and Saenger 1987). Mangroves typically grow in saline coastal soils, or muds, which are poorly drained because of their fine texture, the flatness of the landscape, and their regular inundation by the tides. These soils can remain saturated, or nearly so, even at low tide, such that their interstitial water is poorly flushed. One of the most important stress factors in mangrove ecosystems is salinity. Several studies have attempted to correlate salinity to the standing crop of vegetation and productivity (Adams 1963; Good 1972; Lugo 1980). Walsh (1974), however, maintained that the correlations often fail to identify directly the differences in primary production of mangroves due to overlapping responses of species to salinity, more so as mangrove swamps vary in terms of tidal and freshwater inputs, duration of dry spells and concentrations of soluble salts in the water and soil (Chapman 1976; Cintron et al 1978). Gradients in salinity are ubiquitous features of wetland habitats along the northern Gulf of Mexico coast along with flooding depth as major influences on plant zonation in coastal marshes (Mitsch and Gosselink 1993; Odum 1988; Earle and Kershaw 1989). At least 35 woody species have been found in mangrove forests of Northern Australia, and maximal in areas where moderate salinities and high soil water contents prevail in the late dry season. Three ecophysiological hypotheses are suggested which might provide mechanistic bases for understanding local variation in species richness along natural salinity gradients (Wells 1985; Wightman 1989). These species are distributed differentially along tidally maintained environmental gradients, such that a banded zonation pattern often forms normal to the shore (Semeniuk 1985; Semeniuk and Wurm 1987), as is typical of mangrove forests worldwide (Hutchings and Saenger 1987). It has long been recognized that such banded zonation patterns are well correlated with the frequency and duration of tidal immersion (Chapman 1975). Some forest tree species adapted to mangrove saline soils are: *Avicennia africana, Rhizophora racemosa, Rhizophora mangle, Rhizophora harrisonii, Laguncularia racemosa, Rhizophora apiculata, Rhizophora lamarckii, Ceriops australis, Excoecaria ovalis, Bruguiera exaristata, Avicennia integra, Camptostemon schultzii, Sonneratia alba, Bruguiera gymnorrhiza, Bruguiera parviflora, Avicennia marina, Diospyros littorea, Rhizophora stylosa, Heritiera fomes, Excoecaria agallocha, Xylocarpus mekongensis, Sonneratia apetala, and Avicennia officinalis.*

Ayyad and El-Gharee (1982) studied the extensive salt marshes in the plains and depressions of the western Mediterranean desert of Egypt. They classified the habitat types into three categories: (A) habitat with shallow water table and high salinity, (B) habitat with relatively deep water table and high salinity, and (C) transitional habitats in which salinity and water table are not the controlling factors. Fourteen vegetation types were distinguished; each dominated by one or two species. The dominating life forms are chamaephytes (woody plants with perennating buds borne close to the ground, no more than 25 cm above the soil surface) in sites of high salinity, and therophytes (annual plants which survive the unfavorable season in the form of seeds and complete their life cycle during favorable seasons) in sites of low salinity. There was a general trend of increasing salinity and concentration of different ions from habitat type A to habitat type C through habitat type B. Salt-desert shrub communities of the Intermountain West cover some 16 million ha of rangeland within the Great Basin Desert (Holmgren and Hutchings 1972). Many of these salt-desert plant communities are in pure stands of single species while others occur as a mixture of species. There are many factors that develop the often distinct boundaries of salt-desert plant communities. West (1982) attributes this to the fact that the salinity and aridity tolerance levels of various species cause a sorting effect along a moisture and salinity gradient. Branson et al.

(1967) observed the soil–moisture relationship and soil salts as the primary cause of different plant communities. There were different amounts of salt and sodium between the soils of each species. Billings (1949) found salt-desert shrub soils to have carbonates accumulating just below the surface and the presence of salts in the subsoil.

The ability of a plant to grow and complete its life cycle on saline substrates that contain high concentrations of salt is called salt tolerance (Jeschke 1984). Salt-tolerant plants can withstand the toxic and osmotic effects of increased ion concentration caused by the presence of excess salts (especially Na^+, Cl^-, and SO_4^{2-}) without serious impairment of vital function (Larcher 1980). Salt tolerance may involve salt avoidance. For example, in woody species, exclusion of sodium or especially chlorine ions from plant roots or shoots is the most important mechanism for salt tissue tolerance (Allen et al. 1994). Species with salt-tolerant protoplasts can survive 4–8 % NaCl, whereas salt-sensitive protoplasts are destroyed in solutions of 1.0–1.5 % NaCl (Larcher 1980). Salt tolerance is often measured by plant growth or yield (Shannon 1985). Other factors indicating plant salinity tolerance are germination rates and changes in net photosynthesis, water status, or stomatal conductance (Pezeshki and Chambers 1986). Salinity tolerance may be affected by soil calcium concentrations (Rengel 1992). Marcar and Crawford (2004) and Jackson and Bird (2008) proposed salt tolerance of some tree species for plantation in saline landscapes. The list is summarized as follows:

Low salinity (2–4 dS m^1) Acacia mearnsii (Black Wattle), Corymbia maculata (Spotted Gum), Eucalyptus globulus (Tasmanian Blue Gum).

Medium salinity (4–8 dS m^1) Acacia melanoxylon (Blackwood), Allocasuarina verticillata (Drooping Sheoak), Eucalyptus leucoxylon (Yellow Gum), Eucalyptus melliodora (Yellow Box), Eucalyptus microcarpa (Grey Box), Eucalyptus ovata (Swamp Gum), Eucalyptus viminalis (Manna Gum), Casuarina cunninghamiana (River Sheoak), Eucalyptus cladocalyx (Sugar Gum), Eucalyptus kitsoniana (Gippsland Mallee), Eucalyptus polyanthemos (Red Box), and Eucalyptus tricarpa (Red Ironbark).

High salinity (8–16 dS m^1) Eucalyptus camaldulensis (River Red Gum), Eucalypt hybrids (e.g., E. camal. x E. grandis), Eucalyptus largiflorens (Black Box), and Eucalyptus halophylla (Salt Lake Mallee).

Extreme salinity (>16 dS m^1) Melaleuca decussata (Totem Poles) and Melaleuca halmaturorum (Salt Paperbark).

Sodic soils are more problematic than saline soils. Sodic soils have high exchangeable sodium, high pH (between 8 and 10), and poor structural quality. These soils are usually dispersed and impervious. As a consequence, the movement of water and air is restricted (Singh and Garg 2007). In general, soil sodicity is a more widely spread form of land degradation occupying >2 billion ha throughout the world (Grainger 1988). Sodic soils are highly alkaline in reaction (pH 8.0–10.0). Sodic soils have nutrient limitations and are deficient in zinc, iron, phosphorus, and occasionally calcium, potassium, and magnesium. Zinc deficiency in alkaline soil has been well documented with many crop species and some species develop severe symptoms at pH 8.2, whereas for other species the symptoms are lacking. In sodic soils, boron may be at phytotoxic concentrations due to negligible leaching (Marschner 1986). Bingham and Nelson (1971) subjected adult avocado trees to different soil sodium saturation levels (SAR from 0 to 12) and observed the performance of the trees. Fruit production after one season of sodium treatment was eliminated from trees under SAR 8 and SAR 12 treatments. These treatments severely damaged the trees, producing profuse leaf burn, high leaf contents of Na (0.8–1.0 % Na), followed by defoliation. All fruits were shed by the trees at the onset of leaf defoliation. However, the SAR 0- and SAR 4-treated trees remained normal.

The growth responses of *Dalbergia sissoo* and *A. nilotica* seedlings were compared by Singh and Garg (2007) on different levels of soil sodicity. The growth and dry weight of 1-year old seedlings of both the tree species decreased as the level of sodicity increased. However, the suppression in growth caused by sodicity was relatively greater in *D. sissoo* than in *A. nilotica*. Singh and Garg (2007) observed that *Derris indica*, *D. sissoo*, *Azadirachta indica*, *Cassia siamea*, *Terminalia arjuna*, and *Syzigiumcumini* were the dominant species. *A. nilotica* is a leguminous tree species commonly found in interior Karnataka on sodic soils. This species has been successfully used in the reclamation of sodic soils in India (Basavaraja et al. 2011). Other species used for sodic soil reclamation are Prosopis juliflora, D. sissoo, and E. tereticornis (Mishra and Sharma 2010). A list of some deciduous and evergreen tree species suitable for sodic soils is given as follows:

Deciduous trees Ash (*Fraxinus* sp.), Chokecherry (*Prunus virginiana*), Hackberry (*Celtis sp.*), Japanese Tree Lilac (*Syringa reticulate*), Maple-Hedge (*Acer campestre*), Maple-Norway varieties (*Acer latanoides*), Mulberry (*Morus alba*), Oak-Bur (*Quercus macrocarpa*), and Oak-Swamp (*Quercus bicolor*).

Evergreen trees Arizona Cypress (*Cupressus glabra*), Austrian Pine (*Pinus nigra*), Black Hills Spruce (*Picea glauca densata*), Blue Atlas Cedar (*Cedrus atlantica glauca*), Bos-

Table 3.4 Forest tree species suitability for sodic soils. (Source: Gupta et al. 1995; Dagar and Singh 1994)

pH	Suitable species
>10	*Acacia nilotica, Cassuarina equisetifolia, Prosopis juliflora, Prosopis cineraria, Tamarix articulate*
9.0–10.0	*Pithecellobium dulce, Salvadora persica, Salvadora oleoides, Capparis decidua, Terminalia arjuna, Cordia rothu, Albizzia lebbeck, Pongamia pinnata, Sesbania sesban, Eucalyptus tereticornis, Parkinsonia aculeata, Cassia siamea*
8.6–9.0	*Butea monosperma, Acacia auriculiformis, Azadirachta indica, Grevillia robusta, Populus delteoides, Melia azaderach, Delbergia sissoo, Hardwickia binata, Morus alba, Moringa olifera, Tectona grandis*

nian Pine (*Pinus leucodermis*), Deodar Cedar (*Cedrus deodara*), and Juniper-upright (*Juniperus sp*).

The suitability of different forest tree species in different types of sodic soils is given in Table 3.4.

3.9 Free Carbonates

Some soils have a high content of free carbonates, i.e., carbonates of the alkaline earth metals calcium and magnesium. They either have originated from lime-rich or calcareous parent materials or have undergone calcification during soil formation. Calcareous soils may contain some carbonates of iron, zinc, and manganese as well (Brown et al. 1978). Calcite and aragonite ($CaCO_3$), dolomite [$CaMg(CO_3)_2$], and magnesite ($MgCO_3$) are the calcium- and magnesium-containing minerals found in calcareous soils. The carbonate minerals may be primary carbonates originating from carbonate-rich parent materials, or secondary carbonates formed by reprecipitation. The common field test for the presence of carbonates in soil is the addition of dilute hydrochloric acid solution (10% HCl). Calcareous soils effervesce upon addition of dilute hydrochloric acid. The actual carbonate content, expressed as $CaCO_3$ equivalent, can be determined in the laboratory (Day 1983). Calcareous soils have a pH between 7.0 and 8.5 (Yaalon 1957). Calcium and magnesium carbonates are very reactive soil constituents. Carbonates are found in sand, silt, and clay fractions; carbonates in the silt and clay fractions are the most reactive. The chemistry of carbonate-rich soils is dominated by particulate calcium and magnesium carbonates, dissolved carbonates, and exchangeable calcium ions. The high pH of the calcareous soils often creates Fe and Mn deficiency to conifers such as *Pinus resinosa, Pinus radiata, Pinus sylvestris, Picea excelsa, Populus tremuloides*, etc. in calcareous soils. However, some coniferous forest tree species including *Picea glauca* and *Picea englemannii* thrive well in calcareous soils (Kishchuk 2000).

Calcareous soils cover more than 30% of the earth's surface (Marschner 1995) occurring in Brazil, China, Indo-China, Sumatra, Malaya, Sarawak, Java, and New Guinea. The forests over limestone have several characteristic species occurring in habitats normally drier than the surrounding areas (Aider et al. 2001). LA CWCS (2005) describes a forest community on calcareous soil in the uplands of Central, western, and northwest Louisiana. It characteristically occurs on hills and slopes on either side of small creeks, at times in a mosaic with calcareous prairies. Soils are stiff calcareous clays (surface pH ~ 6.5–7.5) with very high shrink–swell characteristics. Common overstory species include *Quercus stellata* (post oak, often dominant), *Q. shumardii* (shumard oak), *Q. alba*(white oak), *Q. muhlenbergii* (chinkapin oak), *Carya myristiciformis* (nutmeg hickory), *C. ovata* (shagbark hickory), *C. tomentosa* (mockernut hickory), *Pinus echinata* (shortleaf pine), *P. taeda* (loblolly pine), *Fraxinus americana* (white ash), *Diospyros virginiana* (persimmon), *Liquidambar styraciflua* (sweetgum), *Celtis* spp. (hackberries), *Gleditsia triacanthos* (honey locust), *Morus rubra* (red mulberry), *Fagus grandifolia* (beech), *Ulmus rubra* (slippery elm), *U. americana* (American elm), *U. alata* (winged elm), *U. crassifolia* (rock elm), and *Acer rubrum* (red maple). *Quercus sinuata*-var. *sinuata* (Durand oak) and *Q. oglethorpensis* (Oglethorp oak) may rarely be present. Trees, especially pines, are often stunted and/or crooked due to extreme physical soil properties. Midstory and understory shrubs typically include *Viburnum rufidulum* (rusty blackhaw), *Crataegus* spp. (hawthorns), *Prunus mexicana* (Mexican plum), *Cercis canadensis* (red bud), *Chionanthus virginicus* (fringe-tree), *Asimina triloba* (paw-paw), *Ilex decidua* (deciduous holly), *Vaccinium arboreum* (winter huckleberry), *Rhamnus caroliniana* (Indian cherry), *Rhus copallina* (flame-leaf sumac), *Ostrya virginica* (hop-hornbeam), and *Aesculus pavia* (red buckeye). *Maclura pomifera* (osage-orange) may occur sporadically, especially in northwest Louisiana. The herbaceous layer may contain *Symphyotrichum drummondii* (Drummond's aster), *Solidago auriculata* (auricled goldenrod), *Cynoglossum virginianum* (hound's-tounge), *Antennaria plantaginifolia* (plantain-leaf pussy-toes), *Lithospermum tuberosum* (tuberous puccoon), *Pedicularis canadensis* (Canadian lousewort), *Podophyllum peltatum* (may-apple), *Phlox divaricata* (phlox), *Elephantopus* spp. (elephant-foot), *Viola* spp. (violets), *Chasmanthium* spp. (spangle-grasses), *Bromus* spp. (brome grasses), *Onosmodium hispidissimum* (false-gromwell), *Sanicula canadensis* (snakeroot), *Zizia aurea* (golden alexanders), *Tipularia discolor* (crane-fly orchid), *Agrimonia* spp. (agrimony), *Galium* spp. (bedstraws), and others. Mixtures of sugar maple (Acer saccharum var. saccharum), black maple (Acer nigrum), white oak (Quercus alba), northern red oak (Quercus rubra), black oak (Quercus velutina), and hickories (Carya spp.) are typical of calcareous soils in Arizona. Another variant features codominance by white oak, chikapin oak (Quercus muhlenbergii), white ash (Fraxinus americana), and hickories (Anon 2012).

Soddy calcareous soils are found to develop on carbonate rocks (limestone, chalk, and dolomite) beneath coniferous, deciduous–coniferous, or broad-leaved forests. Such soils are permeable. A typical soddy calcareous soil consists of a humus layer 10–15 to 30–40 cm thick and the carbonate rock underlying it; it is dark gray and effervesces because of acid from the surface. Typical properties of soddy calcareous soils are a weak alkaline or close to neutral reaction in the humus layer and a weak alkaline reaction in layers B and C; a high humus content (6–15 %); complete saturation of the adsorbing complex with bases (Ca and Mg); a water-resistant granular or nutlike-granular structure; high biological and microbiological activity; and significant reserves of phosphorus, potassium, and nitrogen (Rudneva 2010).

3.10 Redox Potential (E_h)

There are a variety of electron acceptors and electron donors in soil. An electron acceptor is a chemical entity that accepts electrons from another chemical entity. An electron acceptor is an oxidizing agent that, by virtue of its accepting electrons, is itself reduced in the process. Prominent oxidizing agents in soil are O_2, NO_3^-, Fe (III), Mn (IV), SO_4^{2-}, and CO_2. An electron donor donates electron during an oxidation–reduction reaction, and by donating electrons it becomes oxidized. Electron donors are reducing agents and prominent reducing agents in soil are H_2, S, S^{2-}, Fe (II), Mn (II), CH_4, and organic compounds. However, oxidation is the loss of electrons or an increase in oxidation state by a molecule, atom, or ion. Reduction is the gain of electrons or a decrease in oxidation state. A common redox reaction in soil is given as follows:

$$Fe(OH)_3 + e^- + 3H^+ \Leftrightarrow Fe^{2+} + 3H_2O.$$

This reaction proceeds to the left (oxidation) if the soil is aerobic, and it proceeds to the right (reduction) if the O_2 supply in soil is low. This reaction or any oxidation–reduction reaction can be divided into two half reactions: one in which a chemical species undergoes oxidation and one in which another chemical species undergoes reduction. The driving force is the oxidation–reduction potential which is a measure of the affinity of a substance for electrons. It indicates the electronegativity of substances. Substances less electronegative than hydrogen have negative redox potentials and substances more electronegative than hydrogen have positive redox potentials. The redox potential is denoted by E_h and expressed in volts (V) or millivolts (mV). Redox potential varies within a wide range depending on the aeration in the soil. It may range from 100 to−300 mV in O_2 depleted soils, and may be as high as 800 mV in very well-aerated soils. According to Jackson (2005), well-oxidized soils, moderately well-oxidized soils, poorly oxidized soils, much reduced soils, and extremely reduced soils may have redox potential values around 500, 300, 100,−200, and−300 mV, respectively.

Redox potential is an integrated soil measurement that reflects several environmental conditions in the soil associated with aeration, moisture, and organic matter dynamics. It is related to water table fluctuations, precipitation and landscape gradients, organic matter decomposition rates, nutrient dynamics, biological diversity, and plant species distribution. It is an excellent indicator of soil biological processes. Thus, redox serves as an ecological indicator of site productivity at the ecosystem scale and may be used for management purposes as its magnitude can be altered by activities such as harvesting and drainage. A threshold value of 300 mV has been documented as the critical value below which anaerobic conditions in the soil develop (Toktam et al. 2010).

A change in redox status of a soil indicates changing availability of electron acceptors, requiring fundamental changes in microbial metabolic lifestyles. Under rapidly fluctuating conditions, microbial populations can be periodically activated and inactivated, which in turn quickly alters the nature and rate of key biogeochemical transformations (Pett-Ridge et al. 2006). Dynamic redox conditions are characteristic of tropical forest ecosystems, where soils cycle between oxic and anoxic on a scale of hours to days (Silver et al. 2001). Constant high temperature and soil moisture coupled with high rates of net primary productivity and substantial labile carbon pools fuel considerable biological oxygen demand, leading to frequent episodes of anoxia (Schuur and Matson 2001). Under fluctuating redox conditions, aerobic heterotrophic respiration is considered to be responsible for most soil CO_2 emissions (Subke et al. 2006). Low or fluctuating redox conditions drive N_2O, CH_4, and CO_2 production via anaerobic microbial respiration and fermentation (Teh et al. 2005; Liptzin and Silver 2009). Recent research has highlighted the possibility of spatially separated biogeochemical processes in sediments under fluctuating redox conditions (Nielsen et al. 2010), suggesting that enhanced spatial compartmentalization occurs in tropical forest soils as well. Brettar et al. (2002) studied nitrogen transformation and redox potential in floodplain soil of two depth profiles from a hardwood forest of Upper Rhine valley. The reduction of nitrate was followed and compared with redox potential and organic matter content. The redox potential under denitrifying condition ranged from 10 to 300 mV. Redox potential values decreased with increasing nitrate reduction rates and increasing organic matter content.

In high-E_h, well-oxygenated forest soils, obligate aerobes depend on oxidative respiration and phosphorylation to maintain their energy and nutrient requirements. While these organisms generally do not find anoxic conditions toxic, many are inactivated or starved as O_2 becomes limiting (Fenchel and Finlay 1995). Conversely, many obligate anaerobes such as methanogens, iron and sulfate reducers, find O_2 toxic because they lack superoxide dismutase and catalase enzymes necessary to degrade O_2 and peroxides (Imlay 2002). Free O_2 toxicity may also occur due to the disruption of enzymes essential to anaerobic energy metabolism, such as hydrogenases and pyruvate:ferredoxin oxidoreductase. Because of the importance of soil oxygen availability to microbial metabolism, spatial shifts in microbial physiology are traditionally thought to occur along gradients of soil redox potential. As soil redox potential decreases, the dominance of functional groups shifts.

Mangrove-forest tree species may alter sediment redox conditions through either rhizosphere oxidation (McKee 1993) or nutrient cycling processes (Sherman et al. 1998). Since mangrove soils are typically waterlogged, and hence anaerobic, microbial decomposition takes place through a series of oxidation–reduction (redox) processes. Totally anoxic sediments have redox potentials below −200 mV, while typical oxygenated soils have potentials of above +300 mV. The measurement of Eh has been used as a rapid means of assessing the potential impact of additional organic input to marine sediment. Reliable measurements of redox require great care to minimize exposure of the soil sample to air (English et al. 1997).

Study Questions

1. What are the important chemical properties of soil? Give an account of soil solution. Why the composition of soil solution is so variable?
2. What is soil reaction? Define soil pH and narrate the effects of soil pH on the growth and distribution of forest tree species.
3. How do charges on soil colloids develop? Why there are variations in the cation-exchange capacities of different soil colloids? What are factors that affect cation exchange in forest soil?
4. Explain the significance of electrical conductivity and sodium saturation of soil. How does soil salinity affect growth of forest trees? Give an account of mangrove soil salinity.
5. What is the significance of free carbonates in forest soil? Why are calcareous soils not ideal habitats for conifers? Discuss the impact of Eh on the growth of forest trees.

References

Abrahamsen G (1994) Soil chemical effects of liming-considerations related to forest liming. In: Nilsen P (ed) Proc. from an International seminar on counteractions against in forest ecosystems. March 3–4 1994, Mastemyr, Norway

Adams CD (1963) Factors influencing vascular plant zonation in North Carolina salt marshes. Ecology 45:479–494

Adams ML HOJ, Nilsson NHS, Powell KJ (2000) The relationship between soil solution pH and Al^{3+} concentrations in a range of South Island (New Zealand) soils. Aust J Soil Res 38(1):141–154

Agboola AA, Ayodele O (1987) Soil test calibration for Nigerian soils. Fertilizer research 14:227–234

Aidar MPM, De Godoy JRL, Bergmann J, Joly CA (2001) Atlantic forest succession over calcareous soil, Parque Estadual Turístico do Alto Ribeira–PETAR, SP Rev Bras Bot 24(4)

Allen JA, Chambers JL, Stine M (1994) Prospects for increasing the salt tolerance of forest trees: a review. Tree Physiol 14:843–853

Anon (2012) The natural communities of Virginia classification of ecological community groups. Department of conservation and recreation. Natural heritage program, 217 Governor Street, Richmond

Ayyad MA, El-Gharee REM (1982) Salt marsh vegetation of the western mediterranean desert of Egypt. Vegetatio 49(1):3–19

Bailey SW, Horsley SB, Long RP, Hallett RA (2004) Influence of edaphic factors on sugar maple nutrition and health on the Allegheny Plateau. Soil Sci Soc Am J 68:243–252

Bakker MR, Kerisit R, Verbist K, Nys C, Stokes A (1999) Effects of liming on rhizosphere chemistry and growth of fine roots and of shoots of sessile oak (*Quercus petrea*). The supporting roots–structure and function. Plant Soil 217:243–255

Basavaraja PK, Sharma SD, Dhananjaya BN, Badrinath MS (2011) *Acacia nilotica*: A Tree Species for Amelioration of Sodic Soils in Central Dry Zone of Karnataka, India. Res J Agri Sciences 2(2):311–316

Billings WD (1949) The shadscale vegetation zone of Nevada and eastern California in relation to climate and soils. American Midlands Naturalist 42:87–109

Bingham FT, Nelson CO (1971) The effects of sodium on mature avocado trees. California Avocado Society 1970-71 Yearbook 54:75–78

Bloom PR, McBride MB, Weaver RM (1979) Aluminum organic matter in acid soils: Buffering and solution aluminum activity. Soil Sci Soc Am J 43:488–493

Bonneau M (2005) Evolution of mineral fertility of an acidic soil during a period of ten years in the Vosges Mountains (France). Impact of humus mineralisation. Annals of Forest Science 62:253–260

Branson FA, Miller RF, McQueen IS (1967) Geographic distribution and factors affecting the distribution of salt desert shrubs in the United States. J Range Manage 20:287–296

Brettar I, Sanchez-Perez JM, Trémolières M (2002) Nitrate elimination by denitrification in hardwood forest soils of the Upper Rhine floodplain—correlation with redox potential and organic matter. Hydrobiologia 469(1-3):11–21

Brown G, Newman ACD, Rayner JH, Weir AH (1978) The structures and chemistry of soil clay minerals. In: Greenland DJ, Hays MHB (eds) The chemistry of soil constituents. Wiley, New York

Campbell JL, Hornbeck JW, McDowell WH, Buso DC, Shanley JB, Likens GE (2000) Dissolved organic nitrogen budgets for upland, forested ecosystems in New England. Biogeochem 49:123–142

Carou S, Dennis I, Aherne J, Ouimet R, Arp PA, Watmough SA, DeMerchant I, Shaw M, Vet B, Bouchet V, Moran M (2008) A National picture of acid deposition critical loads for forest soils in Canada. A report prepared for the CCME Acid Rain Task Group, Canadian Council of Ministers of the Environment

Chapman HD (1965) Cation exchange capacity. In: Black CA (ed) Methods of soil analysis; American Society of Agronomy: Madison

Chapman VJ (1976). Mangrove vegetation. J. Cramer, Vaduz, Germany

Cintron G, Lugo AE, Pool DJ, Morris G (1978) Mangroves of arid environments in Puerto Rico and adjacent islands. Biotropica 10:110–121

Clark JS (1965) The lime potential and percent base saturation of some representative Podzolic and Brunisolic Soils in Canada. Soil Sci Soc Am J 30(1):93–97

Comerford NB, Harris WG, Lucas D (1990) Release of non-exchangeable potassium from a highly weathered, forested Quartzipsamment. Soil Sci Soc Am J 54:1421–1426

Cooper A (1997) Plant species coexistence in cliff habitats. J Biogeogr 24:483–494
Cronan CS, April R, Bartlett RT, et al (1989) Aluminum toxicity in forests exposed to acidic deposition: the ALBIOS results. Water Air Soil Pollut 48:181–192
Dagar JC, Singh NT (1994) Agroforestry options in the reclamation of problem soils. In: Thampan PK (ed) Trees and tree farming. Peekay Tree Crops Development Corporation, Cochin
Day JH (1983) The Canadian soil information system: Mannual for describing Canada soils in the field. Agriculture Canada Expert Committee on Soil Survey
De Silva BLT (1934) The distribution of "calcicole" and "calcifuge" species in relation to the content of the soil in calcium carbonate and exchangeable calcium, and to soil reaction. J Ecol 22:532–553
Don A, Kalbitz K (2005) Amounts and degradability of dissolved organic carbon from foliar litter at different decomposition stages. Soil Biology and. BioChemistry 37:2171–2179
Earle JC, Kershaw KA (1989) Vegetation patterns in James Bay coastal marshes. III. Salinity and elevation as factors influencing plant zonations. Can J Bot 67:2967–2974
Egbuchua CN, Bosah BO (2011) Characterization of some forest soil under teak (*Tectona grandis*) plantation in a humid forest region of Nigeria. Int J Sci Nature 2(3):821–825
English S, Wilkinson C, Basker V (1997) Survey manual for tropical marine resources (2nd Ed). Australian Institute of Marine Science, Townsville
Ewald J (2003) The calcareous riddle: Why are there so many calciphilous species in the Central European flora? Folia Geobot 38:357–366
Fahey TJ, Yavitt JB (1988) Soil solution chemistry in lodgepole pine (*Pinus contorta* ssp. Latifolia) Ecosystems, southeastern Wyoming, USA. Biogeochemistry 6:91–118
Falkengren-Grerup U (1994) Importance of soil solution chemistry to field performance of *Galium odoratum* and *Stellaria nemorum*. J Appl Ecol 31(182-):192
Federer CA, Hornbeck JW (1985) The buffer capacity of forest soils in new England water, air, & soil pollution 26(2):163–173
Fenchel T, Finlay BJ (1995) Ecology and evolution in anoxic worlds. Oxford University Press, Oxford, United Kingdom
Fenn ME, Huntington TG, McLaughlin SB, Eagar C, Gomez A, Cook RB (2006) Status of soil acidification in North America. J For Sci 52 (Special Issue): 3–13
Formanek P, Vranova V (2002) A contribution to the effect of liming on forest soils: review of literature. J For Sci 48(4):182–190
Fox TR, Comerford NB (1990) Low-molecular-weight organic acids in selected forest soils of the south-eastern USA. Soil Sci Soc Am J 54:1139–1144
Fox TR, Comerford NB, McFee WW (1990) Kinetics of phosphorus release from spodosols: effects of oxalate and formate. Soil Sci Soc Am J 54:1441–1447
Gauthier C (2002) Contribution to the study of fractionation of free aluminum in solutions of forest soils. Influence of quality and nature of organic matter. These N°dordre: 38-2002, Faculté des Sciences et Techniques, Université de Limoges, 156 p. (in French)
Gillman GP, Bruce RC, Davey BG, Kimble JM, Searle PL, Skjemstad JO (1983) A comparison of methods used for determination of cation exchange capacity. Commun Soil Sci Plant. Anal 14(11):1005–1014
Good RE (1972) Salt marsh production and salinity. Bull Ecol Soc Am 53:22
Grainger A (1998) Estimating areas of degraded tropical lands requiring replishment of forest cover. Int Tree Crops J 5:31–61
Gupta RK, Tomar OS, Minash PS (1995) Managing salt affected soils and waters for afforestation. Bull 7.CSSRI, Karnal
Heisey RM (1995) Growth trends and nutritional status of sugar maple stands on the Appalachian Plateau of Pennsylvania. USA Water Air Soil Pollut 82:675–693
Herbert BE, Bertsch PM (1995) Characterization of dissolved and colloidal organic matter in soil solution: a review. In: McFee WW, Kelly JM (eds). Carbon forms and functions in forest soils. Madison, WI: Soil Science Society of America 41:1696–702
Hinrichsen D (1986) Multiple pollutants and forest decline. Ambio 15(5):258–265
Holmgren RC, Hutchings SS (1972) Salt desert shrub response to grazing use. In: Wildland shrubs-their history and utilization. Gen. Tech. Rep. INT-1. USDA Forest Serv. Intermountain Forest and Range Exp Sta Ogden, Utah
Hoover KL, Rightnour TA (2002) Mosquito Creek phase I-atmospheric acidification abatement demonstration project. Mosquito creek sportsman association-Pennsylvania growing greener program final report. Pennsylvania department of conservation and natural resources, Harrisburg, Pennsylvania, USA
Hue NV, Craddock GR, Adams F (1986) Effect of organic acids on aluminum toxicity in subsoils. Soil Sci Soc Am J 50:28–34
Huntington TG (2000) The potential for calcium depletion in forest ecosystems of southeastern United States: review and analysis. Global Biogeochem Cycles 14:623–638
Hutchings P, Saenger P (1987) Ecology of Mangroves. University of Queensland Press, St Lucia
Imlay JA (2002) How oxygen damages microbes: oxygen tolerance and obligate anaerobiosis. Adv Microb Physiol 41:111–153
Jackson ML (2005) Soil chemical analysis–advanced course. 2nd revised edition. Parallel Press, University of Wisconsin. Madison Libraries, Madison
Jackson T, Bird R (2008) Trees and shrubs for saline land: research in south west Victoria, Department of primary industries, Hamilton, Victoria
James BR, Riha SJ (1985) pH buffering in forest soil organic horizons: relevance to acid precipitation. J Envir Quality 15(3):229–234
Jeschke WD (1984) K+-Na+exchange at cellular membranes, intracellular compartmentation of cations, and salt tolerance. In: Staples RC, Toenniessen GH (eds) Salinity tolerance in plants. Strategies for crop improvement. Wiley, New York
Johnson DW, Cole DW, Horng FW, Van Miegroet H, Todd DE (1982) Chemical characteristics of two forested ultisols and two forested inceptisols relevant to anion production and mobility. Publ. No. 1670, Oak Ridge National Laboratory Environmental Science Division: Oak Ridge, TN
Johnson CE, Romanowicz RB, Siccama TG (1997) Conservation of exchangeable cations After clear-cutting of a northern hardwood forest. Can J Forest Res 25:859–868
Johnson DW, Todd DE (1998) Harvesting effects on long-term changes in nutrient pools of mixed oak forest. Soil Sci Soc Am J 62:1725–1735
Johnson CE (2002) Cation exchange properties of acid forest soils of the northeastern USA. European Journal of. Soil Sci 53(2):271–282
Kalisz PJ, Stone EL (1980) Cation exchange capacity of acid forest humus layers. Soil Sci Soc Am J 44(2):407–413
Khan GS, Khan ZH, Imran J, Quraishi MU, Yaqoob S, Khan SH (2009) Effect of salinity on germination and growth of some forest tree species at seedling stage. J Agri Res 47(3):271–279
Kimble JM, Ping CL, Sumner ME, Wilding LP (2000) Andisols. In: Sumner ME (ed) Handbook of soil science. CRC Press, New York
Kishchuk BE (2000) Calcareous soils, their properties and potential limitations to conifer growth in southern British Columbia and western Alberta: a literature review. Ministry for Invermere, British Columbia Inf Rep NOR-X–370
Klnzel H (1983) Influence of limestone, silicates and soil pH on vegetation. In: Lange O, Nobel PS, Osmund CB, Ziegler HJ (eds) Physiological plant ecology III. Springer, Berlin
Kolb TE, McCormick LH (1993) Etiology of sugar maple decline in four Pennsylvania stands. Can J For Res 23:2395–2402
Kozlowski TT (1997) Responses of woody plants to flooding and salinity. Tree Physiology Monograph No. 1
Kreutzer K (1995) Effects of forest liming on soil processes. Plant Soil 168-169:447–470
Krug EC, Charles R (1983) Acid rain on acid soil: a new perspective. Science New Series 221(4610):520–525

LA CWCS (2005) Conservation habitatas and species assessments. The Louisiana comprehensive wildlife conservation Strategy (LA CWCS). National council for science and the environment

Larcher W (1980) Physiological plant ecology, 2nd edn. Springer-Verlag, Berlin

Lawrence GW, David MB, Baily SW, Shortle WC (1997) Assessment of calcium status in soils of red spruce forests in the northeastern United States. Biogeochemistry 38:19–39

Lawrence GB, David MB, Lovett GM, Murdoch PS, Burns DA, Stoddard JL, Baldigo BP, Porter JH, Thompson AW (1999) Soil calcium status and the response of stream chemistry to changing acidic deposition rates in the Catskill Mountains of New York. Ecol Appl 9:1059–1072

Leinonen L, Juntto S (1992) (eds) Air quality measurements (1991) Finnish Meteorological Institute, Helsinki

Lindberg SE, Johnson DW, Lovett FM, VanMiergroet H, Taylor GE Jr, Owens JG (1989) Sampling and analysis protocols and project description for the integrated forest soil study. Oak Ridge National laboratory, Environ Sci Publ 3321, Oak Ridge, TN

Liptzin D, Silver WL (2009) Effects of carbon additions on iron reduction and phosphorus availability in a humid tropical forest soil. Soil biology biochemistry 41:1969–1702

Lochman V, Sebkova V (1998) The development of air pollutant depositions and soil chemistry on the research plots in the eastern part of the Ore Mts. Lesnictví-Forestry 45:549–560

Long PL, Horsley SB, Lilja PR (1997) Impact of forest liming on growth and crown vigor of sugar maple and associated hardwoods. Can J For Res 27:1560–1573

Lövblad G, Amann M, Andersen B, Hovmand M, Joffre S, Pedersen U (1992) Deposition of sulfur and nitrogen in the Nordic countries: present and future. Ambio 21:339–347

Lugo AE (1980) Mangrove ecosystems: successional or steady state? In: Tropical succession. Biotropica 65–72

Lynch JA (1998) Atmospheric deposition in Pennsylvania. In: Sharpe SW, Drohan JR (eds) The effects of acidic deposition on Pennsylvania's forests. Proceedings of the 1998 Pennsylvania acidic deposition conference, Volume 1. Environmental Resources Research Institute, Pennsylvania State University, University Park, Pennsylvania, USA

Makeschin F (1991) Auswirkungen von saurer Beregnung und Kalkung auf die Regenwurmfauna (*Lumbricidae: Oligochaeta*) im Fichtenaltbestand Höglwald. Ökosystemforschung Höglwald. In: Kreutzer K, Göttlein A. (eds) Forstwiss Forsch *39*: 117–128

Marcar NE, Crawford DF (2004) Trees for saline landscapes, Publication No. 03/108, RIRDC, Canberra

Mares V (1992) Vliv melioračních opatření na růstové podmínky výsadeb v imisních oblastech. Zpr. Lesn. Výzk., XXXVII (2): 5–7

Marschner H (1986) Mineral Nutrition of Higher Plants. Academic Press, London

Marschner H (1995) Mineral nutrition of higher plants. Academic Press, London

Matzner E (1992) Acidification of forests and forest soils: Current status. Stud Env Sci 50:77–86

McDaniel PA, Wilson MA (2007) Physical and chemical characteristics of ash-influenced soils of inland Northwest forests In: Page-Dumroese, Deborah;Miller, Richard;Mital, Jim;McDaniel, Paul;Miller, Dan (eds) 2007. Volcanic-Ash-derived forest soils of the inland Northwest: Properties and implications for management and restoration. Proceedings RMRS-P-44; Fort Collins, USDA Forest Service, Rocky Mountain Research Station

McKee KL (1993) Soil physicochemical patterns and mangrove species distribution-reciprocal effects? J Ecol 81:477–487

Meiwes KJ (1995) Application of lime and wood ash to decrease acidification of forest soils. J For Sci 49:182–190

Menendez I, Moreno G, Gallardo JF, Saavedra J (1995) Soil solution composition in forest soils of Siera de Gata Mountains, Central-Western Spain: Relationship with soil water content. Arid Soil Res Rehab 9:495–502

Michalzik B, Kalbitz K, Park JH, Solinger S, Matzner E (2001) Fluxes and concentrations of dissolved organic carbon and nitrogen—a synthesis for temperate forests. Biogeochem 52:173–205

Mishra A, Sharma SD (2010) Influence of forest tree species on reclamation of semiarid sodic soils. Soil Use Manag 26(4):445–454

Mitsch WJ, Gosselink JG (1993) Wetlands 2d ed. Van Nostrand Reinhold, New York, NY

Mladkova L, Bordvka L, Drabek O (2005) Soil properties and selected aluminum forms in acid forest soils as influenced by the type of stand factors. Soil Sci Plant Nutr 51(5):741–744

Montagnini F, Haines B, Swank WT (1991) Soil-solution chemistry in black locust, pine/mixed-hardwoods and oak/hickory forest stands in the southern Appalachians, USA. For Ecol Manage 40:199–208

Moore JD, Camiré C, Ouimet R (2000) Effects of liming on the nutrition, vigor and growth of sugar maple at the Lake Clair Watershed, Quebec, Canada. Can J For Res 30:725–732

Neff JC, Hooper DU (2002) Vegetation and climate controls on potential CO_2, DOC and DON production in northern latitude soils. Global Change Biology 8:872–884

Nielsen UN, Osler GHR, Campbell CD, Burslem DFRP, Van Der Wal R (2010) The influence of vegetation type, soil properties and precipitation on the composition of soil mite and microbial communities at the landscape scale. J Biogeogr 37:1317–1328

Nilsson SI, Tyler G (1995) Acidification-induced chemical changes of forest soils during recent decades—a review.—Ecol Bull 44: 54–64

Nour ME, Khalil AAM, Abdelmajid E (2006) Effect of salinity on seed germination characteristics of five arid zone tree species. University of Khartoum J Agric Sci.14(1), 23–31

Odum WE (1988) Comparative ecology of tidal freshwater and salt marshes. Annual Review of Ecology and Systemnatics 19:147–176

Osman KT (2013) Soils: principles, properties and management. Springer

Ouimet R, Camiré C (1995) Foliar deficiencies of sugar maple stands associated with soil cation imbalances in the Quebec Appalachians. Can J Soil Sci 75:169–175

Pabian SE, Brittingham MC (2007) Terrestrial liming benefits birds in an acidified forest in the Northeast. Ecol Appl 17(8):2184–2194

Perina V, Podrazsky V (1988) Účinnost vápnění v Jizerských horách. Lesn. Práce 67:12–17

Pett-Ridge J, Silver WI, Firestone MK (2006) Redox fluctuations frame microbial community impacts on N-cycling rates in a humid tropical forest soil. Biogeochemistry 81(1):95–110

Pezeshki SR, Chambers JL (1986) Effect of soil salinity on stomatal conductance and photosynthesis of green ash (*Fraxinus pennsylvanica*). Can J For Res 16:569–573

Pierzynski GM SJT, Vance GF (2000) Soils and environmental quality, CRC Press LLC, USA

Podrazsky V (1992) Změny chemismu skarifikovaného a neskarifikovaného horského humusového podzolu po vápnění. Zpr. Lesn. Výzk., XXXVII (1): 22–25

Pohlman AA, McColl JG (1986) Kinetics of metal dissolution from forest soils by soluble organic acids. J Environ Qual 15:86–92

Rengel Z (1992) The role of calcium in salt toxicity. Plant Cell and. Environment 15:625–632

Reuss JO, Roswall EC, Hopper RWE (1990) Aluminum solubility, calcium-aluminum exchange, and pH in acid forest soils. Soil Sci Soc Am J 54:374–380

Rhoades JD (1982) Cation exchange capacity. In: Page AL (ed) Methods of soil analysis. Part 2. Chemical and microbiological properties; American Society of Agronomy: Madison, WI

Richards LE (1954) Diagnosis and improvement of saline and alkali soils. USDA Salinity Laboratory Staff, Handbook 60, USDA

Richter DD, Johnson DW, Dai KH (1992) Cation exchange reactions in acid forested soils: effects of atmospheric pollutant deposition. In: Johnson DW, Lindberg ED (eds) Cation Exchange and Al Mobilization. Chapter 9: Atmospheric Deposition and Forest Nutrient Cycling; Ecological Studies, 91. Springer-Verlag, New York, 707

Richter AK, Hirano Y, Luster J, Frossard E, Brunner I (2011) Soil base saturation affects root growth of European beech seedlings. J Plant Nutr Soil Sci 174:408–419

Rodhe H, Langner J, Gallardo L, Kjellström E (1995) Global scale transport of acidifying pollutants. Water, Air and Soil Pollution 85: 37–50

Rorison IH (1960) Some experimental aspects of the calcicole-calcifuge problem: I. The effects of competition and mineral nutrition upon seedling growth in the field. J Ecol 48:585–599

Roskams P (1997) Part II National Inventories: Belgium. In: Vanmechelen L, Groenemans R, and Van Ranst E (eds) Forest Soil Condition in Europe. Results of a Large-Scale Soil Survey. EC, UN/ECE, ICP Forests and the Ministry of the Flemish Community

Ross DS, Matschonat G, Skyllberg U (2008) Cation exchange in forest soils: the need for a new perspective. European Journal of. Soil Sci 59(6):1141–1159

Rudneva E (2010) The Great Soviet Encyclopedia, 3rd Edition (1970-1979). The Gale Group, Inc

Sanchez PA (1976) Properties and management of soils in the tropics. A wiley international publication. Wiley and sons NY

Schmidt IK, Tietema A, Williams D, Gundersen P, Beier C, Emmett BA, Estiarte M (2004) Soil solution chemistry and element fluxes in three european heathlands and their responses to warming and drought. Ecosystems 7:638–649

Schutt P, Cowling E (1985) Plant Disease 69:448

Schuur EAG, Matson PA (2001) Net primary productivity and nutrient cycling across a mesic to wet precipitation gradient in Hawaiian montane forest. Oecologia 128:431–442

Semeniuk V (1985) Mangrove environments of Port Darwin, Northern Territory: the physical framework and habitats. J. R. Soc. Western Austr 67:81–97

Semeniuk V, Wurm PAS (1987) The mangroves of the Dampier Archipelago, Western Australia. J. R. Soc. Western Austr 69:29–87

Shannon MC (1985) Principles and strategies in breeding for higher salt tolerance. Plant Soil 89:227–241

Sharpe WE, Sunderland TL (1995) Acid-base status of upper rooting zone soil in declining and non-declining Sugar maple (*Acer saccharum* marsh) stands in Pennsylvania. 10th Central Hardwood Forest Conference

Sharpe WE, Sunderland TL (1995) Acid-base status of upper rooting zone soil in declining and non-declining sugar maple (*Acer saccharum*) stands in Pennsylvania. USDA For Serv Res Gen Tech Rep NE–197

Sharpe WE, Brittingham MC, Tzilkowski WM, Swistock BR, Bohnenblust AK, Donaldson LD, Kim HJ, McNeal LJ, Pabian SE, Rummel SM, Voorhees CK (2006) Evaluation of whole watershed and riparian wetland liming to mitigate acidity. Grant no. 41000155481ME352917. Final report to the Pennsylvania department of environmental protection, Harrisburg, Pennsylvania, USA

Sherman RE, Fahey TJ, Howarth RW (1998) Soil-plant interactions in a neotropical mangrove forest: iron, phosphorus and sulfur dynamics. Oecologia 115:553–563

Silver WL, Herman DJ, Firestone MK (2001) Dissimilatory nitrate reduction to ammonium in upland tropical forest soils. Ecology 82:2410–2416

Simonsson M, Berggren D (1998) Aluminum solubility related to secondary solid phases in upper B horizons with spodic characteristics. European Journal of. Soil Sci 49:317–326

Singh B, Garg VK (2007) Phytoremediation of a sodic forest ecosystem: Plant community response to restoration process Not Bot Hort Agrobot Cluj 35(1):71–84

Skinner MF, Zabowski D, Harrison R, Lowe A, Xue D (2001) Measuring the cation exchange capacity of forest soils. Commun Soil Sci Plant Anal 32(11 & 12):1751–1764

Skyllberg U, Borggaard O (1998) Proton surface charge determination in Spodosol horizons with organically bound aluminum. Geochim Cosmochim Acta 62:1677–1689

Skyllberg U (1999) pH and solubility of aluminum in acidic forest soils: a consequence of reactions between organic acidity and aluminum alkalinity. Eur J Soil Sci 50:95–106

Skyllberg U, Raulund-Rasmussen K, Borggaard OK (2001) pH buffering in acidic soils developed under *Picea abies* and *Quercus robur*—effects of soil organic matter, adsorbed cations and soil solution ionic strength. Biogeochemistry 56:51–74

Sollins P, McCorison FM (1981) Nitrogen and carbon solution chemistry of an old growth coniferous forests watershed before and after cutting. Water Resour Res 17:1409–1418

Sparks DL (2003) Environmental soil chemistry. Academic Press, Amsterdam

Sposito G (1996) The Environmetal Chemistry of Aluminum, CRC Press LLC, USA

Subke JR, Inglima I, Cotrufo MF (2006) Trends and methodological impacts of soil CO_2 efflux partitioning: a meta analytical review. Glob Change Biol 12:921–943

Sukardjo S (1994) Soils in the mangrove forests of the Apar Nature Reserve, Tanah Grogot, East Kalimantan, Indonesia. Southeast Asian Studies 32(3):385–398

Sun D, Dickinson GR (1995) Salinity effects on tree growth, root distribution and transpiration of *Casuarina cunninghamiana* and *Eucalyptus camaldulensis* planted on a saline site in tropical north Australia. For Ecol Manag 77(1-3):127–138

Teh YA, Silver WL, Conrad ME (2005) Oxygen effects on methane production and oxidation in humid tropical forest soils. Global Change Biology 11:1283–1297

Tisdale SL, Nelson WL, Beaton JD, Havlin JL (1993) Soil acidity and basicity. In Soil fertility and fertilizers. 5th ed, Macmillan Publ, New York

Toktam S, Cindy P, Les L (2010) Redox potential: An indicator of site productivity in forest management. EGU General Assembly 2-7 May, 2010 in Vienna, Austria

Trettin CC, Johnson DW, Todd DE Jr (1999) Forest nutrient and carbon pools at walker branch watershed: Changes during a 21-Year Period. Soil Sci Soc Am J 63:1436–1448

Walsh GE (1974) Mangroves: A review. In: Reimold R, Queen H (eds) Ecology of halophytes. Academic Press, New York

Wells CG, Davey CB (1966) Cation-exchange characteristics of forest floor materials. Soil Sci Sco Am J 30(3):399–402

Wells AG (1985) Grouping of tidal systems in the Northern Territory and Kimberley region of Western Australia on presence/absence of mangrove species. In: Bardsley KN, Davie JDS, Woodroffe CD (eds) Coasts and tidal wetlands of the Australian monsoon region. North Australian Research Unit, Darwin

West NE (1982) Intermountain salt-desert shrublands. Ecosystems of the world. Amsterdam: Elsevier Scientific. 5:375–398

Westerhoff P, Mash H (2002) Dissolved organic nitrogen in drinking water supplies: a review. J Wat Supply Res Techn-Aqua 51:415–448

Wightman G (1989) Mangroves of the Northern territory. Northern territory botanical Bull. No. 7. Conservation Commission of the Northern Territory, Palmerston

Williston HL, LaFayette R (1978) Species suitability and pH of soils in southern forests. USDA Forest Service. Southeastern Area, State and Private Forestry. Forest Management Bulletin

Wilmot TR, Ellsworth DS, Tyree MT (1995) Relationships among crown condition, growth, and stand nutrition in seven northern Vermont sugarbushes. Can J For Res 25:386–397

Wolt JD (*1994)* Soil solution chemistry: applications in environmental science and agriculture. Wiley, New York

Yaalon DH (1957) Problems on soil testing on calcareous soils. Plant Soil 8:275–288

Yu Z, Zhang Q, Kraus TEC, Dahlgren RA, Anastasio C, Zasoski RJ (2002) Contribution of amino compounds to dissolved organic nitrogen in forest soils. Biogeochemistry 61:173–198

Organic Matter of Forest Soils

4

Organic matter is a major component of soil. It has tremendous ecological significance; it determines soil health, influences ecosystem productivity, and affects climate quality. Soil organic matter (SOM) has three main fractions, fresh undecomposed residues, decomposing and partially decomposed materials, and a highly decomposed and stable product humus (it is not completely decomposed organic matter; complete decomposition of organic matter produces carbon dioxide, water, and other inorganics). Plant litters including dead leaves, stems, barks, flowers, fruits, and logs are the major sources of forest SOM. Soil animals, microorganisms, and roots also contribute to the SOM. The threshold value for organic matter in agricultural soil is 2% by weight, beyond which soil quality does not remain sustainable, but no threshold level for forest soils has so far been established. Forest mineral soils have generally 1–5% organic matter by weight. Forest soils usually have higher organic matter than agricultural soils. There are two types of organic matter in soil, active or labile and passive or stable. Humus is the stable fraction of SOM. Forest ecologists identify three types of humus: mull humus, mor humus, and moder humus depending on the degree of decomposition and integration with mineral matter, acidity, and base contents. This categorization is more pronounced in temperate and boreal forest soils. A deep O horizon also develops in these types of forests. SOM performs a variety of physical, chemical, and biological functions including aggregation, soil reaction and ion exchange and nutrient cycling in addition to supply of food and energy to soil biota.

4.1 Soil Organic Matter

SOM was defined by Baldock and Skjemstad (1999) as "all organic materials found in soils irrespective of origin or state of decomposition." Since SOM exists in various complex states in soil, it is actually difficult to measure the SOM content; most methods determine the soil organic carbon (SOC) content and estimate SOM through a conversion factor. The amount of SOM that exists in any given soil is determined by the balance between inputs from biota (vegetation, roots, microorganisms, animals) and output (CO_2 from microbial decomposition). Soil type, climate, topography, soil biota, mineral composition, management, and their interactions are modifying factors that will affect the total amount of SOM in a soil and its distribution within the profile. Any changes in the natural state of the soil systems (conversion to agriculture, deforestation, and afforestation) will lead to change in the amount and state of SOM. Mineral soils may contain on an average 5% organic matter by volume. Organic soils may have more than 80% organic matter by volume.

SOM concentrations are major indicators of soil quality. Janzen et al. (1992) observed a relationship between SOC in the uppermost 15 cm and soil productivity in agricultural point of view. They found an upper threshold level of SOC beyond which no further increases in productivity were achieved. The threshold value for SOC for these dryland sites in Alberta, Canada, was at 2% SOC. Howard and Howard (1990) suggested that the threshold value for most soils was at 2% SOC (equivalent to 3.4% SOM), below which most soils are prone to structural destabilization and crop yields are reduced. No such threshold value for SOM in forest soils could be established, although SOM influences almost all physical, chemical, and biological behavior of the forest soils.

4.2 Sources of Organic Matter in Forest Soil

There are various sources of organic matter in a forest soil. As forest soils are not usually manured except in nursery seedling development processes, all sources of organic matter in forest soils are native. Forest vegetations continuously cast litters from the above- and belowground parts. Litters are dead parts of the forest vegetation, including leaves, branches, twigs, flowers, fruits, barks, and even large logs, existing in various stages of decomposition above the soil surface. Litters may also include dead bodies of soil animals,

K. T. Osman, *Forest Soils,* DOI 10.1007/978-3-319-02541-4_4, © Springer International Publishing Switzerland 2013

insects, etc. Plant leachate, sloughed cells, and root exudates are other sources of organic matter from vegetation. To all these are added microbial cell substances. Dead aboveground and burrowing animals also significantly contribute to the forest SOM since forests are habitats of a large group of big and small animals.

Fisher and Binkley (2000) reported that there were wide variations in the amount of organic matter additions in forest soils; and all factors that affect the growth and biomass production of ecosystems influenced this addition. Important factors were climate (rainfall and temperature), topography, soil, age of forest (in case of plantations), altitude, latitude, etc. They suggested that a tropical forest may shed on an average 10 t ha^{-1} aboveground litters annually in comparison to 5 t ha^{-1} in temperate and 3 t ha^{-1} in boreal forests. Persson (2012) studied organic matter in six boreal forest types and observed that substantial amounts of organic matter from dead fine roots are continuously accumulated in the soil. Pregitzer et al. (1995) suggested that large amounts of C contained in plants recycle back to the soil through rhizosphere processes. Some investigators have quantified the root production (Vogt et al. 1986; Cheng et al. 1990; Eissenstat and Yanai 1997) and total rhizodeposition (Kuzyakov and Domanski 2000). Ludmila (2010) observed that plant residue carbon was mainly allocated on the soil surface, in the forest litter. Plant residues in the soil (dead roots + other mortmass) were estimated to contribute 10–30 % of the plant residues carbon, or 2.5–15.1 % of the total soil carbon. Soil surface and in-soil dead plant material included 60–95 % of heavily decomposed residues. In an Amazon forest on an Oxisol in Venezuela, Cuevas (1983) found that annual fine root production in the first 10 cm of soil and the root mat above the mineral soil was 8 t ha^{-1} per year. Sanford (1985) found that fine root turnover in the upper 10 cm of the same Oxisol was 25 % per month. Biomass and nutrient fluxes in litterfall for the same forest amounted to 7.6 t dry matter ha^{-1} per $year^{-1}$ (Cuevaz and Medina 1986), indicating that fine root production and turnover in forests on very infertile soils may be more important for nutrient cycling than litterfall.

4.3 Categories of SOM

There are three categories of SOM on the basis of the stages of decomposition. The categories are fresh or undecomposed organic matter, partially decomposed organic matter, and fully decomposed organic matter.

Fresh or undecomposed organic matter Fresh SOM includes freshly fallen leaves, twigs, branches, flowers, fruits, dead animal tissue, etc., if they have been incorporated into the soil. These materials begin to decompose immediately after their incorporation. It often becomes very difficult to separate decomposed and undecomposed organic matter because all organic matter in soil is subjected to decomposition immediately after their incorporation. For all practical purposes, fresh organic matter includes materials the origin and structure of which can be identified. It comprises about 1–10 % of total SOM (Lickacz and Penny 2001).

Partially decomposed organic matter It includes organic residues in soil that have undergone decomposition to a considerable extent. The structure and origin of the materials may or may not be identified. It consists of gradually decomposing detritus and partially broken down plant and animal tissues. It comprises about 10–40 % of total SOM.

Fresh, undecomposed, and partially decomposed organic matter together constitute the "active soil organic matter." They are so called because they are easily attacked by soil organisms, they are decomposed very fast, and they provide food and energy to heterotrophic soil organisms.

Fully decomposed organic matter or humus Humus is said to be "completely decomposed organic matter." It is not actually "completely" decomposed in the sense that it is still organic and complete decomposition of organic matter produces CO_2 and water with some bases. It is "fully" decomposed in the sense that it has undergone decomposition for a considerable time, say some hundreds or thousands of years, and that further decomposition proceeds extremely slow. Humus may be defined as a brown to black, amorphous, colloidal organic matter that has undergone decomposition to such an extent that it has become stabilized with soil constituents. Humus is composed of stable natural organic complexes formed by the products of biological decomposition and resynthesis. Humus is fairly resistant to decomposition due to its physical state and chemical nature. Humus is present as coating on soil particles, as gum within aggregates, and as organo-mineral complexes, which make them inaccessible to decomposing organisms. Again, humus contains refractory substances such as lignin and polyphenols. Because of these refractory substances, humus is called passive SOM. It comprises about 40–60 % of total SOM.

4.4 Functions of Organic Matter in Forest Soil

SOM performs a variety of physical, chemical, and biological functions in forest soil (Fig. 4.1). The most striking physical benefit the forest soils get from the organic matter is in the aggregation of soil particles. The effect is physical, but the processes of aggregation are all complex interactions of physical, chemical, and biological processes. Aggregation in forest soil is a natural process and forest soils are usually better structured than agricultural soils. There are plenty of litters and their decomposition products to stabilize

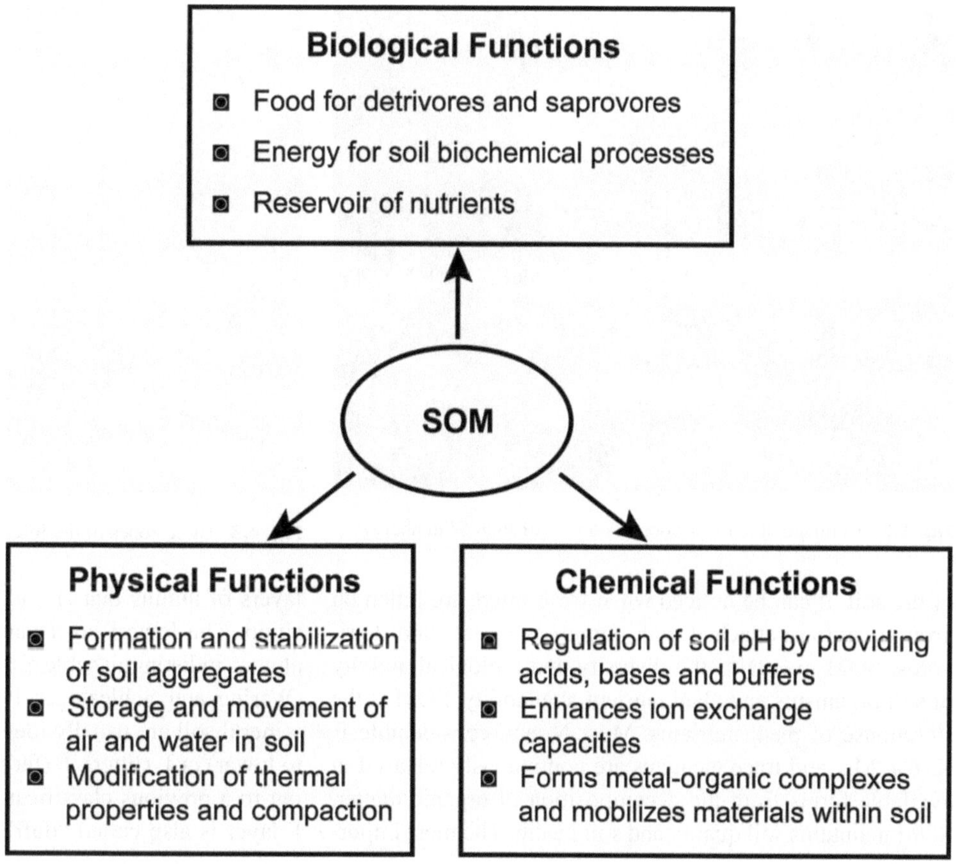

Fig. 4.1 Schematic summary of functions of soil organic matter (SOM) in forest soil

peds. In soil, charged colloid surfaces, organic molecules, cations, and water molecules bridge each other and flocculate. At the microscopic level, clay and humic substances act as bridges between the sand and silt particles, producing microaggregates of soil (McLaren and Cameron 1996). The soil microaggregates are bound together into macroaggregates by fungal hyphae, plant roots, and other stabilizing agents at the macroscopic level (Oades 1984). Carter and Gregorich (1996) reported that the turnover time of organic matter in aggregate could be much longer than that of litter remained without aggregation. So, organic matter stabilizes aggregates, and aggregates stabilize the organic matter. It is one process of carbon sequestration in soil. In agricultural soils, tillage and other soil disturbances destroy or modify soil structure. In coniferous and broadleaf forests, SOC content was positively correlated with silt, clay, and water stable aggregates, but negatively with bulk density. Among different land-use systems studied, bulk density was the least in broadleaf forest and the highest in paddy rice fields (Han et al. 2010). Admixing of organic matter in soil increases porosity and decreases bulk density. SOM influences the hydrological properties of forest soils; it increases infiltration and percolation and decreases evapotranspiration and runoff. It increases the water-holding capacity and hydraulic conductivity.

There are a few chemical properties of soil that are not affected by soil organic matter. SOM provides chemical reactants, including carbohydrates, proteins, lignins, acids, alcohols, enzymes, etc. that take part in chemical reactions and that provide energy for other reactions to occur. SOM influences pH, redox potential (Eh), and cation exchange capacity (CEC). Forest soils are commonly acidic in character due to the continuous production of organic acids and CO_2 from decomposing litters.

Litters of some forest vegetation such as broadleaf forests of the temperate region may release substantial quantities of bases to make the underlying soils neutral to slightly alkaline in reaction. The intermediate decomposition of organic matter in waterlogged conditions may produce acid peat soils. Humus has a great surface area and a large number of electrical charges on their surfaces. So, humus has a very high CEC, 10–30 $cmol_c$ per gram of colloids. Forest soils are rich in humus and CEC.

Soil is a living and dynamic ecosystem. A healthy soil is teeming with microscopic and larger organisms that perform many vital functions including converting dead and decaying matter as well as minerals to plant nutrients. Different soil organisms feed on different organic substrates. Their biological activity depends on the organic matter supply. Organic matter addition to soils fuels the biotic population

Fig. 4.2 An image of a forest floor. (Courtesy of John Wittenberg)

Fig. 4.3 Three zones of the forest floor

of the soil. It can be noticed when fresh litters are fallen on ground, and are attached with numerous insects and detrivores. SOM maintains the diversity and biological activity of soil organisms and plant nutrient availability. SOM is the storehouse of plant nutrients. Most N, and considerable P, K, Ca, Mg, and trace elements are continuously released in available forms from the decomposition of organic matter. SOM maintains soil quality and soil health. The most important environmental role of SOM is the carbon sequestration. Recognition of the importance of SOM in the global C balance has spurred many recent attempts to estimate the global SOM pool (Paustian et al. 1997). Increasingly, there is agreement about estimates of the global SOM pool (Batjes 1996) at approximately 1,500 Pg of organic C in the top 1 m of soil.

4.5 Forest Floor

The continuous layer of organic residues developed on the surface of a forest soil by the accumulation of fallen litters is popularly known as forest floor in forestry literature. It is actually the organic O horizon, most frequently developed on the surface of temperate and boreal forests and rarely in humid tropical forests. According to Smith and Heath (2002), the forest floor includes nonliving plant mass in various states of mechanical and chemical decay on the surface of the forest soil. Forest floor is formed by the deposition of dead organic residues from both floral and faunal components. Large vertebrate detritus may sometimes form a small part; the large part is composed of dead plant materials including leaves, twigs, barks, and woody stems (Fig. 4.2). There may also be a significant amount of fine roots in the forest floor, mushrooms, and lichens. The deposition of organic residues can form identifiable layers above the mineral soil. In a well-developed forest floor, layering can be identified, ranging from freshly fallen leaves and twigs to lower layers of humus that are at an advanced state of decomposition. The lowest layer can be amorphous and sometimes almost indistinguishable from upper layers of mineral soil (Waring and Schlesinger 1985). Organic layers above the mineral soil are usually identified as Oi, Oe, and Oa (upper to lower) or L (litter), F (fermented), and H (humus) according to a previous classification system (Federer 1982). The F layer is also called "duff" (Fig. 4.3). Distinctions among pools of coarse woody debris, the forest floor, and SOM are necessary to avoid confusion (Smith and Heath 2002). Large woody material may be excluded from forest floor. Small woody material that is considered part of the forest floor is based on diameters of recently fallen branches and stems. Diameters ranging from less than a centimeter to more than 7 cm are reported as the distinction between coarse woody debris and forest floor (Rollinger et al. 1998). Typical approximate figures for the accumulation of organic matter in forest floors are 15–100 Mg ha^{-1} in northern boreal forests, 7.5–12.5 Mg ha^{-1} in temperate broadleaved forests, and 1–2.5 Mg ha^{-1} in tropical forests (Vogt et al. 1986). The state of decomposition is a basis for distinguishing layers of the forest floor. The distinction between the lowest layers of forest floor and upper (generally A horizon) layers of mineral soil is sometimes clear. However, purely organic layers of forest floor and organic-rich layers of the mineral soil are sometimes difficult to discern. This can be even more difficult when mechanical mixing of layers has occurred (Ryan et al. 1992). Field classifications are usually based on morphology and are closely related to an individual researcher's experience (Federer 1982).

The L, F, and H layers correspond to Oi, Oe, and Oa sub-horizons, respectively, of the Pedology terminology (Fig. 4.3).

The accumulation of forest floor mass depends on rates of inputs such as litterfall and loss such as litter decomposition which are spatially (Simmons et al. 1996) and temporally

(Nemeth 1973) heterogeneous. Processes can be influenced by season, climate, and edaphic factors, as well as stand composition, age, and disturbance history (Schlesinger 1977; Mattson and Smith 1993; Binkley 1995). Decomposition rates are principally affected by quality and quantity of material as well as a favorable microenvironment (Schlesinger 1977; Hendrickson et al. 1989; Berg et al. 1996). Though quantity and quality of inputs affect carbon mass stored in forest floors, factors controlling carbon loss appear to be somewhat more important in controlling the carbon level in temperate forests (Vogt et al. 1996).

Numerous reviews report forest location and composition are associated with an average amount of forest floor mass (Schlesinger 1977; Vogt et al. 1986; Vogt et al. 1995; Vogt et al. 1996; Rollinger et al. 1998). For example, Vogt et al. (1995) reviewed a number of broad influences on soil and forest floor carbon. They demonstrated a trend of greater forest floor accumulation along a climatic gradient from tropical to temperate to boreal forests. The result of adding evergreen and deciduous classifications suggested their usefulness for predicting forest floor mass. Evidence further suggests that additional predictor variables such as soil type, stocking, productivity, or management intensity can affect forest floors, but these variables are not as consistently useful as region or forest type for making large-scale estimates (Vogt et al. 1995; Vogt et al. 1996). Climate affects accumulation and loss of forest floor mass. Longer growing seasons and warmer temperatures are associated with greater productivity and litterfall, but they also are associated with more rapid decomposition, especially if litter retains sufficient moisture and aeration (Smith and Heath 2002).

Fisher and Binkley (2000) reported physical and chemical properties of the forest floor. The L layer has freshly fallen leaves, branches, twigs, flowers, fruits, etc. They are unaltered, and their original structure is easily recognizable. The layer is loose and has a very low bulk density (even <0.2 g cm^{-3}). Bulk density increases with the stage of decomposition. Therefore, the F layer has intermediate bulk density and the H layer has the highest bulk density. The F layer is partially decomposed and the original structure of materials has largely been retained. The sources of materials are identifiable. The H layer is the humus layer which is brown to black, amorphous, and finely divided material that is relatively stable and colloidal material formed by the decomposition of overlying litter materials. It may be intimately mixed with mineral matter. Forest floor materials have high water-holding capacity. Litters have high C/N ratio which decreases with advancement of decomposition. The C/N ratio of L layer may be around 60:1 and that of the H layer may be close to 15:1. The undecomposed loose organic debris on the surface of the forest floor is not strictly included in SOM. The part of organic residues incorporated in the soil is the actual SOM.

Osman (2013) describes the significance of the forest floor in the functioning of forest ecosystems. The forest floor is a natural organic mulch of the forest soil. It reduces raindrop impact, velocity of run-off, concentration of water in narrow channels, and detachment of soil particles. It soaks water, retains water, and enhances infiltration. It improves soil structure, aeration, and drainage and reduces soil erosion. The forest floor is a spongy material and it may hold several hundred times of its own weight of moisture (Fisher and Binkley 2000). Thus, the forest floor acts as a storage tank of water which gets infiltrated gradually within the soil. The forest floor acts as a barrier to the diffusion of water vapor from the soil to the air. Evaporation of water through a forest floor is slow. Forest trees absorb nutrients from deeper parts of the soil and return much of the absorbed nutrients to the soil surface in litters. Huge nutrients are locked up in the standing forest floor materials. If these nutrients were released all at a time (as by accidental fires), they would all have been lost by washing and leaching. Litters are slowly and gradually decomposed, available nutrients are released slowly, and nutrients are absorbed by plant roots. So, there is little loss of nutrients. A large number of soil-dwelling organisms belonging to invertebrates, fungi, algae, bacteria, and archaea inhabit the forest floor. Forest floor materials provide a habitat for all of them. These organisms perform many different functions, including burrowing, pedoturbation, mixing, aggregating, and nutrient cycling. Organic residues in the forest floor are food for detrivores and saprovores. Soil organisms use organic residues as food and mediate their fragmentation, mixing, and decomposition. By this way, litter materials become soil materials. Fallen seeds are trapped in the forest floor and are preserved in it until germination.

4.6 Chemical Composition of SOM

SOM is a highly variable chemical entity because of the variable composition of different sources of organic residues. A variety of organic compounds predominate at different stages of decomposition as well. So, at any given time, SOM includes materials that are added in plant and animal residues, compounds that are formed during decomposition, substances synthesized by microorganisms during their metabolism, and compounds in root exudates and microbial secretions. In fact, SOM contains most, if not all, of the organic compounds synthesized by living organisms. SOM includes:

1. Simpler substances such as sugars, amino acids, and other small molecules
2. High-molecular-weight organic materials such as polysaccharides, proteins, waxes, lipids, lignins (Figs. 4.4 and 4.5)
3. Humus

So, SOM contains two types of substances: humic substances and non-humic substances.

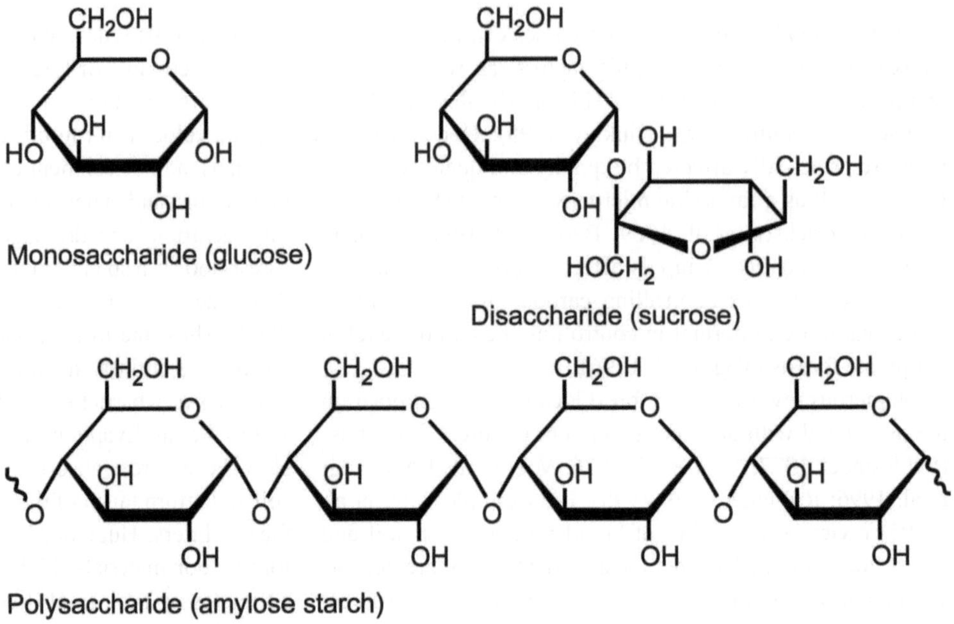

Fig. 4.4 Structures of monosaccharides and polysaccharides. (Source: http://chemistry2.csudh.edu/rpendarvis/monosacch.html)

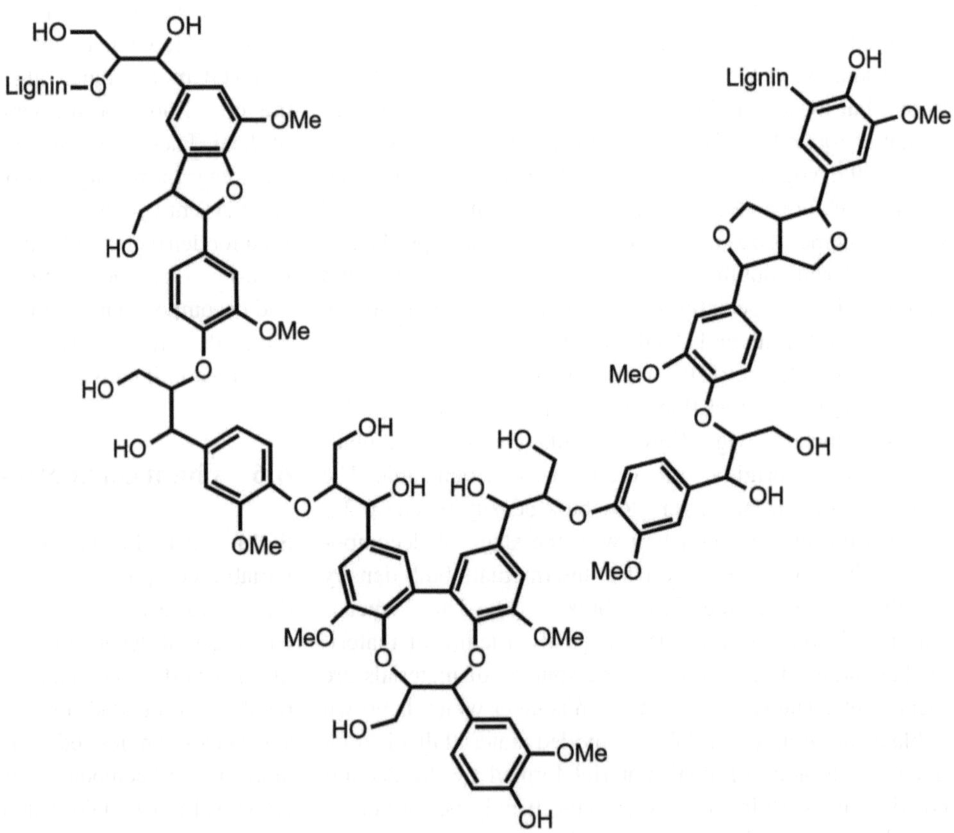

Fig. 4.5 Structure of lignin. (Source: http://www.polypompholyx.com/2012/09/far-from-the-light-of-day/)

Plant residues contain 15–40 % cellulose, 10–43 % hemicellulose, 25–40 % lignin, 1–10 % waxes and lipids, and 1–15 % proteins (Schulze et al. 2005). Microbial decomposition proceeds very fast after the addition of organic residues and slows down gradually. Susceptible or "labile" substances such as sugars (Cheshire 1979), amino acids (Stevenson 1982), other organic acids (Stevenson 1967), celluloses, etc. are quickly lost and "stable" or resistant substances such as lignin increase in proportion. Some carbohydrates and proteins are newly synthesized by microorganisms for their body

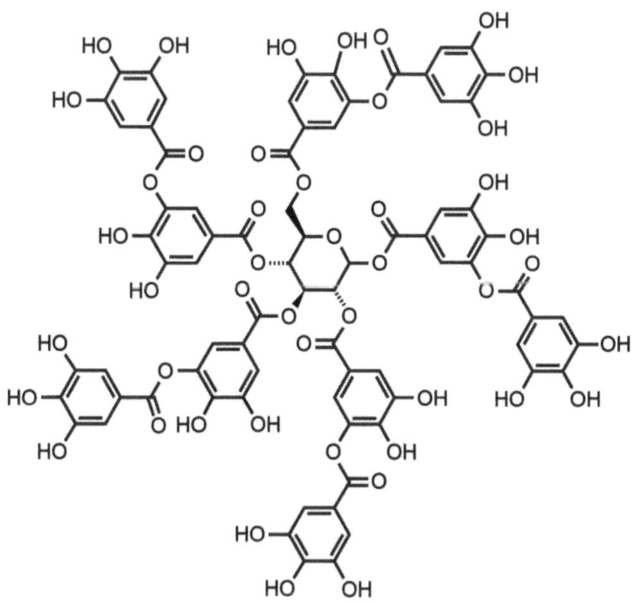

Fig. 4.6 Structure of polyphenol. (Source: http://en.wikipedia.org/wiki/Polyphenol)

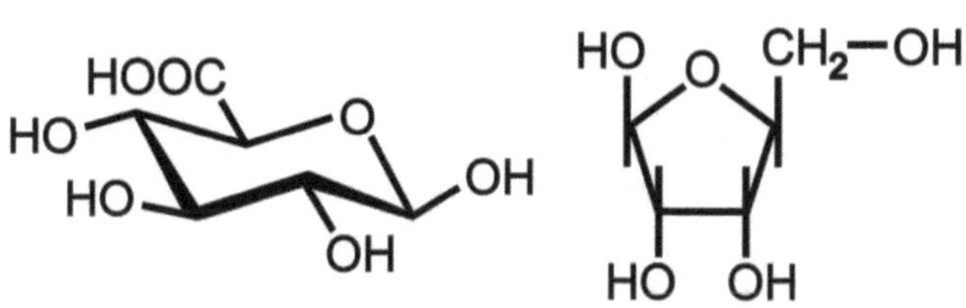

a b

Fig. 4.7 Structure of uronic acid (a) and pentose sugar (b). (Source: http://en.wikipedia.org/wiki/Uronic_acid)

materials. These substances are also added to soils after their death. Microbial polysaccharides and proteins form complexes with lignin and other compounds rich in polyphenols and accumulate in humus (Fig. 4.6) (Coleman et al. 2004).

Some carbohydrates may be detrital, in that they originally entered the soil in plant debris, but most carbohydrates are thought to be products of microbial synthesis in situ. Hydrolyzed products of soil polysaccharides are pentoses, hexoses, deoxyhexoses, hexosamines, and uronic acids (Fig. 4.7) (Jenkinson 1988).

There is a considerable amount of amino acid polymers, such as peptides, and a small quantity of amino sugars and free amino acids. Peptides and polypeptides generally remain linked with other organic fractions, and with clay and polyvalent cations. They are stable substances which predominate with lignins in humus (Fig. 4.8).

Humic substances are defined as condensed polymers of aromatic and aliphatic compounds produced by the decomposition of plant and animal residues and by microbial synthesis. They are amorphous, dark-colored, and hydrophilic, with a wide range in molecular weight from a few hundreds to several thousands (Evangelou 1998). Three fractions of materials may be obtained during the acid–alkali fractionation of humus. They are humic acid, fulvic acid, and humin (Kononova 1966). But these fractions do not represent any discrete group of compounds; they are actually products obtained in different steps of the fractionation procedure. These fractions, however, provide a convenient way of describing the composition of humus. Chemical characteristics of these fractions are mentioned as follows (Figs. 4.9, 4.10 and 4.11):

1. Humic acid—soluble in dilute alkali, but precipitates in acid solution
2. Fulvic acid—soluble both in alkali and acid solutions
3. Humin—insoluble both in alkali and acid solutions

Generally, humic and fulvic acids have a similar structure, but they differ in molecular weight and elemental and functional group contents. Fulvic acid is lower in molecular weight and contains more oxygen-containing functional groups and elemental oxygen, but less nitrogen and carbon per unit weight than humic acid (Schnitzer and Khan 1972; Sposito et al. 1976; Schnitzer 1991). According to Aswathanarayana (1999), humic acid contains 560, 47, and 355 g kg^{-1} C, H, and O, respectively, while fulvic acid contains 457, 54, and 448 g kg^{-1}, respectively.

4.7 Dissolved Organic Matter in Forest Soil

Part of the SOM is water soluble and is encountered in the soil solution as dissolved organic matter (DOM). DOM has an important role in soil solution chemistry, soil forming processes, elemental cycling, and plant nutrition (Stevenson 1983; Duchaufour 1991), and in the transport of elements within the soil (Qualls et al. 1991). Additionally, DOM has been shown to affect the stabilization of aggregates (Kuiters 1993). DOM is present in all ecosystems. It occurs in forms that range in size from simple amino acids to complex high-molecular-weight DOM. The origin, function, and fate of these compounds in terrestrial ecosystems are only partially understood (Neff and Asner 2001).

Huang et al. (1998) found that DOM was the most oxidized and microbially processed fraction of SOM. Kaiser and Guggenberger (2000) showed that the carbohydrate and lignin profiles of DOM most resembled that of organic matter found in the clay fraction of the mineral soil, which is generally considered to be a highly altered biological reaction product. Thus, DOM released from fresh plant material and litter would seem to undergo transformations, transport reactions, and lag times prior to exiting mineral soil profiles (Sanderman et al. 2008). While these studies point to the significance of highly altered OM found within the mineral soil as a major source of DOM, other investigations (Aufdenkampe et al. 2001; Kaiser et al. 2001; Hernes et al. 2007)

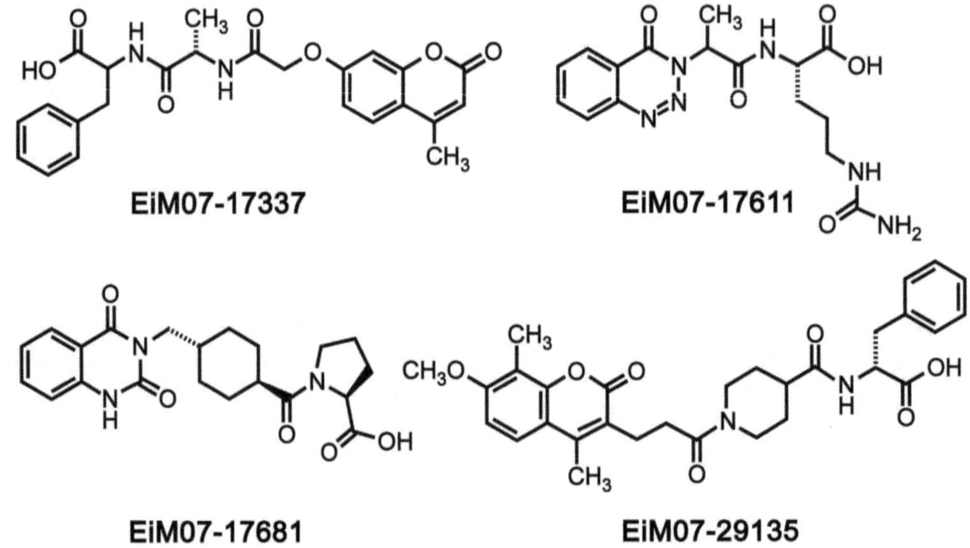

Fig. 4.8 Structure of peptides. (Source:http://www.google.com.bd/imgres?imgurl=http://blogbiosyn.files.wordpress.com/2013/09/peptides-1)

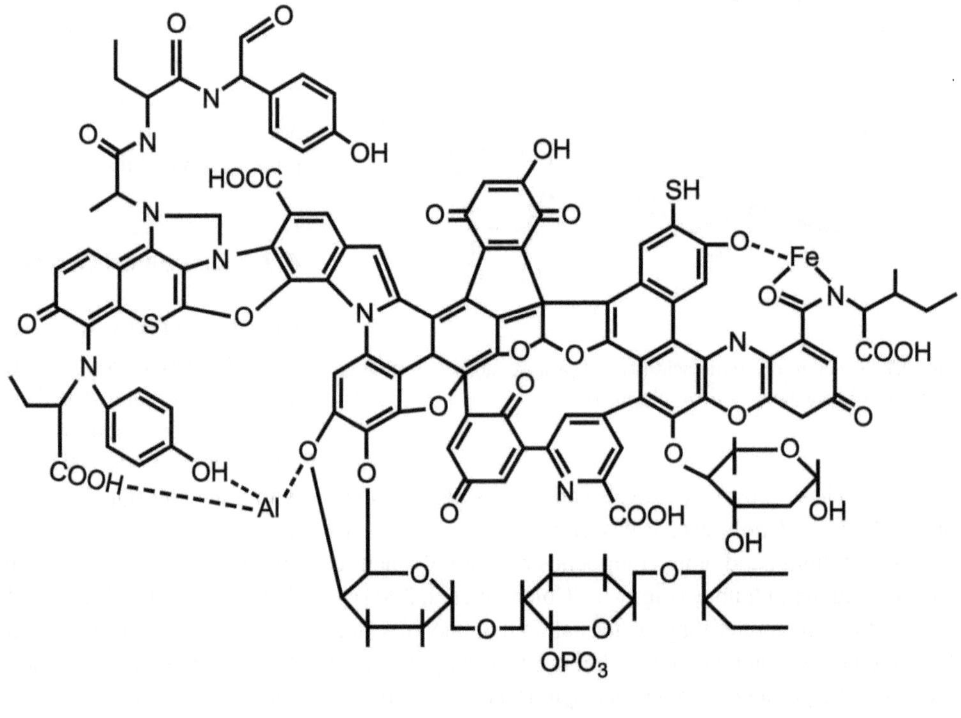

Fig. 4.9 Structural model of humic acid. (Source: http://eximedlab.com/index.php?p=research&sp=1)

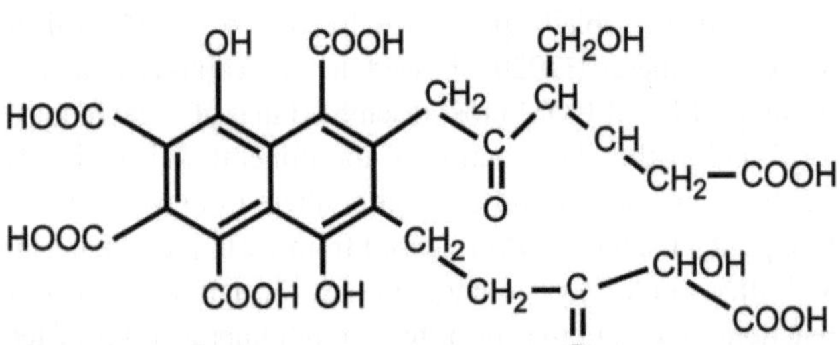

Fig. 4.10 Model structure of fulvic acid. (Source: http://www.humet.com/acatalog/humifulvatescience.html)

have stressed the importance of preferential or selective adsorption of surficially derived DOM in controlling the DOM composition in mineral soils. Despite the more dynamic nature of the potential DOM pool relative to bulk SOM, the field-collected DOM, even within the O horizon, is rarely dominated by the low-molecular-weight, hydrophilic, and highly labile organic compounds which are suggested by studies that examine water-extractable OM from fresh litter material (Kalbitz et al. 2003). Studies with nuclear magnetic resonance (NMR) reveal that DOM resembles highly altered and chemically recalcitrant organic matter rather than fresh plant residues.

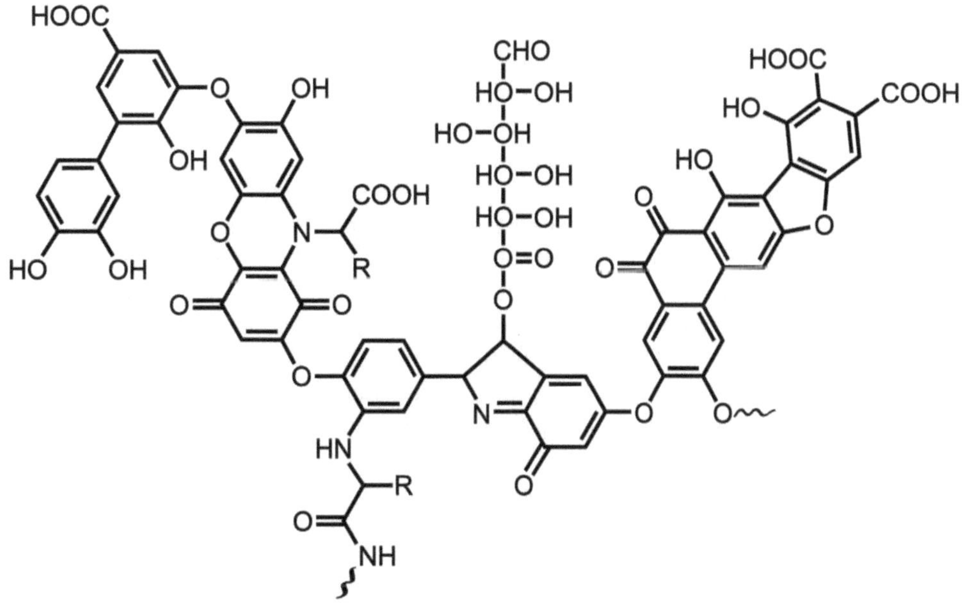

Fig. 4.11 Structure of humin. (Source: http://en.wikipedia.org/wiki/Humic_acid)

DOM is a small fraction but it strongly concerns with carbon and nitrogen dynamics in forest ecosystems (Kalbitz et al. 2000). Biodegradation of DOM, which is probably most active in surface organic horizons, is a process to release gaseous carbon from soils to atmosphere by preferential metabolization of carbohydrates leaving recalcitrant constituents mainly composed of lignin-derived phenols (Kalbitz et al. 2003). The sorption of DOM onto soil minerals is one of the most important geochemical processes for preservation of carbon and nitrogen included in lignin-derived and microbial synthesized acidic compounds in soils (Guggenberger and Kaiser 2003). After passing through soil systems, DOM might be discharged into aquifers as allochthonous DOM in natural water systems, stream, river, ocean, and so on (Lobbes et al. 2000; Amon and Benner 2003).

DOM is probably the most bioavailable fraction of SOM (Marschner and Kalbitz 2003) and contributes significantly to nutrient cycles (Qualls 2000). Pizzeghello et al. (2006) suggested that DOM plays an important role in forest ecosystems in the following ways:

1. It is a major mode of nitrogen and phosphorus export in forest ecosystems.
2. It determines the balance and accumulation of nitrogen, phosphorus, and possibly of carbon, in long-term soil development.
3. Being normally acidic, and displaying a powerful metal-complexing activity, DOM is deeply involved in mineral weathering, metal toxicity mediation, and metal export to water bodies.
4. It provides a potential source of carbon for microbial growth.

Many authors suggest that tree species and ground vegetation affect the quality and quantity of DOM in the soil (Hongve et al. 2000; Kaiser et al. 2001; Park et al. 2002), but others believe that tree species have little effect on the DOM composition and chemical reactivity (Souminen et al. 2003). According to Qualls (2000), 22–41 % of the C in the freshly fallen autumn leaf litter of a deciduous forest is water soluble.

The constituents of DOM may influence the growth of forest trees. For example, Pizzeghello et al. (2006) observed that the growth of *Picea abies* seedlings was influenced by the phenolic acid content of soil. At concentrations between 1 and 100 μM, phenylacetic and protocatechuic acids inhibited root growth to the same extent as indoleacetic acid, while p-hydroxybenzoic acid had a stimulating effect comparable to that of gibberellic acid. The aliphatic and phenolic acids appear to be related to plant strategies that influence soil fertility affecting plant growth through rhizodeposition. The role of low-molecular-weight aliphatic and phenolic acids as molecular markers of ecosystem function is also important.

4.8 Types of Humus in Forest Soils

Forest soil humus is one of the most interesting components of the forest environment. Humus is responsive: Humus type and characteristics are indicators of forest treatment and, to some extent, of site quality. Foresters find great differences in the humus beneath various kinds of forests and suggest that these differences influence the physical, chemical, and biological properties of forest soils. There are three major types of forest humus: mull humus, mor humus, and duff mull or moder humus (Hoover and Lunt 1952). These humus types differ in the presence or absence of a humified layer; the degree of incorporation of organic matter into the upper mineral soil layer; and the structure, thickness, and organic matter content of the humus layer (Fisher and Binkley 2000).

Mull humus is the brown to black, amorphous, colloidal, decomposed organic matter that is intimately mixed with the mineral matter of the surface soil. Mull humus is less acid, has higher CEC, has a crumb structure, and is rich in bases (Barrat 1964). It generally develops under broadleaf forests on fertile calcium rich soils. Mull is a type of humus characteristic for chestnut soils, phaeozems, rendzinas, and others soils (Alfisols, Spodosols, Mollisols). This type of humus may also arise under grass vegetation. Mull is a well-humified organic matter, which is produced in a very biologically active habitat. This type of humus is characterized by a neutral pH, C/N ratio nearing to 10, and ability to create stable mineral–organic complexes. It represents true humus.

According to Fisher and Binkley (2000), mor humus is a superficial deposit of fibrous and partially decomposed organic matter that is acidic in reaction and poor in bases. It generally develops under coniferous forests and moorland forests in poorly fertile soils. This humus arises under conditions of low biological activity in soil. The mineralization of organic matter proceeds slowly and creates layers, which maintain a structure of vegetable material. It resembles peat. Acidophilic fungi and low active invertebrates participate in transformations of plant residues. Acidic forest soils tend to have a buildup of mor humus on the surface caused by the absence of earthworms below pH 4.8 and reduction in fungal and bacterial activity. It is low in nitrogen and wide in C/N ratio, 20–40. Mor humus is also called raw humus. In mor humus, the H layer rests on the surface soil with practically no mixing with the underlying mineral soil. The biochemical composition of litters is important in the formation of different types of humus. For example, stabilized leaf proteins play an important role in the formation of mor humus. These proteins, stabilized in the dying leaf by tannin-like materials, are so resistant to decomposition that the tissues in which they occur accumulate on the surface of the mineral soil. Withholding of supplies of available nitrogen in these proteins may also delay decomposition of other materials.

Duff mull or moder humus is intermediate or transitional between mor and mull, characteristic of fairly acidic soils often under trees with hard litter such as beech or oak. It is frequently formed in sod-podzolic soils, loesses, and mountain grassland soils. The organic horizons with moder humus consist of thin litter (2–3 cm), which gradually, without bounds, passes on to humus-accumulative horizons. Moder is a type of medium humified humus. C/N ratio is within 15–25. Here, earthworms are usually absent and the litter is mainly decomposed by fungi and arthropods (ants, mites, springtails, millipedes, woodlice/sawbugs, etc.), leaving a mix of plant fragments and mineral particles held together by arthropod fecal material. This gives a loose, crumbly texture to the humus (Thomas and Packham 2007).

According to Kononova (1966), there are three types of humus:

First type of humus is characteristic of podzolic soils, gray brown soils, and lateric soils (Spodosols and Alfisols) under forest communities. In this humus predominate humic acids, thus humic acid/fulvic acid ratio is below 1. Humic acids indicate a small extent of aromatic rings condensation and they are approximate to fulvic acids. Considerable hydrophilic properties of humic acids favor the creation of chelates with polyvalent cations and ability to displacement deep into profile of soil. Considerable mobility of this humus favors the process of podzolization.

Second type of humus is characteristic of phaeozems, rendzinas, black earths, and brown soils (Alfisols and Mollisols). The humic acid/fulvic acid ratio is greater than 1. The extent of aromatic rings condensation is high in humic acids, which cause their hydrophobic properties and inability to create chelates. Humic acids are strongly connected with the mineral portion of soil in this type of humus.

Third type of humus is characteristic of semidesert soils. In this humus predominate fulvic acids fraction, whereas arise of humic acids is limited. Beyond this, humic acids are largely bounded with mineral portion of soil.

4.9 Organic Matter Content in Forest Soils

The forest soil has high organic matter content at the surface, where litterfall has contributed. Subsoil gets organic matter from the occasional death and decay of tree roots. However, SOM in a forest soil decreases rapidly with depth. In the native grassland, on the other hand, organic matter content is also high at the surface, but it drops off slowly with depth. The organic matter content in soil depends on soil conditions, vegetation cover, topography, hydrological conditions, elevation, and soil management. The two major natural variables which affect forest SOM are temperature and moisture. Temperature affects organic matter accumulation in two ways. First, plants tend to grow faster and produce more total mass as temperatures increase. Second, however, and usually overcoming the first point is that, as temperatures increase, microbial activity, including organic matter decomposition, also increases. For this reason, relatively little organic matter is found to persist in the forest floor of a tropical rain forest. Microbial activity is just too intense to allow organic matter to accumulate there. On the other hand, soil moisture has a little more predictable effect on SOM. As rainfall increases, the total plant production of organic matter increases, and so SOM increases. It becomes complex, however, when temperature and moisture effects are put together. In the continental USA, east of the Rocky Mountains, the general trend is that SOM content increases from west to east and from south to north. The west-to-east trend is because of increasing rainfall, while the south-to-north trend is because of lower temperatures preserving the organic matter that is produced. One natural exception occurs where water accumulates to a degree that the soil is flooded for long periods of the year. In swamps or bogs, the excess water produces a shortage of oxygen as a consequence of which organic matter builds up regardless of the temperature until the swamp is drained.

Total grassland area of the world is 3,040 × 10^6 ha, while total forest area is 4,050 × 10^6 ha. There is almost no grassland in Histosols, and no forest in Aridisols or Mollisols. Most of the potentially arable land not yet cultivated is grassland, and the greater part of the grazeable land not yet used is now savannah or forest. A large part of mountain soil is barren land. The soils of humid regions, such as Oxisols, Alfisols, Ultisols, and most of the Inceptisols and Entisols, are forested if they are not used for crops or grazing. These various types of soils have tremendous variation in their organic matter contents. Some examples are given subsequently.

Organic matter content is usually 1–5% of the dry weight in forest soil and it decreases with depth. The carbon content is about 58% of the organic matter content. There are some forest soils which either have higher organic matter content or are organic soil in nature. Under natural conditions, the content of organic matter in soils of stable forest ecosystems is constant; the rate of decomposition equals the rate of supply of organic matter from vegetation. The equilibrium is disturbed when forests are burned or cleared, and the land is used for agriculture or otherwise. The decline is rapid in the first few years after deforestation and gradually slows down over the next 10–50 years (Buringh 1984). A soil, with a bulk density of 1.5 and a carbon content of 3% in the 0–25 cm layer, 1% in the 25–50 cm layer, 0.3% in the 50–75 cm layer, and 0.1% in the 75–100 cm layer, contains 165 t ha^{-1} total organic carbon within 1 m soil depth. The 0–25 cm layer contains 113 t ha^{-1} or 68% of the total. In Mollisols (grassland soils), the total soil carbon is more than 200 t ha^{-1} and the surface layer of 0–25 cm contains only 25% of the carbon because the humus layer is very deep. The average percentage of organic carbon of 1 m depth in forest soils belonging to different soil orders is given in Table 4.1.

Organic matter contents of 1 m upper soil range from 0.87 to 2.50%. If 70% of organic matter concentrates in upper 25 cm depth, then the range becomes 2.60–7.5%. In Table 4.1, no figures for Aridisols and Mollisols are given for forestlands, because Aridisols are soils usually found in deserts and Mollisols are usually found in prairies. Histosols represent the typical organic or peat soils of coastal swamps, which generally have not been reclaimed except for some small areas in Western Europe and North America (Buringh 1984).

Table 4.1 Estimated forest areas in different soil orders and their average organic matter content up to 1 m depth from surface. (Data extracted and calculated from Buringh 1984)

Soil orders	Estimated forest land (10^6 ha)	Average soil organic matter content			
		Virgin forest		Secondary forest	
		t ha^{-1}	%	t ha^{-1}	%
Alfisols	800	270	1.80	200	1.33
Aridisols	0	–	–		–
Entisols	550	230	1.53	170	1.13
Histosols	100	375	2.50		–
Inceptisols	650	270	1.80	200	1.33
Mollisols	0		–	–	–
Oxisols	450	240	1.60	180	1.20
Spodosols	250	130	0.87	100	0.67
Ultisols	400	240	1.60	180	1.20
Vertisols	50	190	1.27	140	0.93
Mountain soils	800	200	1.33	150	1.00
Total	3,040		–		

4.10 SOM Turnover

The turnover of an element in a pool is generally determined by the balance between its inputs and outputs to and from the pool. Turnover is most often quantified as the element's mean residence time (MRT) or its half-life (T1/2). The MRT of an element in a pool is defined as 1) the average time the element resides in the pool at steady state or 2) the average time required to completely renew the content of the pool at steady state. The term half-life of SOM is the time required for half of the currently existing stock to decompose. MRT values can be estimated by different methods and different values are obtained with these methods. Also, variations occur due to differences in vegetation, climate, soil type, and other factors. The ^{13}C method is generally used in medium-term observations or experiments (5–50 years); hence, this method gives an estimate of turnover dominated by relatively recent inputs and C pools. In contrast, the oldest and most recalcitrant C pools dominate estimates by radiocarbon dating because of the long-term time frame (200–40,000 years) that this method measures (Goh 1991).

SOM is the major reservoir of carbon in terrestrial ecosystems, storing some 1,500 Pg (1 Pg = 10^{15} g) of carbon in the upper meter of mineral soils (Jobbagy and Jackson 2000). This is slightly more than twice the amount of carbon present in the atmosphere as CO_2. However, knowing the size of the soil carbon storage in soils is enough alone for predicting its potential to influence atmospheric CO_2 concentrations. An important area of understanding is the role of SOM in carbon cycling. Not all of the carbon in soils interacts with atmospheric CO_2 on the same time scale. For this reason, SOM may be categorized into several pools with characteristic turnover times of year, years to decades, and over several hundred years (Parton et al. 1987). Most of the roughly 80–160 Pg C in surface litters (Matthews 1997) and an estimated 200–300 Pg C in SOM are in forms that accumulate and decompose on time scales of a century or less (Schimel 1995; Potter and Klooster 1997). The remainder, constituting the majority of carbon stored in mineral soils, is stable on time scales of centuries to millennia. In general, forest clearing is associated with soil C loss and regrowth is associated with soil C gain in temperate and tropical forest ecosystems (Camargo et al. 1999).

Primary production and microbial decomposition are the overall biological processes governing inputs and outputs and, hence, SOM turnover. These two processes and their balance are controlled by complex underlying biotic and abiotic interactions and feedbacks linked with the state factor model of soil formation (Jenny 1980). Climate, especially temperature and precipitation, affects both production and decomposition of SOM. Vegetation type affects production rates and the types and quality of organic inputs (belowground vs. aboveground, amounts of structural tissue, C/N and lignin/N ratios), as well as the rates of water and nutrient uptake—influence decomposition rates. Disturbance or management practices also exert considerable influence on SOM turnover via direct effects on inputs and outputs and through indirect effects on the factors controlling these fluxes.

Study Questions

1. What is soil organic matter? What are the sources of organic matter in forest soils? Give an account of the functions of organic matter in forest soil.
2. Explain the structure and properties of a forest floor. Narrate its significance. Why are tropical forests usually lacking a well-developed forest floor?
3. Define humus. Explain why humus is relatively stable in soil. Discuss different types of humus in forest soils. Discuss the turnover of SOM with emphasis on humus.
4. What do you mean by DOM? Discuss the chemical nature of DOM and explain the significance of DOM.
5. Give an account of the chemical composition of SOM. Explain that humic acid, fulvic acid, and humin are products of acid–alkali fractionation procedure; they are not discrete chemical compounds.

References

Amon RMW, Benner R (2003) Combined neutral sugars as indicators of the diagenetic state of dissolved organic matter in Arctic Ocean. Deep Sea Res I 50:151–169

Aswathanarayana U (1999) Soil Resources and the Environment. Oxford and IBH Publishing Co Pvt Ltd, New Delhi

Aufdenkampe AK, Hedges JI, Richey JE, Krusche AV, Llerena CA (2001) Sorptive fractionation of dissolved organic nitrogen and amino acids onto fine sediments within the Amazon Basin. Limnol Oceanogr 46:1921–1935

Baldock JA, Skjemstad JO (1999) Soil organic carbon/soil organic matter. In: Peverill KI, Sparrow LA, Reuter DJ (eds) Soil analysis: An interpretation manual. CSIRO Publishing, Collingwood

Barrat BC (1964) A classification of humus foms and micro-fabrics of temperate grasslands. Eur J Soil Sci 15(2):342–356

Batjes NH (1996) Total carbon and nitrogen in the soils of the world. Eur J Soil Sci 47:151–163

Berg B, Ekbohm G, Johansson MB, McClaugherty C, Rutigliano F, De Santo AV (1996) Maximum decomposition limits of forest litter types: a synthesis. Can J Bot 74:659–672

Binkley D (1995) The influence of tree species on forest soils: processes and patterns. Agronomy Society of New Zealand Special Publication 10, Wellington, New Zealand

Buringh P (1984) Organic Carbon in Soils of the World. In: Woodwell GM (ed) The Role of Terrestrial Vegetation in the Global Carbon Cycle: Measurement by Remote Sensing. John Wiley & Sons Ltd

Camargo P, Trumbore SE, Martinelli LA, Davidson EA, Nepstad DC, Victoria R (1999) Carbon dynamics in regrowing forest of eastern Amazonia. Glob Chang Biol 5:693–702

Carter MR, Gregorich EG (1996) Methods to characterize and quantify organic matter storage in soil fractions and aggregates. In: Carter MR, Stewart BA (eds) Structure and organic matter storage in agricultural soils. Lewis Publ. CRC Press, Boca Raton

Cheng W, Coleman DC, Box JE Jr (1990) Root dynamics, production and distribution in agroecosystems on the Georgia Piedmont using minirhizotrons. J Appl Ecol 27:592–604

Cheshire MW (1979) Nature and Origin of Carbohydrates in Soil. Elsevier, London

Coleman DC, Crossley DA Jr, Hendrix PF (2004) Fundamentals of Soil Ecology, 2nd edn. Elsevier-Academic Press, USA

Cuevas E (1983) Crecimiento de raices finas y su rclaci6n con los procesos de descomposici6n de materia orginica y liberaci6n de nutrientes en bosques del Alto Rio Negro en el Territorio Federal Amazonas. Ph.D. dissertation. Instituto Venezolano de Investigaciones Cientfficas, Caracas, Venezuela

Cuevas E, Medina E (1986) Nutrient dynamics within Amazonian forest ecosystems. I. Nutrient flux in fine litter fall and efficiency of nutrient utilization. Oecologia (Berlin) 68:466–472

Duchaufour P (1991) Pedologie: Sol, Vegetation, Environnement. Masson, Paris

Eissenstat DM, Yanai RD (1997) The ecology of root lifespan. Adv Ecol Res 27:1–60

Evangelou VP (1998) Environmental soil and water chemistry: principles and applications. Wiley, New York

Federer CA (1982) Subjectivity in the separation of organic horizons of the forest floor. Soil Sci Soc Am J 46:1090–1093

Fisher RF, Binkley D (2000) Ecology and Management of Forest Soils Wiley, New York

Goh KM (1991) Carbon Dating. In: Coleman DC, Fry B (eds) Carbon isotope techniques. Academic Press, San Diego

Guggenberger G, Kaiser K (2003) Dissolved organic matter in soil: challenging the paradigm of sorptive preservation. Geoderma 113:293–310

Han KH, Ha SG, Jang BC (2010) Aggregate Stability and Soil Carbon Storage as Affected by Different Land Use Practices. Proceedings of International Workshop on Evaluation and Sustainable Management of Soil Carbon Sequestration in Asian Countries. Bogor, Indonesia Sept 28-29, 2010

Hendrickson OQ, Chatarpaul L, Burgess D (1989) Nutrient cycling following whole-tree and conventional harvest in northern mixed forest. Can J For Res 19:725–735

Hernes PJ, Robinson AC, Aufdenkampe AK (2007) Fractionation of lignin during leaching and sorption and implications for organic matter ""freshness"". Geophys Res Lett 34:L17401

Hongve D, Hees PAW van, Lundstrom US (2000) Dissolved components in precipitation water percolated through forest litter. Eur J Soil Sci 51:667–677

Hoover MD, Lunt HA (1952) A key for the classification of forest humus types. Soil Sci Soc Amer Proc 16:368

Howard PJA, Howard DM (1990) Use of organic carbon and loss-on-ignition to estimate soil organic matter in different soil types and horizons. Biology and Fertility of soils 9:306–310

Huang Y, Eglinton G, Van der Hage ERE, Boon JJ, Bol R, Ineson P (1998) Dissolved organic matter and its parent organic matter in grass upland soil horizons studied by analytical pyrolysis techniques. Eur J Soil Sci 49:1–15

Janzen HH, Larney FJ, Olson BM (1992) Soil quality factors of problem soils in Alberta. Proceedings of the 29th Annual Alberta Soil Science Workshop, Feb. 19-20, 1992, Lethbridge, Alberta

Jenkinson DS (1988) Soil organic matter and its dynamics. In: Wild A (ed) Russel's Soil Condition and Plant Growth, 11th edn, English Language Book Society, London

Jenny H (1980) The Soil Resource-Origin and Behavior. Springer, New York

Jobbagy EG, Jackson RB (2000) The vertical distribution of soil organic carbon and its relation to climate and vegetation. Ecol Appl 10:423–436

Kaiser K, Guggenberger G (2000) The role of DOM sorption to mineral surfaces in the preservation of organic matter in soils. Org Geochem 31:711–725

Kaiser K, Guggenberger G, Zech W (2001) Isotopic fractionation of dissolved organic carbon in shallow forest soils as affected by sorption. Eur J Soil Sci 52:585–597

Kalbitz K, Solinger S, Park JH, Michalzik B, Matzner E (2000) Controls on the dynamics of dissolved organic matter in soils: a review. Soil Sci 165:277–304

Kalbitz K, Schmerwitz J, Schwesig D, Matzner E (2003) Biodegradation of soil-derived dissolved organic matter as related to its properties. Geoderma 113:273–291

Kononova MM (1966) Soil Organic Matter. Pergamon, Oxford

Kuiters AT (1993) Dissolved organic matter in forest soils: Sources, complexing properties and action of herbaceous plants. Chemistry and. Ecology 8:171–184

Kuzyakov Y and Domanski G (2000) Carbon input by plants into the soil. Review J Plant Nutr Soil Sci 163: 421–431

Lickacz J, Penny D (2001) Soil organic matter. Plant Industry Divn, Alberta. http://www1.agric.gov.ab.ca/$department/deptdocs.nsf/all/agdex890?opendocument

Lobbes JM, Fitznar HP, Kattner G (2000) Biogeochemical characteristics of dissolved and particulate organic matter in Russian rivers entering the Arctic Ocean. Geochim Cosmochim Acta 64:2973–2983

Ludmila M (2010) Soil Organic Matter in Forest Ecosystems of the Forest-tundra zone of Central Siberia. EGU General Assembly 2010, held 2-7 May, 2010 in Vienna, Austria

Marschner B, Kalbitz K (2003) Controls of bioavailability and biodegradability of dissolved organic matter in soils. Geoderma 113:211–235

Matthews E (1997) Global litter production, pools, and turnover times: Estimates from measurement data and regression models. J Geophys Res 102:18771–18800

Mattson KG, Smith HC (1993) Detrital organic matter and soil CO_2 efflux in forests regenerating from cutting in West Virginia. Soil Biology and. BioChemistry 25:1241–1248

McLaren RG, Cameron KC (1996) Soil Science, 2nd edition. Oxford University Press

Neff JC, Asner GP (2001) Dissolved Organic Carbon in Terrestrial Ecosystems: Synthesis and a Model. Ecosystems 4:29–48

Nemeth JC (1973) Dry matter production in young loblolly (*Pinus teada* L.) and slash pine (*Pinus elliottii* Engelm.) plantations. Ecol Monogr 43:21–41

Oades JM (1984) Soil organic matter and structural stability: mechanisms and implications for management. Plant Soil 76:319–337

Osman KT (2013) Soils: principles, properties and management. Springer

Park JH, Kalbitz K, Matzner E (2002) Resource control on the production of dissolved organic carbon and nitrogen in a deciduous forest floor. Soil Biol Biochem 34:1391

Parton WJ, Schimel DS, Cole CV, Ojima DS (1987) Analysis of factors controlling soil organic matter levels in Great Plain Grasslands. Soil Sci Soc Am J 51:1173–1179

Paustian K, Levine E, Post WM, Ryzhova IM (1997) The use of models to integrate information and understanding of soil C at the regional scale. Geoderma 79:227–260

Persson HA (2012) The High Input of Soil Organic Matter from Dead Tree Fine Roots into the Forest Soil. Int J Forest Res 2012 Article ID 217402, 9 pages

Pizzeghello D, Zanella A, Carletti P, Nardi S (2006) Chemical and biological characterization of dissolved organic matter from silver fir and beech forest soils. Chemosphere 65:190–200

Potter CS, Klooster SA (1997) Global model estimates of carbon and nitrogen storage in litter and soil pools-response to changes in vegetation quality and biomass allocation. Tellus Series B-Chemical and Physical Meteorology 49B:1–17

Pregitzer KS, Zak DR, Curtis PS, Kubiske ME, Teeri JA, Vogel CS (1995) Atmospheric CO_2, soil nitrogen and turnover of fine roots. New Phytol 129:579–585

Qualls RG (2000) Comparison of the behaviour of soluble organic and inorganic nutrients in forest soils. Forest Ecol Manag 138:29–50

Qualls RG, Haines BL, Swank WT (1991) Fluxes of dissolved organic nutrients and humic substances in a deciduous forest. Ecology 72:254–266

Rollinger JL, Strong TF, Grigal DF (1998) Forested soil carbon storage in landscapes of the Northern Great Lakes Region. In: Lal R, Kimble JM, Follett RF, Stewart BA (eds) Management of carbon sequestration in soil. CRC Press, Boca Raton

Ryan DF, Huntington TG, Martin CW (1992) Redistribution of soil nitrogen, carbon and organic matter by mechanical disturbance during whole tree harvesting in northern hardwoods. For Ecol Manag 49:87–99

Sanderman J, Baldock JA, Amundson R (2008) Dissolved organic carbon chemistry and dynamics in contrasting forest and grassland soils. Biogeochemistry (2008) 89:181–198

Sanford RL Jr (1985) Root ecology of mature and successional forests. Ph.D. dissertation. University of California, Berkeley

Schimel DS (1995) Terrestrial ecosystems and the carbon cycle. Global Change Biology 1:77–91

Schlesinger WH (1977) Carbon balance in terrestrial detritus. Annu Rev Ecol Syst 8:51–81

Schnitzer M (1991) Soil organic matter-The next 75 years. Soil Sci 151:41

Schnitzer M, Khan SU (1972) Humic Substances in the Environment. Dekker, New York

Schulze WX, Gleixner G, Kaiser K, Guggenberger G, Mann M, Schulze E-D (2005) A proteomic fingerprint of dissolved organic carbon and of soil particles. Oecologia 142:335–343

Simmons JA, Fernandez IJ, Briggs RD, Delaney MT (1996) Forest floor carbon pools and fluxes along a regional climate gradient in Maine, USA. For Ecol Manag 84:81–95

Smith JE, Heath LS (2002) A Model of Forest Floor Carbon Mass for United States Forest Types. Research Paper NE-722, United States Department of Agriculture Forest Service, Northeastern Research Station

Souminen K, Kitunen V, Smolander A (2003) Characteristics of dissolved organic matter and phenolic compounds in forest soils under silver birch (Betula pendula), Norway spruce (Picea abies) and Scots pine (Pinus sylvestris). Eur J Soil Sci 54:287–293

Sposito GK, Holtzclaw M, Baham J (1976) Analytical properties of the soluble, metal complexing fractions in sludge-soil mixtures. II. Comparative structural chemistry of the fulvic acid. Soil Sci Soc Am 40:691

Stevenson FJ (1967) In: McLaren AD, Peterson GH (eds) Soil Biochemistry vol 1. Marcel Dekker, USA

Stevenson FJ (1982) Nitrogen in agricultural soils. Agron Monogr No 22:67

Stevenson FJ (1983) Trace metal-organic matter interactions in geologic environments. In: The significance of trace elements in solving petrogenetic problems and controversies. Theophrastus Publications, Athens

Thomas PA, Packham JR (2007) Ecology of Woodlands and Forests, Description, Dynamics and Diversity. Cambridge University Press, Cambridge

Vogt KA, Grier CC, Vogt DJ (1986) Production, turnover and nutrient dynamics of above- and belowground detritus of world forests. Adv Ecol Res 15:303–378

Vogt KA, Vogt DJ, Brown S, Tilley JP, Edmonds RL, Silver WL, Siccama TG (1995) Dynamics of forest floor and soil organic matter accumulation in boreal, temperate, and tropical forests. In: Lal R, Kimble J, Levine E, Stewart BA (eds) Soil management and greenhouse effect. CRC Lewis Publishers, Boca Raton

Vogt KA, Vogt DJ, Palmiotto PA, Boon P, O'Hara J, Asbjornsen H (1996) Review of root dynamics in forest ecosystems grouped by climate, climatic forest type and species. Plant Soil 187:159–219

Waring R, Schlesinger W (1985) Forest ecosystems: concepts and management. Academic Press, Orlando

5 Biological Properties of Forest Soils

Many organisms, both macro- and micro- flora and fauna, live in the forest floor and mineral layers of the forest soil. Macrofloras include plant roots, which function in water storage, drainage, aeration, and nutrient cycling in soil by their proliferation, death, and decay. In addition, they influence biological and biochemical environment of the soil by their rhizoplane and rhizosphere effects. Soil microflora includes algae, bacteria, and fungi, which act on organic matter and mineral transformations. Organic matter decomposition, humification, mineralization, nitrification, denitrification, nitrogen fixation, etc., are some of the important transformations mediated by forest soil microflora. Forest soil harbors plenty of macro-, meso-, and microfauna, including rotifers, nematodes, protozoa, acari, collembolan, protura, diplura, symphyla, enchytraedae, isoptera, opiliones, isopoda, amphipoda, chilopoda, diplopoda, megadrili, coleoptera, mollusca, earthworms, etc. Soil faunas fragment litters and mix them with the mineral soils through their burrows. They regulate the structure and composition of soil biotic communities and their functions.

5.1 Forest Soil Organisms

The ultimate health of a forest depends to a large extent on the biological dynamics in soil that regulate organic matter acquisition and decomposition, and nutrient recycling. Microorganisms—fungi and bacteria—influence, by their metabolic activity, the acquisition of nutrients and water through tree roots in the soil (Moldenke 1990). Associations of mycorrhizal fungi facilitate uptake of nutrients by plant roots from the soil (Feldman 1988). Soil invertebrates such as mites, centipedes, and millipedes play an irreplaceable role in soil genesis, physical structure, and chemical content. Arthropod feces make up the bulk of particulate matter in the forest floor, while arthropod burrowing activities aerate the soil, making it accessible to root penetration and water absorption (Rusek 1986). Examining the influence of soil arthropods on coniferous litter decomposition, Teuben and Roelofsma (1990) found that isopods and collembolla, two very common and abundant soil arthropods, enhanced microbial activity and concentrations of exchangeable nitrate, ammonium, and phosphate. The presence of soil fauna is generally assumed to benefit plants by increasing the amount of available nutrients for root uptake. Setala and Huhta (1991) have shown in laboratory experiments that soil fauna exerts a positive influence on birch seedling growth. Root biomass of the seedlings was as much as 70 % greater in the presence of nematodes and microarthropods. Soil invertebrates affect soil formation and change the forest environment as a whole. Soil ecosystems, and especially their faunal components, are highly sensitive to successional stage changes following forest utilization and management practices, such as logging and burning (Torgersen 1993). Soil arthropods function as biological indicators of forest type (Moldenke 1990). Coniferous forest soils in western Oregon contain some of the richest and most diverse terrestrial complexes of soil-dwelling arthropods in the world (Petersen and Luxton 1982). Oribatid mites (150,000 m^{-2}) and springtails (50,000 m^{-2}) are the most abundant soil arthropods in Oregon's western forests. One square foot of old-growth forest soil can contain more than 200 species (Moldenke and Lattin 1990).

For convenience, soil organisms can be divided into two categories—flora and fauna. Plant roots are the major forest soil flora. The most important portion of soil flora is the microflora, including bacteria, fungi, and actinomycetes. Soil fauna is divided into microfauna (body width < 0.1 mm), mesofauna (body width 0.1–2.0 mm) and macrofauna (body width > 2 mm) on the basis of their sizes (Swift et al. 1979). Examples of these categories are microfauna (protozoa and some nematodes), mesofauna (microarthropods and enchytraeids), and macrofauna (earthworms, termites, and millipedes). The following categorization of forest soil organisms may be taken for convenience of discussion (Fig. 5.1).

Nature Education (2012) presents abundance of some groups of soil organisms in agricultural, grassland, and forest soils. There is little difference in the number of bacteria (100 million to 1 billion per gram soil) among the land-use

K. T. Osman, *Forest Soils*, DOI 10.1007/978-3-319-02541-4_5, © Springer International Publishing Switzerland 2013

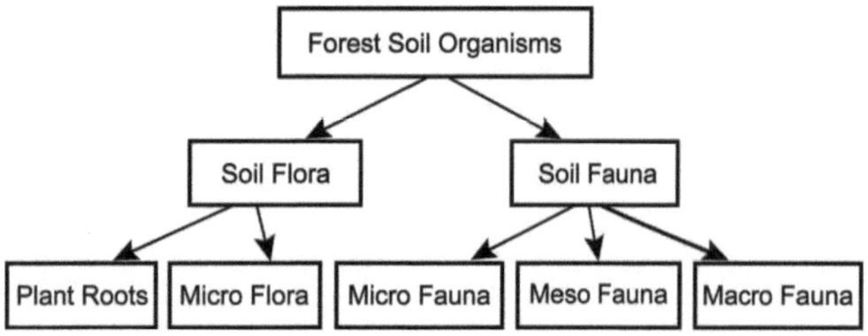

Fig. 5.1 Categories of forest soil biota

Table 5.1 Abundance of soil organisms in agricultural, prairie, and forest soils

Organisms	Agricultural soil	Prairie soil	Forest soil
2 *Number of organisms per gram soil*			
Bacteria	100 million to 1 billion	100 million to 1 billion	100 million to 1 billion
Fungi	Several meters	10s to 100s m	1–60 km
Protozoa	1,000s	1,000s	100,000s
Nematodes	10–20	10s to 100s	100s
3 *Number of organisms per square meter*			
Arthropods	<10	45–200	900–2,300
Earthworms	4–25	8–42	8–42 (0 in conifers)

Table 5.2 Dominant functional groups of soil organisms in different forests

Forest	Dominant functional groups of soil organisms
Pine, Fir, Cedar	Fungi, ectomycorrhizal fungi, ectomycorrhizal mats, fungal-feeding nematodes, fungal feeding microarthropods
Mesquite	Fungi, ectomycorrhizal fungi, fungal-feeding nematodes, fungal feeding microarthropods
Aspen, Alder Wilow	Bacteria, fungi, protozoa, nematodes, rotifers, microarthropods
Oak, Madrone	Fungi, ectomycorrhizal fungi, ectomycorrhizal mats, fungal-feeding nematodes, fungal feeding microarthropods
Sub-alpine herbaceous woodland	Bacteria, VAM, protozoa, bacteria-feeding nematodes
Alpine tundra	Fungi, bacteria, protozoa, nematodes, rotifers
Glacial tundra	Bacteria, protozoa, nematodes

classes. Fungi, protozoa, and arthropods are much more abundant in forest soils. Earthworm population is similar between prairie and forest soils but coniferous forests are devoid of earthworms. The abundance of these groups of organisms is shown in Table 5.1.

Ingham (1995) suggested "dominant functional groups of soil organisms" in different types of managed and natural ecosystems. Table 5.2 is constructed taking his data for various forests. Fungi, bacteria, protozoa, and nematodes are the most predominant functional groups. Ectomycorrhizal fungi are dominant in soils of Pine, Fir, Cedar, Mesquite, Oak, and Madrone forests.

5.2 Plant Roots

Roots are primarily the anchoring and absorbing organs of vascular plants. By their death and decay, roots leave channels for movement of water and oxygen through the soil. Roots also circulate water, organic matter, and nutrients within the soil profile. The most important ecological function, the roots perform, is perhaps the modulation of biological activity in soils of their vicinity. Fine and small roots (<5 mm) and coarse roots (>5 mm) are two major components of belowground biomass. Respiration of these roots is a major source of carbon dioxide in soil. Roots release metabolites and their decay products through rhizodeposition. Rhizodeposition is a process of transport of fixed carbon in the canopy through the roots and release in deeper soil. Rhizodeposition releases exudates of water-soluble, low-molecular-weight compounds leached from the roots and mucilages secreted by golgi organelles in the root cap, hydrolysates of the polysaccharides of the primary cell wall and sloughed root cap cells, mucilage secreted by epidermal cells and root hairs, and those produced by bacterial degradation of dead epidermal cells. There are lysates and sloughed cells from the epidermis and cortex of roots (Lavelle and Spain 2003).

Clark (1949) proposed the term "rhizoplane" to refer to the immediate surface of plant roots together with any closely adhering particles of soil or debris. The term "rhizosphere" was introduced by the German scientist Hiltner in 1904 for the region of the soil which is influenced by the activity of plant roots. Floral and faunal communities in the rhizoplane (surface of the root) and in the rhizosphere (around the root) are different from other zones of the soil. Some microorganisms preferably colonize the rhizoplane and some rhizoplane bacteria can promote the growth of plants. These plant growth-promoting bacteria include *Pseudomonas*, *Azospirillum*, *Azotobacter*, *Klebsiella*, *Enterobacter*, *Alcaligens*, *Arthobacter*, *Burkholderia*, *Bacillus*, and *Serratia* (Han and Lee 2005). Some fungi inhabit the root surface in a mycelial state. They belong to the genera *Mortierella*, *Cephalosporium*, *Trichoderma*, *Penicillium*, *Gliocladium*, *Gliomastix*, *Fusarium*, *Cylindrocarpon*, *Botrytis*, *Coniothyrium*, *Mucor*, *Phoma*, *Pythium*, and *Aspergillus*.

The rhizosphere soil differs in physical and chemical properties from the bulk soil (Whalley et al. 2005), although it is difficult to make a sharp demarcation in field between rhizosphere and non-rhizosphere zone of soil. The rhizosphere zone is characterized by intense biological and chemical activity. Since plant roots exude many biochemical compounds, including amino acids, other organic acids, carbohydrates, sugars, vitamins, mucilage (polysaccharides), proteins, flavones, enzymes, hydrocyanic acid, glycosides, auxins, and saponins (Gupta and Mukerji 2002), growth of some microorganisms is favored and that of some others is inhibited in the rhizosphere. Root exudates stimulate the growth of many bacteria, including free-living nitrogen fixers of the genera *Azotobactor*, *Azospirillum*, and *Azoarcus*, the symbiotic nitrogen fixers of the genus *Rhizobium*, and several other bacteria and fungi, including the mycorrhizal fungi.

5.3 Mycorrhizae

The German forest pathologist A.B. Frank in 1885 first applied the term mycorrhiza (*fungus–root*) to fungus–tree associations (Jhonson 2009). Mycorrhiza refers to the symbiotic association between a fungus and a root of a vascular plant. In this mutualistic association, the fungus gets carbohydrates synthesized by the plant and the plant gets the benefits of the huge mycelial network that absorbs water and nutrients from a larger volume of soil. Plant partners in nutrient-poor soils are especially benefitted from mycorrhiza. There are arbuscular, ecto-, ectendo-, arbutoid, monotropoid, ericoid, and orchidaceous mycorrhizas. The arbuscular mycorrhiza and ectomycorrhiza are the most abundant and widespread (Siddiqui and Pichtel 2008). Except ectomycorrhizas, the others were earlier taken together as endomycorriza. The hyphae of ectomycorrhizal fungi do not penetrate individual cells within the root but constitute a hyphal sheath, or mantle, covering the root tip and a hartig net of hyphae surrounding the plant cells within the root cortex. On the other hand, the hyphae of endomycorrhizal fungi penetrate the cell wall and invaginate the cell membrane. Ectomycorrhizas are found to form association between the roots of woody plants, such as birch, dipterocarps, eucalyptus, oak, and pine, and the fungi that belong to the Basidiomycota, Ascomycota, and Zygomycota.

5.4 Soil Microflora

Bacteria Bacteria are unicellular prokaryotes ranging in size from 0.3 to more than 3 μm. They can be classified on various bases such as cell morphology (cocci, bacilli, and spirilli), cell-wall structure, specific staining (Gram-positive and Gram-negative), presence of endospores, mobility of cells, and shape and position of flagella. There are cocci (sphere, 0.5 μm), bacilli (rod, 0.5–0.3 μm), or spirilli in soil, the bacilli being more common and spirilli being rare (Baudoin et al. 2002). There are again autochthonous bacteria which get their nutrition from native soil organic or mineral matter (*Arthrobacteria* and *Nocardia*), and zymogenous bacteria which require additional external substrates (*Pseudomonas* and *Bacillus*). The number of zymogenous bacteria increases when appropriate substrate is added and gradually declines when the added substrate is exhausted. Examples of zymogenous bacteria include cellulose decomposers and nitrogen transformers (Giri et al. 2005). Based on energy source and nature of the electron donors, bacteria are divided into four categories: (i) photolithotrophic bacteria—using energy from light and inorganic reduced substances (essentially sulfides) as electron donors; (ii) photo-organotrophic bacteria—photosynthetic organisms using oxidizable organic substrates as electron donors; (iii) chemo-lithotrophic bacteria—using the energy produced by redox reactions and four kinds of mineral substrates as electron donors (reduced nitrogen, sulfur or iron compounds, and hydrogen); and (iv) chemo-organotrophic bacteria—typical heterotrophic organisms which take their energy from redox reactions and use organic compounds as electron donors. There are also obligate chemoautotrophs such as *Nitrobacter* utilizing nitrite and *Nitrosomonas* utilizing ammonium, while *Thiobacillus* converts inorganic sulfur compounds to sulfate and *Ferrobacillus* converts ferrous ions to ferric ions (Baudoin et al. 2002). There are also functional categories of bacteria such as cellulolytic, chitinolytic, nitrifiers, N-fixers, denitrifiers, etc.

The most common soil bacteria belong to the genera Pseudomonas, Arthrobacter, Clostridium, Achromobacter, Bacillus, Micrococcus, Flavobacterium, Corynebacterium, Sarcina, Azosprillium, and Mycobacteria. Myxobacteria belonging to the genera Myxococcus, Chondrococcus, Archangium, Polyangium, Cytophaga, and Sporocytophaga are also common. The latter two genera are cellulolytic and, hence, are dominant in cellulose-rich litter. Based on the temperature tolerance, bacteria are grouped as psychrophilous (below 20 °C), mesophyllous (15–45 °C), and thermophilous (45–65 °C). However, mesophyllous bacteria constitute the bulk of forest soil bacteria. The cyanobacteria are Gram-negative eubacteria characterized by their ability to perform oxygenic photosynthesis. The dominant cyanobacteria belong to the genera Chrococcus, Aphanocapsa, Lyngbya, Oscillatoria, Phormidium, Microcoleus, Cylindrospermum, Anabaena, Nostoc, Scytonema, and Fischerella (Benizri et al. 2002).

Actinobacteria The Actinobacteria were earlier called Actinomycetes. They are filamentous bacteria and are mostly Gram positive. They have ramified pseudomycelium whose diameter (0.5–1 μm) is much smaller than that of fungi. Soil acidity and waterlogging are unfavorable soil conditions for Actinobacteria. They cannot grow in soils more acid than pH 5. The most favorable range of pH for Actinobacteria is 6.5–8.0. The commonest genera of Actinobacteria are *Streptomyces*. In contrast, *Nocardia*, *Micromonspora*,

Actinomyces, *Actinoplanes*, and *Streptosporangium* are only encountered occasionally. Although there is evidence that actinobacteria are abundant in soils, it is generally concluded that they are not as important as bacteria and fungi as decomposers (Foth 1990).

Fungi Fungi are heterotrophic organisms; they can be saprophytic living on dead organic matter, or parasitic living on live plant or animal tissues. Some fungi are deadly pathogens of plants and animals. Fungi have filamentous mycelium composed of individual hyphae. The hyphae may be uni-, bi-, or multinucleate and septate or nonseptate (Hawksworth 1991). The carbon source of saprophytes is organic detritus, which they decompose in the course of their growth and nutrition. Content and quality of soil organic matter affect the population and activity of saprophytic fungi in soils.

Fungi are classified into Phycomycetes, Ascomycetes, Basidiomycetes, and Fungi imperfecti (Alexander 1977). Phycomycetes constitute an important class of parasitic or saprophytic fungi. The fungal body may be an undifferentiated mass of protoplasm to a well-developed and much-branched mycelium. Reproduction is mainly sexual, by the formation of conidia or sporangia. Ascomycetes include fungi which produce spores inside a sac called an ascus. Each ascus usually contains four to eight spores depending on the species. The Basidiomycetes fungi constitute the most conspicuous group of fungi in the environment and include mushrooms, puffballs, and bracket fungi. Basidiomycetes produce a highly specialized sporangium, the basidia. Many fungi, commonly isolated from soils, come under the class Fungi imperfecti because they produce abundant asexual spores but lack sexual stages (Lynch 1987). Many members of Asco- and Basidiomycetes can degrade very complex organic compounds such as cellulose or lignin. Many others live as mycorrhizas on roots of higher plants and obtain simple sugars from their plant partners. The most common genera of fungi in soils are: *Acrostalagmus, Aspergillus, Botrytis, Cephalosporium, Gliocladium, Monilia, Penicillium, Scopulariopsis, Spicaria, Trichoderma, Trichothecium, Verticillium, Alternaria, Cladosporium, Pillularia, Cylindrocarpon, Fusarium, Absidia, Cunninghamella, Mortierella, Mucor, Rhizopus, Zygorynchus, Pythium, Chaetomium,* and *Rhizoctonia* (Hawksworth 1991). Some fungi such as *Alternaria, Aspergillus, Cladosporium, Dematium, Gliocladium, Helminthosporium, Humicola,* and *Metarhizium* produce substances similar to humic substance in soil and, hence, may be important in the maintenance of soil organic matter (Hawksworth 1991).

Fungi dominate the microbial community in acid soils; they are also present in neutral or alkaline soils, and some can tolerate a pH more than 9.0. Fungi are strictly aerobic and their numbers decrease in waterlogged soils. However, moist organic detritus such as leaf litter or wood are ideal habitat for fungi.

Algae Algae are less numerous than fungi in soil. Algae may be unicellular (*Chlamydomonas*) or filamentous (*Spirogyra* and *Ulothrix*). Algae contain chlorophyll and they are photoautotrophic organisms. They use CO_2 from the atmosphere and produce O_2. Some of the common green algae occurring in most soils belong to the genera *Chlorella, Chlamydomonas, Chlrococcum, Oedogonium, Chlorochytrium,* and *Protosiphone* (Lynch 1990). Algae require light and since light under a dense closed forest is low, population of algae is not high in forest soil.

5.4.1 Functions of Microflora in Forest Soil

In view of the microbial diversity and soil functionality, it may be considered that 80–90% of the processes in soil are reactions mediated by microbes (Coleman and Crossley 1996; Nannipieri and Badalucco 2003). Indeed, bacteria and fungi are highly versatile; they can carry out almost all known biological reactions. Algae are autotrophic and are capable of performing photosynthesis. Algae are most active and abundant in moist open forest soils. Because they need light for photosynthesis, algae are most common at or very near the soil surface. Some tropical rainforests are very dense and closed and allow little light penetration on the soil surface. Algal growth may be restricted in those soils. Some algae make symbiotic associations with fungi forming lichens that are important in colonizing bare rocks, litter, and tree barks. The most common groups of algae present in soil are green, yellow–green, and diatoms. In addition to producing a substantial amount of organic matter in some fertile soils, certain algae excrete polysaccharides that have very favorable effects on soil aggregation. Bacteria belong to the kingdom Monera, can be either autotrophic or heterotrophic, and can be both aerobic and anaerobic. They are single-celled, prokaryotic organisms and are the most diverse group of soil organisms. A gram of soil may contain 20,000 different species. Due to their diversity, they play many roles in the soil and are involved in all of the organic transactions. Their most important role is in decomposition of dead organic matter and mineralization of nutrients such as N, P, and S to forms available to plants. Another critical process in which bacteria are prominent is N-fixation. Cyanobacteria are either autotrophic or heterotrophic. They are also prokaryotic and are among the most ancient of organisms. Many genera of cyanobacteria including *Anabaena* sp. can fix atmospheric N_2. They are found in plenty in moist soils. Fungi belong to eukaryotic and heterotrophic microflora. They depend on living or dead organic materials for their C and energy. They may be divided into three groups including (1) yeasts (single-celled organisms that live in moist soils), (2) molds, and (3) mushroom fungi. Fungi play an important role in the decomposition of various organic substances (cellulose, hemicellulose, and lignin), they may form symbiotic relationships with plants (mycorrhizae) or act as pathogens, and they play a major role in humus formation and aggregate stabilization. Actinomycetes are single-celled, prokaryotic, filamentous, and often profoundly branched organisms. They are of great importance in the decomposition of soil organic matter and the mineralization of nutrients, especially in alkaline soils. Many actinomycetes produce antibiotic compounds that kill other microorganisms.

Soil bacteria are important regulators of biogeochemical cycles in forest soils, and plant—bacteria interactions in the rhizosphere are the determinants of plant health and soil fertility. There are plant growth-promoting rhizobacteria (PGPR), plant health-promoting rhizobacteria (PHPR), or nodule-promoting rhizobacteria (NPR). Generally, PGPR function in three different ways: synthesizing particular compounds for the plants, facilitating the uptake of certain nutrients from the soil, and lessening or preventing the plants from diseases (Guo et al. 2004). Plant growth promotion and development can be facilitated both directly and indirectly. Indirect plant growth promotion includes the prevention of the deleterious effects of phytopathogenic organisms. This can be achieved by the production of siderophores, i.e., small metal-binding molecules. Biological control of soil-borne plant pathogens and the synthesis of antibiotics have also been reported in several bacterial species. Another mechanism by which PGPR can inhibit phytopathogens is the production of hydrogen cyanide (HCN) and/or fungal cell-wall degrading enzymes, such as chitinase and β-1,3-glucanase. Direct plant growth promotion includes production of plant hormones such as auxins, cytokinins, gibberellins, ethylene, and abscisic acid. Production of indole-3-ethanol or indole-3-acetic acid (IAA), the compounds belonging to auxins, have been reported for several bacterial genera (Zahir et al. 2004). PGPR also help solubilization of mineral phosphates and other nutrients, enhance resistance to stress, stabilize soil aggregates, and improve soil structure and organic matter content.

Important biochemical reactions of BNF occur mainly through symbiotic association of N_2-fixing microorganisms with legumes that converts atmospheric elemental nitrogen (N_2) into ammonia (NH_3) (Shiferaw et al. 2004). Symbiotic nitrogen-fixing bacteria include the genera *Rhizobium, Bradyrhizobium, Azorhizobium, Allorhizobium, Sinorhizobium*, and *Mesorhizobium*. These bacteria form nodules in roots of forest tree legumes and fix atmospheric nitrogen. Effective nitrogen-fixing leguminous trees of tropical forests include *Acacia angustissima, Acacia auriculiformis, Acacia mangium, Acacia mangium, Albizia lebbeck, Albizia saman, Calliandra calothyrsus, Dalbergia sissoo, Desmodium rensonii, Enterolobium cyclocarpum, Erythrina poeppigiana, Erythrina variegate, Flemingia macrophylla, Gliricidia sepium, Inga edulis, Intsia bijuga, Leucaena diversifolia, Leucaena leucocephala, Paraserianthes falcataria, Pongamia pinnata, Pterocarpus indicus, Sesbania grandiflora, Sesbania sesban*, and *Tephrosia vogelii.* Leguminous trees of the arid and semiarid regions are *Acacia holosericea, Acacia confuse, Albizia guachapele, Albizia lebbeck, Albizia saman, Chamaecytisus palmensis, Cajanus cajan, Dalbergia sissoo, Dalbergia retusa, Enterolobium cyclocarpum, Leucaena leucocephala*, and *Sesbania sesban.* Important nitrogen-fixing leguminous trees of temperate forests are *Acacia angustissima, Albizzia julibrisin, Robnia pseudoacacia, Amorpha fruticosa, Gymnocladus dioica, Laburnum anagyroides*, and *Spartium junceum*. Lupines (genus: *Lupinus*) are tree legumes found in boreal forests. Recently, several non-rhizobial species, belonging to α and β subgroup of Proteobacteria such as *Methylobacterium, Blastobacter, Devosia, Phyllobacterium, Ochrobactrum, Agrobacterium, Cupriavidus, Herbaspirillum, Burkholderia*, and some γ-Proteobacteria have been reported to form nodules and fix nitrogen in legume roots (Balachandar et al. 2007). The actinomycetes from the genus *Frankia* nodulate plants from eight additional plant families (Benson and Silvester 1993). Important tree species that develop nodules with *Frankia* and attain the capacity of fixing nitrogen are *Alnus incana, Alnus cordata, Alnus rugosa, Alnus serrulata, Coriaria arborea, Casuarina cunninghamiana, Ceanothus americanus, Cercocarpus montanus, Elaeagnus angustifolia, Elaeagnus commutata, Elaeagnus multiflora, Elaeagnus umbellate, Myrica cerifera, Myrica pensylvanica, Myrica gale*, and *Shepherdia canadensis*. These are important tree species of the temperate forests.

Cleveland et al. (1999) revealed average rates of N-fixation by symbiotic understory legumes in temperate coniferous forest ecosystems of 7.3 kg N ha^{-1} per year. Depending on the species present, percent cover, and geographical location, this N-fixation rate could range widely between 1 and 160 kg N ha^{-1} per year. *Ceanothus* spp. were reported to add 7–75 kg N ha^{-1} per year in Oregon. Youngberg and Wollum (1976) reported rates on average of 70 kg N ha^{-1} per year. *Lupinus* and *Shepherdia* may fix only about 3 kg N ha^{-1} per year on similar sites in Oregon (Jurgensen et al. 1997). In logged lodgepole pine stands in southern British Columbia, *Lupinus arcticus* was found to fix only around 2 kg N ha^{-1} per year, while *Shepherdia canadensis* fixed just under 1 kg N ha^{-1} per year (Hendrickson and Burgess 1989). *Alnus* spp. have been estimated to fix between 60 and 200 kg N ha^{-1} per year in dense stands.

Coarse woody debris, litter, and soil can provide a carbon source for free-living N-fixing bacteria as they decompose these substrates (Hicks et al. 2003) because the C to N ratio in these substrates is high (Vitousek et al. 2002). In large woody debris in coastal sites in Oregon, Crawford et al. (1997) found N-fixation rates of less than 1 kg N ha^{-1} per year. These rates are similar to those found in forests on Vancouver Island, where N-fixation rates in coarse woody debris were between 1 and 2.1 kg N ha^{-1} per year (Brunner and Kimmins 2003). Forest floor litter in the Pacific Northwest was found to fix trace amounts of N, usually less than 1 kg N ha^{-1} per year. In old-growth Douglas-fir forests in western Montana, Jurgensen et al. (1991) found N-fixation rates of 0.8 kg N ha^{-1} per year in woody residue, forest floor material, and mineral soil. Free-living nitrogen-fixing bacteria or associative nitrogen-fixing bacteria include *Azotobacter*, *Azospirillum*, *Bacillus*, *Clostridium*, *Desulfovibrio*, *Enterobacter*, *Klebsiella*, *Pseudomonas*, *Rhodospirillum*, and the Cyanobacteria *Anabaena* and *Nostoc* (Ishizuka 1992).

Mineralization of nitrogen from soil organic matter is the principal source of nitrogen to plants in most forest ecosystems (Vernimmen et al. 2007). Rates of N mineralization can regulate the productivity of many forests (Pastor et al. 1984; Miller et al. 2003). The rate of ammonification (transformation of organic N to NH_4–N) and nitrification (oxidation of NH_4–N to NO_3–N) play a key role on the availability, mobility, and loss of N in forest soils. Factors affecting nitrification in soils are pH, moisture, temperature, C to N ratio of the litter, the presence of plant-produced allelochemicals, and the supply of essential nutrients, notably phosphorus (Tietema et al. 1992; Nugroho et al. 2007). Studies of nitrification in undisturbed tropical forest soils have reported a wide range of nitrification rates (0.1–6 mg N $g^{-1}d^{-1}$) (Hafkenscheid 2000; Van Dam 2001). Low to very low nitrification and nitrogen mineralization rates have been reported for heath forest soils and in wet montane tropical soils (Hafkenscheid 2000), while high rates have been found in fertile volcanic soils (Marrs et al. 1988) and high to intermediate rates in lowland Oxisols and Ultisols (Vitousek and Matson 1988; Montagnini and Buschbacher 1989). Neill et al. (1997) examined soil inorganic N concentrations, net mineralization and nitrification rates in six forests and 11 pastures 3 years old or older on Ultisols and Oxisols that encompassed a wide variety of soil textures and spanned a 700 km geographical range in the southwestern Brazilian Amazon Basin. Forest soils had higher extractable NO_3–N and total inorganic N concentrations than pasture soils, but substantial NO_3–N occurred in both forest and pasture soils. Rates of net N mineralization and net nitrification were higher in forest soils. Greater concentrations of soil organic matter were associated with greater rates of net N mineralization and net nitrification. Perakis and Sinkhorn (2011) observed surface mineral soil N (0–10 cm) in a temperate forest to range from 0.29 to 0.78% was linearly related to net N mineralization ranging from 8 to 82 kg N ha^{-1} per year. Net N mineralization rate was not related to soil C to N ratio, soil texture, precipitation, and temperature differences. Net nitrification was negatively related to soil pH and accounted for $<20\%$ of net N mineralization at low-N sites, increasing to 85–100% of net N mineralization at intermediate- and high-N sites. Boreal forest soils usually contain large amounts of organically bound nitrogen (Viro 1969), but the rate of mineralization is relatively slow and the amount of mineralized nitrogen is low (Nommik 1982). It has recently been demonstrated that conifers such as *Pinus sylvestris* and *Picea abies* can take up some organic forms of N via mycorrhiza (Nasholm et al. 1998), but their chief form is inorganic. The slow pace of nitrogen mineralization in boreal forest soils is due to low soil pH, low temperature, and poor litter quality. In Norway spruce forest soil (litter, humus, and mineral soil layer to a depth of 50 cm) in Sweden, annual net N mineralization was estimated to be 0.5–5.0% of the total amount of N, 35–105 kg N ha^{-1} (Persson and Wiren 1995). The organic horizons (litter and humus layer) accounted for 32–74% of the annual mineralization. In the humus layer of Finnish coniferous forest soils, the nitrogen in the microbial biomass accounts for about 4–6% of the total amount of nitrogen (Martikainen and Palojarvi 1990; Smolander et al. 1994).

The chief nitrifying organisms are (Robertson and Groffman. 2007)—Autotrophic nitrifiers; Ammonia oxidizers: bacteria of the genera *Nitrosomonas*, *Nitrosolobus*, *Nitroso*, *Nitrosovibrio;* Nitrite oxidizers: bacteria of the genera *Nitrobacter*, *Nitrospira*, *Nitrosococcus*, *Nitrospina;* and Heterotrophic nitrifiers: bacteria such as *Arthrobacter globiformis, Aerobacter aerogenes, Thiosphaera pantotropha, Streptomyces grisens*, *Pseudomonas*. The fungi *Aspergillus flavus* also has the capacity of nitrification.

Denitrification is carried out by facultative anaerobes, predominantly heterotrophic bacteria, the most common being species of the genera *Pseudomonas* and *Alcaligenes* (Focht and Verstraete 1977). Most of denitrifying bacteria require anaerobic conditions, but some species continue to denitrify at varying levels of dissolved oxygen (Lloyd et al. 1987; Jetten et al. 1997).

Soil microflora has important roles in the transformations of sulfur in forest soil as well. Sulfur deficiencies have been found, although less frequently, in forest soils of some regions, especially where atmospheric inputs are low or where there is excess N. Examples of such forests include *Pinus radiata* in Australia and *Pseudotsuga menziesii* (Douglas fir) fertilized with N along the Pacific Coast of North America. Sulfur in soils can be grouped into two broad areas: organic and inorganic forms. Most of the total S pool in forest soils is composed of organic fractions (Mitchell and Fuller 1988), primarily due to leaf litter inputs (Pregitzer et al. 1992). Organic S in soil organic matter occurs in two primary forms: ester sulfates, which have C–O–SO_3 linkages, and carbon-bonded S, which has direct C–S linkages. A few other organic forms also exist, but they are of comparatively minor importance. Ester sulfates include compounds such as choline sulfate, phenolic sulfates, and sulfated polysaccharides. Carbon-bonded S is comprised principally of amino acids, such as methionine and cysteine, and sulfolipids (Neptune et al. 1975). Sulfur mineralization has been described as a microbially driven decomposition; however, mineralization of ester sulfates does not require direct microbial activity (McGill and Cole 1981) due to extracellular enzymes. Organic matter decomposition rates vary with chemical makeup and complexities, and whether net S mineralization occurs depends upon the S supply and microbial demand. If more S is available to the system than needed to meet S requirements of sulfur-using microorganisms, mineralization will occur. If less S is available than needed, all S entering the system will become immobilized because microbial needs must be satisfied first.

Microorganisms also mediate S oxidation and reduction which are key processes in S transformations. Completely mineralized S compounds can be transformed to SO_4^{2-} or H_2S by microbes. Hydrogen sulfide is a common product of mineralization. The reduction step from SO_4^{2-} to H_2S is microbially mediated, primarily by bacterial species in the genera of *Desulfovibrio*, *Desulfotomaculum*, and *Desulfomas* (Kennedy 1986) under anaerobic conditions. Reduction can also occur in anaerobic microniches that are present within a larger aerobic system (Jorgenson 1977). Hydrogen sulfide can be lost through volatilization, join with metal ions to form insoluble sulfide precipitates, or be oxidized to elemental S. Sulfur-oxidizing microorganisms are ubiquitous in most forest soils. Heterotrophic oxidizers, such as *Arthrobacter* and *Pseudomonas*, carry out the majority of S oxidation (Janzen and Bettany 1987) and require organic C to satisfy energy and C needs. In the absence of sufficient organic C, autotrophic microbes, including those in the genus *Thiobacillus*, become more important. They obtain their energy from inorganic S and their C from CO_2. Oxidation by *Thiobacillus* has been reported to occur in soils from pH <2 to pH >9 (Kennedy 1986). Dynamics of oxidation of inorganic sulfur compounds to sulfate by spruce forest soil was investigated by Lettl et al. (1981). Sulfide, sulfite, and thiosulfate are oxidized to sulfate at a maximal rate at the beginning of the reaction, oxidation of elemental sulfur exhibits a lag phase. There was a linear relationship between the amounts of sulfate produced and concentrations of substrates in the soil. They determined the species representation of Thiobacilli in F, H, and A horizons at ten localities of the spruce forest. F horizon exhibited a high oxidative activity, a lower activity was found in horizon H, and the lowest one was detected in horizon A.

5.5 Soil Fauna

Forest soils harbor a large number of fauna including herbivores, carnivores, parasites, or predators. For example, in a single square meter of soil of a European beech forest, as many as 1,000 species of invertebrates may be collected (Schaefer and Schauermann 1990). Richness differs greatly between taxa. Microarthropod communities may comprise hundreds of species with a diversity of up to 400–500 species of Acari and 60–80 species of Collembola. Under comparable conditions, nematodes may comprise up to 90 species, Protozoa 60, Enchytraeidae 22, earthworms 15–17, Diplopoda 15 (Lavelle 1996). In African savannas, termite communities may comprise up to 60 species. Villalobos and Lavelle (1990) collected 113 species of soil Coleoptera. Major members of fauna of the forest soils are as follows:

1. Microfauna: Rotifer, Nematoda, and Protozoa
2. Mesofauna: Acari, Collembolan, Protura, Diplura, Symphyla, Enchytraedae, and Isoptera
3. Macrofauna: Opiliones, Isopoda, Amphipoda, Chilopoda, Diplopoda, Megadrili, Coleoptera, Araneida, and Mollusca

Forest soil organisms occupy different microhabitats for their accommodation, movement, and functions. The microfauna (protozoa and small nematodes) inhabit water films. The mesofauna inhabit air-filled pore spaces; they cannot create spaces for themselves. The macrofauna make their own spaces through their burrowing activities and, like the megafauna, can have large influences on gross soil structure (van Vliet and Hendrix 2003). Some of the faunas alter the physical structure of the soil, influencing rates of nutrient and energy flow. Different organisms perform different functions in forest soil.

The Microfauna Rotifer, Nematoda, and Protozoa represent the microfauna of the soil and litter. Protozoa are the most abundant and they belong to four ecological groups: flagellates, naked amoebae, testate amoebae, and ciliates (Lousier and Bamforth 1990). Flagellates have one or more whiplike propulsive organs known as flagella. They play a significant role in nutrient turnover by their feeding activities, with bacteria as their principal prey items. Naked amoebae are very numerous and active in a wide range of agricultural, grassland, and forested soils. Testate amoebae are less numerous than naked amoebae, except in moist, forested systems where they thrive. They perform functions similar to naked amoebae. Ciliates tend to be restricted to very moist or seasonally moist habitats. They are less numerous than others, with a general range of 10–500/ g of litter or soil. Figure 5.2 shows a testate amoebae (a) and a flagellate protozoa (b).

Microfaunas in the rhizosphere play an important role in release of plant-available nutrients, accumulation, and stabilization of soil organic carbon, hormonal effects on roots, microbial diversity and functional stability, multi-trophic interactions above ground, and bioremediation of contaminated soils. Selective grazing, active dispersal and excretion by the microfauna not only benefit the rhizosphere ecologi-

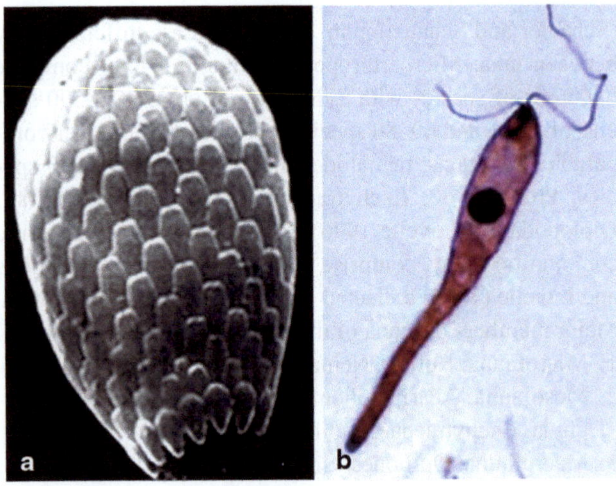

Fig. 5.2 Protozoa (**a**) Testate amoebae and (**b**) flagellate protozoa. (Image courtesy of Dr. Wilhelm Foissner)

cal functions but also have an impact on the whole soil and aboveground community.

The Mesofauna

Rotifera Rotifera are considered to be aquatic organisms and are found in water films and water-filled pores. There are very common rotifers not only in aquatic but also in terrestrial habitats (Donner 1965; Ricci and Balsamo 2000; Fontaneto and Melone 2003). Rotifers are present in all types of soils (Sohlenius 1979; 1982). These rotifers are hydrobionts of semiaquatic living mode and are able to undergo anhydrobiosis to withstand drought (Ricci 2001). The role and importance of rotifers in soil communities have not been entirely appreciated yet, but they probably impact soil bacterial communities through their filtration activity (Pourriot 1979; Donner 1965).

Nematoda Nematodes are non-segmented worms typically 50 µm in diameter and 1 mm in length. An incredible variety of nematodes function at several trophic levels of the soil food web. Some feed on the plants and algae (first trophic level); others are grazers that feed on bacteria and fungi (second trophic level); and some feed on other nematodes (higher trophic levels). Ingham et al. (1985) found up to 70 % of the bacterial and fungal-feeding nematodes in the rhizosphere. Nematodes seek areas of concentrated organic matter.

Tardigrada The Tardigrada are bilaterally symmetrical micrometazoans with four pairs of lobopod legs terminating in claws or sucking disks. They occupy a diversity of niches in marine, freshwater, and terrestrial environments throughout the world. Some have a cosmopolitan distribution, while others are endemic. About 900 species have been described so far (Nelson 2002). In one of the ecological studies for soil-inhabiting targigrades in Japan, Ito (1995) conducted faunal research focused on five forests in the southern part of Kanto District, Central Japan. He showed that the generic composition of targigrades was linked to the status of the forests and pointed out that it could be used as a good indicator of the forests. Fig. 5.3 (a) and (b) shows images of nematode and tardigrada.

Microarthropods Soil microarthropods usually comprise the mite orders Oribatida, Acaridida, Actinedida, Tarsonemida, and Mesostigmata, with occasionally some Metastigmata. Also included are the insect order Collembola and Protura, Diplura and the myriapod order Symphyla. Other arthropod groups are too rare, although many soil insect species fit in the size range of microarthropods. Soil microarthropods are very numerous; species numbers up to 45 per 500 cm^2 have been found in Central European forests (Moritz 1965) and up to 26 per 5 cm^2 in Finnish forests (Karppinen 1958) for oribatid units only. Total numbers of microarthropod species must be even higher. Such a group

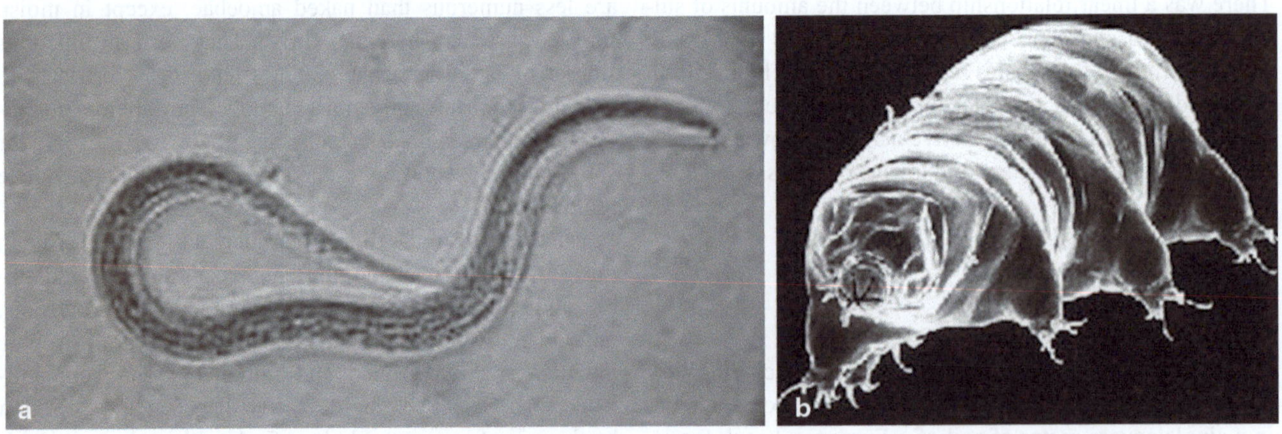

Fig. 5.3 (**a**) Nematode and (**b**) Tardigrada. (Image courtesy of University of Minnesota)

Fig. 5.4 Mite. (Courtesy of Tom Muray)

Fig. 5.5 Protura. (Courtesy of Dr. Karen Meusemann)

Collembola Collembola are also called "springtails." They are wingless soft-bodied hexapods that are usually between 1 and 3 millimeters in length and occur in varying habits, such as soil surface and litter, under rocks or the bark of trees. Collembolans occur in great numbers throughout the upper layer of forest soils, where their major food is fungi associated with decaying vegetation. They are often the most numerous of the microarthropods in the rhizosphere. They have varied food habits, with species being detritivores, microbivores, coprovores, and saprovores. Collembola together with other soil arthropods such as acari constitute an important component of soil mesofauna in almost all forest ecosystems and are indispensable in decomposition of organic matter and efficient nutrient cycling.

Protura Protura are wingless soil microarthropods—usually 0.2–2.0 mm in length. Their body is divided into a head, a thorax with three pairs of legs, and a segmented abdomen. Eyes or antennae are absent and the head is quite small and narrow. Protura are white, but the end of the abdomen and the tips of the forelegs are often light yellow. Protura are one of the oldest groups of terrestrial arthropods. They are distributed in all terrestrial regions and climate zones, except the Arctic and Antarctica. Proturans are found almost universally in localities where decaying organic matter and sufficient moisture are present. They occur abundantly in moist woodland humus, among dead leaves, in moldering timber and in soil. They are most abundant in the topsoil, especially the moist humus layer (Nosek 1977). They can be found as deep as 50 cm under pasture soil (Haarlov 1960). Proturans occur frequently in nests of soil-dwelling small mammals, such as voles and moles (Nosek and Vysotskaya 1976). Figures 5.4 and 5.5 show images of a mite and a protura, respectively.

Fig. 5.6 Diplura. (Courtesy of John of Vladimir Motycka)

Diplura Diplurans are small (usually less than 10 mm; rarely 40–50 mm) slender creatures with relatively long antennae with two pronged tails and six legs (Fig. 5.6). Diplurans have strong legs and flexible bodies suited for pushing through soil. The head is globose with long, multisegmented antennae and no eyes; the mouthparts are adapted to gnawing. Diplura inhabit the soil, the forest floor, anthills, or rotten wood. The majority are predators or feed on animal remains. More than 200 species of Diplura are known; they are divided into three families—Campodeidae, Japygidae, and Projapygidae. Diplurans principally live in soil but can also be found in compost heaps, decaying vegetation and under wood and stones. Some species live in caves and others deep in soil. In the food chain, diplurans are predacious and feed on tiny creatures such as arthropods. Diplura occur in tropical and temperate soils in low densities. They are predators on mites and other small arthropods but also ingest fungal mycelia and detritus.

Pseudoscorpionida Pseudoscorpions are small (typically less than 5 mm in length) predacious arachnids, with a general similarity to true scorpions, but without the tail and stingers (Fig. 5.7). They are generally light tan to reddish brown to black in color and are typically compressed dorsoventrally. They have two main body parts, divided into an anterior prosoma or cephalothorax and a posterior opistho-

Fig. 5.7 Pseudoscorpionida. (Courtesy Meyer of Vladimir Motycka)

Fig. 5.8 Symphyla. (Courtesy of Washington State University)

soma or abdomen. The dorsal surface of the cephalothorax shows little signs of segmentation, whereas the abdomen is divided into 11 or 12 clearly defined segments. Pseudoscorpions are cryptic animals, living amongst leaf litter, under rocks, within compost piles, in the forest floor, under bark and within decaying wood, in caves, and in various vertebrate nests.

Symphyla Symphylids are small, white, eyeless, elongate, many legged invertebrates that resemble tiny centipedes (Fig. 5.8). Symphylans are about ¼ inch long. They differ from centipedes in several characteristics, but superficially, symphylids have 12 body segments and 12 pairs of legs, whereas centipedes have at least 15 pairs of legs, the first pair modified as fangs. Some species are pests feeding on roots of seedlings. Symphylans thrive in very moist soils high in organic matter and live on other insects.

Pauropoda Pauropods are small (less than 5 mm), pale, terrestrial arthropods. Adult pauropods have 11 (or sometimes

Fig. 5.9 Pauropoda. (Courtesy of Tom Muray)

Fig. 5.10 Enchytraeidae. (Courtesy of Walter Pfliegler)

12) body segments and 9 (or sometimes 10 or 11) pairs of legs (Fig. 5.9). They also possess unique forked antennae and a distinctive locomotory pattern characterized by rapid burst of movement and frequent abrupt changes in direction. Most pauropods lack eyes and a tracheal system. Pauropods can be found in soil, decaying wood, leaf litter, forest floor, and other moist places, where they feed on fungi and decaying organic matter. Over 500 species of pauropods have been described so far.

Enchytraeidae One group of the Oligochaeta is the Enchytraeidae, also known as pot worms and/or white earthworms due to their pale color and small size (Fig. 5.10). The Enchytraeidae family contains 27 genera and about 500 species; however, new species are still being described each year (Chalupsky 1991). Most species are hermaphrodites but also parthenogenetic species and a few species reproducing by fragmentation (fissiparous), such as *Cognettia sphagnetorum* and *Enchytraeus bigeminus,* exist (Christensen 1964). Enchytraeids are widely distributed from the Arctic to tropi-

Fig. 5.11 Isopoda. (Courtesy of Dr. David A Kendall)

Fig. 5.12 Pill Millipede. (Diplopoda; courtesy of Martin Wolf)

cal areas (Nurminen 1965; Petersen and Luxton 1982; Standen 1988). Different species also have specific preferences of habitat, and enchytraeids can be divided into euedaphic (soil-living species), like *Achaeta* sp. and *Cernosvitoviella* sp., which are, generally, found deeper in the soil compared to epedaphic species, like *Mesenchytraeus glandulosus* and *Cognettia sphagnetorum,* which live in the topsoil but can also, occasionally, be found above the soil (Abrahamsen 1972; Dozsa–Farkas 1992). Enchytraeids feed on various items but are, generally, said to be 80% microbivorous and 20% saprovorous (Didden 1993). Enchytraeids are involved both directly and indirectly in the decomposition processes and the nutrient mineralization (Williams and Griffiths 1989).

The Macrofauna The importance of macrofauna to the promotion of soil fertility has been stressed in some reviews (Fragoso et al. 1993; Lavelle et al. 1997; Garnier–Sillam and Harry 1995; Nash and Whitford 1995; Brussaard and Jumas 1996; Wood 1996). The distribution, protection, and stabilization of organic matter, the genesis of soil structure (macroaggregates), humification, the release of immobilized N and P, the improvement of drainage and aeration, and the increase in exchangeable cations have all been demonstrated in soils modified by termites and earthworms (Mulongoy and Bedoret 1989; Lavelle et al. 1998). Soil ants and other macrofauna represent predators, herbivores, and bioturbators, bringing about important changes in the physical and chemical properties of soils, as well as dispersing plant propagules. Networks of galleries and chambers increase the porosity of the soil, increasing drainage and aeration (Cherrett 1989) and reducing bulk density (Baxter and Hole 1967).

Macroarthropods Macroarthropods are a group of larger insects, spiders, myriapods, and others. Their typical body lengths range from 10 mm to 15 cm (Shelley 2002). Many macroarthropods are cryptozoans; they dwell beneath stones, logs, under bark, or in cracks and crevices. Macroarthropods have direct effects on soil structure, porosity, aeration, and water movement. Termites and ants are important movers and mixers of soil; they bring deeper soil to the surface and on top of the litter layer. Emerging nymphal stages of cicadas may disturb soil structure. Scarabaeid beetles larvae sometimes churn the soil in grasslands. Many macroarthropods are temporary soil residents. Macroarthropods may have a major influence on the microarthropod community. For example, collembola are important food items for spiders.

Isopoda Isopods are relatively small crustaceans with seven pairs of legs of similar size and form, ranging in size from 300 μm to nearly 50 cm (Fig. 5.11). They are typically flattened dorsoventrally. Soil isopods are crustaceans; they occur under rocks and in similar habitats. They are generally saprovores. They can also feed upon roots or foliage of seedlings. Isopods possess heavy, sclerotized mandibles and are capable of fragmentation of decaying plant residues. In the laboratory, terrestrial isopods feed upon fecal pellets dropped by themselves or by any other isopods.

Diplopoda The Diplopoda (millipedes) (Fig. 5.12) is the third largest class of terrestrial Arthropoda with more than 12,000 described species in 145 families and 16 orders (Adis 2002; Shelley 2002, Sierwald and Bond 2007, Shelley 2007). A major component of terrestrial ecosystems throughout the temperate, subtropical, and tropical zones of the world, they are ecologically important as detritivores and are important biogeographical indicators because of their profound diversity and geological age as well as low vagility (Hopkin and Read 1992). Millipedes, whose population density in forest soil sometimes exceeds 1,000 individuals per square meter, are estimated to consume 10–15% of the annual leaf fall in temperate forest. They can be important in calcium cycling. Fig. 5.13 shows a cylindrical diplopoda.

Chilopoda Chilopoda, or centipedes, have an elongated, flattened multi-segmented body, many legs, and a wriggling running gait (Fig. 5.14). They are generally 2–5 cm in size,

Fig. 5.13 Cylidrical Millipede. (Diplopoda; courtesy Martin Wolf)

Fig. 5.14 Centipede. (Chilopoda; courtesy of Phil Bendle)

Fig. 5.15 Scorpionida. (Courtesy of Cris Huh)

although some tropical forms reach more than 20 cm. Each segment of the body bears a single pair of walking legs; the total number of segments and pairs of legs can vary from 15 to more than a 100. The segments of the body are covered with lightly sclerotized, leathery plates, connected to one another by the soft pliable cuticle. The last legs differ in structure and function; often they are modified as genital appendages (gonopodes). Centipedes are predators, able to tackle relatively large and active prey, grasping and paralyzing their victims with their poison claws. Centipedes attack a large variety of invertebrate prey—other soil arthropods, earthworms, mollusks, etc. Predatory ground beetles and large spiders may hunt centipedes, particularly the juveniles. Lithobiids are the common brown, flat centipedes of litter in hardwood forests. Centipedes attack a large variety of invertebrate prey—other soil arthropods, earthworms, mollusks, etc.

Scorpionida The scorpions (Arachnid) have long legs, segmented, stinger-bearing abdomen, and chelate palpi (Fig. 5.15). They are inhabitants of warm, dry, tropical, and temperate regions. Scorpions commonly occur in deserts but there are also many other found in other habitats, including grasslands and savannahs, deciduous forests, montane pine forests, rain forest, and caves. Scorpions are distinctive, predatory arthropods that are extremely hardy and adaptable creatures. Scorpions vary enormously in length, from around 1 cm to around 20 cm. They are nocturnal creatures; during the day, they burrow under the soil to rest. All scorpions are venomous, the venom being used mainly to kill or paralyze prey. The venom is a mixture of compounds, each of which has a different effect depending on lifestyle and prey species. The venom is stored in glandular sacs and released in quantities regulated by the scorpion. Of all the known species of scorpions, only 30–40 have venom that is strong enough to harm a person and all of these belong to the family Buthidae.

Araneae (Spiders) Spiders, belonging to the class Arachnida, are a diverse group, with much variation in size, body form, and coloration (Fig. 5.16). Most ground-dwelling spiders are of various shades of back, gray, or brown, but a few are brightly colored. The body of a spider consists of two major parts—the anterior cephalothorax and the posterior abdomen—which are connected by a narrow waist. The cephalothorax, which is a fused head and thorax, is covered by dorsal and ventral sclerotized plates; the abdomen is much softer and more flexible by comparison. At the posterior end of the abdomen are the spinnerets which the spider uses to exude and spin threads of silk. The cephalothorax bears the sensory organs, mouthparts, and the four pairs of walking legs. Spiders possess up to eight simple eyes or ocelli, each

Fig. 5.16 Spider. (Araneae)

of which consists of a single lens. Ground-dwelling spiders can be found in a variety of habitats from coastal sand dunes, to pastures, to native forests. Some of the small litter-inhabiting spiders could be considered microarthropods.

Opiliones Opiliones are delicate, shy and the largest arachnids in woodlands. Their bodies are small but their legs are unusually long which suggest that their habitat is litter surface or exposed areas. Some species occur high in foliage, others in subcanopy, some on soil surface, and some in litter layers. They are mainly active during the night, when the native forest species can be observed in the open on the surface of trees and on the forest floor. The short-legged Laniatores are common among damp leaf litter, under the loose bark of rotting logs on the ground, in moss, and in similar habitats. Many species of harvestmen are cave dwellers.

The Pterygote insects Many winged insects (Pterygota) are residents of soils. Some are permanent soil inhabitants spending whole of their life in or on soil. Immature stages of other species are true soil dwellers but their adults are flying insects. All major winged insect orders—the Coleoptera (beetles), Lepidoptera (butterflies and moths), Hymenoptera (bees, wasps and ants), and Diptera (flies)—include soil-dwelling species. Termites belong to Isoptera and they are saprophages. The Homoptera (aphids, cicadas), Orthoptera

Fig. 5.17 Coleoptera.. (Courtesy of Jim McClarin)

(grasshoppers and crickets), and minor orders such as the Dermaptera (earwigs) contain soil-dwelling species. Of 26 pterygote insect orders, all but seven contain at least some soil-dwelling species.

Coleoptera Coleoptera (beetles) is the largest order in the class Insecta. Adult beetles have a hard, dense exoskeleton that covers and protects most of their body surface (Fig. 5.17). The front wings, known as elytra, are just as hard as the rest of the exoskeleton. At rest, both elytra meet along the middle of the back, forming a straight line that is probably the most distinctive characteristics of the order. Beetles have wide feeding habits and inhabit all terrestrial environments. Many species are herbivores, feeding on roots, stems, leaves, or reproductive structures of plants. Some species live on fungi, others burrow into plant tissues, and still others excavate tunnels in wood or under bark. Many beetles are predators. They live in the soil or on vegetation and attack a wide variety of invertebrate hosts. Some beetles are scavengers, feeding primarily on carrion, fecal material, decaying wood, or other dead organic matter.

Hymenoptera One of the largest orders of insects is the Hymenoptera. They have two groups of great importance: the ants and the ground-dwelling wasps. The ants are the most significant family of soil insects for their large influence on soil structure. Some bees and wasps nest in soil and also have some impacts on soils. Ants are widely distributed, numerous, and diverse.

Diptera Many of the true flies are soil dwellers in some stages of their life cycles. Many species that live in aboveground habitats pupate in the soil. Many species of fly larvae are important saprovores in soils. They are restricted to moist soils rich in organic matter. Fly larvae have a major impact on decomposition rates of carrion. Maggots of various types hasten the decomposition rate significantly.

Isoptera The Isoptera (termites) are among the most important of forest soil fauna for their impact on soil structure and

Fig. 5.18 Termites in a subterranean colony. (Courtesy of ORKIN)

Fig. 5.19 Gastropod. (Courtesy of Ryan Durand)

on decomposition of detritus. Some termites possess a gut flora of protozoa, which enable them to digest cellulose. Their normal food is wood that has come into contact with soil. Most species of termites construct spectacular mounds. Termitidae do not have protozoan symbionts but possess an array of microbial symbionts (bacteria and fungi) that enable them to digest the humified organic matter in tropical soils. There are wood-feeding, plant-feeding, and humus-feeding termites (Fig. 5.18).

Other Pterygota The Orthoptera, grasshoppers and crickets, lay eggs in soils and some are active on the soil surface. The Psocoptera, psocids, are a small order of insects that occasionally become abundant in leaf litter. They feed on organic detritus, algae, lichens, and fungus. The order Homoptera, cicadas, aphids, and others, has members important as belowground herbivores and as soil movers.

Gastropoda The Gastropoda or gastropods, more commonly known as snails and slugs, are a large taxonomic class within the phylum Mollusca (Fig. 5.19). They can be microscopic to large in size. Soil gastropods are typically herbivores, fungivores, and detritivores (Burch and Pearce 1990) that exhibit weak levels of intraspecific competition (Barker and Mayhill 1999). Snails and slugs are essential components of forest ecosystems. They decompose forest litter, recycle nutrients, build soils, and provide food and calcium for birds, amphibians, reptiles, small mammals, and invertebrates. As land snail communities can consume less than 0.5 % of annual litter input per year (Mason 1970), some speculate that few resources, beyond $CaCO_3$ (Boycott 1934) and appropriate resting site availability (Pearce 1997), will limit their distribution. At landscape scales, land snail population size and faunistic composition have been suggested to vary with habitat and vegetation types (Nekola 2002). Abele (2010) found that in boreal forests, gastropod assemblages were influenced by canopy composition, with most gastropods of the mixedwood showing a strong affinity for broadleaf dominated forests. Tree species mixture influenced gastropod distribution; basal tree area of either conifer or broadleaf trees was generally associated with gastropod distribution within a stand.

Oligochaeta—earthworms The most familiar and often the most important of the soil fauna with respect to soil processes are the earthworms. They fragment plant residues, bury them in, and mix them well with the soil. More than 3,500 earthworm species have been identified so far. Earthworms are classified within the phylum Annelida, class Oligochaeta, and order Opisthophora. Ten of the 16 families consist of the terrestrial forms commonly known as earthworms. Species within the families Lumbricidae and Megascolecidae are ecologically important.

Earthworms are soft-bodied, segmented animals (Fig. 5.20). In length, they range from a few millimeters to more than a meter. They consist of a simple, tube-within-a-tube body plan, the outer tube constituting the body proper and the internal tube comprising the alimentary canal. Soil material is ingested and drawn through the mouth into a muscular buccal cavity and then through the pharynx into the esophagus. Many species have a muscular esophageal gizzard that grinds and mixes food material as it passes through. Earthworms are called ecosystem engineers because they have pronounced effects on soil structure, aeration, and water movement by their burrowing activities as well as their ingestion of soil and production of castings (Vliet and Hendrix 2003). Casts are produced after earthworms ingest mineral soil and particulate organic matter, mix them together, and enrich them with organic secretions in the gut, and then egests the material as slurry or as discrete fecal pellets within or upon the soil. Turnover rates of soil through earthworm casting range from 40–70 t ha^{-1} per year in temperate grasslands to 500–1,000 t ha^{-1} per year in tropical savannas. Earthworm burrowing in soil creates macropores of various sizes, depths, lengths, and orientations, depending on their species and soil

Fig. 5.20 Earthworm. (Courtesy of Bil Leonard)

type. Continuous macropores resulting from earthworm burrowing may enhance water infiltration by functioning as bypass flow pathways through saturated soils. Earthworms affect ecosystem structure and function directly by ingesting, altering, and mixing organic matter and mineral soil (Lavelle et al. 1998), thereby changing soil structure, chemistry, and biology (Devliegher and Verstraete 1997). Exotic earthworm invasions have been associated with decreasing forest-floor thickness and the development of thick A horizons (Bohlen et al. 2004).

Earthworms may have some detrimental effects. These include removing and burying of surface residues that would otherwise protect soil surfaces from erosion; fresh casts increase erosion and surface sealing; transmit plant or animal pathogens; increase losses of soil nitrogen through leaching and denitrification; and increase soil carbon loss through enhanced microbial respiration.

5.5.1 Macrofauna in Soils of Different Types of Forests

The abundance, biomass, vertical distribution, and taxonomic composition of soil invertebrates (springtails, macrofauna, and termites) were studied by Anichkin et al. (2007) in some tropical forests differing in edaphic and climatic conditions: lowland forests dominated by *Lagerstroemia* spp. or *Dipterocarpus* spp. in the Cat Tien National Park and in a mountain pine (*Pinus kesiya*) forest on the Da Lat Plateau, southern Vietnam. In the lowland forests, springtails had a relatively low density (10,000–12,000 individual per square meter), but their diversity was high (41–43 species in each forest). The density of large soil invertebrates reached 500–700 individual per square meter at a biomass of approximately 30 g m^{-2} (with earthworms accounting for up to 230 individual per square meter and 19–28 g m^{-2}). Among termites, species of the genera *Macrotermes* and *Odontotermes* were dominant. Their total biomass in some areas exceeded 15–20 g m^{-2}. In the mountain pine forest, the total biomass of soil macrofauna was approximately 11 g m^{-2}, the abundance and diversity of springtails were low (7,500 individual per square meter, 28 species), and wood-destroying species of the genera *Schedorhinotermes* sp. and *Coptotermes* sp. dominated among termites. In temperate forests, woodlice (Isopoda) and millipedes (Diplopoda) are the most abundant macrofauna and they have been found to play a significant role in decomposition and nitrogen release (Irmler 2000; David, and Handa 2010). Woodlice and millipedes provide important indirect effects through the breakdown of leaf litter into smaller particles, mixing of the leaf litter into a homogenous state, and moving it to more favorable microclimate conditions (Sutton 1980; Hassall et al. 1987; David and Handa 2010). Collembola (Springtails) and symphylans are two small invertebrates found in the leaf litter on the forest floor of temperate forests. Better known are earthworms, which constantly burrow through the decaying material and move organic material down into the mineral soil, and bring mineral soil to the surface. This mixing, combined with the aerating effect of the earthworm burrows, makes forest soils much richer. The species richness and total biomass of soil organisms are significantly lower in the boreal forest than at lower latitudes. Dominant soil organisms are protozoa, nematodes, rotifers, and tardigrades. These organisms live primarily in soil water film and soil pore water. The soil fauna of the boreal forest region is distinctive because it generally lacks large invertebrates such as millipedes, isopods (springtails), and earthworms, especially in the middle and northern taiga.

5.5.2 Functions of Macrofauna

Soil macrofaunas are an important component of the biodiversity of forest ecosystems. They include ants, earthworms, termites, amphipods, centipedes, millipedes, snails, slugs, and to a small extent mole rats. Soil macrofaunas have a role in nutrient release and physical parameters, e.g., bulk density, porosity, and water availability (Lee and Foster 1991; Lavelle et al. 1992; Brussard et al. 1993; Tinzara and Tukahirwa 1995; Black and Okwakol 1997; Beare et al. 1997). Certain groups of macrofauna, particularly termites and earthworms, can substantially modify soil structure through formation of macropores and aggregates (Lee and Foster 1991). The influence of soil fauna on soil structural properties has been considered to be the best long-term indicator of soil quality (Linden et al. 1994). Despite their role in maintenance of structure and function of the belowground ecosystems, their importance is often overlooked (Crossley et al. 1992). The principal functions of soil macrofauna in forest soils are described subsequently.

Soil Mixing Ants, termites, earthworms, and ground beetles can move huge quantities of soil, bringing back to the surface mineral matters from deeper horizons and burying the organic matter from the surface horizons, from litter and from excrements. For example, a large nest of *Atta* ants in a tropical forest comprises several million individuals. It forms a cavity in soil with numerous chambers. The excavated earth is deposited on the soil surface surrounding the nest. The macromixing activity of earthworms is of major importance to soils. It can be measured by the quantity of casts found on the soil surface. Earthworms can produce casts amounting to 40–250 t ha^{-1} per year. Some can produce up to 2,500 t ha^{-1} per year. Some beetles (especially those of the subfamily Scarabeidae) are coprophagous—they are very efficient at incorporating and removing excrements that are on the soil surface. For example, just a couple of *Heliocopris dilloni*, a large African species, can bury a piece of dung in one night (Waterhouse 1974). Some soil macrofaunas influence soil structure in a less spectacular way, but the micromixing that they realize is as important as macromixing. These organisms, mainly represented by Diptera larvae, have a more limited capacity to dig the soil. They stay on the soil surface where they realize a fundamental task for the incorporation of organic matter to soil.

Gallery construction Gallery (burrow) formation is very important for soil aeration and water flux. For example, earthworms and termites develop networks of galleries that improve large spaces in the soil macroporosity by 20–100 % (Edwards and Bohlen 1996). These galleries are denser in the top 40 cm and can represent up to 3 % of the total soil volume. In these conditions, the water-holding capacity of soil can increase by 80 % and water flux can be from four to ten times faster.

Litter fragmentation The fragmentation of dead wood (high lignin material), carcass, and litter is one of the most important activities of soil fauna. It has a major effect on organic matter evolution in soil, conditioning the activity of bacteria, fungi, and microfauna populations. Fragmentation is performed by phytosaprophagous animals (i.e., animals feeding on decayed plant material and dead animals).

Aggregation Earthworms, termites, millipedes, centipedes, and woodlice ingest soil particles with their food and contribute to aggregate formation by mixing organic and mineral matter in their gut. Food and Agriculture Organization of the United Nations (FAO) (2008) summarized the functions of soil macrofauna as:

Maintenance of soil structure: Bioturbating invertebrates and plant roots

Regulation of hydrological processes: Bioturbating invertebrates and plant roots

Nutrient cycling: Soil and litter-feeding invertebrates

Decomposition of organic matter: Saprophytic and litter-feeding invertebrate detritivores

Suppression of pests and parasites: Nematodes, earthworms, and various predators

Study Questions

1. Mention important forest soil flora and fauna. Explain roots as living components of forest ecosystems. Describe functions of mycorrhiza.
2. Give a brief account of forest soil flora. Narrate functions of bacteria and fungi in a forest soil.
3. Classify forest soil fauna on the basis of body width. Explain functions of soil microfauna.
4. What are mesofaunas? Discuss mesofaunas of the forest soils and their functions.
5. Describe macrofaunas of different forest ecosystem and explain their functions.

References

Abele SE (2010) Gastropod diversity in the boreal mixedwood forest of northern Alberta-variation among forest types and response to partial harvesting. MS Thesis, University of Alberta, Edmonton

Abrahamsen G (1972) Ecological study of Enchytraeidae (Oligochaeta) in Norwegian coniferous forest soils. Pedobiologia 12:26–82

Adis J (2002) Taxonomical classification and biodiversity. In: Adis J (ed) Amazonian Arachnida and Myriapoda. Pensoft Publishers, Sofia

Alexander M (1977) Introduction to soil microbiology 2nd edn. Academic Press, New York

Anichkin AE, Belyaeva NV, Dovgobrod IG, Shveenkova YB, Tiunov AV (2007) Soil microarthropods and macrofauna in monsoon tropical forests of Cat Tien and Bi Dup-Nui Ba National parks, southern Vietnam. Biology bulletin 34(5):498

Balachandar D, Raja P, Kumar K, Sundaram SP (2007) Non-rhizobial nodulation in legumes. Biotechnology and molecular biology review 2(2):49–57

Barker GM, Mayhill PC (1999) Patterns of diversity and habitat relationships in terrestrial mollusc communities of the Pukeamaru ecological district, northeastern New Zealand. J Biogeogr 26:215–238

Baudoin E, Benizri E, Guckert A (2002) Impact of growth stages on bacterial community structure along corn (maize) roots by metabolic and genetic fingerprinting. Appl Soil Ecol 19:135–145

Baxter PF, Hole H (1967) Ant (*Formica cinerea*) pedoturbation in a prairie soil. Soil science society of America proceedings 31:425–428

Beare MH, Reddy MV, Tian G, Scrivasta SC (1997) Agricultural intensification, soil biodiversity and agroecosystem function in the tropics: the role of decomposer biota. Applied Soil Biology 6:87–108

Benizri E, Dedourge O, Di Battista-Leboeuf C, Nguyen CS, Piutti GA (2002) Effect of maize rhizodeposits on soil microbial community structure. Appl Soil Ecol 21:261–265

Benson DR, Silvester WB (1993) Biology of Frankia strains, actinomycete symbionts of actinorhizal plants. Microbiol Rev 57:293–319

Black HIJ, Okwakol MJN (1997) Agricultural intensification, soil biodiversity and agroecosystem function in the tropics: the role of decomposer biota. Applied Soil Biology 6:37–53

Bohlen PJ, Pelletier D, Groffman PM, Fahey TJ, Fisk MC (2004) Influence of earthworm invasion on redistribution and retention of soil carbon and nitrogen in northern temperate forests. Ecosystems 7:13–27

Boycott AE (1934) The habitats of land mollusca in Britain. J Ecol 22:1–38

Brunner A, Kimmins JP (2003) Nitrogen-fixation in coarse woody debris of *Thuja plicata* and *Tsuga heterophylla* forests on northern Vancouver Island. Can J For Res 33:1670–1682

Brussaard L, Jumas NG (1996) Organisms and humus in soils. In: Piccolo A (ed) Humic substances in terrestrial ecosystems. Elsevier, Amsterdam

Brussard L, Hauser S, Tian G (1993) Soil fauna activity in relation to the sustainability of agricultural systems in the humid tropics. In: Mulongoy K, Merckx R (eds) Soil organic matter dynamics and sustainability of tropical agriculture. A Wiley-Sayce co-publication

Burch JB, Pearce TA (1990) Terrestrial Gastropoda. Soil Biology Guide. Wiley , New York

Chalupsky J (1991) Comprehensive guide to the enchytraeid taxonomy. The Swedish University of Agricultural Sciences, Uppsala

Cherrett JM (1989) Leaf-cutting ants: biogeographical, and ecological studies. In: Leith H, Werger MJ (eds) Ecosystems of the World: Tropical Rainforest Ecosystem. Elsevier, New York

Christensen B (1964) Regeneration of a new anterior end in *Enchytraeus bigeminus* (Enchytraeidae, Oligochaeta). Videnskabelige Meddelelser fra Dansk Naturhistorisk Forening i Kjøbenhavn

Clark FE (1949) Soil microorganisms and plant roots. Adv Agron 1:241–288

Cleveland CC, Townsend AR, Schimel DS, Fischer H, Howarth RW, Hedin LO, Perakis SS, Latty EF, Fisher JCV, Elseroad A, Wasson MF (1999) Global patterns of terrestrial biological nitrogen (N_2) fixation in natural ecosystems. Global Biogeochem Cycles 13:623–645

Coleman DC, Crossley DA (1996) Fundamentals of soil ecology. Academic Press, London

Crawford RH, Li CY, Floyd M (1997) Nitrogen-fixation in root-colonized large woody residue of Oregon coastal forests. For Ecol Manag 92:229–234

Crossley DA, Mueller BR, Perdue JC (1992) Biodiversity of microarthropod in agricultural soils: relations to processes. Agriculture. Ecosystems Environment 40:37–46

David JF, Handa IT (2010) The ecology of saprophagous macroarthropods (millipedes, woodlice) in the context of global change. Biological Reviews 85:881–895

Devliegher W, Verstraete W (1997) The effect of Lumbricus terrestris on soil in relation to plant growth: effects of nutrient enrichment processes (NEP) and gut-associated processes (GAP). Soil Biol Biochem 29:341–346

Didden WAM (1993) Ecology of terrestrial Enchytraeidae. Pedobiologia 37:2–29

Donner J (1965) Ordnung Bdelloidea (Rotatoria). Akademie Verlag, Berlin

Dozsa-Farkas K (1977) Beobachtungen über die Trockenheitstoleranze von *Fridericia galba* (Oligochaets, Enchytraeidea). Opuscula Zoologica Budapest 14:77–83

Edwards CA, Bohlen PJ (1996) Biology and ecology of earthworms. 3rd edn. Chapman and Hall, London

FAO (2008) Soil macrofauna field manual technical level. Food and agriculture organization of the United Nations, Rome

Feldman LJ (1988) The habits of roots. BioScience 38(9):612–618

Focht DD, Verstraete W (1977) Biochemical ecology of nitrification and denitrification. In: Alexander M (ed) Advances in Microbial Ecology. Plenum Press, New York

Fontaneto D, Melone G (2003) On some rotifers new for the Italian fauna. Ital J Zool 70:253–259

Foth HD (1990) Fundamentals of soil science, 8th edn. Wiley, New York

Fragoso C, Barois I, Gonzales C, Arteaga C, Patron JC (1993) Relationship between earthworms and soil organic matter levels in natural and managed ecosystems in the Mexican tropics. In: Mulongoy K, Mecchx R (eds) Soil organic matter dynamics and sustainability of tropical agriculture. Wiley, Chichester, UK

Garnier-Sillam E, Harry M (1995) Distribution of humic compounds in mounds of some soil-feeding termite species of tropical rainforests: its influence on soil structure stability. Insect Soc 42:167–185

Giri B, Giang PH, Kumari R, Sachdev RM, Garg AP, Oelmuller R, Varma A (2005) Mycorrhizosphere: strategies and functions. In: Buscot F, Varma A (eds) Soil biology, Volume 3. microorganisms in soils: roles in genesis and functions. Springer-Verlag, Berlin

Guo JH, Qi HY, Guo YH, Ge HL, Gong LY, Zhang LX (2004) Biocontrol of tomato wilt by plant growth promoting rhizobacteria. Biol Control 29:66–72

Gupta R, Mukerji KG (2002) Root exudate-biology. In: Mukerji KG, Manaharachary C, Chamola BP (eds) Techniques in mychorrhizal studies. Kluwer Academic Publishers, Dordrecht

Haarløv N (1960): Microarthropods from Danish soils: ecology, phenology. Oikos Supplement 3:1–176

Hafkenscheid RLLJ (2000) Hydrology and biochemistry of tropical montane rainforests of contrasting stature in the blue mountains, Jamaica. Ph.D. Thesis, Faculty of Earth sciences, Vrije Universiteit, Amsterdam

Han HS Lee KD (2005) Plant growth promoting rhizobacteria effect on antioxidant status, photosynthesis, mineral uptake and growth of lettuce under soil salinity. Res J Agric Biol Sci 1(3):210–215

Hassall M, Turner JG, Rands MRW (1987) Effects of terrestrial isopods on the decomposition of woodland leaf litter. Oecologia 72:597–604

Hawksworth DL (1991) The fungal dimension of diversity: magnitude, significance, and conservation. Mycol Res 95:641–655

Hendrickson OQ, Burgess D (1989) Nitrogen-fixing plants in a cutover lodgepole pine stand of southern British Columbia. Can J For Res 19:936–939

Hicks WT, Harmon ME, Myrold DD (2003) Substrate controls on nitrogen fixation and respiration in woody debris from the Pacific Northwest, USA. Forest Ecol Manage 176:25–36

Hopkin SP, Read HJ (1992) The biology of millipedes. Oxford University Press, Oxford

Ingham RE, Trofymow JA, Ingham ER, Coleman DC (1985) Interactions of bacteria, fungi, and their nematode grazers: effects on nutrient cycling and plant growth. Ecol Monogr 55:119–140

Ingham ER (1995) Soil organisms, bacteria, fungi, protozoa, nematodes and rotifers. Interior Columbia Basin Ecosystem Management Project Report

Irmler U (2000) Changes in the fauna and its contribution to mass loss and N release during leaf litter decomposition in two deciduous forests. Pedobiologia 44:105–118

Ishizuka J (1992) Trends in biological nitrogen jixation research and application. Plant Soil 141:197–209

Ito M (1995) Taxonomic study on the Eutardigrada from the northern slope of Mt. Fuji, central Japan, II. Family Hypsibiidae. Proc Japan Soc Syst Zool 53:18–39

Janzen HH, Bettany JR (1987) Measurement of sulfur oxidation in soils. Soil Sci 143:444–452

Jetten MSM, Logemann S, Muyzer G, Robertson LA, de Vries S , van Loosdrecht MCM, Kuenen JG (1997) Novel principles in the microbial conversion of nitrogen compounds. Antonie Van Leeuwenhoek 71:75–93

Jhonson C (2009) Biology of soil science. Oxford Book Company, Jaipur, India

Jorgenson BB (1977) Bacterial sulfate reduction within reduced microniches of oxidized marine sediments. Mar Biol 41:7–17

Jurgensen MF, Tonn JR, Graham RT, Harvey AE, Geier-Hayes K (1991) Nitrogen-fixation in forest soils of the Inland Northwest. Proceedings-Management and Productivity of Western-Montane Forest Soils USDA FS Gen Tech Rep INT–280

Jurgensen MF, Harvey AE, Graham RT, Page-Dumroese DS, Tonn JR, Larsen MJ, Jain TB (1997) Impacts of timber harvesting on soil organic matter, nitrogen, productivity, and health of Inland Northwest forests. Forest Science 43:234–251

Karppinen E (1958) Ober die Oribatiden (Acar) der finnischen Waldb-Sden. Ann. Zool. Soc. Vanamo 19:1–43

Kennedy IR (1986) Acid soil and acid rain: The impact on the environment of nitrogen and sulphur cycling. Wiley. New York

Lavelle P, Spain AV, Blanchart E, Martin A, Martin PS (1992) Impact of soil fauna on the properties of soils in the humid tropics. In: Sanchez PA, Lal R (eds) Myths and Science of soils in the tropics. Soil Science Society of America, Special publication. No. 29, Madison, Wisconsin

Lavelle P (1996) Diversity of soil fauna and ecosystem function. Biology International 33:3–16

Lavelle P, Bignell DE, Lepage M (1997) Soil function in a changing world: the role of invertebrate ecosystem engineers. European Journal of Soil Biology 33:159–193

Lavelle P, Pashanasi B, Charpentier F, Gilot C, Rossi JP, Derouard L, Andre J, Ponge JF, Bernier N (1998) Large-scale effects of earthworms on soil organic matter and nutrient dynamics. In: Edwards CA (ed) Earthworm ecology. St. Lucie Press, Boca Raton

Lavelle P, Spain AV (2003) Soil ecology. Kluwer Academic Publishers, New York

Lee KE, Foster RC (1991) Soil fauna and structure. Aust J Soil Res 9:754–760

Lettl A, Langkramer O, Lochman V (1981) Dynamics of oxidation of inorganic sulphur compounds in upper soil horizons of spruce forests. Folia Microbiologica 26(1):24–28

Linden DR, Hendrix PF, Coleman DC, Petra CJ, van Vliet (1994) Faunal indicators of soil quality. In: Doran JW, Coleman DC, Bezdicek DF, Stewart BA (eds) Defining soil quality for sustainable environment. SSSA Special Publication, No. 35

Lloyd D, Boddy L, Davies KJP (1987) Persistence of bacterial denitrification capacity under aerobic conditions: the rule rather than exception. FEMS Microbiol Ecol 45:185–190

Lousier JD, Bamforth SS (1990) Soil protozoa. In: Dindal D (ed) Soil biology guide. Wiley, New York

Lynch JM (1987) Microbial interactions in the rhizosphere. Soil microorganisms 30:33–41

Lynch JM (1990) The Rhizosphere. Wiley, New York

Marrs RH, Proctor J, Heaney A, Mountford MD (1988) Changes in soil nitrogen-mineralization and nitrification along an altitudinal transect in tropical rain forest in Costa Rica. J Ecol 76:466–482

Martikainen PJ, Palojarvi A (1990) Evaluation of the fumigation-extraction method for the determination of microbial C and N in a range of forest soils. Soil Biology and. BioChemistry 22:797–802

Mason CF (1970) Snail populations, beech litter production, and the role of snails in litter decomposition. Oecologia 5:215–239

McGill WB, Cole CV (1981) Comparative aspects of cycling of organic C, N, S and P through soil organic matter. Geoderma 26:267–286

Miller EK, Johnson AH, Vann DR, Joshi AB (2003) Nitrogen availability and forest productivity along a climosequence on Whiteface Mountain, New York. Can J For Res 33:1880–1891

Mitchell MJ, Fuller RD (1988) Models of sulfur dynamics in forest and grassland ecosystems with emphasis on soil processes. Biogeochemistry 5:133–163

Moldenke AR (1990) One hundred twenty thousand little legs. Essays onInvertebrate Conservation, The Xerces Society. Wings (Summer): 11–14

Moldenke AR, Lattin JD (1990) Density and diversity of soil arthropods as "biological robes' of complex soil phenomena. Northwest Environ J 6(2):409–410

Montagnini F, Buschbacher R (1989) Nitrification rates in two undisturbed tropical rain forests and three slash-and-burn sites of the Venezuelan Amazon. Biotropica 21:9–14

Moritz M (1965) Untersuchungen tiber den EinfluB von KahlschlagmaB-nahmen auf die Zusammensetzung von Hornmilbengemeinschaften (Acari: Oribatei) norddeutscher Laubund Kiefernmischw~ilder. Pedobiologia 5:65–101

Mulongoy K, Bedoret A (1989) Properties of worm castes and surface soils under various plant covers in the humid tropics. Soil Biology and. BioChemistry 21:197–203

Nannipieri P, Badalucco L (2003) Biological processes In: Bembi DK, Nieder R (eds) Processes in the soil-plant system: modelling concepts and applications. The Haworth Press, Binghamton, NY

Nash MH, Whitford WG (1995) Subterranean termites: regulators of soil organic matter in the Chihuaharan desert. Biol Fertil Soils 19:15–18

Nasholm T, Ekblad A, Nordin A, Giesler R, Högberg M, Hogberg P (1998) Boreal forest plants take up organic nitrogen. Nature 392:914–916

Nature E (2012) Soil, Agriculture, and Agricultural Biotechology http://www.nature.com/scitable/knowledge/library/the-soil-biota-84078125

Neill C, Piccolo MC, Cerri CC, Steudler PA, Melillo JM, Brito M (1997) Net nitrogen mineralization and net nitrification rates in soils following deforestation for pasture across the southwestern Brazilian Amazon Basin landscape. Oecologia 110:243 -252

Nekola JC (2002) Effects of fire management on the richness and abundance of central North American grassland land snail faunas. Animal biodiversity and conservation 25: 53–66

Nelson DR (2002) Current status of the Tardigrada: evolution and ecology. Integr Comp Biol 42(3):652–659

Neptune AML, Tabatabai MA, Hanaway JJ (1975) Sulfur fractions and carbon-nitrogen-phosphorussulfur relationships in some Brazilian and Iowa soils. Soil Sci Soc Am Proc 39:51–55

Nommik H (1982) Nitrogen cycling, leaching and denitrification in forest soils. Second National symposium on biological nitrogen fixation, Helsinki

Nosek J, Vysotskaya SO (1976): Study of Protura from the nests of small mammals of Transkarpathian region. Entomologicheskoe Obozrenie 55: 808–812

Nosek J (1977): Proturan synusies and niche separation in the soil. Ecological Bullentins 25: 138–142

Nugroho RA, Roling WFM, Laverman AM, Verhoef HA (2007) Low nitrification rates in acid Scots pine forest soils are due to pH related factors. Microb Ecol 53:89–97

Nurminen M (1965) Enchytraeid and lumbricid records (Oligochaeta) from Spitsbergen. Enchytraeids (Oligochaeta) from northern Norway and western Lapland. Priliminary notes on the Enchytraeids (Oligochaeta) of the Ahvenanmaa Island, South Finland. Ann Zool Fenn 2:1–17

Pastor J, Aber JD, McClaugherty CA, Melillo JM (1984) Aboveground production and N and P cycling along a nitrogen mineralization gradient on Blackhawk Island, Wisconsin. Ecology 65:256–268

Pearce TA (1997) Interference and resource competition in two land snails: adults inhibit conspecific juvenile growth in field and laboratory. Journal of Molluscan Studies 63:389–399

Perakis SS, Sinkhorn ER (2011) Biogeochemistry of a temperate forest nitrogen gradient. Ecology 92(7):1481–1491

Persson T, Wiren A (1995) Nitrogen mineralization and potential nitrification at different depths in acid forest soils. Plant Soil 168-169:55–65

Petersen H, Luxton M (1982) A comparative analysis of soil fauna populations and their role in decomposition processes. Oikos 39:287–388

Pourriot R (1979) Rotiferes du sol. Rev Ecol Biol Sol 16:279–312

Pregitzer KS, Burton AJ, Mroz GD, Liechty HO, MacDonald NW (1992) Foliar sulfur and nitrogen along a 800-km pollution gradient. Can J For Res 22:1761–1769

Ricci C, Balsamo M (2000) The biology and ecology of lotic rotifers and gastrotrichs. Freshwater Biol 44:15–28

Ricci C (2001) Dormancy patterns in rotifers. Hydrobiologia 446:1–11

Robertson GP, Groffman PM (2007) Nitrogen transformation. In: Paul EA (ed) Soil microbiology, biochemistry, and ecology. Springer, New York

Rusek J (1986) Soil microstructures: contributions of specific soil organisms. Quaestiones Entomologicae 21:497–514

Schaefer M, Schauermann J (1990) The soil fauna of beech forests: comparison between a mull and a moder soil. Pedobiologia 34(5):299–314

Setala H, Hubta V (1991) Soil fauna increase *Betula pendula* growth: Laboratory experiments with coniferous forest floor. Ecology 72(2):665–671

Shelley RM (2002) A revised, annotated, family-level classification of the Diplopoda. Arthropoda Selecta 11(3):187–207

Shelley RM (2007) Taxonomy of extant Diplopoda (millipeds) in the modern era: Perspectives for future advancements and observations on the global diplopod community (Arthropoda: Diplopoda). Zootaxa 1668: 343–362

Shiferaw B, Bantilan MCS, Serraj R (2004) Harnessing the potential of BNF for poor farmers: technological policy and institutional-constraints and research need. In: Serraj R (ed) Symbiotic nitrogen fixation; prospects for enhanced application in tropical agriculture. Oxford & IBH, New Delhi

Siddiqui ZA, Pichtel J (2008) Myidantcorrhizae: An overview. In: Siddiqui ZA, Akhtar MS, Futai K (eds) Mycorrhizae: sustainable agriculture and forestry. Springer, Berlin

Siepel H (1996) Biodiversity of soil microarthropods: the filtering of species. Biodivers Conserv 5:251–260

Sierwald P, Bond JE (2007) Current status of the myriapod class Diplopoda (millipedes): Taxonomic diversity and phylogeny. Annu Rev Entomol 52:401–420

Smolander A, Kurka A, Kitunen V, Mälkönen (1994) Microbial biomass C and N, and respiratory activity in soil of repeatedly limed and N- and P-fertilized Norway spruce stands. Soil Biol Biochem 26:957–962

Sohlenius B (1979) A carbon budget for nematodes, rotifers and tardigrades in a Swedish coniferous forest soil. Holarctic Ecology 2:30–40

Sohlenius B (1982) Short-term influence of clear-cutting on abundance of soil-microfauna (Nematoda, Rotatoria and Tardigrada) in a Swedish pine forest soil. J Appl Ecol 19:349–359

Standen V (1988) Oligochaetes in fi re climax grassland and conifer plantations in Papua New Guinea. J Trop Ecol 4:39–48

Sutton SL (1980) Woodlice. Pergamon Press, New York

Swift MJ, Heal OW, Anderson JM (1979) Decomposition in terrestrial ecosystems. Studies in Ecology vol. 5. Blackwell, Oxford

Teuben A, Roejofsma TAPI (1990) Dynamic interactions between functional groups of soil arthropods and microorganisms during-decomposition of coniferous litter in microcosm experiments. Biol Fertil Soils 9:145–151

Tietema A, Warmerdam B, Lenting E, Riemer L (1992) Abiotic factors regulating nitrogen transformations in the organic layer of acid forest soils-moisture and pH. Plant Soil 147:69–78

Tinzara W, Tukahirwa EM (1995) The effects of soil macrofauna on soil properties in a banana cropping system, Masaka District, Uganda, TSBF, Afnet V R

Torgersen CE (1993) Spatial variability of soil organisms, pH, moisture, O-horizon depth, and temperature indifferentiated conifer stands in the Western Cascades, Oregon. A Bachelor Thesis, Department of Geography, University of Oregon

Van Dam O (2001) Forest filled with gaps: effects of gap size on water and nutrient cycling in tropical rain forest. A study in Guyana. Ph.D. Thesis, Tropenbos-Guyana Series 10. Georgetown, Guyana

Van Vliet PCJ, Hendrix PF (2003) Role of fauna in soil physical processes. In: Abbott LK, Murphy DV (eds) Soil biological fertility-A key to sustainable land use in agriculture. Klewer Publishers

Vernimmen RRE, Verhoef HA, Verstraten JM, Bruijnzeel LA, Klomp NS, Zoomer HR, Wartenbergh PE (2007) Nitrogen mineralization, nitrification and denitrification potential in contrasting lowland rain forest types in Central Kalimantan, Indonesia. Soil Biology Biochemistry 39:2992–3003

Villalobos FJ, Lavelle P (1990) The soil coleoptera community of a tropical grassland from Laguna Verde, Veracruz (Mexico). Revue d'Ecologie et Biologie du Sol. 27(1):73–93

Viro PJ (1969) Prescribed burning in forestry. Communicationes Instituti Forestalis Fenniae 67:49

Vitousek PM, Matson PA (1988) Nitrogen transformations in a range of tropical forest soils. Soil Biology Biochemistry 20:361–367

Vitousek P, Cassman K, Cleveland CC, Crews TE, Field CB, Grimm NB, Howarth RW, Marino R, Martinelli L, Rastetter EB, Sprent JI (2002) Towards an ecological understanding of biological nitrogen-fixation. Biogeochemistry 57/58:1–45

Waterhouse DF (1974) The biological control of dung. Sci Am 230:100–109

Whalley WR, Riseley B, Leeds-Harrison PB, Bird NRA, Leech PK, Adderley WP (2005) Structural differences between bulk and rhizosphere soil. Eur J Soil Sci 56(353-):360

Williams BL, Griffiths BS (1989) Enhanced nutrient mineralization and leaching from decomposing sitka spruce litter by enchytraeid worms. Soil Biology and. BioChemistry 21:183–188

Wood TG (1996) The agricultural importance of termites in the tropics. Agricultural zoology reviews 7:117–155

Youngberg CT, Wollum IIAG (1976) Nitrogen accretion in developing *Ceanothus velutinus* stands. Soil Sci Soc Am J 40

Zahir AZ, Arshad M, Frankenberger WT Jr (2004) Plant growth promoting rhizobacteria: application and perspectives in Agriculture. Adv Agron 81:97–168

6 Nutrient Dynamics in Forest Soil

Forest stands produce 100–500 t ha^{-1} total standing biomass (dry weight) including 60–80 % in stemwood, 10–20 % in branches, 5–10 % in barks, 5–10 % in roots, and 1–5 % in leaves. About 58 % of this biomass is carbon. To produce such huge amounts of biomass, forest plants use 200–2,500 kg ha^{-1} N, similar amounts of K and Ca, and 20–200 kg ha^{-1} P. The annual uptake of nutrients by forest trees varies considerably depending on tree species and soil fertility. In one study, annual uptake rates of N, P, K, Ca, and Mg were 178.59, 27.48, 80.46, 349.39, and 63.70 kg ha^{-1}, respectively. Some absorbed nutrients are resorbed from senescing leaves and meet the urgent needs of new flush. A part of the absorbed nutrients are retained in biomass and used for further biomass accretion. A considerable part of absorbed nutrients is returned annually in the form of litterfall, stemflow, throughfall, root decay, and death and decay of soil organisms. Litterfall is the primary route of nutrient returns in forests and annually 3–15 t ha^{-1} litters fall to the forest floor. Litterfall brings about 30–225 kg N ha^{-1} per year, 1.2–15 kg P ha^{-1} per year, 6–68 kg K ha^{-1} per year, 6.5–290 kg Ca ha^{-1} per year, and 1–64 kg Mg ha^{-1} per year in some instances. These nutrients are released to the soil nutrient pool by the decomposition of litter. Usually K, Ca and Mg and sometimes P are released from the beginning of the decomposition process, and in most of the instances, nitrogen is immobilized by decomposing organisms up to the 50 % of the mass loss, and becomes liberated afterwards.

6.1 Biomass and Nutrients

6.1.1 Biomass Production

One way of estimating the productivity of a forest is the determination of its biomass. Currently, the aims of studies addressing biomass are to elucidate energy and nutrient cycles (Salazar et al. 2010). Such studies are also used to observe the effect of the vegetation on the global CO_2 cycle. Some CO_2 models include the estimation of biomass as the volume, its components, or some related parameters in order to establish the flows of gas between the vegetation, soil, and the atmosphere. Studies of biomass have been carried out in tropical, temperate, and Mediterranean forests and in semiarid areas (Rapp et al. 1999; Santa-Regina and Tarazona 2000, 2001; Navar et al. 2002; Zianis and Mencuccini 2003; Nadezhdina et al. 2004; Rubilar et al. 2005; Segura 2005). However, no robust methods of estimation of biomass are still available (Murali et al. 2005); the methods developed so far include total harvesting of trees or shrubs and sample harvesting (Usotsev and Hoffman 1997). Indirect ways of estimation of stand biomass are based on measurable parameters such as basal area, height, and stem density (Whitmore 1984; Palm et al. 1986; Brown et al. 1991). Further, some reports have estimated biomass using tree inventory data on stems falling under different size classes and wood density (specific gravity) and production (Brown and Lugo 1982; Gillespie et al. 1992).

In this account, the amounts of biomass produced by natural and man-made forests in different regions will be cited from the literature. Dry biomass may vary from 50 to 550 t ha^{-1}, and the factors behind this wide variation in biomass production are soil, climate, species, stand density, stand age, and management. Comeau and Kimmins (1989) measured the distribution of tree biomass and allocation of production in four 70–78-year-old stands of lodgepole pine (*Pinus contorta* var. *latifolia* Engelm.) differing in soil moisture conditions. Aboveground biomass ranged from 116.5 t ha^{-1} on one xeric site to 313.1 t ha^{-1} on one mesic site. Stem biomass represented 68 % and 73 % of total tree biomass on xeric and mesic sites, respectively, whereas total root biomass represented between 20 and 28 % of total lodgepole pine biomass. Fine and small roots represented 4 and 1.5 % of total tree biomass on the xeric and mesic sites, respectively. On the other hand, according to Vitousek and Sanford (1986), biomass does not differ strongly among sites with different soils, except that it is lower in the most infertile white-sand soils. Additional information on tropical forest biomass has been reviewed by Brown and Lugo (1982),

K. T. Osman, *Forest Soils*, DOI 10.1007/978-3-319-02541-4_6, © Springer International Publishing Switzerland 2013

and further calculations based on the merchantable volume of forests are summarized in Brown and Lugo (1984). Strong patterns relating soil fertility to biomass are not apparent in their analysis either. Total aboveground biomass is strongly dependent on stand age (successional status) as well as climate and soils. In their review, Vitousek and Sanford (1986) reported aboveground biomass of 233–402 t ha^{-1} in moderately fertile tropical soils, 182–510 t ha^{-1} in infertile Oxisols/Ultisols, 180–185 t ha^{-1} in Spodosols, and 176–348 t ha^{-1} in montane soils.

Son et al. (2004) investigated biomass production in one pure *Quercus variabilis* Bl. stand (stand 1) and two mixed *Q. variabilis–Q. mongolica* Fisch. stands (stands 2 and 3) in central Korea. Total biomass values, including overstory and understory vegetation, were 137.8, 216.2, and 253.3 t ha^{-1} for stands 1, 2, and 3, respectively. The biomass of overstory vegetation was distributed for the three stands as follows: stemwood > live branch > stem bark > roots > foliage > dead branch > current twig. Annual biomass production was 11.1 t ha^{-1} per year for stand 1, 16.6 t ha^{-1} per year for stand 2, and 19.7 t ha^{-1} per year for stand 3. Stand density influenced the differences in biomass and annual production. Lugo et al. (1988) reported tropical species aboveground biomass accumulation rates of 1.6–29.8 t ha^{-1} per year, with most in the range of 6–15 t ha^{-1} per year. For temperate forest systems reported in Reichle (1981), the range goes from 5 to 15 t ha^{-1} per year. Liu et al. (2002) determined biomass production in a montane evergreen broadleaf forest of *Lithocarpus xylocarpus* in the Ailao Mountains of China. They observed total live biomass of 503 t ha^{-1} and 285 t ha^{-1} in high-density and low-density stands, respectively.

Biomass in various aboveground tree components (stemwood, stembark, branches, and leaves) was quantified in an age sequence of pure *Larix olgensis* plantations (20, 35, 53, and 69 years old) in Northeast China by Zhao et al. (2011). The results show that the aboveground biomass allocation in various tree components was in the order of stemwood (62–83 %), branches (9–21 %), stembark (7–11 %), and leaves (1–6 %) for all stands (Fig. 6.1). Biomass in all components increased almost linearly with stand age. Yang et al. (2011) also suggested that biomass significantly increased in older stands of *Pinus massoniana* plantations in southern China.

Biomass production potentials of tree species are not equal. For example, Fang et al. (1998) reported biomass values for main forest types of China. These forests included evergreen and deciduous, broadleaf and coniferous, boreal, and temperate and tropical forests. Biomass values were: *Picea abies*, 136 t ha^{-1}; *Cunninghamia lanceolata* (CUL), 86 t ha^{-1}; Cypress, 213 t ha^{-1}; *Larix*, 140 t ha^{-1}; Pines, 50–90 t ha^{-1}; *Tsuga*, 209 t ha^{-1}; *Betula*, 123 t ha^{-1}; *Casuarina*, 89 t ha^{-1}; oak, 90 t ha^{-1}; *Eucalyptus*, 178 t ha^{-1}; mixed deciduous forests, 313 t ha^{-1}; *Populus*, 109 t ha^{-1}; and tropical forests, 324 t ha^{-1}. Biomass and productivity in some tropical dry deciduous disturbed teak (*Tectona grandis*) forests of the Satpura plateau were estimated in three communities identified as *Tectona grandis–Lagerstroemia parviflora–Sterculia urens* (site I); *T. grandis–Lannea coromandalica–Diospyros melanoxylon–Butea monosperma* (site II); *T. grandis–Chloroxylon swietenia–L. parviflora–D. melanoxylon* (site III); and a young plantation of *T. grandis* (site IV). The allocation of aboveground biomass ranged from 79 to 86 % of total biomass (IV). There were 28.1–85.3 t ha^{-1} aboveground biomass, 9.1–15.6 t ha^{-1} belowground biomass, and 39.1–100.9 t ha^{-1} total biomass. These values are on the lower side of range reported for tropical/subtropical forests of India (38.6–239.8 t ha^{-1}) (Pande 2005). Nascimentoa and Laurance (2001) estimated total aboveground biomass in 15 undisturbed central Amazonian rainforests, based on estimates of all live and dead plant material within 20 1-ha plots spanning an extensive study area of about 1,000 km^2. Biomass values were very high, averaging 397 t ha^{-1}. The most important component of aboveground biomass was large trees (10 cm diameter-at-breast-height trees, which comprised 81.9 % of total aboveground biomass, followed by downed wood debris (7.0 %); small trees, saplings, and seedlings (< 10 cm diameter at breast height (DBH); 19 5.3 %); lianas (2.1 %); litter (1.9 %); snags (1.5 %); and stemless palms (0.3 %). Liu et al. (2004) measured biomass of plant compartments in *Picea likiangensis* (Franchet) E. Pritzel var. *balfouriana* trees growing in a plantation in West Sichuan, Southwest China, and found total biomass to be 114.83 t ha^{-1}. Tree, shrub, herb, bryophyte, and litter layers accounted for 93.9, 0.9, 0.02, 0.04, and 5.2 %, respectively. The total biomass of tree layers was 107.82 t ha^{-1}. Needles, branches, stemwood, stem bark, and roots accounted for

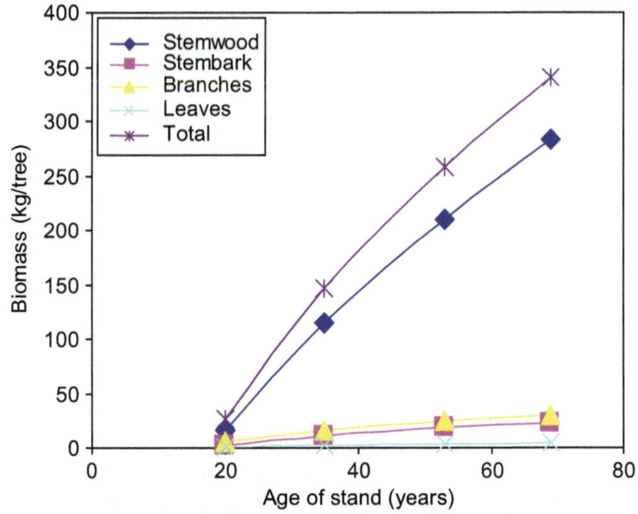

Fig. 6.1 Variation in biomass production per tree with stand age. (Constructed from data of Zhao et al. 2011)

13.2, 19.7, 42.3, 10.0, and 14.8%, respectively. Rutkowski and Stottlemyer (1993) found ten canopy species in a boreal forest stand which contained an aboveground biomass of 108 t ha^{-1}, while the northern hardwood stand had 15 canopy species and a biomass of 267 t ha^{-1}. The boreal understory had a larger number of saplings (10,440 stems ha^{-1}) and much greater biomass (2,275 kg ha^{-1}) than did the northern hardwood stand (2,938 stems ha^{-1}, 85 kg ha^{-1}). Its vascular ground cover also contained more biomass (1,180 stem ha^{-1} to 34 kg ha^{-1} for northern hardwoods). Whittaker et al. (1974) investigated a small watershed in the White Mountains of New Hampshire bearing mesophytic, cool—temperate, broadleaf—deciduous forests in Hubbard Brook. *Acer saccharum*, *Betula lutea*, and *Fagus grandifolia* are dominant, but toward higher elevations *Picea rubens* and *Abies balsamea* also occur. The stands were young following cutting in 1909–1917 but contain older trees; stand composition is thought reasonably representative of the climax. A total of 93 sample trees of major species were cut and roots excavated. Mean dimensions of sample trees, and the constants for the system of logarithmic regressions relating volume, surface, mass, and growth to DBH and other independent variables, show decrease in tree sizes and height/diameter ratios toward higher elevations. Aboveground biomass (dry matter) varied from 102 to 162 t ha^{-1} and aboveground net primary productivity (1956–1960) from 790 to 1,127 g m^{-2} per year. Biomass of root systems is 18–21% of aboveground biomass. Different estimations suggest that a mean climax aboveground biomass for the watershed may be around 350 t ha^{-1}.

Salazar et al. (2010) observed total biomass at a paraclimax chestnut stand of 125 t ha^{-1} and for a climax oak stand of 74 t ha^{-1}, which lie within the range reported by other authors for both species in different geographic locations. Thus, in chestnut coppices devoted to wood production in Spain, Santa-Regina (2000) reported values of 120.4 t ha^{-1}; in France, Santa-Regina et al. (2000), Ranger et al. (1990), and Ranger and Colin-Belgrand (1996) established values of 153.3, 120, and 120 t ha^{-1}; in Italy, values of 107 (La Marca 1984) and 108 t ha^{-1} (Cutini 2000) have been reported. In Slovakia, Tokarár and Krekulova (2004) found biomass values between 95.1 and 174.21 t ha^{-1}. The root biomass was 58 t of dry matter, which was 18% of the total stand biomass. Nutrient concentration in the belowground compartment was lower than in the aerial biomass (Ranger and Gelhaye 2001). Vitousek et al. (1988) reported aboveground biomass of different forest types as: northern coniferous forest, 233 t ha^{-1}; temperate broadleaf deciduous forest, 286 t ha^{-1}; giant temperate coniferous forest, 624 t ha^{-1}; temperate broadleaf evergreen forest, 315 t ha^{-1}; and tropical closed forest, 494 t ha^{-1}. In a study to estimate the aboveground tree biomass in forests of Cameroon (Ibrahima et al. 2010), 112 trees were sampled in three undisturbed plots of 10 × 10 m. Understory biomass was estimated by the harvesting method. The total aboveground tree biomass was 583 t ha^{-1}.

Various factors may affect the relative biomass allocation between roots and aboveground plant parts, as expressed by the root to shoot ratio (R/S). R/S varies with stand or tree age (Gerhardt and Fredriksson 1995). It also depends on tree species and differs between gymnosperms and angiosperms (Cuevas et al. 1991). Several abiotic factors may also influence biomass allocation to roots. For example, decreasing soil moisture produces higher R/S (Brown and Lugo 1982). Soil characteristics, such as nutrient availability (Cavelier 1992) and texture (Vitousek and Sanford 1986; Gerhardt and Fredriksson 1995), are also purported to have a significant influence on root biomass. Whittaker et al. (1974) estimated root biomass in southern Appalachian forests to be 51.4 t ha^{-1}, with 27.2 t ha^{-1} representing roots <2.5 cm in diameter. McGinty (1976) reported root biomass estimates in the temperate deciduous forest of eastern North America to range from 28.3 t ha^{-1} to 51.4 t ha^{-1} for the Coweeta forests.

6.1.1.1 Nutrient Concentrations in Biomass

The distribution of nutrients in several biomass components is determined on the basis of their nutrient concentrations and mass values. Nutrient concentration differs in biomass compartments depending on species, soils, climate, stand age, etc. Mean nutrient concentrations in the biomass of *Vaccinium angustifolium*, *Pinus rigida*, *Quercus alba*, *Q. ilicifolia*, *Q. coccinea*, *Gaylussacia baccata*, *V. vacillans*, *P. rigida*, and *Kalinia angustifolia* in Bookhaven pine–oak forest were measured by Woodwell et al. (1975). Nutrient concentrations varied considerably depending on species, biomass component, and the elements concerned. With the exception of *Kalinia*, the species of the lower strata and earlier woody successional stages of the forest contained higher mean nutrient concentrations in the biomass than did the upper strata and climax tree species. For most species and most nutrients, the patterns of tissue concentrations were similar. Leaves, flowers, and fruits contained the highest concentrations, heartwood the lowest, while twigs, bark, branches, roots, and sapwood had intermediate concentrations. In stands of the main forest types of China, the concentration of N, P, K, and Mg was the highest in leaves and the lowest in stemwood and followed the order of leaves > branches > stembark > stemwood. By contrast, Ca concentration was the highest in stembark. Variations in nutrient concentrations among stand ages depended on both tree components and nutrient elements. With increasing stand age, the concentration of all elements (except Ca) in stemwood decreased significantly. The concentration of stembark Ca and branch N, P, and K significantly increased, while leaf K concentration significantly decreased and leaf Ca concentration increased (Fang et al. 1998). The ordering of major element concentrations was K>N>Mg>Ca>P in branch, stem, and root tissues but was N>K>Mg>Ca>P in leaves (Shanmughavel et al. 2001). In the tree species *Quercus coccinea, Q. alba, Q. ilicifolia, Pinus rigida, Gaylussacia baccata, Vaccinium vacillans, V. angustifolium,* and *Kalmia angustifolia*, weighted means for standing crop nutrient concentrations were within the ranges: 2.02–5.15% N, 0.26–1.26% P, 1.11–2.04% K, 1.11–4.80% Ca, and 0.32–0.74% Mg (Woodwell et al. 1975). In this report, there were extra-normal concentrations of N, P, and Ca in some sites. Svoboda et al. (2006) studied biomass and element pools of spruce trees in the catchments of Plesne and Certovo in the Sumava Mountains, the Czech Republic. Nitrogen had the highest mean concentrations in tree components in all cases. The concentrations of P, Ca, Mg, and K showed similar patterns. Generally, the highest concentrations of these elements were found in foliage, fine branches, fine roots, and bark of stem and branches. Fe, Na, Al, and Mn showed the lowest mean concentrations in tree components for all the analyzed elements.

According to Vitousek and Sanford (1986), concentrations of all of the major nutrients in leaves are significantly elevated on the more fertile tropical soils. The infertile oxisols and ultisols have intermediate nitrogen and low phosphorus and calcium concentrations. Sandy soils support vegetation with low foliar nitrogen and phosphorus, intermediate major cation concentrations, and high specific leaf weights. Forest tree leaves on oxisols had higher nitrogen but lower potassium concentrations than did leaves on the sandy soils. The values of nutrient concentrations given as follows from Vitousek and Sanford (1986) showed a significant bearing of soil fertility on nutrient concentration in biomass of forest soils.

Moderately fertile soil: 2.08–2.45% N, 0.12–0.15% P, 0.85–1.92% K, 0.70–2.29% Ca, 0.26–0.88% Mg.

Infertile oxisols/ultisols: 1.27–1.93% N, 0.05–0.07% P, 0.34–0.54% K, 0.11–0.50% Ca, 0.10–0.29% Mg.

Lowland and Montane sites: 0.6–1.78% N, 0.02–0.12% P, 0.43–1.17% K, 0.46–1.17% Ca, 0.16–0.29% Mg.

Augusto et al. (2000) studied biomass and nutrient dynamics of several woody species at the stand scale. Allometric relationships were built from 13 stands of a woody species of moderate height (European gorse: *Ulex europaeus* L.). In eight other stands, the aboveground biomass of the species was estimated using allometric relationships. The mean concentrations of nutrients among the species were ranked as follows: mean concentration of N and K, European beech > Douglas fir = Norway spruce = Scots pine; Ca, European beech = Norway spruce ≥ Scots pine ≥ Douglas fir; and Mg, European beech ≥ Scots pine ≥ Norway spruce ≥ Douglas fir. For P, no significant difference was found for the tested species. Nutrient concentrations were consistently higher in the bark than in the wood of sessile oak and beeches (Andre et al. 2010).

A good example of the effect of stand density on the concentration of nutrients in leaf, stem, and fine roots was provided by Barron-Gafford et al. (2003). They studied the relationship between nutrient acquisition and stem biomass increment on a stand basis in 4-year-old pine plantations planted at different densities. The stands were intensively managed (bedded site preparation, a high level of fertilization, and nearly complete weed control) loblolly pine (*Pinus taeda* L.) and slash pine (*Pinus elliottii* Engelm.) in the lower Coastal Plain of the southeastern United States. Stand densities were 740, 2,220, and 3,700 trees ha^{-1} on three sites, each with a different soil type. Foliar N and P, stemwood N and K, and fine root N concentrations decreased with increased stand density indicating competition at higher densities. For loblolly pine, foliar N concentrations fell from 1.31% at 740 trees ha^{-1} to 1.09% at 3,700 trees ha^{-1}. Slash pine foliar N concentrations followed a similar pattern, decreasing from 1.12 to 0.91%.

Foliar nutrient concentrations in a study of eight most common canopy tree species tended to decrease with increasing precipitation across a precipitation gradient. Mean foliar P ranged from 0.076% at the 3,100-mm precipitation site to 0.14% at the 1,800-mm site and decreased significantly with increasing precipitation. Mean foliar K ranged from 0.76 at the 2,300-mm site to 1.13% at the 1,800-mm site and showed a weak decreasing trend with increasing precipitation. Mean foliar Ca decreased significantly with increasing precipitation and ranged from 0.82% at the 3,500-mm site to 1.45% at the 2,300-mm site. Mean foliar Mg varied between 0.32% at the 3,500-mm site and 0.52% at the 1,800-mm site and also decreased significantly with increasing precipitation (Santiago et al. 2005).

6.1.1.2 Nutrient Contents in Biomass

Accurate estimates of the amounts of nutrients immobilized in the organs and tissues of different tree species are of prime importance to make appropriate tree species selection in relation to site conditions and to determine the harvesting regime that will ensure forest sustainability (Andre et al. 2010). In a study, Son et al. (2004) estimated nutrient contents in pure and mixed stands of *Quercus variabilis* in forests of central Korea. Nutrient contents in the vegetation were distributed as follows: 471–839 kg N ha^{-1}, 37–71 kg P ha^{-1}, 478–860 kg K ha^{-1}, 428–791 kg Ca ha^{-1}, and 72–125 kg Mg ha^{-1}, with considerable variation within stands. Nutrient allocation, accumulation, and aboveground biomass in 4-year-old grey alder (*Alnus incana* (L.) Moench) and hybrid alder (*Alnus incana* (L.) Moench and *Alnus glutinosa* (L.) Gaertn.) plantations were estimated by Uri et al. (2003). When leaf mass was at its maximum, the amount of N accumulated in the aboveground biomass of the grey alder stand was 142.0 kg ha^{-1}, the amount of P was 16.3 kg ha^{-1}, and the amount of K was 49.5 kg ha^{-1}.

The amount of nitrogen accumulated in a hybrid alder stand totaled 76.8 kg ha^{-1}, that of phosphorus 6.2 kg ha^{-1}, and that of potassium 28.2 kg ha^{-1}. The smaller amounts of N, P, and K bound in the hybrid alder plantation are related to the smaller biomass of the stand. Still, the amounts of N, P, and K consumed for the production of 1 t of biomass were similar in the case of these very young grey alder and hybrid alder stands. The amounts of nutrients consumed to produce 1 t of biomass were: 16.0 kg N, 1.6 kg P, and 5.4 kg K for grey alder and 14.6 kg N, 1.1 kg P, and 5.2 kg K for hybrid alder. In a study in the forests of Cameroon, there were 8,888 kg N ha^{-1}, 6,953 kg Ca ha^{-1}, 2,337 kg K ha^{-1}, 436 kg P ha^{-1}, 311 kg Mg ha^{-1}, and 30 kg Na ha^{-1} in the aboveground biomass (Ibrahima et al. 2010). These values seem to be very high in comparison to other reports. In a review of reports from more than 30 tropical countries, Vitousek and Sanford (1986) observed differences in nutrient contents in the vegetation in soils of variable fertility. Nitrogen in per hectare biomass varied from 1,685 to 1,980 kg in moderately fertile soils, from 741 to 2,430 kg in oxisols/ultisols and from 212 to 618 kg in spodosols. Such differences were also observed in other nutrients. Biomass P ranged from 112 to 290 kg in moderately fertile soils and from 28 to 62 kg ha^{-1} in spodosols.

The N content in a 41-year-old loblolly pine in South Carolina was 123.2 kg ha^{-1} and the P content was 10.4 kg ha^{-1} (Van et al. 1984). In spruce trees of the catchments of Plesne and Certovo in the Sumava Mountains, the total element pools per tree were highly variable because of the differences in total biomass between the individual trees. Generally, stemwood and bark, foliage, and roots contained the highest proportion of the elements. However, there were differences between individual elements. Concerning the important nutrients, while the highest proportion of Ca and Mg was contained in stemwood and bark, the highest proportion of P was contained in foliage. The foliage contained a relatively high proportion of P and K, but a relatively low proportion of Ca and Mg (Svoboda et al. 2006). The relative proportion of foliage in total tree biomass was only 5.4 %, but it contained 33.7 % P, 24.6 % N, 24.3 % K, 14.7 % Mg, and 12.2 % Ca. Salazar et al. (2010) studied nutrient accumulation in the biomass of a paraclimax chestnut and a climax oak stand. They found that leaves were quantitatively lower in nutrient accumulation than branches and trunks because leaf biomass only represented 2.9 and 3 % of the total biomass. However, these smaller amounts of nutrients have greater ecological significance because the leaves of trees are subject to annual shedding cycles, with which the organic matter (OM) and nutrients return to the soil.

6.2 Nutrient Retranslocation

It has extensively been found that dead leaves contain a much less concentration of the major nutrients, notably nitrogen, phosphorus, and potassium, than fresh leaves. This is due to the withdrawal of some nutrients from senescing leaves into new growth or storage parts. The phenomenon is called internal nutrient translocation, nutrient retranslocation, nutrient resorption, or internal nutrient cycling. Internal nutrient translocation may be defined as the total amount of an element withdrawn from older tissues and transferred to new and growing tissues of perennial plants (Lim and Cousens 1986). Each senescing tissue becomes a nutrient source for growing sink organs. The physiological processes involved could be similar to the seasonal withdrawal from previously accumulated reserves in storage tissues (Titus 1989), for example, in fruit trees (Titus and Kang 1980). The most studied mechanism in woody species is the transfer of nutrients from senescing leaves to storage organs (Proe and Millard 1995). The ability of a tree to translocate nutrients prior to foliage senescence allows the tree to decrease the need for external nutrients and directly reuse the nutrient through its internal recycling mechanism.

The factors that affect nutrient resorption include species (Killingbeck 1985; Escudero and del Arco 1987; Jonasson 1990; Escudero et al. 1991), successional stage (Ryan and Bormann 1981; Gholz et al. 1985; Potter et al. 1987; Baker and Attiwill 1985; Helmisaari 1995; Lodhiyal et al. 1995), and site characteristics (Fife and Nambiar 1982; Chapin and Kedrowski 1983; Stachurski and Zimka 1975; Staaf 1982; Nambiar and Fife 1987; Del Arco et al. 1991). However, few studies have considered foliar resorption at a stand level in natural ecosystems. These studies have been limited to conifer ecosystems (Helmisaari 1992, 1995). There are major differences between tree species and between nutrients in the amounts of nutrient retranslocation that takes place. Nutrients that remain in an ionic form in foliage, such as potassium, are very mobile and can be retranslocated easily (Colin-Belgrand et al. 1996). Lal et al. (2001) observed that 51 % of N and 53 % of P were retranslocated or resorbed from the senescing leaves in eight forest tree species. Aerts (1996) reported that mean nutrient resorption efficiency was 50 % for N (n=287) and 52 % for P (n=226). N resorption efficiency of evergreens (47 %) was significantly lower than that of deciduous shrubs and trees (54 %). There was, however, little difference in P resorption efficiency between these growth forms (51 and 50 %, respectively). Aerts (1996) also observed that the mean N and P concentrations in leaves of deciduous shrubs and trees were about 60 % higher than in evergreen species. The nutrient resorption coefficient (PR) for the other nutrients ranged from 48 % for K to 7 % for Ca. There was a considerable interspecific variation in the magnitude of retranslocation for all nutrients with the largest PR values in the species *Ficus racemosa* (for N), *Syzygium cumini* (for P), *Adina cordifolia* (for K), and *Terminalia alata* (for Ca). On an average, across the species, the fraction of nutrient mass retranslocated from leaves was in the order: P>N>K>Na>Ca. This order closely coincides with the physiological mobility of plant nutrients. As retranslocated nutrients from senescing leaves are re-used by new growths, it becomes a major nutrient-conservation mechanism (Chapin 1980; Chabot and Hicks 1982; Aerts 1990), which has important implications at both population and ecosystem levels. It has been postulated that low nutrient loss (with senescing leaves) can help the plant populations adapt to nutrient-poor environments (Grime 1979; Chapin 1980; Berendse and Aerts 1987; Aerts 1990). Resorbed nutrients may be immediately used in supporting growth of new leaves, whereas in the deciduous species, they may be temporarily stored. When sinks in growing shoots are absent, translocated nutrients accumulate in stems and roots (Jonasson 1995). Considerable amounts of nutrients may also be stored in the leaves where they also act as a buffer between the asynchrony of supply and demand (Bloom et al. 1985).

Witherspoon (1962), working with ^{134}Cs (an analog of K) in white oak, estimated that 51 % of the leaf Cs returned to woody tissue prior to leaf fall while 36 % was lost as leaf litter and 13 % was leached. Wells and Metz (1963) estimated that about one-half of the N, P, and K was removed from loblolly pine needles prior to leaf fall. An estimate of the amount of resorption from leaves to woody tissue prior to leaf fall can be obtained by the difference in nutrient content of leaf net primary production in the growing season (Day and Monk 1977) and the nutrient content in leaf litter (Cromack 1973; Cromack and Monk 1975). From this difference, Monk and Day (1988) obtained 1.4 kg ha^{-1} per year for Ca, 28.3 kg ha^{-1} per year for K, 4.6 kg ha^{-1} per year for Mg, 3.3 kg ha^{-1} per year P, and 59.3 kg ha^{-1} per year for N. Adjusting nutrient leaching between the growing season and period of leaf fall, the realistic resorption estimates would be −0.5 kg Ca ha^{-1} per year, 13.5 kg K ha^{-1} per year, 3.4 kg Mg ha^{-1} per year, 3.1 kg P ha^{-1} per year, and 56.5 kg N ha^{-1} per year.

6.3 Nutrient Recycling

Forests are self-nourishing ecosystems which depend on native soil nutrients and on those retained in their own biomass. Natural soil nutrients coming from atmospheric deposition and weathering of soil minerals are not often sufficient for their nourishment. The extra nutrients required are supplied by the mineralization of litter materials which bring back a major part of the nutrients absorbed by live vegetation. On decomposition of litter, nutrients return to the available soil nutrient pool, where they are reused by plants. It is known as nutrient recycling, often considered as a closed system, the biological cycling. Nutrient recycling has four major steps: uptake, retention, return, and release (Fig. 6.2).

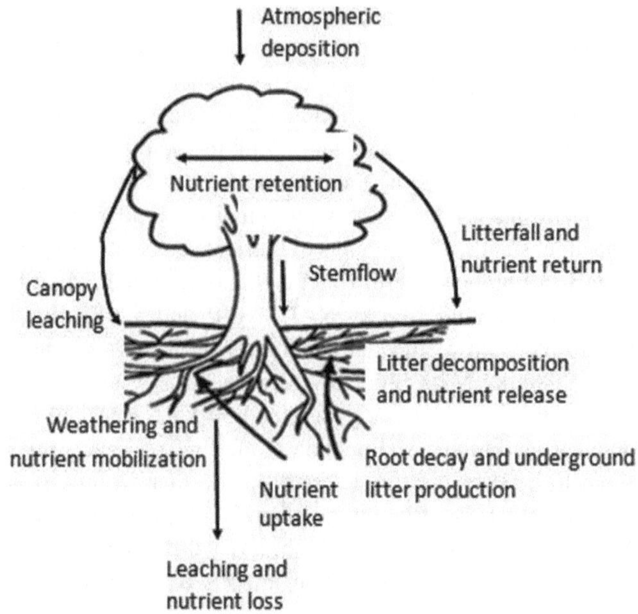

Fig. 6.2 Schematic presentation of the nutrient recycling

6.3.1 Nutrient Uptake

Uptake is defined as the amount of annual nutrient absorption by vegetation from the soil through roots and absorption from rainfall and dryfall through foliage. Forests differ in their nutrient uptake; there are high- and low-demanding tree species. Again, tree age, soil fertility, and climate are important determinants of nutrient requirements. The annual requirement of 115.6 kg N, 12.3 kg P, 67.3 kg K, 62.2 kg Ca, and 9.5 kg Mg ha^{-1} for hardwood temperate forest at Hubbard Brook, New Hampshire, was estimated by Schlesinger (1991). On the other hand, annual uptake per hectare of 34–123 kg N, 2–12 kg P, and 5–75 kg K were reported from several other temperate forests. Thus, nitrogen uptake is ten times that of phosphorus and three times that of potassium (Pritchett and Fisher 1987). However, the greater part of the annual nutrient uptake is returned to the soil in litterfall. The annual uptake for N, P, and K was 61.04, 14.28, and 97.46 kg ha^{-1} in an *Avicennia marina* mangrove forest in Vamleshwar, Gujarat, India (Nirmal Kumar et al. 2011). Maximum annual uptake was 101, 9.3, 44, 37, and 13 kg ha^{-1} per year of N, P, K, and Mg, respectively, in an

8-year-old loblolly pine (*Pinus taeda* L.) stand growing on a nutrient-poor sandy soil in Scotland County, North Carolina, USA (Albaugh et al. 2008).

Meier (1981) reported annual N uptake rates of 10 kg ha^{-1} and 25 kg ha^{-1} by 22- and 175-year-old silver fir (*Abies amabilis*), respectively. Cole et al. (1978) observed annual N uptake of 115 kg ha^{-1} by 35-year-old red alder (*Alnus rubra*) and most of this uptake was met by symbiotic nitrogen fixation. Generally, deciduous tree species take up more nutrients than conifers in the temperate region (Cole and Rapp 1981). Cole and Rapp (1981) reported almost twofold higher nutrient uptake by deciduous trees and they concluded that this was due to annual replacement of the canopy foliage. Their values (kg ha^{-1} per year) are given as follows:

Deciduous forest: 70 kg N, 6 kg P, 48 kg K, 84 kg Ca, and 13 kg Mg ha^{-1} per year

Coniferous forest: 39 kg N, 5 kg P, 25 kg K, 35 kg Ca, and 6 kg Mg ha^{-1} per year

Glumphabutr et al. (2007) estimated the amount of nutrients taken up by trees by the summation of nutrients accumulated in the aboveground biomass of trees and nutrients which returned to the soil in the form of annual litterfall in natural evergreen forests of Thailand. Trees in the dry evergreen forest showed the highest uptake rate with N, P, K, Ca, and Mg at 178.59, 27.48, 80.46, 349.39, and 63.70 kg ha^{-1} per year, respectively. These rates were higher than those of the dry evergreen forest at Namphrom Dam, Chaiyaphum, which were 89.0, 8.0, 58.0, 169.0, and 20.0 kg ha^{-1} per year, respectively. Moist evergreen forests had uptake rates of 118.95, 7.41, 132.83, 120.17, and 26.47 kg ha^{-1} per year (Tsutsumi et al.1983).

6.3.2 Nutrient Retention

Trees use a part of the nutrients absorbed annually for their physiological functions. The remaining part is returned to the soil as litterfall. Annual retention of 5–44 kg N, 1–6 kg P, and 2–20 kg K was found in some temperate forests. These values represented 20–50 % of the annual nutrient uptake (Pritchett and Fisher 1987).

Poggiani (1985) studied the distribution and movement of mineral nutrients in an 11-year-old stand of *Eucalyptus saligna* and in a 14-year-old stand of *Pinus caribaea* var. *hondurensis*, located respectively near Piracicaba and near Agudos (State of São Paulo -Brazil). Results showed that a major portion of nutrients taken up annually by the aboveground components of trees is returned to the soil in litterfall and canopy wash. The values of annual nutrient retention in the *E. saligna* stand were 19.90 kg N, 5.20 kg P, 17.20 kg K, 86.70 kg Ca, and 7.40 kg Mg ha^{-1} against an annual uptake of 47.20 kg N, 7.40 kg P, 39.90 kg K, 130.70 kg Ca, and 16.70 kg Mg ha^{-1}. The percentage retention values were 42.3, 70.2, 50.7, 66.3, and 44.3, respectively. In the *P. caribaea* stand, the annual retention values were 21.70 kg N, 1.10 kg P, 10.70 kg K, 7.30 kg kg Ca, and 2.40 kg Mg ha^{-1} and the uptake values were 65.40 kg N, 3.30 kg P, 32.90 kg K, 27.70 kg Ca, and 8.80 kg Mg ha^{-1}. Thus, uptake and retention of calcium are both considerably higher in *E. saligna* than in *P. caribaea*. *P. caribaea* shows a lower requirement and retention of phosphorus, calcium, and magnesium than *E. saligna*. In a study of nutrient dynamics in the evergreen forests of Thailand, Glumphabutr et al. (2007) showed that the annual retention of N, P, K, Ca, and Mg in a dry evergreen forest was the highest at 103.79, 8.07, 47.03, 209.65, and 31.24 kg ha^{-1} per year, respectively. This was followed by the moist evergreen forest at 46.77, 2.51, 94.01, 39.02, and 8.98 kg ha^{-1} per year, respectively. Annual retention in the hill evergreen forest was the lowest with 25.25, 1.31, 7.70, 10.86, and 2.03 kg ha^{-1} per year, respectively. In an agroforestry ecosystem, the retention of nitrogen and potassium in trees as percentage of total uptake was lower (15.44–20.89, respectively), whereas the retention of calcium, magnesium, and phosphorus ranged from 31.28 to 41.83 %. This indicated that a relatively higher proportion of N and K returned to the soil as litterfall (Bhardwaj et al. 2005).

Nutrient retention in an ecosystem depends on soil and vegetation (Vitousek and Reiners 1975). In some tree species, a relatively large amount of nutrient is used to produce a given amount of biomass (Vitousek 1982). Nutrient-use efficiency is then low. Vegetation that is most effective at retaining nutrients should produce large amounts of woody biomass with high nutrient concentrations. It was hypothesized that early successional tree species may be especially effective as nutrient sinks because of more rapid production and nutrient uptake and greater retention of nutrients in perennial tissues than late successional species (Boring et al. 1981). This hypothesis has been tested for pin cherry (*Prunus pensylvanica* L.) and for several Appalachian species (*Robinia pseudoacacia* L. and *Liriodendron tulipifera*) (Pastor and Bockheim 1984).

6.3.3 Nutrient Return

A major portion of the nutrients taken up annually by forest trees is returned to the soil through the following processes: litterfall, stemflow, throughfall, and exudation and death of roots. Some nutrients are also added to forest soils by the death of soil flora and fauna. However, most of the returns of nutrients in forest ecosystems take place by litterfall.

6.3.3.1 Litterfall

Dead parts of vegetation, including leaves, boles, barks, branches, flowers, fruits, etc., which are shed continuously on the forest floor from standing vegetation are collectively called litters. Litterfall represents a major biological pathway for element transfer from vegetation to soils (Yang et al. 2005) leading to OM replenishment and nutrient cycling (Bhat and Jan 2010). Amount of litterfall, distribution of litter components, concentration of nutrients in litter components, and amount of nutrients returned in litters are all very important aspects of nutrient recycling in forest ecosystems. A significant proportion of terrestrial net primary production is recycled from the trees as litterfall to the forest floor and, subsequently into the detritus food web. In forest ecosystems, litterfall replenishes the pool of nutrients in soil depleted by nutrient uptake and leaching (Gosz et al. 1976; Parker 1983; Morrison 1991). Therefore, litters are natural fertilizers in natural ecosystems. Litterfall leads to the development of forest floor.

Amounts of Litterfall

The amount of litterfall varies widely among forests. Some examples are cited here to show the extent of global variation in OM and nutrient returns in forest ecosystems. Oziegbe et al. (2011) studied litterfall in secondary lowland forests of Nigeria. The total litterfall was 9.3 t ha^{-1} per year for the period of November 2001 to October 2002. They considered the value to be well within the range of 7.0–14.1 t ha^{-1} per year reported for other West African tropical lowland rainforests (Swift et al. 1981; Muoghalu et al. 1993; Songwe et al. 1995; Odiwe and Muoghalu 2003). The value also fell within 5.8–12.0 t ha^{-1} per year reported by Anderson and Swift (1983) and Proctor (1987) for lowland tropical rain forests. Chave et al. (2010) showed that across old-growth tropical rainforests, litterfall averages 8.61 t ha^{-1} per year. Secondary forests have a lower annual litterfall than old-growth tropical forests with a mean of 8.01 t ha^{-1} per year. The outlying secondary forest at the edge of the Mata de Piedade site, the Atlantic rain forest of Brazil, produced 14.74 t ha^{-1} per year litter. Montane forests and low forests had lower mean annual litterfall (7.06 t ha^{-1} per year and 3.01 ± 1.67 t ha^{-1} per year, respectively). Gonzalez (2012) reported a wide range, from 1.99 to 9.16 t ha^{-1} per year, of total litterfall in the floodplain forests of a large Mediterranean river. Li et al. (2005) observed that litterfall in evergreen broadleaf forests in the subtropical region of China ranged from 7.1 to 11.0 t ha^{-1} per year with a mean value of 9.0 t ha^{-1} per year; but the average annual litterfall in the pine forest was only 2.67 t ha^{-1} per year. Yang et al. (2005) estimated litterfall in seven natural forests of *Schima superba* (SCS), *Castanopsis fabri* (CAF), *Tsoongiodendron odorum* (TSO), *Cinnamomum chekiangense* (CIC), *Altingia gracilipes* (ALG), *Castanopsis carlesii* (Hemsl.) (CAC), and *P. massoniana* (PIM) and compared it with that of an adjacent 29-year-old plantation of Chinese fir (CUL) in Jianou, Fujian, China. Mean annual total litterfall over 3 years of observations varied from 4.63 t ha^{-1} in the CUL forest to 8.85 t ha^{-1} in the PIM forest. Litterfall in disturbed and undisturbed teak forests of the Satpura Plateau in India was studied by Pande et al. (2002). They found total annual litterfall to range from 2.58 t ha^{-1} in the disturbed to 4.15 t ha^{-1} in the undisturbed teak forest. Mangrove in Quintana Roo, Mexico, produced a total litter fall of 2.61 t ha^{-1} per year (Navarette 2002). In a *Quercus suber* forest ecosystem of Montseny, Spain, the average annual litter production of 4.77 t ha^{-1} per year was similar to those found in other Mediterranean evergreen forests with relatively high rainfall (Caritat et al. 2006). Siddiqui and Qasim (1990) observed that total amount of litter production in an *Avicenia marina* mangrove forest of Pakistan was 3.71 t ha^{-1} per year. Vitousek and Sanford (1986) reported 3.6–12.6 t ha^{-1} per year litterfall with a mean value of 6.4 t ha^{-1} per year in tropical forests of a number of different countries. Fine litter production was investigated at monthly intervals from January to December 2003 in two tropical dry evergreen forest sites, namely Kuzhanthaikuppam (KK) and Oorani (OR) on the Coromandel Coast of South India. Fine litter production amounted to 13.51 and 13.27 t ha^{-1} per year at KK and OR, respectively (Pragasan and Parthasarathy 2005). From January 1992 to December 1994, Shanmughavel et al. (2000) investigated the litter production and nutrient cycling behavior of *Bambusa bambos* plantations of different ages. The annual litter production was 15.4 t ha^{-1}, 17 t ha^{-1}, and 20.3 t ha^{-1} in 4-, 5-, and 6-year-old plantations. Annual rates of litter production varied between 8 and 10.5 t ha^{-1} per year in a tropical semi-deciduous forest of the southern Amazon Basin (Sanches et al. 2008). Wang et al. (2007) compared litter production in pure and mixed species plantations, taking one pure CUL plantation (PC) and two mixed-species plantations of *C. lanceolata* with *Alnus cremastogyne* (MCA) and *Kalopanax septemlobus* (MCK) in subtropical China. Mean annual litter production values were 4.97 t ha^{-1} for MCA, 3.97 t ha^{-1} for MCK, and 3.46 t ha^{-1} for PC. Broadleaved trees contributed 42 % of the total litter production in MCA and 31 % in MCK. Litterfall in some evergreen and deciduous forests in Chile ranged from 3.5 to 5.8 t ha^{-1} per year and was temporarily lower in the managed than in the unmanaged deciduous stand (Staelens et al. 2011). Average yearly litter production in 450-year-old Douglas fir (*Pseudotsuga menziesii*) stands were studied by Abee and Lavender (1972), which represented six forest communities common to the western Cascades of Oregon. Annual litterfall was 5.89 t ha^{-1} per year. This is approximately one and

Table 6.1 Amounts of litterfall in some tropical, temperate, and Mediterranean forests

Forests/species	Location	Litterfall t ha−1 per year	Reference
Tropical forests			
Natural tropical forests of variable species diversity	Ghana	10.7	Vitousek and Sanford (1986)
	Zaire	12.4	
	Panama	11.4	
	Costa Rica	8.1	
	Guatemala	9.0	
	China	11.6	
	Australia	9.0	
Natural forest	Indonesia	13.67	Triadiati et al. (2011)
Broad-leaved forest	China	3.56	Zhou et al. (2007)
Mixed pine and broad-leaved forest	China	8.61	Zhou et al. (2007)
Evergreen broad-leaved forest	China	8.16	Zhou et al. (2007)
Temperate forests			
Quercus petraea	England	3.86	Carlisle et al. (1966)
Quercus robur	Sweden	5.28	Anderson (1970)
Acer–Fagus–Quercus	Indiana, USA	5.23	Vitousek et al. (1982)
Alnus rubra	USA	7.80	Turner et al. (1976)
Costanea sativa	Spain	6.18	Gallardo et al. (1989)
Fagus sylvatica	Southern Sweden	5.70	Nihlgard (1972)
Pinus sylvestris	Eastern Pyrenees	2.70–4.40	Pausas (1997)
Mediterranean forests			
Quercus coccifera	Southern France	2.30	Rapp (1969)
Quercus ilex	Southern France	3.80–7.0	Rapp (1969)
Quercus ilex	Spain	2.28	Bellot et al. (1992)
Quercus suber	Iberian Peninsula	2.88–4.3	Robert et al. (1996)

a half times the average 3.5 t ha^{-1} per year reported by Bray and Gorham (1964) for cool, temperate forests. Litterfall, litter turnover, and soil respiration were studied for 1 year in two forest plantations of *Pinus merkusii,* one on the slopes of Mount Merapi (at 800 m) and another on the slopes of Mount Merbabu (at 1,050 m) in Central Java. At the Merapi site, total litterfall was 9 t ha^{-1} per year, and at the Merbabu site total litterfall was 4 t ha^{-1} per year (Gunadi 1994). Williams and Gray (1974) estimated annual litterfall at 0.6–1.5 t ha^{-1} per year for arctic regions, 1.0–6.9 t ha^{-1} per year for cool temperate forests, 2.9–8.1 t ha^{-1} per year for warm temperate forests, and 5.5–15.3 t ha^{-1} per year for equatorial rainforests. Litterfall was studied in five even-aged stands of Norway spruce, Sitka spruce, Douglas fir, European beech, and common oak at three different locations in Denmark; two sandy sites, Ulborg and Lindet in Jutland; and one loamy site, Frederiksborg on Zealand. Litterfall was collected during 3 years from 1994 to 1996 in all five species and during 6 years from 1994 to 1999 in Norway spruce, Sitka spruce, and European beech. The average total litterfall was in the range of 3.20–3.70 t ha^{-1} per year (Hansen et al. 2009). Table 6.1 shows the amounts of litterfall in forests of tropical, temperate, and Mediterranean regions.

Factors Affecting Litterfall

The major factors for the variation in litterfall in forests are latitude, altitude, vegetation type, soil type, climate, and seasons. Actually, elements that enhance biomass production also improve litter production and OM inputs in forest ecosystems. Litterfall varies from place to place and from time to time. Accordingly, there are two types of variation in litterfall—spatial variation and temporal variation. Differences in the amount of litterfall in forests of different latitude, altitude, soil type, forest type, and climatic regions are examples of spatial variation.

Hansen et al. (2009) studied litterfall in five even-aged stands of Norway spruce, Sitka spruce, Douglas fir, European beech, and common oak at three different locations in Denmark. They observed no significant differences in total litterfall among sites during the short period, but during the longer period the more fertile sites had significantly higher total and foliar litterfall amounts compared to the less fertile sites. Vitousek and Sanford (1986) reviewed the literature on litterfall in different tropical forests and observed that soil fertility is a determining factor of the amount of litterfall. They collected litterfall data for a large number of soil types throughout the tropics. Average litterfall values for different soil orders of the soil taxonomy were given. Their litterfall

Table 6.2 Periods of peak litterfall in different tree species

Tree species	Peak period of litterfall	Reference
Pseudotsuga	October–November	Dimock (1958)
Pinus caribaea	March and April	Kadeba and Aduayi (1985)
Cunninghamia lancelata	November–March	Wang et al. (2007)
Picea purpurea	Major peak in October, a smaller peak in February	Yang et al. (2005)
Leuceana leucocephala	September–November	Oladoye et al. (2010)
Gmelina arborea	January–February	Ojo et al. (2010)
Quercus suber	May–June	Caritat et al. (2006)

values in different soil fertility categories can be summarized as:

Soil type	Amount of litterfall, t ha^{-1} yr^{-1}
Moderately fertile soils	8.1 – 12.6
Infertile soils	5.8 – 12.0
Sandy soils	5.6 – 9.2
Montane forests	3.6 – 11.0

The distribution of litterfall data shows a definite trend of variation with latitude. Tropical, temperate, and boreal forests represent three distinct latitudinal vis-à-vis geographical positions, and litterfall follows the trend: tropical forest > temperate forest > boreal forest. The corresponding forest types produce average annual litterfall around 10 t ha^{-1}, 5 t ha^{-1}, and < 3 t ha^{-1}. The ranges are as follows:

Forest type	Annual litterfall ranges, t ha^{-1}	Equator ↓ North pole
Tropical forest	8 – 14	Latitude increases
Temperate forest	3 – 6	Litterfall decreases
Boreal forest	2 – 3	

Saenger and Snedaker (1993) stated that a major paradigm in biosphere ecology is that organic production, carbon turnover and, perhaps, species diversity are the highest at tropical latitudes, and that each decrease toward the higher latitudes. A similar trend of variation is observed due to differences in altitude. Generally, the higher the altitude, the lower is the litterfall value. Montane forests produce less litter than lowland forests. Litterfall is also affected by species composition, stand age, basal area, and stand density. Generally, broadleaf species produce higher amounts of litterfall than coniferous species, while evergreen species produce more litter than deciduous species. There is higher litterfall in vigorously growing stands, increasing during the course of stand development. Litterfall increases with canopy closure. There is more litterfall in closed forests than in open forests.

Climate and soil moisture contents are other variables that affect litterfall. Moister climate and soil produce higher litterfall than forests affected by aridity.

Deciduous trees shed all of their leaves in a definite season, preferably winter or autumn. In evergreen forests, litterfall occurs throughout the year, but the maximum litterfall may occur in any one season or uniformly in all seasons. Litterfall increases as the environment becomes drier and decreases further with increasing moistness. The highest litterfall is called peak litterfall. If there is one peak, it is called unimodal litterfall; if there is more than one peak it is termed multimodal litterfall. Yang et al. (2005) observed a unimodal litterfall distribution pattern in CAF, *Altingia gracilipes*, and CAC forests, while multimodal in SCS, TSO, CIC, and PIM forests. The highest senescent needle litterfall in an old-growth temperate rainforest occurred from July to October, but the highest woody litterfall was from January to April (Edmonds and Murray 2002). Litterfall in 4–6-year-old bamboo plantations was found to follow a bimodal pattern with a principal peak in the winter and another in the late summer in all ages (Shanmughavel et al. 2000). Leaf litter production peaked at the end of the May–August dry season, while stem litter production peaked during the wet season, and reproductive litter production peaked during the dry–wet season transition in a tropical semi-deciduous forest of the southern Amazon Basin (Sanches et al. 2008). Usually, peak litterfall occurs in the period of maximum soil moisture depletion. The times of peak litterfall in some forest tree species are given in Table 6.2.

Components of Litter

Leaf litter constitutes the major proportion of litterfall in most forests. Branch litter, flower, fruits, and barks are taken together as miscellaneous litter. Usually, more than 60% of total litter is leaf, although there is variation among forests. Some examples are given. In seven natural forests and in one 29-year-old planted forest of China, Yang et al. (2005) demonstrated that leaf litterfall contributed from 62 to 73% of total litterfall. In disturbed and undisturbed teak forests of the Satpura Plateau in India, leaf litter ranged between 79 and 91% of total litterfall (Pande et al. 2002). Caritat et al. (2006) observed that the main components of litters in a *Quercus suber* forest ecosystem of Spain were the leaves (55 % of the total biomass), followed by acorns (22 %), and twigs (16 %). In an *Avicenia marina* mangrove forest of Pakistan, 74 % of the total production was contributed by leaves and the remaining 26 % by twigs, fruits, and flowers (Siddiqui and Qasim 1990). In an old-growth temperate rainforest watershed in the Hoh River valley, Olympic National Park, Washington, litterfall was collected and sorted into green needles, senescent needles, fine wood, reproductive, and miscellaneous litter. Needles provided the greatest amount of litterfall (60 %), with woody litter and other material averaging 18 and 22 %, respectively (Edmonds and Murray 2002). Of the total annual litter production in bamboo plantations, leaf litter accounted for 58 % and twig litter for 42 % (Shanmughavel et al. 2000). Needle litter in four mature stands of *Pinus sylvestris* corresponded to 40–60 % of the total litter fall. From worldwide data, Bray and Gorham (1964) reported that non-leaf litter averaged from 27 % to 31 % of total litter production. Gunadi (1994) observed that in two pine forest plantations in Central Java, Indonesia, needle litter contributed 63 % of total litter. Leaf litter also contains the highest concentration of nutrients.

A little studied component of forest litter is the down woody debris. Some investigators (Harmon et al. 1986; Krajick 2001) suggested that down woody debris is an important component of forest ecosystems and has two interrelated roles. First, it is a substrate of a variety of decaying organisms whose life cycles are closely tied to decaying wood (deMaynadier and Hunter 1995; McComb and Lindenmayer 1999; Siitonen 2001; Jonsson et al. 2005). Woody debris is important for the regeneration of various tree species. Woody debris acts as "nurse logs" (Cornett et al. 2001; Svoboda et al. 2010). Second, down woody debris is a functional component of the forest, having important influences on soil biology, soil hydrology, geomorphology, and nutrient cycling (Harmon et al. 1986; Keenan et al. 1993; Duvall and Grigal 1999). Therefore, altering woody debris volume, size, and/or spatial distribution by silvicultural treatments affects forest processes such as nutrient flow and respiration (Harmon et al. 1986; McComb and Lindenmayer 1999; Janowiak and Webster 2010). Saunders et al. (2011) suggested that the importance of woody debris, particularly regarding nutrient cycling, remains somewhat poorly understood, probably because of its variation by region and forest type. Laiho and Prescott (1999) reported that woody debris contributed less than 5 % of the N and P released during cycling in conifer forests of southwestern Alberta. On the other hand, Arthur and Fahey (1990) reported a substantial component of a stand's nutrient capital in woody debris. In northeastern North America, there has been little research on the nutrient concentration of down woody debris except for work in mid-to-high-elevation, *Abies balsamea–Picea rubens* forests of the White Mountains of New Hampshire (Lambert et al. 1980; Foster and Lang 1982).

Concentration of Nutrients in Litter

Concentrations of nutrients in litter components are important considerations for estimating nutrient returns in a forest ecosystem from aboveground vegetation. It is the most significant nutrient input in forests because forests are commonly self-nourished. Nutrient concentration is much higher in leaf litter, but leaf litter concentration may also differ due to site and species characteristics. For example, average litter N concentrations, unweighted for seasonal variation in litter production, ranged from 0.44 % in *Eucryphia cordifolia* to 0.84 % in *Laurelia philippiana*. P concentrations ranged from 0.04 % in *E. cordifolia* to 0.12 % in *L. philippiana* and *Aextoxicon punctatum* (Wood et al. 2006). Leaf litter N, P, and K were significantly higher in the inceptisols than in both ultisol plateaus and ultisol slopes. However, topography did not significantly influence leaf litter N, P, and K. Mean leaf litter P for inceptisols was 0.08 % and for ultisol plateaus 0.07 %. Leaf litter N values varied from 1.6 to 1.9 %. The order of nutrient concentrations in litterfall by *T. catappa* stands was Ca > N > K > Mg > P > Na, while that of the adjoining rainforest was N > Ca > K > Mg > P > Na. The concentration of plant nutrients in all the litter components of *Leucaena* litter followed this general trend: N > P > Mg > Ca > K > Na. However, leaf litter and seed components maintain that order, while twigs and pods follow the order N > Mg > P > Ca > K > Na, and flowers show a separate order, i.e., K > N > P > Mg > Ca > Na. This suggests that P and K are concentrated in the reproductive structure (Reinaldo and Philip 1995). The concentration of macro- and micronutrients contributed by litter showed the following order: Ca > N > K > Mg > P > S and Fe > Mn > Zn > B > Cu. The results indicated the important and significant role played by riparian forests in the cycling of nutrients and in restoration of soil fertility, conducting to the equilibrium and sustainability of the natural ecosystems (Fior et al. 2010). In Populus deltoides forests at Kuruskhetra, India, Bhardwaj et al. (2005) observed the following nutrient concentrations in leaf litter: N = 1.28–1.76 %; P = 0.071–0.091 %; K = 0.57–0.92 %; Ca = 0.99–1.04 %; and Mg = 0.36–0.56 %. Nutrient concentration in twig litter was low except for the fact that calcium concentrations were relatively higher in twigs. The nutrient concentration in *Leucaena leucocephala* in Nigeria followed the trend N > P > Mg > Ca > K > Na for leaf litter and seed components. Nutrient concentration in twigs and pods ranked N > Mg > P > Ca > K > Na (Oladoye et al. 2010). Table 6.3 gives some values of nutrient concentrations of different forest types. The ranges of nitrogen, phosphorus, potassium, calcium, and magnesium were: 0.4–2.1, 0.03–0.50, 0.05–2.0, 0.04–1.96, and 0.20–0.56 %, respectively. This wide variation can be caused by species difference, soil dif-

Table 6.3 Nutrient concentration in leaf litter of different forests

Place	Forest type	Percent					Reference
		N	P	K	Ca	Mg	
Malaysia	Dipterocarp rain forest	1.0	0.45	0.15	0.11		Proctor (1984)
Malaysia	Calcareous	0.4	0.16	0.32	0.33		Proctor (1984)
India	Lowland rain forest	1.5	0.06	0.66	1.38	0.39	Swamy and Proctor (1994)
India	*Populus deltoides*	1.9	0.13	0.34	1.35	0.24	Bhardwaj et al. (2005)
Ghana	High forest	1.9	0.07	0.07	1.96		Nye & Greenland (1960)
Congo	Plateau forest	0.6	0.39	0.85	0.43		Laudelot and Meyer (1954)
Ivory Coast	Rain forest	2.0	0.50	0.81	0.32	0.29	Bernhard-Reversat (1972)
Ivory Coast	Rain forest	0.5	0.49	1.36	0.32		Bernhard-Reversat (1972)
Trinidad	*Mom excelsa*	0.9	0.04	0.16	0.91	0.22	Cornforth (1970)
Guatemala	Secondary forest	0.7	0.24	0.12	0.35		Ewel (1976)
Columbia	Lower Montane	1.1	0.07	0.89	–	–	Veneklaas (1991)
Costa Rica	Rain forest	1.9	0.09	0.33	0.08	0.20	Wood et al. (2006)
Costa Rica	*Pinus pinea*	0.5		0.48	0.27		Wood et al. (2006)
Finland	*Pinus sylvestris*	0.8	0.09	0.23	0.42	0.07	Wood et al. (2006)
Ghana	Moderately fertile	2.1	0.09	2.00			Vitousek and Sandford (1986)
Venezuela	Spodosol	0.7	0.05	0.77			Vitousek and Sandford (1986)
Columbia	Oxisol	1.6	0.03	0.17			Vitousek and Sandford (1986)
India	*Tectona grandis*	1.4	0.32	0.38	1.61	0.56	Pande et al. (2002)
Finland	*Picea abies*	1.0	0.09	0.27	0.62	0.11	Ukonmaanaho et al.(2008)
Turkey	*Quercus cerris*	0.8	0.05	0.72	1.07	0.04	Kutbay and Horuz (2001)

ference, and climatic and other environmental variations. Readers are referred to Vitousek and Sandford (1986) to observe how litter nutrient concentration varied within more than 40 different tropical forests and how soil fertility was responsible for this variation.

There is also variation in concentration of nutrients in other litter components as well. Ukonmaanaho et al. (2008) reported that miscellaneous litter (branches, fruits, barks, etc.) of *Pinus syslvestris* contained 0.77% N, 0.06% P, 0.10% K, 0.24% Ca, and 0.04% Mg. Kutbay and Horuz (2001) reported 0.38% N, 0.02% P, 0.3% K. 1.56% Ca, and 0.17% Mg in wood litter; 0.43% N, 0.05% P, 1.32% K. 1.83% Ca, and 0.25% Mg in reproductive litter; and 0.44% N, 0.03% P, 0.93% K, 1.24% Ca, and 0.06% Mg in miscellaneous litter.

Amounts of Nutrients Returned in Litter

Only a small part of the assimilated nutrients of forest trees is used for physiological requirement and stored in biomass, while a large part returns to the soil (Alifragis and Papamichos 1990). The leaves, branches, seeds, bark, and the dead roots may be the means of nutrient return to the soil by the standing biomass (Sollins et al. 1980). These plant residues differ between broad-leaved and conifer species, regarding the quantities that are annually added to the soil surface. Bray and Gorham (1964) reported that conifers produce 17% more plant residues than the broad-leaved species.

Let us take an arbitrary example for estimating annual nutrient returns in a tropical forest. Let there be a total litterfall of 12 t ha^{-1} per year, containing 70% in leaf and 30% in miscellaneous litters. Let the leaf litter contain 1.6% N, 0.12% P, 0.65% K, 1.1% Ca, and 0.25% Mg, and the miscellaneous litter contain 1.2% N, 0.08% P, 0.8% K, 1.7% Ca, and 0.2% Mg. Let us calculate the amounts of nutrients returned annually in litterfall from vegetation to the soil surface.

Total annual litterfall	12 000 kg ha^{-1}
Leaf litter	12000 × 70/100 = 8,400 kg ha−1 per year
Miscellaneous litter	12000 × 30/100 = 3,600 kg ha^{-1} per year
Annual return of N in leaf litter	8,400 × 1.6/100 = 134.4 kg ha^{-1}
Annual return of P in leaf litter	8,400 × 0.12/100 = 10.08 kg ha^{-1}
Annual return of K in leaf litter	8,400 × 0.65/100 = 54.6 kg ha^{-1}
Annual return of Ca in leaf litter	8,400 × 1.1/100 = 92.40 kg ha^{-1}
Annual return of Mg in leaf litter	8,400 × 0.25/100 = 21.0 kg ha^{-1}

Annual return of N in miscellaneous litter 3,600 × 1.2/100 = 43.2 kg ha^{-1}

Annual return of P in miscellaneous litter 3,600 × 0.08/100 = 0.23 kg ha^{-1}

Annual return of K in miscellaneous litter 3,600 × 0.8/100 = 28.8 kg ha^{-1}

Annual return of Ca in miscellaneous litter 3,600 × 1.7/100 = 61.2 kg ha^{-1}

Annual return of Mg in miscellaneous litter 3,600 × 0.2/100 = 7.2 kg ha^{-1}

Annual return of N in total litter 134.4 + 43.2 = 177.6 kg ha^{-1}

Annual return of P in total litter 10.08 + 0.23 = 10.31 kg ha^{-1}

Annual return of K in total litter 54.6 + 28.8 = 83.4 kg ha^{-1}

Annual return of Ca in total litter 92.40 + 61.2 = 153.6 kg ha^{-1}

Annual return of Mg in total litter 21.0 + 7.2 = 28.2 kg ha^{-1}

Therefore, this hypothetical tropical forest would return annually 177.6 kg N, 10.31 kg P, 83.4 kg K, 153.6 kg Ca, and 28.2 kg Mg per hectare. Amounts of actual annual nutrient returns to the forest soils reported so far are comparable. For example, mean annual nutrient returns in litter in a rain forest on an Inceptisol of Costa Rica were 133 kg N, 6 kg P, 23 kg K, 56 kg Ca, and 14 kg Mg per hectare. Somewhat less NPK and higher Ca and Mg were returned in an Ultisol (Wood et al. 2006). The order of nutrient elements return via litterfall by *Terminalia catappa* stands was Ca > N > K > Mg > P > Na, while that of the adjoining rainforest was N > Ca > K > Mg > P > Na. In a 5–7-year-old *Populus deltoides* agroforestry system at Kurukshetra, annual return of nutrients (kg ha^{-1}) in litterfall and fine roots was: 95.96 N, 8.85 P, 48.08 K, 64.73 Ca, and 27.42 Mg (Bhardwaj et al. 2002). Annual return of N in litter of CUL in China ranged from only 4.3 to 7.0 kg ha^{-1} per year and Ca from 3.0 to kg ha^{-1} per year. Morellato (1992) observed an annual addition through litter of 206 kg N ha^{-1}, 11.2 kg P ha^{-1}, 37.8 kg K ha^{-1}, 269.2 kg Ca ha^{-1}, and 29.9 kg Mg ha^{-1}. Fior et al. (2010) obtained the following ranges of N, P, K, Ca, and Mg, consecutively: 159–278, 11.4–15.5, 27.8–43.9, 164.6–265.2, and 20.7–30.5 kg ha^{-1} per year. Meanwhile, the returns of P, K, and Mg through leaf litter were much smaller than those for N and Ca (Wang et al. 2007). Litters return substantial amounts of micronutrients as well. For example, Vital et al. (2004) observed annual addition through litter of 18.3 kg Fe ha^{-1}, 5.6 kg Mn ha^{-1}, 0.33 kg B ha^{-1}, 0.22 kg Zn ha^{-1}, and 0.15 kg Cu ha^{-1}. Fior et al. (2010) observed higher inputs of B, Fe, Zn, and Cu in riparian forests of Brazil. Pande et al. (2002) observed annual returns 25.71 kg N, 6.85 kg P, 9.45 kg K, 69.92 kg Ca, and 11.53 kg Mg ha^{-1} in litters of dry deciduous teak forests of the Satpuras, India. Vitousek and Sanford (1986) reviewed data on nutrient cycling in more than 30 tropical forests. They categorized forests on different soil fertility categories and showed that litterfall, nutrient concentration in litters, and nutrient returns varied among these categories and within a category. They found ranges of nutrient returns in litters of tropical forests are: 34–224 kg N ha^{-1} per year, 1.2–15 kg P ha^{-1} per year, 6–68 kg K ha^{-1} per year, 6.5–290 kg Ca ha^{-1} per year, and 1–64 kg Mg ha^{-1} per year. Nutrient returns to the forest floor of a sub-alpine coniferous forest were 18, 2, 23, 45, 4, and 5 kg ha^{-1} per year for N, P, K, Ca, Mg, and Mn, respectively, mostly in the form of litter, except in the case of K and Mg where throughfall was important (Turner and Singer 1976). Distribution and movement of nutrients were studied by Poggiani (1985) in an 11-year-old stand of *Eucalyptus saligna* and in a 14-year-old stand of *Pinus caribaea* var. *hondurensis*, located respectively near Piracicaba and near Agudos (State of Sao Paulo- Brazil). In the *E. saligna* stand annual leaf litter returned 27.3 kg of N, 2.2 kg of P, 16.7 kg of K, 44 kg of Ca, and 9.3 kg of Mg ha^{-1}. In the stand of *P. caribaea*, the corresponding nutrient returns were 43.7, 2.2, 22.2, 20.4, and 6.4 kg ha^{-1} per year. A study on nutrient cycling in two forest ecosystems in the Sierra de la Demanda, a climax beech forest (*Fagus sylvatica* L.) and a reforested plantation of Scots pine (*Pinus sylvestris* L.), on land suitable for beech was done by Santa-Regina and Teresa Tarazona (2000). They concluded that nutrient return to the soil by the litter was higher under beech compared to pine for all analyzed elements (P: 0.23 and 0.11 kg ha^{-1} per year; Ca: 17.4 and 12.8 kg ha^{-1} per year; Mg: 4.0 and 2.0 kg ha^{-1} per year, and K: 11.5 and 5.7 kg ha^{-1} per year) except N (19.8 and 23.3 kg ha^{-1} per year).

6.3.3.2 Nutrient Returns in Throughfall, Stemflow, and Canopy Drip

Stemflow is the flow of rainwater along plant branches and stems. The amount of stemflow is determined by leaf size and shape and stem and branch architecture and by rainfall characteristics. In general, broadleaf trees have more stemflow than coniferous vegetation. Some rainwater also falls as canopy drip. This is especially true of coniferous vegetation. Several studies of the elemental content of precipitation under forest stands (Tamm 1951; Madgwick and Ovington 1959; Will 1959; Voigt 1960) have demonstrated that throughfall (rainwater which has passed through tree crowns) contains significantly higher quantities of many nutrient elements than rainfall collected in adjacent openings. This process is controlled by factors like plant leaf and stem density, type of the precipitation, intensity of the precipitation, and duration of the precipitation event (Parker 1983). The amount of precipitation passing through varies greatly with vegetation type. Some nutrients from vegetation are leached with stemflow, canopy drip, and throughfall. These pathways generally represent a relatively minor vector for nitrogen, phosphorus, and calcium transfer in tropical forests, although it is the major pathway of potassium transfer. Nutrients deposited on foliar surfaces through rainfall or dry fallout may be carried downward in throughfall (Carlisle et al. 1966). Nutrients may also be leached from foliar surfaces by precipitation (Stenlid 1958) and, together with atmospheric input, deposited as enriched solutions of throughfall on forest floors (Tamm 1953). The contribution of throughfall and stemflow as pathways of the nutrient recycling within the forested Hubbard Brook ecosystem was investigated by Eaton et al. (1973). They observed concentrations of Ca, Mg, K, Na, NO_3, SO_4, NH_4, Cl, PO_4, organic N, and OM in throughfall. Variation in throughfall and stemflow chemistry was determined under American beech, sugar maple, and yellow birch. Nutrients more commonly found in an ionic form (like K) were found to move very rapidly from the forest canopy in throughfall and stemflow. The annual nutrient return to the forest soil for the elements K and S is predominantly via throughfall and stemflow and little because of litterfall. Stemflow transfers 5–20 % of the total in precipitation-borne solutes (Parker 1983). Reiners (1972) reported quantities of nutrients returned in throughfall in three forests of Minnesota. In oak, fen, and swamp forests, respectively, there were 5.5, 5.5, and 6 kg ha^{-1} per year N; 0.6, 0.7, and 0.5 kg ha^{-1} per year P; 7.6, 10.5, and 10.7 kg ha^{-1} per year Ca; and 3.1, 3.8, and 3.7 kg ha^{-1} per year Mg. Throughfall deposition of major base cations was studied by means of throughfall analysis in a deciduous beech (*Fagus sylvatica* L.) forest edge in Belgium over a period of 1 year by Devlaeminck et al. (2005). Throughfall fluxes of Cl^-, NH_4^+, and Na^+ were significantly elevated at the forest edge compared to the forest interior. As no edge effect on throughfall water volume could be detected, the observed edge enhancement effects were mainly due to dry deposition and canopy exchange patterns. Borong et al. (1993) studied nutrient elements in precipitation and throughfall in a pine–spruce–fir forest of Changbai Mountain Natural Reserve. The amount of nutrient elements in throughfall was higher than that in precipitation, and their leached amount from canopy was in the order of K > Mg > Ca > N. They suggested that from the viewpoint of nutrient return, throughfall plays a more important role in comparison to litterfall. Zhang (1999) measured concentrations of N, P, K, Ca, and Mg in open rainfall and in throughfall, stemflow, and litterfall from various tree species. Average annual nutrient inputs (kg- ha^{-1}) for total-N, NO_3–N, NH_4–N, P, K, Ca, and Mg were 7.49, 1.04, 1.11, 2.44. 8.53, 19.09, and 4.76, respectively. Litterfall from the plantation canopy was the major contributor of N (90.5 % of all nitrogen inputs), whereas throughfall contributed the most K (62.7 % of all potassium inputs), and stemflow accounted for a very small proportion of the nutrients but contributed the most P (50.3 % of all phosphorus inputs).

6.3.3.3 Nutrient Returns Through Root Decay

Very little information is available on nutrient turnover by forest tree roots. However, the data of Sanford (1985) from a Venezuelan Oxisol are mentioned here as an example. Estimated turnover of roots less than 2 mm in diameter in the upper 10 cm was about 25 % per month; it yielded an annual fine-root production of 15.4 Mg ha^{-1}. Nitrogen and phosphorus concentrations in roots less than 2 mm in diameter were 2.23 and 0.073 %, respectively. Therefore, the quantities of nitrogen and phosphorus added to the soil by turnover of these roots were 343 kg ha^{-1} per year and 11 kg ha^{-1} per year, respectively, in contrast to 61 kg kg.ha^{-1} per year ^{N} and 0.8 kg ha^{-1} per year P added by fine litterfall.

According to Gordon and Jackson (2000), fine roots are important sources and sinks for nutrients in terrestrial ecosystems. Across a range of ecosystems, net primary production can be greater belowground than above (Caldwell 1987), and nutrient concentrations in fine roots may be higher than those in foliage (Meier et al. 1985) and their life spans are considerably shorter (Vogt et al. 1983). Nutrient release from decomposing roots is a pathway of significant nutrient flux in terrestrial ecosystems (Joslin and Henderson 1987; Fahey et al. 1988). In forests, the amount of carbon and nutrients returned to the soil from fine-root turnover may equal or exceed that from leaf litter (Joslin and Henderson 1987; Raich and Nadelhoffer 1989). Minimal retranslocation of nutrients from roots upon senescence may also contribute to the importance of fine roots in nutrient cycling (Aerts 1990; Nambiar and Fife 1991).

6.3.4 Nutrient Release

Absorbed and biomass-bound nutrients are released to forest soil primarily through the decomposition of litters fallen from aboveground vegetation. Some nutrients may also be released by canopy leaching and exudation of roots and death and decay of roots and soil organisms.

6.3.4.1 Litter Decomposition

Litter decomposition is a complex process; actually, it is a combination of a series of different processes operating simultaneously under given environmental conditions. According to Berg (1986), litter decomposition involves two simultaneous and fundamental sets of processes: (1) the concomitant mineralization and humification of lignin, cellulose, and other compounds by a succession of microorganisms; and (2) the leaching downward in the soil of soluble compounds whose C and nitrogen (N) are progressively mineralized or immobilized. These processes are controlled by abiotic factors, such as climate, and by biotic factors, such as litter chemical composition and soil organisms. Litter decomposition may be rapid or slow depending on litter quality and environment, and it proceeds at different rates at different stages in a particular environment (Satchell 1974). The rate of decomposition also varies for different litter components; leaf litter decomposes faster than wood litter. The major groups of litter-decomposing organisms are bacteria, actinomycetes, fungi, protozoa, nematodes, microarthropods, macroarthropods, enchytraeid worms, and lumbricid worms. The accepted pathway of decomposition of plant debris in the soil can be summarized as follows: (1) development of phylloplane microflora, (2) colonization by saprophytic microorganisms, (3) comminution and ingestion by invertebrates, resulting in incorporation of OM into the soil and a great increase in surface area but little chemical change, (4) microbial colonization and utilization of feces and comminuted litter resulting in chemical degradation of tissues and production of complex phenolic polymers (humic acids), and (5) the formation of stable organo-mineral complexes (Burges 1967; Wood 1976).

Methods of Studying Rates of Litter Decomposition

The decomposition of plant litter is one of the most crucial processes in the biogeochemical cycle of the forest ecosystems (DeCatanzaro and Kimmins 1985). It has been studied in almost all forest types, and investigators primarily adopted two different approaches to estimate the rate of litter decomposition in a forest stand: (1) litter disappearance from forest floor and (2) mass loss of confined litter.

Litter Disappearance from Forest Floor

Here, the rate of litter decomposition is estimated by the difference in standing litter mass between the initial and the time of observation. The understory in a patch of forest, say 1 m^2, is cleared and a net is fixed overhead to prevent further litter addition to the place. Let the mass of standing litter on the forest floor be M_0 (g) at time t_0, and M_t (g) at time t. Then, the rate of litter decomposition R becomes

$$R = \left(\frac{M_0 - M_t}{t - t_0} \right),\ gm^{-2} \ldots.$$

Wiegert and Evans (1964) proposed a paired-plot technique for the determination of litter decomposition rates over short periods of time in old fields of upland and swale vegetation. An area of approximately 2 m was selected on one side, having relatively homogeneous vegetation in the field. Two quadrats of 0.5 m^2 were located within this area so that they shared one border and also had similar amounts and species composition of dead material. The living vegetation was removed from within both the quadrats. One quadrat was then selected at random and the dead material was removed from it. After an interval, the dead material was removed from the second plot. The difference between the weights of the first plot and the second plot gave the amount of material that had decomposed during the interval.

Mass Loss of Confined Litter

The patterns of decomposition, nature of changes in chemical composition, release of nutrients, and biology of decomposition have been studied by subjecting confined leaf litter of various plant species to decomposition in natural environments. Various containers of wire screening (Falconer et al. 1933; Lunt 1935; Coldwell and DeLong 1950), containers of nylon hairnets, and nylon litterbags (Bocock and Gilbert 1957; Bocock et al. 1960) have employed to study rates of litter decomposition. Yamane and Sato (1971) used plastic wire cages for studying decomposition of leaf blades and culms of *Miscanthus sinensis* and *Sasa palmata*. Ashton (1975) used fly-wire baskets for studying decomposition of *Eucalyptus* leaf litter. The most widely used method in decomposition studies is the litterbag technique (Singh and Gupta 1977). The flexibility of nylon permits the enclosed leaves to become embedded in the natural litter layer. In addition, nylon has the advantage of being biologically inert. Besides, mesh sizes may be selected to allow the entry of all but the largest litter-inhabiting invertebrates. The periodic weight loss in these enclosed litter samples represents decomposition rates. Litterbags with measured amounts of litter are placed in the forest floor and observed periodically for their mass loss to estimate rates of decomposition.

Most studies revealed that the litterbag method yields data which are pertinent for understanding the role of different groups of soil invertebrates in decomposition. Besides, the litterbag method minimizes the loss due to fragmentation. This technique, however, may underestimate the rate of decomposition because the litterbags restrict the entry of large invertebrates, scavengers, and decomposers that could be important detritivores (Singh and Gupta 1977). Roots and fungal hyphae might enter the litterbags during exposure in the field and hamper the interpretation of changes in C, N, P, and K concentrations. However, despite the disadvantages, the litterbag method at present is the most appropriate technique available to study OM breakdown in decomposition studies under field conditions (Knacker et al. 2003)

Mathematical Models

Some mathematical models were developed for the explanation of the process of litter decomposition (Jenny et al. 1949; Olson 1963; Witkamp and Olson 1963). According to Jenny et al. (1949), the annual input of OM must be equal to the annual outputs at the steady state which may or may not be reached in a forest depending on the successional stage and the nature of disturbance. The rate of OM decomposing annually could be estimated by calculating a community parameter k'. The parameter k' is calculated using the formula:

$$k^{l} = \frac{A}{Fe + A},$$

where Fe is the amount of material present on the forest floor just before annual leaf fall or the minimum value for the litter and A is the annual increment in this material. The time required for the amount of litter to reach an equilibrium state may be calculated by the formula:

$$t = \frac{-\ln(Fe - F)}{Fe.k^{-1}},$$

where t is the time in years, F is the percentage of equilibrium value, and k is the decomposition rate. However, according to Olson (1963), the natural communities do not exactly fit the above model. He proposed alternative models. At a steady-state condition, the rate of loss from the system can be estimated by using the equation:

$$k = \frac{L}{X_{ss}},$$

where k is the rate parameter, L is the annual litter production, and X_{ss} is the annual litter accumulation of forest floor in steady state. In cases where the material is confined in mesh bags, the measuring of weight loss is a special case with essentially no addition of material, i.e., litter input is zero or L=0. The model for the constant potential weight loss in this case can be represented by the equation:

$$\frac{X}{X^0} = e^{-kt},$$

where X is the weight remaining at time t, X^0 is the original weight, e is the base of natural logarithm, and k is the decomposition constant.

The rates of litter decomposition in various types of forest ecosystems have been studied using the equations given by Jenny et al. (1949) and Olson (1963). Under tropical conditions, the decomposition parameters for the forests have been estimated by Laudelout and Meyer (1954), Nye (1961), Madge (1965), Hopkins (1966), Cornforth (1970), etc. In temperate conditions, the decomposition parameters have been determined for several forest communities (Crosby 1961; Dahlman and Kucera 1965; Koelling and Kucera 1965; Attiwill 1968; Old 1969; Reiners and Reiners 1970; Lang 1974; Rochow 1974; Redmann 1975), and others.

Rate of Litter Decomposition

Litter decomposition proceeds at a very slow pace in some forests. For example, Jenny et al. (1949) reported litter decomposition rate of 1–3% per annum for Californian pine forests, and Olson (1963) observed 6.25% per annum in Minnesota pine forests. Some workers (Mikola 1954; Kendrick 1959; Crosby 1961; Witkamp 1966; Thomas 1968; Woodwell and Marples 1968) reported 11–44% litter decomposition per annum in temperate forests. Decomposition rates in temperate forests are lower than tropical forests. It may take only 0.1–0.6 years for complete litter decomposition in moist semi-deciduous, moist evergreen, and perhumid forests in the tropical environment (Jenny et al. 1949; Laudelout and Meyer 1954; Nye 1961; Madge 1965; Hopkins 1966), <1 year for temperate deciduous forests (Ovington 1962), 1.6 years for subtropical forests (Jenny et al. 1949), 3–5 years for coniferous forests in Britain (Ovington 1962), and as many as 28 years for high mountain oak and to 60 years for pine forests of California (Jenny et al. 1949). In temperate forests, the rates are comparatively higher for broad-leaf litter than conifer litter. Most litter decomposition studies have been conducted with leaf litter. Branch and wood litters decompose much slower than leaf litter, and litter of all tree species do not decompose at equal rates. Litter decomposition studies have revealed that litter decomposition in a forest floor occurs at different rates at different stages of decomposition.

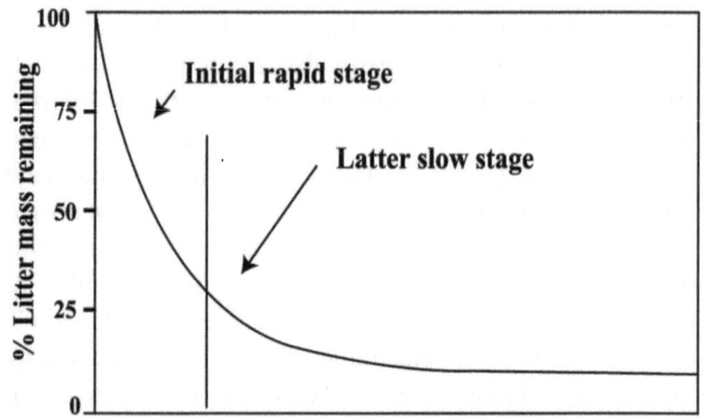

Fig. 6.3 An initial rapid decomposition stage is followed by a latter slow stage

It has been found to be a two-stage phenomenon (Berg and Staaf 1980); the initial stage is rapid which is followed by a latter slow stage (Fig. 6.3). At one time, the rate diminishes to almost zero; the time required to achieve this stage is highly variable from forest to forest, depending on litter quality and environment. Melillo et al. (1989) presented a general model of the decomposition process from litter to humus in two phases. During the early stage, there is rapid loss of water-soluble components followed by rapid loss of cellulose from the litter. There is little loss, or possibly even a gain of insoluble decay products (collectively referred to as "lignin"). During the early phase, carbon is relatively available and nutrients are limiting, and there is immobilization of the limiting nutrient (usually N). Once the litter reaches the late stage of decomposition, it is considered to be humus and it is distinguished by a stabilized content and slow decay of all components. The late stage of decay is characterized by a net loss of lignin and net mineralization of N (Prescott 2005)

Under the influence of the prevailing climatic environment, different litter species have their own specific rate of decomposition which governs the rate at which nutrients are released from litter (Das and Chaturvedi 2003). Bocock and Gilbert (1957) reported the highest rate of decomposition of *Betula verrucosa* (82.9–91.5 % per annum on mull) and the lowest for *Quercus petraea* and *Q. robur* (26.2–17.4 % per annum on mull). Bocock et al. (1960) compared mass loss of confined leaf litter in litterbags of *Q. petraea* and *Fraxinus excelsior* and found higher rates for *Fraxinus*. Only 5 % of the *Fraxinus* leaf material remained in the bags at the end of the investigation, while 85 % of the leaf material of *Quercus* remained undecomposed within that period. Bocock (1964) observed higher rates of decomposition for *Alnus glutinosa* as compared to *Betula pendula*. Moore (1984) conducted a litterbag experiment to examine the decomposition of six litter types (*Picea mariana, Cladina stellaris, Betula glandulosa,* and *Ledum groenlandicum*) in a mature spruce–lichen woodland in northern Quebec. The decomposition was measured over a 2-year period. Mass losses after 2 years were between 10 and 60 %, with the largest losses occurring in *B. glandulosa* and *L. groenlandicum* leaves and the smallest in *C. stellaris*. Nicholas et al. (1995) observed a clear tendency for litter of *Celtis zenkeri* to decompose faster than that of *Desbordesia glaucescens, Cola lepidota,* and *Ceiba pentand.* Nakane (1995) observed lower rate of decomposition of the leaf litter in the Japanese cedar (*Cryptomeria japonica*) plantation and red pine (*Pinus resinosa*) forest, compared to that of the oak forest. Tree species can affect the decomposition process through the quality of their leaf fall and through the species-specific conditions that they generate in their environment. In an experiment, litterbags containing leaf litter of the winter-deciduous *Quercus canariensis*, the evergreen *Q. suber*, and mixed litter were incubated beneath distinct plant covers. Tree species affected decay dynamics through their litter quality and, to a lesser extent, through the induced environmental conditions (Aponte et al. 2011).

Processes and Magnitude of Nutrient Release

Nutrient release from plant and animal debris is a fundamental process of nutrient recycling in natural ecosystems and it determines the turnover rate and distribution of bioelements in soil (Staaf 1980). The nutrient release phenomenon has considerably been studied but yet mechanisms of nutrient release from litter materials have not been fully understood. Mechanical leaching often causes a rapid loss of some elements like sodium and potassium (Kucera 1959; Frankland 1966), and magnesium is also found to be leached considerably from forest litter (Stachurski and Zimka 1975). Other elements undergo microbial mineralization. Some elements, such as K, are generally released to the environment right from the start of the decomposition process, but some elements are found to be released only at an advanced stage of decomposition. In most instances, nitrogen is released after more than 50% mass loss has occurred; most of the times, nitrogen (Gilbert and Bocock 1960; Bocock 1963; Anderson 1973; Howard and Howard 1974; Staaf and Berg 1977), and sometimes phosphorus (Gosz et al. 1973) are even accumulated in decomposing litter due to microbial immobilization. Accumulation of nitrogen and phosphorous and release of calcium and magnesium in decomposing litter were reported by Day (1982). Potassium was accumulated following initial losses. Accumulations were the result of immobilization by decomposing microbes. In a litterbag study with litters of *Picea mariana, Cladina stellaris, Betula glandulosa,* and *Ledum groenlandicum*), Moore (1984) observed that Ca, Mg, and K were rapidly lost from the decomposing tissues, except for *C. stellaris.* All the tissues, except *B. glandulosa,* showed an accumulation of N, associated with high C to N ratios. Nitrogen tended to remain immobilized in the litter, at least over the first 2 years. Rates of nutrient release (N, P, S, K, Mn, Ca, Zn, Fe, Mn, Cu, and Na) were measured in decomposing leaf litter of yellow birch, sugar maple, and beech. The concentration and absolute weight of N, S, and P in the leaf litter of all species increased with time. The amount of the increase as well as the initiation of nutrient release was influenced by C to element ratios in the leaf tissue. These studies also indicate that P levels can influence the mineralization or immobilization of other important nutrients (Gosz et al. 1973). Nutrient content of decomposing litter was determined in a *Eucalyptus diversicolor* forest in south-western Australia by O'Connell (1988). Changes in nutrient content of five leaf litters, karri (*E. diversicolor*) and four understory species *Trymalium spathulatum, Bossiaea laidlawiana, Acacia urophylla,* and *Casuarina decussate*, were studied over 82 weeks using the litterbag technique. Concentrations of N, P, and Ca increased in most litter components during decomposition, concentrations of Mg and S changed little, and the concentrations of Na, K, and Cl decreased markedly. Moore et al. (2006) examined the patterns of nitrogen and phosphorus gain, retention, or loss in ten foliar tissues in a litterbag experiment over 6 years at 18 upland forest sites in Canada, ranging from subarctic to cool temperate. Nitrogen was usually retained in the decomposing litter until about 50 % of the original C remained. The peak N content in the litter was observed at between 72 and 99 % of the original C remaining. The rate of N release from the litters was not related to the original N concentration in the litters. Phosphorus was immediately lost from all litters, except beech leaves. The rate of P loss was inversely correlated with the original litter P concentration. The soil underlying the litterbags influenced the pattern of N and P dynamics in the litters. There was a trend for higher N and P retention in the litter at sites with lower soil C to N and

N to P quotients, respectively. Xu et al. (2006) investigated the patterns of mineral element dynamics during decomposition of seven common canopy species litters in a subtropical evergreen broad-leaved forest by means of the litterbag technique over 2 years. They observed the element mobility in decomposition litters as: Na=K>Mg=Ca>N=Mn=Zn =P>Cu>>Al>>Fe. Concentrations of K and Na decreased in all species as decomposition proceeded. Calcium and Mg also decreased in concentration but with a temporal increase in the initial phase of decomposition, while the concentrations of other elements (Zn, Cu, Al, and Fe) increased for all species with exception of Mn which revealed a different pattern in different species. In most species, microelements (Cu, Al, and Fe) significantly increased in absolute amounts at the end of the litterbag incubation, which could be ascribed to a large extent to the mechanism of abiotic fixation to humic substances rather than biological immobilization. Nutrient release during decomposition of organic residues displays complex patterns depending on litter types, nutrients, diversity of decomposers, and conditions of decomposition. Therefore, any generalization is difficult.

A significant increase in concentrations of most nutrients in woody litters with advancing decay for both hardwoods and softwoods has been reported (Foster and Lang 1982; Arthur and Fahey 1990; Alban and Pastor 1993; Ganjegunte et al. 2004; Mladenoff et al. 2010), although the observation has not always been statistically significant or consistent across species (Krankina et al. 1999; Laiho and Prescott 2004). Several explanations for these increases have been proposed, including concentration of nutrients due to mass loss, inputs by throughfall and litterfall, and root colonization (Foster and Lang 1982; Arthur and Fahey 1990); presence of nutrient-rich basidiomycetes within litters (Harmon et al. 1994) etc. For Mn, Al, B, and K, however, static–to-decreasing nutrient concentrations were observed with increasing decay class (Saunders 2011). Potassium is known to be highly mobile (Arthur and Fahey 1990; Krankina et al. 1999) and leaching likely reduces its concentration over time (Brown et al. 1996).

6.4 Forest Floor Biomass and Nutrients

The forest floor composed of organic detritus is an important component of a forest. The magnitude and chemical contents of this compartment often represent a significant proportion of the total OM and nutrients in temperate forests (Rodin and Basilevich 1967) and may affect plant regeneration, hydrologic processes, nutrient cycling, and forest productivity (Westman 1975; Youngberg 1966). Forest floor characteristics may be dependent on the climate, soil fertility, successional stage, and species composition (Ovington 1968; Lang and Forman 1978). Many studies have shown (summarized by Lang and Forman 1978) that forest floor detritus, nutrient standing crops, and distribution patterns vary with major changes in species composition among forest communities. The forest floor is a major reservoir of OM and nutrients for the ecosystem and, as such, it influences or regulates most of the functional processes occurring throughout the ecosystem. Patterns of forest floor mass accumulation have been observed across a variety of forest ecosystems globally. The size and structure of the forest floor is a natural function of climate, vegetation, and time (Olson 1963). Soil OM accumulations in mature forests may exceed the living forest biomass (Rodin and Basilevich 1967); however, human interference can markedly alter these natural relationships (Likens et al. 1970; Pierce et al. 1972; Bormann et al. 1974). Ovington (1965) stated that greater forest floor mass accumulation occurs with increasing distance from the equator. Tropical forests have less OM on the forest floor than temperate forests. In tropical regions, forest floor mass has a wide range of variation: 3.3–10.0 Mg ha^{-1} for lowland rain forest and 5.1–16.5 Mg ha^{-1} for montane rain forest (Tanner 1981; Anderson et al. 1983). Zheng et al. (2006) reported forest floor mass in Xishuangbanna, China. They found 5.14 Mg ha^{-1} which is similar to forest floor mass in Ghana (4.9 Mg ha^{-1}) and Colombia (5.0 Mg ha^{-1}), higher than that in Pasoh Malaysia (3.2 Mg ha^{-1}) and Zaire (3.9 Mg ha^{-1}), and lower than in Sarawak Malaysia (5.5–7.1 Mg ha^{-1}) and Panama (11.2 Mg ha^{-1}). Vogt et al. (1986) compared forests worldwide and stated that patterns of forest floor mass accumulation can be distinguished by the evergreenness or deciduousness of forest foliage. Under the same climate conditions, forest floor mass in deciduous forests is generally half of those in evergreen forests. Tanner (1981) and Anderson et al. (1983) reported residence times (year) of forest floor mass in montane and lowland rain forests of 0.7–2.2 and 0.3–1.0, respectively. Zheng et al. (2006) observed the residence time of forest floor mass in Xishuangbanna to be 0.51. Average weight of the forest floor on an undisturbed watershed in Hubbard Brook Forest ranged from 25.50 to 85.50 Mg ha^{-1} (Gosz et al. 1976). The weighted watershed average was 46,800 kg/ha. Although the F and H horizons did not vary significantly with time, the L horizon increased significantly during the period June to August largely as a result of a severe hail storm. The order of abundance of elements in the forest floor was N>Ca<>Fe>S>P>Mn>K>Mg>Na>Zn>Cu. The concentrations of Ca, K, and Mn decreased with depth in the forest floor while N, P, S, Na, Fe, Zn, and Cu concentrations increased. N to P ratios were similar in decomposing leaf tissue, the forest floor, litterfall, and net stemflow plus throughfall, suggesting a similar pattern of cycling. Luvall and Weaver (1985) studied the distribution patterns of forest floor detritus and nutrients in southwestern Illinois oak–hickory forests. In these forests, the forest floor OM (32.8–38.7 Mg ha^{-1}) was equivalent to 16–19% of the total aboveground tree biomass. It also accounted for a significant amount of the aboveground nutrient pools of Ca (26–32 %), Mg (55–61 %), K (21–24 %), Mn (27–29 %), Cu (17–27 %), and Zn (35 %). The wood component in the forest floor accounted for substantial amounts of the total nutrients found there with 36–58 % Ca, 31–38 % Cu, 42–47 % K, 29–35 % Mg, 15 % Mn, 19–37 % P, and 36–40 % Zn. Nasiru (2000) determined forest floor litter biomass in *Tectona grandis* (teak) stands of different ages and adjacent natural forest in the Shasha Forest Reserve of Osun State, south-western Nigeria. The teak plantations were planted in 1965, 1970, 1975, 1980, and 1985. The forest floor litter biomass was the highest for the 1975 plantation (10.20 Mg ha^{-1}). The biomass for forest floor litter under other stands (6.7–7.4 Mg ha^{-1}) was not significantly different. The nitrogen contents in the litter were 10.5 kg ha^{-1}, 15.4 kg ha^{-1}, 11.2 kg ha^{-1}, 6.7 kg ha^{-1}, 11.4 kg ha^{-1}, and 11.8 kg ha^{-1} while the phosphorus contents were 7.2 kg ha^{-1}, 8.1 kg ha^{-1}, 7.9 kg ha^{-1}, 8.1 kg ha^{-1}, 7.1 kg ha^{-1}, and 9.7 kg ha^{-1} for 1965, 1970, 1975, 1980, and 1985 teak plantations and natural forest, respectively. The soil OM content was the highest under natural forest (5.6 %) and this was followed by the 1965 teak plantation (4.6 %). Soil OM content was the lowest in the 1970 plantation (2.5 %). The soil nitrogen content was also the highest under natural forest (0.3 %) followed by the 1965 teak plantation (0.2 %) and the lowest in the 1970 teak plantation (0.1 %).

Study Questions

1. Explain biomass, aboveground biomass, standing biomass, forest floor biomass, and litter biomass. How does the amount of standing biomass production differ among sites of varying fertility?
2. Discuss nutrient concentration and nutrients retained in biomass of forests. How can nutrient use efficiency be increased?
3. Define nutrient cycling and nutrient recycling. Discuss the process of nutrient recycling in forest ecosystems.
4. Narrate the significance of litterfall. Discuss factors affecting litterfall. Explain importance of throughfall and stemflow in nutrient recycling.
5. Explain litter decomposition as a two-stage phenomenon. Discuss factors affecting litter decomposition and nutrient release.

References

Aerts R (1990) Nutrient use efficiency in evergreen and deciduous species from heathlands. Oecologia 84:391–397

Abee A, Lavender D (1972) Nutrient cycling in throughfall and litterfall in 450-year-old Douglas-fir stands. Proceedings-Research on Coniferous Forest Ecosystems-A symposium. Bellingham, Washington-March 23-24, 1972

Aerts R (1996) Nutrient resorption from senescing leaves of perennials: are there general patterns? J Ecol 84:597–608

Alban DH, Pastor J (1993) Decomposition of aspen, spruce, and pine boles on two sites in Minnesota. Can J For Res 23(9):1744–1749

Albaugh TJ, Allen HL, Fox TR (2008) Nutrient use and uptake in *Pinus taeda*. Tree Physiol 28:1083–1098

Alifragis D, Papamichos NT (1990) Conservation of forest soil productivity. Proceedings of the third National Soil Conference: 187–195 (In Greek)

Anderson F (1970) Ecological studies in a Scanian woodland and meadow area, southern Sweden. II. Plant biomass, primary production and turnover of organic matter. Bot Not 123:8–51

Anderson JM (1973) The breakdown and decomposition of sweet chestnut (*Castanea sativa* Mill.) and beech (*Fagus sylvatica*L.) leaf litter in two deciduous woodland soils. Oecologia 12:251–274

Anderson JM, Swift MJ (1983) Decomposition in tropical forests. In: Sutton SL, Whitmore TC, Chadwick AC (eds) Tropical rain forest: ecology and management. Blackwell Publications, Oxford

Anderson JM, Proctor J, Vallack HW (1983) Ecological studies in four contrasting lowland rain forests in Gunung Mulu National Park, Sarawak. III. Decomposition process and nutrient losses from leaf litter. J Ecol 71:503–527

Andre F, Jonard M, Ponette Q (2010) Biomass and nutrient content of sessile oak (*Quercus petraea* (Matt.) Liebl.) and beech (*Fagus sylvatica* L.) stem and branches in a mixed stand in southern Belgium. Sci Total Environ 408(11):2285–2294

Aponte C, García LV, Pérez-Ramos IM, Gutiérrez E, Marañón T (2011) Oak trees and soil interactions: a positive feedback model. J Veg Sci 22:856–867

Arthur MA, Fahey TJ (1990) Mass and nutrient content of decaying boles in an Engelmann spruce subalpine fir forest, Rocky Mountain National Park, Colorado. Can J For Res 20(6):730–737

Ashton DH (1975) Studies of litter in *Eucalyptus regnans* forest. Aust J Bot 23:413–433

Attiwill PM (1968) The loss of elements from decomposing litter. Ecology 49:142–145

Augusto L, Ranger J, Ponette Q, Rapp M (2000) Relationships between forest tree species, stand production and stand nutrient amount quantifying the difference between concentration values from the literature and actuals. Ann For Sci 57:313–324

Baker TG, Attiwill PM (1985) Above-ground nutrient distribution and cycling in *Pinus radiata* D. Don and *Eucalyptus obliqua* L'Hérit. forests in southeastern Australia. For Ecol Manage 13:41–52

Barron-Gafford GA, Will RE, Burkes EC, Shiver B, Teskey RO (2003) Nutrient concentrations and contents, and their relation to stem growth, of intensively managed *Pinus taeda* and *Pinus elliottii* stands of different planting densities. For Sci 49(2):291–300

Bellot J, Sanchez JR, Lledo MJ, Martinez P, Escarre A (1992) Litterfall as a measure of primary production in Medditerranean holm-oak forest. Vegetatio 99–100:69–76

Berendse F, Aerts R (1987) Nitrogen-use-efficiency: a biologically meaningful definition? Funct Ecol 1:293–296

Berg B (1986) Nutrient release from litter and humus in coniferous forest soils: a mini review. Scand J For Res 1:359–369

Berg B, Staaf H (1980) Decomposition rate and chemical changes of Scots pine needle litter. II. Influence of chemical composition. Ecol Bull 32:373–390

Bernhard-Reversat F (1972) Decomposition de la litiere de feuilles en forest ombrophile de Cote d'Ivoire. Oecol. Plant 7:279–300

Bhardwaj HI, Rangappa M, Hamama AA (2002) Planting date and genotype effects on tepary bean productivity. HortScience 37(2):317–318

Bhardwaj BB, Gupta SR, Reena S, Sodhi JS, Singh A (2005) Nutrient dynamics in a *Populus deltoides* agroforestry system at Kurukshetra. Bull Natl Inst Ecol 15:99–108

Bhat NH, Jan S (2010) Litterfall and Nutrient Return in Ulmus villosa Forests of Dachigam National Park, Jammu and Kashmir. Res J Agric Sci 1(4):363–365

Bloom AJ, Chapin FS III, Mooney HA (1985) Resource limitation in plants-an economic analogy. Annu Rev Ecol Syst 16:363–392

Bocock KL (1963) Changes in the amount of nitrogen in decomposing leaf litter of sessile oak (Quercus petraca). J Ecol 51:555–566

Bocock KL (1964) Change in the amounts of dry matter, nitrogen, carbon and energy in decomposing woodland leaf litter in relation to the activities of the soil fauna. J Ecol 52:273–284

Bocock KL, Gilbert OJW (1957) The disappearance of litter under different woodland conditions. Plant Soil 9:179–185

Bocock KL, Gilbert OJW, Capstick CK, Twinn DC, Waid JS, Woodman M (1960) Changes in the leaf litter when placed on the surface of soils with contrasting humus types. 1. Losses in dry weight of oak and ash leaf litters. J Soil Sci 11:1–9

Boring LR, Monk CD, Swank WT (1981) Early regeneration of a clearcut southern Appalachian forest. Ecology 62:1244–1253

Bormann FH, Likens GE, Siccama TG, Pierce RS, Eaton JS (1974) The effect of deforestation on ecosystem export and the steady-state condition at Hubbard Brook. Ecol Monogr 44:255–277

Borong C, Guangshan X, Xiaoyuan G, Guoliang Z (1993) Nutrient input of throughfall in a pine-spruce-fir forest of Changbai Mountain. Chinese J Appl Ecol 4(4):447–449

Bray JR, Gorham E (1964) Litter production in forests of the world. Advan Ecol Res 2:101–157

Brown S, Lugo AE (1982) Structure and production in tropical forest and their role in global carbon cycle. Biotropica 14:161–187

Brown S, Lugo AE (1984) Biomass of tropical forests: a new estimate based on forest volumes. Science 233:1290–1293

Brown S, Gillespie AJR, Lugo AE (1991) Biomass of tropical forests of south and Southeast Asia. Can J For Res 21:111–117

Brown S, Mo J, McPherson JK, Bell DT (1996) Decomposition of woody debris in Western Australian forests. Can J For Res 26(6):954–966

Burges A (1967) The decomposition of organic matter in the soil. In: Burges A, Raw F (eds) Soil biology. Academic Press, London

Caldwell MM (1987) Competition between roots in natural communities. In: Gregory PJ, Lake JV, Rose DA (eds) Root development and function. Cambridge University Press, New York

Caritat A, Berthou EG, Lapena R, Vilar L (2006) Litter production in a *Quercus suber* forest of Montseny (NE Spain) and its relationship to meteorological conditions. Ann For Sci 63:791–800

Carlisle A, Brown AHF, White EJ (1966) Litterfall, leaf production and effects of defoliation by Tortrix viridiana in a sessile oak (Quercus petraea) woodland. J Ecol 54:65–85

Cavelier J (1992) Fine-root biomass and soil properties in a semideciduous and a lower montane rain forest in Panama. Plant Soil 142:187–201

Chabot BF, Hicks DJ (1982) The ecology of leaf life spans. Annu Rev Ecol Syst 13:229–259

Chapin FS (1980) The mineral nutrition of wild plants. Annu Rev Ecol Syst 11:233–260

Chapin FS, Kedrowski RA (1983) Seasonal changes in nitrogen and phosphorus fractions and autumn retranslocation in evergreen and deciduous taiga trees. Ecology 64:376–391

Chave J, Navarrete D, Almeida S et al. (2010) Regional and seasonal patterns of litterfall in tropical South America. Biogeosciences 7:43–55

Coldwell BB, DeLong WA (1950) Studies on the composition of deciduous forest tree leaves before and after partial decomposition. Sci Agric 30:456–466

Cole DW, Rapp M (1981) Elemental cycling in forest ecosystems. In: Reichle DE (ed) Dynamic properties of forest ecosystems. IBP 23. Cambridge University Press, Cambridge

Cole DW, Gessel SP, Turner J (1978) Comparative mineral cycling in red alder and Douglas-fir. In: Briggs DG, DeBell DS, Atkinson WA (eds) Composition, Utilization and Management of alder. USDA For Serv Gen Tech Report. PNW −70. USDA, Washington

Colin-Belgrand M, Ranger J, Bouchon J (1996) Internal Nutrient Translocation in Chestnut Tree Stemwood: III. Dynamics Across an Age Series of *Castanea sativa* (Miller). Ann Bot (Lond) 78:729–740

Comeau PG, Kimmins JP (1989) Above-ground and below-ground biomass and production of lodgepole pine on sites with differing soil-moisture regimes. Can J For Res 19:447–454

Cornett MW, Puettmann KJ, Frelich LE, Reich PB (2001) Comparing the importance of seedbed and canopy type in the restoration of upland Thuja occidentalis forests of northeastern Minnesota. Restor Ecol 9(4):386–396

Cornforth IS (1970) Leaf fall in a tropical rain forest. J Appl Ecol 7:603–608

Cromack K Jr (1973) Litter production and litter decomposition in a mixed hardwood watershed and in a white pine watershed at Coweeta Hydrologic Station, North Carolina. Ph.D. thesis, Univ. Georgia, Athens

Cromack K Jr, Monk CD (1975) Litter production, decomposition, and nutrient cycling in a mixed hardwood watershed and white pine watershed. In: Howell FG, Gentry JB, Smith MH (eds) Mineral Cycling in Southeastern Ecosystems. US Energy Research and Development Admin. Symposium Series CONF- 740613, Washington, DC

Crosby JS (1961) Litter-and-duff fuel in shortleaf pine stands in southeast Missouri. Central States For Exp Stn Tech Paper 178. US For Serv Columbus, Ohio

Cuevas E, Brown S, Lugo AE (1991) Above- and belowground organic matter storage and production in a tropical pine plantation and a paired broadleaf secondary forest. Plant Soil 135:257–268

Cutini A (2000) Biomass, litterfall and productivity in chestnut coppices of various age at Monte Amiata (Central Italy). Ecologia Mediterranea 26(1-2):33–41

Dahlman RC, Kucera CL (1965) Root productivity and turnover in native prairie. Ecology 46:84–89

Das DK, Chaturvedi OP (2003) Litter quality effects on decomposition rates of forestry plantations. Tropical Ecology 44(2):261–264

Day FP Jr (1982) Litter Decomposition Rates in the Seasonally Flooded Great Dismal Swamp. Ecology 63(3):670–678

Day FP, Monk CD (1977) Net primary production and phe-nology on a southern Appalachian watershed. Am J Bot 64:1117–1125

DeCatanzaro JB, Kimmins JP (1985) Changes in the weight and nutrient composition of litterfall in three forest ecosystem types on Coastal British Columbia. Can J Bot 63:1046–1056

Del Arco JM, Escudero A, Garrido MV (1991) Effect of site characteristics on nitrogen retranslocation from senescing leaves. Ecology 72:701–708

deMaynadier PG, Hunter ML Jr (1995) The relationship between forest management and amphibian ecology: a review of the North American literature. Environ Rev 3:230–261

Devlaeminck R, An DS, Hermy M (2005) Variation in throughfall deposition across a deciduous beech (*Fagus sylvatica* L.) forest edge in Flanders. Sci Total Environ 337:241–252

Dimock IIEJ (1958) Litterfall in a Young Stand of Douglas-Fir Northwest Science 32(1):19–29

Duvall MD, Grigal DF (1999) Effects of timber harvesting on coarse woody debris in red pine forests across the Great Lakes states, USA. Can J For Res 29(12):1926–1934

Eaton JS, Likens GE, Bormann FH (1973) Throughfall and Stemflow Chemistry in a Northern Hardwood Forest. J Ecol 61(2):495–508

Edmonds RL, Murray GLD (2002) Overstory litter inputs and nutrient returns in an old-growth temperate forest ecosystem, Olympic National Park, Washington. Can J For Res 32(4):742–750

Escudero A, del Arco JM (1987) Ecological significance of the phenology of leaf abscission. Oikos 49:11–14

Escudero A, del Arco JM, Sanz IC, Ayala J (1991) Effect of leaf longevity and retranslocation efficiency on the retention time of nutrients in the leaf biomass of different woody species. Oecologia 90:80–87

Ewel JJ (1976) Litter fall and leaf decomposition in a tropical forest succession in eastern Guatemala. J Ecol 64:293–308

Fahey TJ, Hughes JW, Pu M, Arthur MA (1988) Root decomposition and nutrient flux following whole-tree harvest of northern hardwood forest. For Sci 34:744–768

Falconer JG, Wright JW, Beall HW (1933) The decomposition of certain types of forest litter under field conditions. Am J Bot 20:196–203

Fang JY, Wang GG, Liu GH, Xu SL (1998) Forest biomass of China: An estimate based on the biomass-volume relationships. Ecol Appl 84(4):1084–1091

Fife DN, Nambiar EKS (1982) Accumulation and retranslocation of mineral nutrients in developing needles in relation to seasonal growth of young radiata pine trees. Ann Bot 50:817–829

Fior RC, Soares MR, Casagrande JC (2010) Nutrient input through litter in riparian forest in different stages of ecological succession. 19th World Congress of Soil Science, Soil Solutions for a Changing World. 1-6 August 2010, Brisbane, Australia. Published on DVD

Foster JR, Lang GE (1982) Decomposition of red spruce and balsam fir boles in the White Mountains of New Hampshire. Can J For Res 12(3):617–626

Frankland JC (1966) Succession of fungi on decaying petioles of *Pteridium aquilinum*. J Ecol 54:41–63

Gallardo JF, Santa-Regina I, San Miguel C (1989) Ciclos bioegeoquimicos en bosques de la Sierra de Bejur (Salamanca, Espana). I. Production de hojarasca. Rev Ecol Biol Sol 26:35–46

Ganjegunte GK, Condron LM, Clinton PW, Davis MR, Mahieu N (2004) Decomposition and nutrient release from radiate pine (Pinus radiata) coarse woody debris. For Ecol Manag 187(2-3):197–211

Gerhardt K, Fredriksson D (1995) Biomass allocation by broadleaf mahogany seedlings, Swietenia macrophylla (King), in abandoned pasture and secondary dry forest in Guanacaste, Costa Rica. Biotropica 27:174–182

Gholz HL, Fisher RF, Pritchet WL (1985) Nutrient dynamics in slash pine plantation ecosystems. Ecology 66:647–659

Gilbert O, Bocock KL (1960) Changes in the leaf litter when placed on the surface of soils with contrasting humus types. II. Changes in the nitrogen content of oak and ash litter. J Soil Sci 11:10–19

Gillespie AJR, Brown S, Lugo AE (1992) Tropical forest biomass estimations from truncated stand tables. For Ecol Manag 48:69–87

Glumphabutr P, Kaitpraneet S, Wachrinrat C (2007) Nutrient Dynamics of Natural Evergreen Forests in the Eastern Region of Thailand. Kasetsart J Nat Sci 41:811–822

Gonzalez E (2012) Seasonal patterns of litterfall in the floodplain forest of a large Mediterranean river. Limnetica 31(1):173–186

Gordon WS, Jackson RB (2000) Nutrient concentrations in fine roots. Ecology 81(1):275–280

Gosz JR, Likens GE, Bormann FH (1973) Nutrient Release From Decomposing Leaf and Branch Litter in the Hubbard Brook Forest, New Hampshire. Ecol Monogr 43(2):173–191

Gosz JR, Likens GE, Bormann FH (1976) Organic matter and nutrient dynamics of the forest and forest floor in the Hubbard Brook Forest. Oecologia 22:305–320

Grime JP (1979) Plant strategies and vegetation processes. Wiley, Chichester

Gunadi B (1994) Litterfall, litter turnover and soil respiration in two pine forest plantations in Central Java, Indonesia. J Trop For Sci 6(3):310–322

Hansen K, Vesterdal L, Schmidt IK, Gundersen P, Sevel L, Birk AB, Pedersen LB, Hansen JB (2009) Litterfall and nutrient return in five tree species in a common garden experiment. For Ecol Manag 257:2133–2144

Harmon ME, Franklin JF, Swanson FJ, Sollins P, Gregory SV, Lattin JD, Anderson NH, Cline SP, Aumen NG, Sedell JR, Lienkaemper GW, Cromack K Jr, Cummins KW (1986) Ecology of coarse woody debris in temperate ecosystems. Adv Ecol Res 15:133–302

Harmon ME, Sexton J, Caldwell BA, Carpenter SE (1994) Fungal sporocarp mediated losses of Ca, Fe, K, Mg, Mn, N, P, and Zn from conifer logs in the early stages of decomposition. Can J For Res 24(9): 1883–1893

Helmisaari HS (1992) Nutrient retranslocation in three Pinus sylvestris stands. For Ecol Manag 51:347–367

Helmisaari HS (1995) Nutrient cycling in Pinus sylvestris stands in eastern Finland. Plant Soil 168-169:327–336

Hopkins B (1966) Vegetation of the Olokemeji Reserve, Nigeria. IV. The litter and soil with special reference to their seasonal changes. J Ecol 54:687–703

Howard PJ, Howard DM (1974) Microbial decomposition of tree and shrub litter. I. Weight loss and chemical composition of decomposing litter. Oikos 25:341–352

Ibrahima A, Mvondo Ze A, Ntonga JC (2010) Aboveground biomass and nutrient accumulation in the tropical rainforests of southern Cameroon: effects of logging. Cameroon J Exptl Biol 6(1):49–62

Janowiak MK, Webster CR (2010) Promoting ecological sustainability in woody biomass harvesting. J For 108(1):16–23

Jenny H, Gessel SP, Bingham FT (1949) Comparative study of decomposition rates of organic matter in temperate and tropical regions. Soil Sci 68:419–432

Jonasson S (1990) Implications of leaf longevity, leaf nutrient reabsorption and translocation for the resource economy of five evergreen plant species. Oikos 56:121–131

Jonasson S (1995) Resource allocation in relation to leaf retention time of the wintergreen *Rhododendron lapponicum*. Ecology 76:475–485

Jonsson BG, Kruys N, Ranius T (2005) Ecology of species living on dead wood—lessons for dead wood management. Silva Fennica 39(2):289–309

Joslin JD, Henderson GS (1987) Organic matter and nutrients associated with fine root turnover in a white oak stand. For Sci 33:330–346

Kadeba O, Aduayi EA (1985) Litter production, nutrient recycling and litter accumulation in*Pinus caribaea* Morelet var.*hondurensis* stands in the northern Guniea savanna of Nigeria. Plant Soil 86(2):197–206

Keenan RJ, Prescott CE, Kimmins JP (1993) Mass and nutrient content of woody debris and forest floor in western red cedar and western hemlock forests on northern Vancouver Island. Can J For Res 23(6):1052–1059

Kendrick WB (1959) The time factor in decomposition of coniferous leaf litter. Can J Bot 37:907–912

Killingbeck KT (1985) Automnal resorption and accretion of trace metals in gallery forest trees. Ecology 66:283–286

Knacker T, Forster B, Rombke J, Frampton GK (2003) Assessing the effects of plant protection products on organic matter breakdown in arable fields-litter decomposition test systems. Soil Biol Biochem 35:1269–1287

Koelling MR, Kucera CL (1965) Production and turnover relationships in native tall grass prairie. Iowa State J Sci 39:387–392

Krajick K (2001) Defending deadwood. Science 293(5535):1579–1581

Krankina ON, Harmon ME, Griazkin AV (1999) Nutrient stores and dynamics of woody detritus in a boreal forest: modeling potential implications at the stand level. Can J For Res 29(1):20–32

Kucera CL (1959) Weathering characteristics of deciduous leaf litter. Ecology 40(3):485–487

Kutbay HG, Horuz A (2001) Litterfall and nutrient returns in *Quercus cerris* L, Var *cerris* forests in the central Black Sea Region of Turkey. Pak J Bot 33(3):293–303

La Marca O (1984) Ricerche sulla biomassa dei cedui di catagno (*Castanea sativa* Mill.) dell Valle dell'Irno (AV e SA). Richerche Spèrimentali di dendrometria e di Auxomatria 8:63–79

Laiho R, Prescott CE (1999) The contribution of coarse woody debris to carbon, nitrogen, and phosphorus cycles in three Rocky Mountain coniferous forests. Can J For Res 29(10):1592–1603

Laiho R, Prescott CE (2004) Decay and nutrient dynamics of coarse woody debris in northern coniferous forests: a synthesis. Can J For Res 34(4):763–777

Lal CB, Annapurna C, Raghubanshi AS, Singh JS (2001) Foliar demand and resource economy of nutrients in dry tropical forest species. J Veg Sci 12:5–14

Lambert RL, Lang GE, Reiners WA (1980) Loss of mass and chemical change in decaying boles of a subalpine balsam fir forest. Ecology 61(6):1460–1473

Lang GE (1974) Litter dynamics in a mixed oak forest on the New Jersey Piedmont. Bull Torrey Bot Club 101:277–286

Lang G, Forman TTR (1978) Detrital dynamics in a mature oak forest: Hutcheson Memorial Forest, New Jersey. Ecol 59(3):580–595

Laudelout H, Meyer I (1954) Les cycles d'elements minerales et de matiere organique en foret equatoriale Congolaise. Trans 5th Int Cong Soil Sci 2:267–272

Li ZA, Zou B, Xia HP, Ren H, Mo JM, Weng H (2005) Litterfall Dynamics of an evergreen broadleaf forest and a pine forest in the subtropical region of China. For Sci 51(6):608–615

Likens GE, Bormann FH, Johnson NM, Fisher DW, Pierce RS (1970) Effects of forest cutting and herbicide treatment on nutrient budgets in the Hubbard Brook Watershed Ecosystem. Ecol Monogr 40:23–47

Lim MT, Cousens JE (1986) The internal transfer of nutrients in a Scots pine stand. 2. The patterns of transfer and the effects of nitrogen availability. Forestry 59:17–27

Liu W, Fox JED, Xu Z (2002) Biomass and nutrient accumulation in montane evergreen borad-leaved forest (Lithocarpus xylocarpus type) in Ailao Mountains, SW China. For Ecol Manage 158:223–235

Liu X, Xu H, Berninger F, Luukkanen O, Li C (2004) Nutrient distribution in *Picea likiangensis* trees growing in a plantation in West Sichuan, Southwest China. Silva Fennica 38(3):235–242

Lodhiyal LS, Singh RP, Singh SP (1995) Structure and function of an age series of poplar plantations in central Himalaya. II. Nutrient dynamics. Ann Bot 76:201–210

Lugo AE, Brown S, Chapman J (1988) An analytical review of production rates and stemwood biomass of tropical forest plantations. For Ecol Manage 23:179–200

Lunt HA (1935) Effect of weathering upon dry matter and composition of hardwood leaves. J For 33:607–609

Luvall JC, Weaver GT (1985) Organic matter and nutrient content of the forest floor of oak-hickory forests in southwestern Illinois. Paper presented at the Fifth Central Hardwood Forest Conference held at Urbana, Illinois, April 15-17, 1985

Madge DS (1965) Leaf fall and litter disappearance in a tropical forest. Pedobiologia 5:273–288

Madgwick HAI, Ovington JD (1959) The chemical composition of precipitation in adjacent forest and open plots. Forestry 32(1):14–22

McComb W, Lindenmayer D (1999) Dying, dead, and down trees. In: Hunter ML Jr (ed) Maintaining biodiversity in forest ecosystems. Cambridge University Press, England

McGinty DT (1976) Comparative root and soil dynamics on a white pine watershed and in the hardwood forest in the Coweeta basin. PhD Thesis, Univ. of Georgia, Athens

Meier CE (1981) The role of fine roots in nitrogen and phosphorus budgets of young and mature *Abies amabilis* ecosystems. Ph D Thesis, University of Washington

Meier CE, Grier CC, Cole DW (1985) Below- and aboveground N and P use by *Abies amabilis* stands. Ecology 66:1928–1942

Melillo JM, Aber JD, Linkins AE, Ricca A, Fry B, Nadelhoffer KJ (1989) Carbon and nitrogen dynamics along the decay continuum: plant litter to organic matter. Plant Soil 115:189–198

Mikola P (1954) Kokeellisia tutkimuksia metsakarikkeiden hajaantumisnopeu-desta. Commun Inst For Fenn 43:1–50

Mladenoff DJ, Forrester JA, Schatz J (2010) Impacts of biomass removal on carbon and nutrient pools in Wisconsin northern hardwood forests: establishment of a long-term study. Final report. State of Wisconsin, Public Service Commission, Madison

Monk CD, Day FP Jr (1988) Biomass, primary production, and selected nutrient budgets for an undisturbed hardwood watershed, p. 151–159. In: Swank WT, Crossley DA Jr (eds) Forest ecology and hydrology at Coweeta. Springer-Verlag, New York

Moore TR (1984) Litter decomposition in a subarctic spruce-lichen woodland, Eastern Canada. Ecology 65(1):299–230

Moore TR, Trofymow JA, Prescott CA, Fyles J, Titus BD, et al (2006) Patterns of carbon, nitrogen and phosphorus dynamics in decomposing foliar litter in Canadian forests. Ecosystems 9(1):46–62

Morellato LPC (1992) Nutrient cycling in two south-east Brazilian forests. I. Litter fall and litter standing crop. J Trop Ecol 8:205–215

Morrison IK (1991) Addition of organic matter and elements to the forest floor of an old-growth *Accer saccharum* forest annual litterfall. Can J For Res 21:461–468

Muoghalu JI, Akanni SO, Eretan OO (1993) Litterfall and nutrient dynamics in a Nigerian rainforest seven years after a ground fire. J Veg Sci 4:303–328

Murali KS, Bhat DM, Ravindranath NH (2005) Biomass estimation equations for tropical deciduous and evergreen forests. Int J Agric Res Gov Ecol 4(1):81–92

Nadezhdina N, Tatarinov F, Ceulemans R (2004) Leaf area and biomass of *Rhododendron* understory in a stand of Scots pine. For Ecol Manage 187:235–246

Nakane K (1995) Soil carbon cycling in a Japanese cedar (*Cryptomeria japonica*) plantation. For Ecol Manag 72:185–197

Nambiar EKS, Fife DN (1987) Growth and nutrient retranslocation in needles of radiata pine in relation to nitrogen supply. Ann Bot 60:147–156

Nambiar EKS, Fife DF (1991) Nutrient retranslocation in temperate conifers. Tree Physiol 9:185–207

Nascimentoa HEM, Laurance WF (2001) Total aboveground biomass in central Amazonian rainforests: a landscape-scale study. Forest Ecology and Management 5793:1–11

Nasiru AA (2000) Litter biomass and soil organic matter content in a chronosequence of tectona grandis (l.f.) stands in shasha forest reserve, Nigeria. M. Tech Thesis. Department of Forestry and Wood Technology, School of Postgraduate Studies, Federal University of Technology, Akure, Nigeria

Navar J, Mendez E, Dale V (2002) Estimating stand biomass in the Tamaulipan thornscrub of northeastrn Mexico. Ann For Sci 59:813–821

Navarrete AJ (2002) Litter Production of *Rhizophora Mangle* at Bacalar Chico, Southern Quintana Roo, Mexico. Universidad y Ciencia 18(36):79–86

Nicholas C, Songwe DU, Okali U, Fasehun FE (1995) Litter decomposition and nutrient release in a tropical rainforest, Southern Bakundu Forest Reserve. Cameroon J Trop Ecol 11(3):333–350

Nihlgard B (1972) Plant biomass, primary production and distribution of chemical elements in a beech and a planted spruce forest in southern Sweden. Oikos 23:69–81

Nirmal Kumar IJ, Sajish PR, Nirmal Kumar R, Basil G, Shailendra V (2011) Nutrient Dynamics in an *Avicennia marina* (Forsk.) Vierh. Mangrove Forest in Vamleshwar, Gujarat, India. Not Sci Biol 3(1):51–56

Nye PH (1961) Organic matter and nutrient cycles under moist tropical forests. Plant Soil 13:333–346

Nye PH, Greenland DJ (1960) The Soil under Shifting Cultivation. Tech. Comm. No. 51. Commonwealth Bureau of Soils, Harpenden, England

O'Connell AM (1988) Nutrient Dynamics in Decomposing Litter in Karri (Eucalyptus Diversicolor F. Muell.) Forests of South-Western Australia. J Ecol 76(4):1186–1203

Odiwe AI, Muoghalu JI (2003) Litterfall dynamics and forest floor litter as influenced by fire in a secondary lowland rainforest in Nigeria. Trop Ecol 44:241–249

Ojo AF, Kadeba TOS, Kayode J (2010) Litter mass and nutrient dynamics in a transformed rainforest ecosystem in Southwestern Nigeria. Bangladesh J Sci Ind Res 45(4):351–358

Oladoye AO, Ola-Adams BA, Adedire MO (2010) Litterfall Dynamics in *Leuceana Leucocephala* (Lam) De Wit Plantation in the Nigerian Derived Savanna. ARPN Journal of Agricultural and Biological Science 5(2):31–38

Old SM (1969) Microclimates, fire and plant production in an Illinois prairie. Ecol Monogr 34:355–384

Olson JS (1963) Energy storage and the balance of producers and decomposers in ecological system. Ecology 44:322–331

Ovington JD (1962) Quantitative ecology and the woodland ecosystem concept. Adv Ecol Res 1:103–192

Ovington JD (1965) Organic production, turnover and mineral cycling in woodlands. Biol Rev 40(3):295–336

Ovington JD (1968) Some factors affecting nutrient distribution within ecosystems. In: Eckardt FE (ed) Functioning of terrestrial ecosystems of the primary production level. Proc Copen Symp, UNESCO, Paris

Oziegbe MB, Muoghalu JI, Okel SO (2011) Litterfall, precipitation and nutrient fluxes in a secondary lowland rain forest in Ile-Ife, Nigeria. Acta Botanica Brasilica 25(3):664–671

Palm CA, Houghton RA, Melillo JM, Skole DL (1986) Atmospheric carbon dioxide from deforestation in southeastern Asia. Biotropica 18(3):177–188

Pande PK (2005) Biomass and productivity in some disturbed tropical dry deciduous teak forests of Satpura plateau, Madhya Pradesh. Trop Ecol 46(2):229–239

Pande PK, Meshram PB, Banerjee SK (2002) Litter production and nutrient return in tropical dry deciduous teak forests of Satpura plateau in central India. Trop Ecol 43(2):337–344

Parker GG (1983) Throughfall and Stemflow in the Forest Nutrient Cycle. Adv Ecol Res 13:57–133

Pastor J, Bockheim JG (1984) Distribution and cycling of nutrients in an aspen-mixed-hardwood-spodosol ecosystem in Northern Wisconsin. Ecology 65(2):339–353

Pausas JG (1997) Litter fall and litter decomposition in *Pinus sylvestris* forests of the eastern Pyrenees. J Veg Sci 8:643–650

Pierce RS, Martin CW, Reeves C, Likens GE, Bormann FH (1972) Nutrient losses from clearcutting in New Hampshire, p. 285–295. In: Natl Symp on Watersheds in Transition, Amer Water Resources Assoc Proc, Series No. 14

Poggiani F (1985) Nutrient cycling in *Eucalyptus* and *Pinus* plantations ecosystems: silvicultural implications. IPEF 31:33–40

Potter CS, Ragsdale HL, Berish CW (1987) Resorption of foliar nutrients in a regenerating southern Appalachian forest. Oecologia 73:268–271

Pragasan LA, Parthasarathy N (2005) Litter production in tropical dry evergreen forests of south India in relation to season, plant life-forms and physiognomic groups. Curr Sci 88(8):1255

Prescott CE (2005) Decomposition and Mineralization of Nutrients from Litter and Humus. In Bassiri Rad H (ed) Nutrient Acquisition by Plants: An Ecological Perspective. Ecological Studies 181:15–41

Pritchett WL, Fisher RF (1987) Properties and management of forest soils. 2d ed. John Wiley & Sons, New York

Proctor J (1984) Tropical forest litterfall. II. The data set. In: Chadwick AC, Sutton SL (eds) Tropical rain forest: ecology and management, supplementary volume. Blackwell Scientific Publications, Oxford

Proctor J (1987) Nutrients cycling in primary and old secondary, rainforests. Appl Geogr 7:135–152

Proe MF, Millard P (1995) Effect of P supply upon seasonal growth and internal cycling of P in Sitka pruce (*Picea sitchensis* (Bong.) Carr.) seedlings. Plant Soil 168-169:313–317

Raich JW, Nadelhoffer KJ (1989) Belowground carbon allocation in forest ecosystems: global trends. Ecology 70:1346–1354

Ranger J, Colin-Belgrand MC (1996) Nutrient dynamics of chestnut tree (*Castanea sativa* Mill.) coppice stands. For Ecol Manag 86:259–277

Ranger J, Gelhaye D (2001) Belowground biomass and nutrient content in a 47-year-old Douglas-fir plantation. Ann For Sci 58:423–430

Ranger J, Nys C, Bouchon J (1990) Les relations entre la fertilite du sol, la productin et l´utilisation d´eléments nutritifs dans les taillis de châtaigniers. Acta Oecol 11:487–501

Rapp M (1969) Production de litiere et apport au d elements mineraux dans deux Ecosystemes mediterraneens de la foret de *Quercus ilex* et la garrigue de *Q. coccifera*. Oecol Plant 4:377–410

Rapp M, Santa-Regina I, Rico M, Gallego HA (1999) Biomass, nutrient content, litterfall and nutrient return to the soil in Mediterranean oak forest. For Ecol Manag 119:39–49

Redmann RE (1975) Production ecology of grassland plant communities in western North Dakota. Ecol Monogr 45:83–106

Reichle DE (1981) Dynamic Properties of Ecosystems. Cambridge University

Reinaldo IB, Philip MF (1995) Carbon and Nutrient Flows in an Amazonian forest. Fine litter production and composition at Apiau, Roraima, Brazil

Reiners WA (1972) Nutrient content of canopy throughfall in three Minnesota forests. Oikos 23:14–22

Reiners WA, Reiners NM (1970) Energy and nutrient dynamics of forest floors in three Minnesota forests. J Ecol 58:497–519

Robert B, Caritat A, Bertoni L, Vilar L, Molinas M (1996) Nutrient content and seasonal fluctuations in the leaf component of cork oak (*Quercus suber* L.) litterfall. Vegetatio 122:29–35

Rochow JI (1974) Litter fall relations in a Missouri forest. Oikos 25:80–85

Rodin LE, Basilevich NI (1967) Production and mineral cycling in terrestrial vegetation (English translation ed. by Fogg GE). Oliver and Boyd, Edinburgh-London

Rubilar RA, Allen HL, Kelting DL (2005) Comparison of biomass and nutrient content equations for successive rotations of loblolly pine plantations on an Upper Coastal Plain Site. For Ecol Manage 28:548–564

Rutkowski DR, Stottlemyer R (1993) Composition, biomass and nutrient distribution in mature northern hardwood and boreal forest stands, Michigan. Am Midl Nat 130:13–30

Ryan DF, Bormann FH (1981) Nutrient resorption in northern hardwood forests. BioScience 32:29–32

Saenger P, Snedaker SC (1993) Pantropical trends in mangrove aboveground biomass and annual litterfall. Oecologia 96(3):293–299

Salazar S, Sanchez LE, Galindo P, Santa-Regina I (2010) Above-ground tree biomass equations and nutrient pools for a paraclimax chestnut stand and for a climax oak stand in the Sierra de Francia Mountains, Salamanca, Spain. Scientific Research and Essays 5(11):1294–1301

Sanches L, Valentini CMA, Junior OBP, Nogueira J, Vourlitis GL, Biudes MS, Silva CJ da, Bambi P, Lobo FA (2008) Seasonal and interannual litter dynamics of a tropical semideciduous forest of the southern Amazon Basin, Brazil. J Geophys Res 113:1–9

Sanford RL Jr. (1985) Root ecology of mature and successional Amazon forests. PhD Thesis. Univ Calif, Berkeley, CA

Santa-Regina I (2000) Biomass estimation and nutrient pools in tour *Quercus pyrenaica* in Sierra de Gata Mountains, Salamanca, Spain. For Ecol Manag 132:127–141

Santa-Regina I, Tarazona T (2000) Nutrient return to the soil through litterfall and throughfall under beech and pine stands of Sierra de la Demanda, Spain. Arid Soil Res Rehab 14(3):239–252

Santa-Regina I, Tarazona T (2001) Organic matter and nitrogen dynamics in a mature forest of common beech in the Sierra de la Demanda, Spain. Ann For Sci 58:301–314

Santa-Regina I, Leonardi S, Rapp M (2000) Organic matter and foliar nutrient dynamics in *Castanea sativa* Mill. Coppice stands of southern Europe. Ecol Med 26(1-2):71–81

Santiago LS, Schuur EAG, Silvera K (2005) Nutrient cycling and plant-soil feedbacks along a precipitation gradient in lowland Panama. J Trop Ecol 21:461–470

Satchell JE (1974) Litter interface of animate/inanimate matter. In: Dickinson CH, Pugh CJF (eds) Biology of Plant Litter Decomposition. Academic Press, New York

Saunders MR, Fraver S, Wagner RG (2011) Nutrient concentration of down woody debris in mixedwood forests in central Maine, USA. Silva Fennica 45(2):197–210

Schlesinger WH (1991) Biogeochemistry: An Analysis of Global Change. Academic Press, New York

Segura M (2005) Allometric models for tree volume and total aboveground biomass in tropical humic forest in Costa Rica. Biotrop 37(1):2–8

Shanmughavel P, Peddappaiah RS, Muthukumar T (2000) Litter production and nutrient return in *Bambusa bambos* plantation. J Sustain For 11(3):71–82

Shanmughavel P, Sha L, Zheng Z, Cao M (2001) Nutrient cycling in a tropical seasonal rain forest of Xishuangbanna, Southeast China. Part I: Tree species:nutrient distribution and uptake. Bioresour Technol 80:163–170

Siddiqui PJA, Qasim R (1990) Litter Production and Physico-Chemical Conditions in Mangrove, Avicennia Marina (Forsk.) Vierh., Swamps At Karachi Back Waters and Bakran Creek. Journal of Islamic Academy of Sciences 3:15–21

Siitonen J (2001) Forest management, coarse woody debris and saproxylic organisms: Fennoscandian boreal forests as an example. Ecological Bulletins

Singh JS, Gupta SR (1977) Plant decomposition and soil respiration in terrestrial ecosystems. Bot Rev 43(4):449–528

Sollins P, Grier CC, McCorison FM, Cromack K, Fogel Jr R, Fredriksen RL (1980) The internal element cycles of an old-growth Douglas-fir ecosystem in Western Oregon. Ecol Mono 50(3):261–285

Son Y, Park IH, Yi MJ, Jin HO, Kim DY, Kim RH, Hwang JK (2004) Biomass, production and nutrient distribution of a natural oak forest in central Korea. Ecol Res 19:21–28

Songwe NC, Okali DUU, Fasehun FE (1995) Litter decomposition and nutrient release in a tropical rainforest, Southern Bakundu Forest Reserve, Cameroon. J Trop Ecol 11:333–350

Staaf H (1980) Release of plant nutrients from decomposing leaf litter in a South Swedish beech forest. Holarct Ecol 3:129–136

Staaf H (1982) Plant nutrient changes in beech leaves during senescenceas influenced by site characteristics. Oecol Plant 3:161–170

Staaf H, Berg B (1977) Mobilization of plant nutrients in a Scots pine forest moor in central Sweden. Silva Fenn 11:210–216

Stachurski A, Zimka JR (1975) Leaf fall and the rate of litter decay in some forest habitats. Ekol Pol 23:103–108

Stachurski A, Zimka JR (1975) Methods of studying forest ecosystems: leaf area, leaf production and withdrawal of nutrients from leaves of trees. Ekol Pol 23:637–648

Staelens J, Ameloot N, Almonacid L, Padilla E, Boeckx P, Huygens D, Verheyen K, Oyarzún C, Godoy R (2011) Litterfall, litter decomposition and nitrogen mineralization in old-growth evergreen and secondary deciduous Nothofagus forests in south-central Chile. Revista Chilena De Historia Natural 84(1):125–141

Stenlid G (1958) Salt losses and redistribution of salts in higher plants. Plant Physiol 4:615–637

Svoboda M, Matejka K, Kopacek J (2006) Biomass and element pools of selected spruce trees in the catchments of Plesne and Certovo Lakes in the Sumava Mts. J For Sci 52(10):482–495

Svoboda M, Fraver S, Janda P, Baèe R, Zenáhlíková J (2010) Natural development and regeneration of a Central European montane spruce forest. For Ecol Manag 260(5):707–714

Swamy HR, Proctor J (1994) Litterfall and nutrient cycling in four rainforests in the Sringeri area of the Indian Western Ghats. Global Ecology and Biogeography Letters 4:155–165

Swift MJ, Smith RA, Perfect JJ (1981) Decomposition and mineral nutrient dynamics of plant litter in a regenerating bush—fallow in subhumid tropical Nigeria. J Ecol 69:981–995

Tamm CO (1951) Removal of plant nutrients from tree crowns by rain. Physiol Plant (4):184–188

Tamm CO (1953) Growth, yield and nutrition in carpets of forest moss (*Hylocomium splendens*). Meddr Stat Skogsforskningsinst 43:1–140

Tanner EVJ (1981) The decomposition of leaf litter in Jamaican montane rain forests. J Ecol 69:263–275

Thomas WA (1968) Decomposition of loblolly pine needles with and without the addition of dogwood leaves. Ecology 49:568–571

Titus JS (1989) Nitrogen recycling in the apple (*Malus domestica* Borkh.). Annales des Sciences Forestie Res 46 (suppl): 654–659

Titus JS, Kang SM (1980) Nitrogen metabolism, translocation and recycling in apple trees. Horticultural Reviews: 204–246

Tokarar F, Krekulova E (2004) Aboveground biomass production and leaf area index in various types of chestnut (*Castanea sativa* Mill.) stands in Slovakia. Ekologia (Bratislava) 23(4):342–352

Triadiati S, Tjitrosemito E, Guhardja, Sudarsono I, Qayim, Leuschner C (2011) Litterfall production and leaf-litter decomposition at natural forest and cacao agroforestry in Central Sulawesi, Indonesia. Asian J Biol Sci 4:221–234

Tsutsumi T, Yoda K, Sahunalu P, Dhanmanonda P, Prachaiyo B (1983) Forest: Felling, burning and regeneration. In: Youngberg CT, Davey CB (eds) Tree Growth and Forest Soils. Proceeding of the Third North America Forest Soils Conference, Oregon State University Press, USA

Turner J, Singer MJ (1976) Nutrient distribution and cycling in a subalpine coniferous forest ecosystem. J Appl Ecol 13:295–301

Turner J, Cole DW, Gessel SP (1976) Mineral nutrient accumulation and cycling in a stand of red alder (Alnus rubra). J Ecol 64:965–974

Ukonmaanaho L, Merila P, Nojd P, Nienminen TM (2008) Litterfall production and nutrient return to the forest floor in Scotspine and Norway spruce stands in Finlanl. Boreal Env Res 13(B):67–91

Uri V, Tullus H, Lohmus K (2003) Nutrient allocation, accumulation and above-ground biomass in grey alder and hybrid alder plantations. Silva Fenn 37(3):301–311

Usotsev VA, Hoffman CW (1997) Combining harvest sample data and inventory data to estimate forest biomass. Scand J For Res 12(3):273–279

Van L, David H, Waide JB, Michael JT (1984) Biomass and nutrient content of a 41-year-old Loblolly Pine (*Pinus taeda* L.) plantation on a poor site in South Carolina. For Sci 30(2):395–404

Veneklaas EJ (1991) Litterfall and nutrient fluxes in two montane tropical rain forests, Colombia. J Trop Ecol 7(3):319–336

Vital ART, Guerrini IA, Franken WK (2004) Produção de serapilheira e ciclagem de nutrientes de uma floresta estacional semidecidual em zona ripária. Revista Árvore 28:793–800

Vitousek (1982) Nutrient cycling and nutrient use efficiency. Am Nat 119:553–572

Vitousek PM, Reiners WA (1975) Ecosystem succession and nutrient retention: a hypothesis. Bio Sci 25:376–380

Vitousek PM, Sanford RL Jr (1986) Nutrient cycling in moist tropical forest. Annu Rev Ecol Syst 17:137–167

Vitousek PM, Gosz JR, Grier CC, Melillo JM, Reiners WA (1982) A comparative analysis of potential nitrification and nitrate mobility in forest ecosystems. Ecol Mongr 52:155–177

Vitousek PM, Fahey T, Johnson DW, Swift MJ (1988) Element interactions in forest ecosystems: succession, allometry, and in-put-output budgets. Biogeochemistry 5:7–34

Vogt KA, Grier CC, Meier CE, Keyes MR (1983) Organic matter and nutrient dynamics in forest floors of young and mature *Abies amabilis* stands in western Washington, as affected by fine-root input. Ecol Monogr 53:139–157

Vogt KA, Grier CC, Vogt DJ (1986) Production, turnover and nutrient dynamics of the above-and belowground detritus of world forests. Adv Ecol Res 15:303–377

Voigt GK (1960) Alternation of the composition of rainwater by trees. Am Mid-land Nat 63(2):321–326

Wang Q, Wang S, Fan B, Yu X (2007) Litter production, leaf litter decomposition and nutrient return in *Cunninghamia lanceolata*

plantations in south China: effect of planting conifers with broad-leaved species. Plant Soil 297(1-2):201–211
Wells CG, Metz LJ (1963) Variation in nutrient content of Loblolly pine needles with season, age, soil and position on the crown. Soil Sci Soc Am Proc 27:90–93
Westman WE (1975) Edaphic climax patterns of the pygmy forest region of California. Ecol. Monog 45:109–135
Whitmore TC (1984) Tropical rainforests of the far east. Oxford University Press, London
Whittaker RH, Bormann FH, Likens GE, Siccama TG (1974) The Hubbard Brook Ecosystem Study: forest biomass and production. Ecol Monogr 44:233–254
Wiegert RG, Evans FC (1964) Primary production and disappearance of dead vegetation on an old field in southeastern Michigan. Ecology 45:49–63
Will GM (1959) Nutrient return in litter and rainfall under some exotic conifer stands in New Zealand. New Zealand J Agric Res 2:719–734
Williams ST, Gray TRG (1974) Decomposition of Litter on the Soil Surface. In: Dickinson CH, Pugh GJF (eds) Biology of Plant Litter Decomposition, Vol. 2. Academic Press, London
Witherspoon JPJr, (1962) Cycling of cesium-134 in White Oak trees on sites of contrasting soil type and moisture. I. 1960 growing season. In: Ist National Symposium Radioecology, Fort Collins, Colorado. Reinhold, New York
Witkamp M (1966) Decomposition of leaf litter in relation to environmental conditions, microflora and microbial respiration. Ecology 47:194–201
Witkamp M, Olson JS (1963) Breakdown of confined and non-confined oak litter. Oikos 14:138–147
Wood TG (1976) The role of termites (Isoptera) in decomposition processes. In: Anderson JM, Macfadyen A (eds) The role of terrestrial and aquatic organisms in decomposition processes, Blackwell Sci Pub Oxford
Wood TE, Lawrence D, Clark DA (2006) Determinants of leaf litter nutrient cycling in a tropical rain forest: soil fertility versus topography. Ecosystems 9:1–11
Woodwell GM, Marples TG (1968) The influence of chronic gamma irradiation on production and decay of litter and humus in an oak-pine forest. Ecology 49:456–464
Woodwell GM, Whittaker RH, Houghton RA (1975) Nutrient concentrations in plants in the Brookhaven oak-pine forest. Ecology 56:318–332
Xu XN, Shibata H, Enoki T (2006) Decomposition patterns of leaf litter of seven common canopy species in a subtropical forest: dynamics of mineral nutrients. J For Res 17(1):1–6
Yamane I, Sato K (1971) Decay of litter of *Miscanthus sinensis* and *Sasa palmata* in Kawatabi IBP area. In: Numata M (ed) Grassland ecosystem studies. Jap Comm Int Biol Prog Grassland Project
Yang H, Wang S, Zhang J, Fan B, Zhang W (2011) Biomass and nutrients of *Pinus massoniana* plantations in southern China: simulations for different managing practices. J Food. Agric Env 9(1):689–693
Yang WQ, Wang KY, Kellomaki S, Gong H (2005) Litter dynamics of three subalpine forests in Western Sichuan. Pedosphere 15(5):653–659
Yang YS, Guo JF, Chen GS, Xie JS, Gao R, Li Z, Jin Z (2005) Litter production, seasonal pattern and nutrient return in seven natural forests compared with a plantation in southern China. Forestry 78(4):403–415
Youngberg CT (1966) Forest floors in Douglas-fir forests: 1. Dry weight and chemical properties. Soil Sci Amer Soc Proc 30:406–409
Zhang P (1999) nutrient inputs from trees via throughfall, stemflow and Litterfall in an intercropping system. MS Thesis, The University of Guelph, Canada
Zhao Q, Liu X, Zeng D (2011) Aboveground biomass and nutrient allocation in an age-sequence of *Larix olgensis* plantations. J For Res 22(1): 71–76
Zheng Z, Shanmughavel P, Sha L, Cao M, Warren M (2006) Litter decomposition and nutrient release in a tropical seasonal rain forest of Xishuangbanna, Southwest China. Biotropica 38(3):342–347
Zhou GY, Guan LL, Wei XH, Zhang DQ, Zhang QM, Yan JH, Wen DZ, Liu JX, Liu SG, Huang ZL, Kong GH, Mo JM, Yu QF (2007) Litterfall production along successional and altitudinal gradients of subtropical monsoon evergreen broadleaved forests in Guangdong, China. Plant Ecol 188:77–89
Zianis D, Mencuccini M (2003) Aboveground biomass relationships for beech (*Fagus moesiaca* Cz.) trees in Vermio Mountain, Northern Greece, and generalized equations for *Fagus* sp. Ann For Sci 60:439–448

Forest Types and Their Associated Soils

7

For variations in climatic, topographic, and edaphic conditions, we find the evolution of different types of forests in different geographic regions of the world. These forests have diverse structures, compositions, and functions. They can be classified on the basis of their regions of occurrence, species composition, habitat type, leaf type, persistence, etc. There are natural and artificial forests, broadleaf and coniferous forests, evergreen and deciduous forests, primary and secondary forests, etc. World forests are classified into three broad types—tropical forests, temperate forests, and boreal forests—mainly on the basis of geographical positions. Besides these types, there are montane forests, mangrove forests, and swamp forests. Likewise, there are many different types of soils in the world. These soils are classified into 12 orders in USDA Soil Taxonomy and 32 Reference Soil Groups (RSGs) by World Reference Base (WRB) for Soil Resources. As vegetation is an active factor of soil formation, and as soil and vegetation evolve together as coupled systems through natural succession, and as particular vegetation is commonly associated with a definite climate, we see widely different types of soils in different forest types. For example, typical soils of tropical forests are the Oxisols (Ferralsols, Plinthosols, and Nitisols), Ultisols (Plinthols, Planosols, and Alisols), and some Alfisols (Albelluvisols, Luvisols, and Planosols); typical soils of temperate forests are Alfisols and Spodosols (Podzols) along with some Histosols; and typical soils of the boreal forests are the Gelisols (Cryosols and Histosols) and Spodosols. There are some Entisols (Regosols, Leptosols, Fluvisols, and Arenosols) and Inceptisols (Cambisols, Gleysols, Umbrisols, and Nitisols) in all of these forests. In the following sections, a brief description of world forest types, world soil classification, and associations of different soils with different forests is given.

7.1 Forest Types of the World

Forests are large tracts of uncultivated lands occupied by trees, shrubs, herbs, and other vegetations along with insects and animals, large and small, and microorganisms interacting together and remaining in a dynamic equilibrium with their abiotic environments. Forests cover almost 30% of the world's ice-free land (FAO 2001) and perform such important ecological functions as the regulation of climate, hydrology and geomorphology, primary productivity, biogeochemical cycling, etc.

There are many different types of forests in the world. Primarily, forests may be natural or artificial. Natural forests have evolved in a place spontaneously by natural succession. On the other hand, artificial forests or man-made forests have been created by human interventions—by sowing and planting. Natural forests contain indigenous trees and other plant species. They are native forests. Artificial forests may consist of indigenous or exotic trees. Exotic trees are those trees that do not naturally grow in the area and that have been imported from another ecological zone. Forests may be evergreen (retaining leaves and remaining green all the year round) or deciduous trees (shedding all the leaves and remaining leafless in a season) depending on leaf persistence of dominant trees. Examples of evergreen trees include Artocarpus heterophyllus, Bischofia javanica, Calophyllum elatum, Euvodia lunuankenda, Hopea ponga, Mesua ferrea, Myristica dactyloides, etc., which are familiar trees of tropical forests. The deciduous trees include *Acrocarpus fraxinifolius*, *Bombax ceiba*, *Chukrasia tabularis*, *Dalbergia latifolia*, *Grewia tiliaefolia*, *Lagerstroemia microcarpa*, *Pterospermum* sp., *Tectona grandis*, *Terminalia bellirica*, *Toona ciliate*, etc., found in the tropical region. A forest may contain evergreen and deciduous trees, both being dominant. These forests are called mixed evergreen forest or semi-evergreen forest if the proportion of the evergreen trees is higher

K. T. Osman, *Forest Soils*, DOI 10.1007/978-3-319-02541-4_7, © Springer International Publishing Switzerland 2013

than the deciduous, and it is mixed deciduous if the proportion of the deciduous trees is higher. Forests are sometimes classified on the moistness of the habitat or the amount of rainfall they receive into moist forests and dry forests. There are moist evergreen, moist deciduous, and dry deciduous forests. Tree species may have either broad, flat, and expanded leaves, or narrow leaves, i.e., needles. On this basis, forests may be broad-leaved forests and coniferous forests. Coniferous forests are mainly found in the continental and upland regions of North America, Northeast Europe, and Russia. Broadleaf forests are generally hardwood forests, while coniferous forests are softwood forests, thus termed on the relative ease of sawing. There are broad-leaved evergreen and broad-leaved deciduous forests; similarly, there are coniferous evergreen and coniferous deciduous forests. Forests are often classified on the basis of the habitat of their occurrence such as montane forests, swamp forests, mangrove forests, etc. Forests that have not yet experienced the human axe are virgin forests, but most forests are disturbed forests. There are three major vegetation zones or forest biomes in the world: tropical, temperate, and boreal forests. Besides these main types, there are some other minor forest types in each of these zones.

7.1.1 Tropical Forests

The area of the earth between latitudes 23.5° N and 23.5° S (at both sides of the equator) is called the tropics. The tropics are characterized by their distinct seasonality where winter is almost absent, and there are two prominent seasons: rainy and dry. Average temperature ranges from 20 to 25 °C and the average temperature difference between the three warmest and the three coldest months is not more than 5 °C. Precipitation is evenly distributed throughout the year, with annual rainfall exceeding 2,000 mm. Tropical forests are characterized by a great diversity of tree species with plenty of orchids, bromeliads, lianas, ferns, mosses, and palms. According to FAO (1994), the tropical forest zone contains 1.76 billion ha, divided into six ecofloristic zones, including the tropical rain forests, the moist deciduous forests, the dry zone forests, and the upland forests (hill and mountain forests).

1. Tropical rainforests are found in areas with more than 2,500 mm of annual rainfall. They are evergreen, luxuriant, and rich in animal and plant species. More than half the world's 718.3 million ha of rain forests are located in two countries: Brazil (41 %) and Indonesia (13 %).
2. Moist deciduous forests occur in areas with an annual rainfall of 1,000–2,000 mm. Forest structure varies depending on the amount and distribution of rain, the type of soil, and the length of the dry season. This forest type is generally less diverse than rain forest.
3. Dry zone forests are found in tropical areas receiving between 500 and 1,000 mm of rainfall per year. They are relatively open and include thornland, shrubland, savannah, and other short and sparse woody vegetation. More than half are in Africa. Dry forest types include oak, mesquite, pinon–juniper, maquis, and acacia.
4. Tropical upland forests are forests above 800 m and include cloud forests (montane rain forests), which are shorter, floristically simpler, and more heavily laden with mosses and lichens than lowland rain forests. Tropical upland species are similar to temperate forest species. The upland zone covers the Himalayas, parts of Myanmar, Thailand, and Vietnam, the highlands of Mexico, the Andes, and the highlands of Ethiopia and mountains around Lake Victoria.

There can be another distinct form—the mixed evergreen (or semi-evergreen) forests.

7.1.1.1 Tropical Rainforests

Tropical rainforests are found between latitudes 10° N and 10° S. Tropical rainforests include the Amazon Basin of South America, the Zaire Basin of Africa, and the islands and peninsulas of Southeast Asia. There are distinct differences in physiognomy and species composition among these forests. The Amazon Basin of South America encompasses five and a half million square kilometers of rainforests. Brazil contains 60 %, Peru 13 %, Colombia 10 %, and minor amounts in Venezuela, Ecuador, Bolivia, Guyana, Suriname, and French Guiana. The Amazon comprises the most species-rich tract of tropical rainforest in the world. Some of the dominant tree species in these forests are (Duivenvoorden 1995):

Well-drained flood plain: *Annona hypoglauca, Cecropia membranacea, Endlicheria cf. pyriformis, Ficus aff. insipida, Inga brachystachys, Iriartea deltoidea, Protium cf. nodulosum, Qualea ingens, Sarcaulusl brasiliensis, Zygia aff. Pilosula, Bocageopsis multiflora, Didymocistus chrysadenius, Euterpe precatoria, Licania apetala, Matisia lasiocalyx, Mollia lepidota, Oenocarpus bataua, Oxandra polyantha, and Virola calophylla.*

Well-drained uplands: *Anibal sp., Chimarrhis sp., Clathrotropis macrocarpa, Dendrobangia aff. boliviana, Dendropanax sp., Eschweilera aff. coriacea, Eschweiler a sp. 1, Eschweilera sp. 2, Eschweilera sp. 3, Eschweilera punctata, Guarea macrophylla, Iryanthera ulei, Micranda spruceana, Monopteryx uaucu, Oenocarpus bataua, Pouteria sp., Protium cf. nodulosum, Protium nitidifolium, Senefeldera sp., Swartzia cf. racemosa, Tachigali aff. paniculata, Unonopsis stipitata, Virola calophylla, Virola calophylloidea, and Virola theiodora.*

Fig. 7.1 A view of a tropical rainforest of Hawaii. (Photo courtesy of Maleische Archipel)

Poorly drained flood plain: *Elaeoluma glabrescens, Mauritia flexuosa, Oxandra polyantha, Rhodognaphalopsis sp. 1, Triplaris weigeltiana, Acosmium nitens, Bocageopsis cf. canescens, Elaeoluma sp., Eschweilera itayensis, Euterpe precatoria, Licania sp., Rhodognaphalopsis sp. 1.*

The Zaire basin contains the largest patch of Tropical Rainforest in Africa and about one-tenth of the rainforests of the world. Madagascar, also a part of Africa, is the fourth largest island in the world and is one of the places in the world that has the richest plant species (Fig. 7.1). The dominant tree species are (Fashing and Gathua 2004):

Acacia abyssinica, Alangium chinense, Albizia gummifera, Albizia sp., Aningeria altissima, Antiaris toxicaria, Bequaertiodendron oblanceolatum, Bersama abyssinica, Bischoffia javanica, Blighia unijugata, Bridelia micrantha, Canthium keniense, Casaeria battiscombei, Cassipourea ruwensorensis, Celtis africana, Celtis gomphophylla, Celtis mildbraedii, Chaetacme aristata, Chrysophyllum albidum, Cordia africana, Cordia millenii, Craibia brownie, Croton macrostachyus, Croton megalocarpus, Croton sylvaticus, Cupressus sp., Diospyros abyssinica, Dovyalis macrocalyx, Ehretia cymosa, Entada abyssinica, Erythrina abyssinica, Fagaropsis angolensis, Ficus sp., Ficus cyathistipula, Ficus exasperate, Ficus lutea, Ficus natalensis, Ficus sansibarica, Ficus sur, Ficus sycomorus, Ficus thonningii, Ficus vallischoudae, Funtumia africana, Harungana madagascariensis, Heinsenia diervillioides, Khaya anthotheca, Kigelia moosa, Lepidotrichilia volkensii, Lepisanthes senegalensis, Maesa, Maesopsis eminii, Manilkara butugi, Margaritaria discoidea, Markhamia lutea, Maytenus senegalensis, Milicia excelsa, Monodora myristica, Morus mesozygia, Nuxia congesta, Olea capensis, Phyllanthus inflatus, Polyscias fulva, Premna, Prunus africana, Psidium guajava, Rinorea brachypetala, Rothmannia urcelliformis, Sapium ellipticum, Spathodea campanulata, Strombosia scheffleri, Strychnos usambarensis, Suregada procera, Teclea nobilis, Terminalia brownie, Toona ciliate, Trema orientalis, Trichilia emetica, Trilepisium madagascariense, Vangueria volkensii, Vitex doniana, Vitex keniensis, Zanthoxylum gillettii, and Zanthoxylum mildbraedii.

In Southeast Asia, the tropical rainforests are found in India, Bangladesh, Sri Lanka, Malaysia, Brunei, Indonesia, Myanmar, and Papua New Guinea. The rainforests found in India, Bangladesh, and Sri Lanka are in small patches and strips, while, on the other hand, Indonesia contains one-tenth of the world's rainforest and 40% of all Asian rainforests. Indonesia is losing its rainforests to commercial logging fast and Malaysia too has lost about two-third of its lowland forest. Papua New Guinea still has areas of rainforest yet to be disturbed. Some of the dominant tree species of these forests are (Hadi et al. 2009):

Alangium ridleyi, Anisophyllea disticha, Antidesma sp., Aporosa benthamiana, Artocarpus heterophyllus, Artocarpus sp., Baccaurea javanica, Baccaurea sp., Baccaurea sumatrana, Bhesa paniculata, Bridelia stipularis, Cannarium megalathum, Cinnamomum iners, Dillenia excelsa, Diospyros ellipticifolia, Dipterocarpus sp., Disoxylum sp., Drypetes sp., Durio sp., Eugenia grandis, Eugenia sp., Excoecaria sp., Horsfieldia irya, Knema sp., Horsfieldia sp., Ixora, Knema sp., Mallotus sp., Phaenathus sp., Psychotria sp., Santiria sp., Shorea pauciflora, Urophyllum griffithianum, and Xanthophyllum sp.

Australia's tropical rainforests are distributed along the coast of Queensland. The rainforests there take up about 10,500 km^2. In these forests, the canopy is often multilayered consisting of an upper, discontinuous layer of emergents, over the main canopy and subcanopy. Below the canopy is an understory of sparse shrubs and seedlings. The upper, discontinuous layer includes canopy emergents that may be 40–50 m tall and have large spreading crowns. This layer is composed of species such as *Araucaria cunninghamii, Ficus* spp., *Lophostemon confertus* and, in some sites, *Eucalyptus* spp. The canopy/subcanopy layer contains a diverse range of species. Representative species include: *Araucaria cunninghamii, Ficus* spp., *Argyrodendron trifoliolatum/Heritiera trifoliolata, Castanospermum australe, Cryptocarya obovata, Dendrocnide excelsa, Diploglottis australis, Dysoxylum fraserianum, Dysoxylum mollissimum, Elattostachys nervosa, Endiandra pubens, Flindersia schottiana, Gmelina leichhardtii, Neolitsea australiensis, Neolitsea dealbata, Sloanea australis, Sloanea woollsii*, and *Toona ciliate* (Floyd 1990).

7.1.1.2 Tropical Mixed Forests

Mixed evergreen and mixed deciduous forests are characterized by the presence of both evergreen (full of leaves throughout the year) and deciduous (remaining leafless in a

Fig. 7.2 Tropical mixed evergreen forests of Andaman. (Photo courtesy of Dr. George Weber)

Fig. 7.3 Tropical deciduous forest of Mexico. (Photo courtesy of Dr. Mark E Olson)

season of the year) tree species at various proportions. However, there are enormous variations in species composition in mixed forests of the tropics. Dittus (1977) reported the dominance of the following tree species in mixed evergreen forests of Sri Lanka:

Adina cordifolia, Alstonia scholaris, Azadirachta indica, Bauhinia racemosa, Canthium coromandelicum, Cassia fistula, Cassia roxburghii, Chloroxylon swietenia, Cordia sp., Drypetes sepiara, Eugenia bracteata, Ficus amplissima, Ficus bengalensis, Ficus (several other species), Garcinia spicata, Gleniea unijuga, Grewia polygama, Holoptelea integrifolia, Ixora arborea, Lanneaz coromandelica, Lepisanthes tetraphylla, Madhuca longifolia, Manilkara hexandra, Mitragyne parvifolia, Phyllanthus polyphyllus, Polyalthia korinthii, Polyalthia longifolia, Premna tomentosa, Sapindus emarginatus, Schleichera oleosa, Stereospermum personatum, Strychnos potatorum, Syzigium cumini, Vitex pinnata, and Walsura piscidia. Fig. 7.2 shows a section of a tropical mixed evergreen forest of Andaman.

Moist tropical deciduous forests in India cover an area of 1,20,000 ha. *Shorea robusta* is the main species there with other associates like *Albizzia procera, Dillenia pentagyna, Cassia fistula, Terminalia belerica, Terminalia chebula, Bauhinia acuminata, Lagerstroemia speciosa, Syzygium* spp., etc. The principal tree species of tropical moist deciduous forests of Kerala, India is teak (*Tectona grandis*). *T. grandis* occurs in some places in India and Myanmar as almost pure stands. Associated tree species include *Terminalia* species, *Lagerstroemia lanceolata*, and *Cassia fistula*. They exhibit low branching in contrast to trees of evergreen forests. Tropical deciduous forests are less diverse than the rainforests or evergreen forests. Sahu et al. (2007) inventoried the dry deciduous forests of Bodhu district, Eastern Ghats of India. They found 91 tree species within a 4-ha sampled area. The predominant tree species are *Shorea robusta, Madhuca indica, Buchanania lanzan, Cleistanthus collinus, Terminali alata, Anogeissus latifolia, Lagerstroemia parviflora,* and *Diospyros melanoxylon.* Fig. 7.3 shows a view of a tropical deciduous forest.

7.1.1.3 Tropical Mangrove Forests

Mangroves are one of the world's dominant coastal ecosystems comprised chiefly of flowering trees and shrubs uniquely adapted to marine and estuarine tidal conditions (Hogarth 1999; Saenger 2002; FAO 2007). They form distinctly vegetated and often densely structured habitat of verdant closed canopies cloaking coastal margins and estuaries of equatorial, tropical, and subtropical regions around the world (Spalding et al. 1997). Mangrove forests occur in the area within latitudes 25° N and 25° S in coastal and estuarine tidal floodplains. Mangrove forests cover an area of 137,760 km^2 spanning 118 countries and territories (Giri et al. 2011) in the tropical and subtropical regions of the world. The largest percentage of mangroves is found between 5° N and 5° S latitudes. Approximately, 75 % of world's mangroves are found in just 15 countries. Asia has the largest amount (42 %) of the world's mangroves, followed by Africa (21 %), North/Central America (15 %), Oceania (12 %), and South America (11 %). The habitat is distinctly saline, although differentially depending on the depth and duration of inundation and position of land in relation to sea level. Mangroves dominate three-quarters of tropical coastlines. The saline conditions tolerated by various mangrove species range from brackish water, through pure seawater (30–40 ppt), to water concentrated by evaporation to over twice the salinity of ocean water (up to 90 ppt). About 110 tree species are found to thrive on mangrove habitats, but only a few of them belong to the typical mangrove family the Rhizophoraceae and the genus *Rhizophora*. There are six functional types of mangrove forests, namely, fringe, riverine, basin, overwash, scrub (dwarf), and hammock forests. Mangroves are well known for their morphological and physiological adaptations coping with salt, saturated soils, and regular tidal inunda-

Fig. 7.4 A section of mangrove forests in Sundarbans, Bangladesh at low tide. (Photo courtesy of Dr. Animesh Biswas)

tion, notably with specialized attributes like exposed breathing roots above ground, extra stem support structures, salt-excreting leaves, low water potentials and high intracellular salt concentrations to maintain favorable water relations in saline environments, and viviparous water-dispersed propagules. Pneumatophores protrude over the water level and exchange gases through lenticels. These "breathing tubes" typically reach heights of up to 30 cm, and in some species, over 3 m. Mangroves have notable aboveground breathing roots and shallow belowground anchoring roots, because their soils are usually saturated and airless. The trees have broad support structures, such as buttresses and sturdy prop roots, because soils are often soft and unconsolidated. Root structures above ground include four types: (1) pneumatophores—pencil-like (*Avicennia*), erect conical (*Xylocarpus moluccensis*), thinly conical (*Sonneratia alba*), and elongate conical (*Sonneratia caseolaris* and *Sonneratia lanceolata*); (2) knee roots—thick and knobbly (*Bruguiera*) and thin and wiry (*Lumnitzera littorea*); (3) stilt roots (*Rhizophora*); and (4) buttresses—sinuous planks (*Xylocarpus granatum*, *Heritiera littoralis*, and *Ceriops* spp., *Pelliciera rhizophorae*).

Mangroves have acknowledged roles in coastal productivity and connectivity (Mumby et al. 2004), often supporting high biodiversity and biomass not possible otherwise. Mangrove ecosystems are key sources of coastal primary production with complex trophic linkages (Robertson et al. 1992), as nursery and breeding sites of marine and arboreal life, and as physical shelter and a buffer from episodic severe storms, river flows, and large waves. Fig. 7.4 shows an image of a tropical mangrove forest.

Some important plant species of mangroves are (Giri et al. 2011):

Avicennia alba, Avicennia marina, Acrostichum aureum, Bruguiera cylindrica, Bruguiera gymnorhiza, Bruguiera parviflora, Camptostemon schultzii, Ceriops tagal, Excoecaria agallocha, Heritiera fomes, Heritiera littoralis, Lumnitzera littorea, Lumnitzera racemosa, Nypa fruticans, Pluchea indica, Rhizophora apiculata, Rhizophora mucronata, Sonneratia ovata, Xylocarpus spp., and Xylocarpus granatum.

7.1.1.4 Tropical Freshwater Swamp Forests

Freshwater swamp forests are found in areas which are seasonally or permanently inundated with freshwater. They normally occur along the lower reaches of rivers and around freshwater lakes. Freshwater swamp forests are also found in low-lying drainage basins. Peat swamp forests occur in submerged soils that prevent woody debris from fully decomposing, which over time creates a thick layer of acidic peat. In the tropics, freshwater swamp forests are distributed in Eastern and Western Congo and Nigeria in Africa; Indonesia, Papua New Guinea, Thailand, Myanmar, Bangladesh, Malaysia, Cambodia, Vietnam, and Sri Lanka in Asia; and Brazil, Peru, Bolivia, Guyana, Venezuela, Suriname, and Mexico of South and Central America. Over long periods of time, freshwater swamp forests may transition into peat swamp forests where the accumulation of organic materials gives rise to high levels of acidity which in turn limits their decomposition. The peat which is made up of partly decayed vegetative material will build up over a long period of time to form a mound several meters thick with its highest point in the center and tapering off near the edges which are usually delimited by significant streams or rivers along with the characteristic riverside vegetation. The species composition varies from place to place. Koponen et al. (2004) observed the dominance of the following tree species in a tropical freshwater swamp forest in French Guiana:

Malouetia tamaquarina, Taralea oppositifolia, Caryocar microcarpum, Licania heteromorpha, Licania macrophylla, Caraipa sp., Diospyros guianensis, Duguetia calycina, Pterocarpus officinalis, Eschweileria pedicellata, Lecythis corrugate, Hydrochorea coiymbosa, and Virola surinamensis.

Theilade et al. (2011) reported the following tree species of variable dominance in swamp forests of Cambodia:

Myristica iners, Litsea sp., Eugenia spp., Mastixia pentandra, Archidendron clypearia ssp., clypearia, Ficus spp., Livistona saribus, Macaranga triloba, Pternandra caerulescens, Eugenia spp., Baccaurea bracteata, Euonymus glaber, Memecylon umbellatum, Parinari annamensis, Callophyllum ssp., and Xerospermum noronhianum. Fig. 7.5 shows a section of a freshwater swamp forest.

7.1.1.5 Tropical Montane Forests

The tropical montane cloud forest is composed of forest ecosystems of distinctive floristic and structured form. It typically occurs as a relatively narrow altitudinal zone where the atmospheric environment is characterized by persistent, frequent or seasonal cloud cover at the vegetation level. Enveloping cloud or wind-driven clouds influence

Fig. 7.5 A freshwater swamp forest in Sunamganj, Bangladesh. (Photo courtesy of Dr. Kamrul Huda)

the atmospheric interaction through reduced solar radiation and vapor deficit, canopy wetting, and general suppression of evapotranspiration. The net precipitation (throughfall) is significantly enhanced through direct canopy interception of cloud water and low water use by the vegetation (Hamilton et al. 1993). In the humid tropics, montane cloud forests are found between 500 and 3,500 m altitude (LaBastille and Pool 1978; FAO 1992) with major occurrence between 1,200 and 2,500 m (Stadtmüller 1987). Some vegetation gradients occur with increasing altitude: diminishing tree height, simplified stratification, smaller leaf size, more open understory, some floristic changes, and more epiphytes, mosses, and lichens (Whitmore 1975; White 1983; Jacobs 1988). Tropical montane forests are usually of low stature (Letouzey 1985), whereas submontane forests have greater height, are richer in species, with more vascular epiphytes, and incorporate species from low and mid-altitude forests (Thomas and Achoundong 1991). They occur in tropical America and Southeast Asia and in tropical Africa (Stadtmüller 1987). The canopy in montane forests of Nilgiris, India is dominated by Lauraceaous species such as *Cryptocareya nessiana, Cinnamomum malabathrum, Beischemedia wightii,* and *Litsea* spp. Other species in the canopy include *Syzygium* spp., *Elaeocarpus* spp., and *Isonandra candolleana* (Suresh and Sukumar 2011). Tropical montane forests are highly threatened ecosystems with a rapidly declining area (Doumenge et al. 1995; Gentry 1995). For the Andean montane forests, Hamilton (1995) estimated that the original forest cover already has been reduced by 90%. Himalayan montane forests are mainly dominated by *Pinus roxburghii* and *Quercus* species. *P. roxburghii* forest is found mainly in mid-altitude (1,000–2,200 m) as pure patch but mostly mixed with broad-leaved forest in the valleys and shaded areas closely associated with *Lyonia ovalifolia, Pyrus pashiya, Mallotus phillipensis,* and *Shorea robusta* at lower altitude and *Cedrus deodara, Myrica esculenta, Q. leucotrichophora, Q. glauca,* and *Rhododendron arboreum* at higher altitudes of this region. *Q. leucotrichophora* grows in north aspect mainly cooler region below 2,000 m and is found either pure or mixed with other broad-leaved and conifer species (Ram et al. 2004; Semwal et al. 2007; Semwal et al. 2010).

7.1.2 Temperate Forests

The Encyclopedia Britannica states that the temperate forests represent a vegetation type with a more or less continuous canopy of broad-leaved trees. Such forests occur between approximately 25° and 50° latitude in both hemispheres. In the Northern Hemisphere, they grade into boreal forests further north, which are dominated by evergreen conifers, so that mixed forests containing both broadleaf and coniferous trees occupy intermediate areas. Well-defined seasons with a distinct winter characterize this forest biome. Temperature ranges from −30 to 30 °C, and precipitation differs from 750 to 1,500 mm with more or less uniform distribution over the year. There is a moderate climate and a growing season of 140–200 days during 4–6 frost-free months. Canopy is moderately dense and allows light to penetrate, resulting in well-developed understory vegetation. Flora is characterized by 3–4 tree species per square kilometer. Trees are mainly broad-leaved and deciduous, and include such species as oak (*Quercus* sp.), hickory (*Carya* sp.), beech (*Fagus* sp.), hemlock (*Conium* sp.), maple (*Acer* sp.), basswood (*Tilia Americana*), cottonwood (*Populus deltoides*), elm (*Ulmus* sp.), and willow (*Salix* sp.). There are some evergreen and coniferous forests as well. Temperate forests occupy much of North America, Europe, Asia, Australia, New Zealand, and a small part of South America, but those in the Southern Hemisphere are almost exclusively evergreen; temperate deciduous forests are generally found in mid-latitudinal belts in the Northern Hemisphere (Reich and Frelich 2002). Deciduous forests are found in regions of the Northern Hemisphere that have moist, warm summers and frosty winters—primarily eastern North America, eastern Asia, and Western Europe. In contrast, evergreen forests typically grow in areas with mild, nearly frost-free winters. Evergreens fall into two subcategories—broad-leaved forests and sclerophyllous forests (trees have small, hard, thick leaves). Broad-leaved forests dominate the natural vegetation of New Zealand; they are significantly represented in South America, eastern Australia, southern China, Korea, and Japan, and they occur in less well-developed form in small areas of southeastern North America and southern Africa. Sclerophyllous forests occur particularly in Australia and in the Mediterranean region.

7.1.2.1 Temperate Deciduous Forest

Temperate deciduous forests are composed of broad-leaved deciduous forest tree species. They are found in areas where warm moist summers alternate with mild winters (Archibold 1995; Reich and Frelich 2002). The three major areas of this forest type occur in the Northern Hemisphere: eastern North America, eastern Asia (Wen 1999), and Europe. Smaller areas

Fig. 7.6 Temperate broadleaf deciduous forest. (Photo courtesy of Daisyzhou)

Fig. 7.7 Temperate coniferous forest. (Photo courtesy of Green Forest)

occur in Australia and southern South America (Archibold 1995). Examples of typical trees include oak (*Quercus* sp.), maple (*Acer* sp.), beech (*Fagus* sp.), and elm (*Ulmus* sp.). Some specific trees of the temperate deciduous forests are *Acer saccharum, Tilia americana, Ostrya virginiana, Ulmus americana, Ulmus rubra*, and *Quercus rubra* (Fig. 7.6).

7.1.2.2 Temperate Mixed Forests

These forests accommodate coniferous trees along with broad-leaved trees in various proportions. Temperate mixed forests are distributed in Austria, Belgium, Belarus, China, Czech Republic, Denmark, France, Germany, Lithuania, Moldova, the Netherlands, North Korea, Poland, Portugal, South Korea, and Turkey. Plant species found in these forests include broadleaf trees such as maple (*Acer* sp.), and birch (*Betula* sp.), as well as conifers such as pines (*Pinus* sp.) and firs (*Abies* sp.). Guan et al. (2006) reported that dominant tree species in an old temperate mixed forest of northeastern China are the Korean pine (*Pinus koriaensis*), Tuan linden (*Tilia amurensis)*, Mono maple (*Acer mimo*), Manchurian ash (*Fraxinus mandshurica*), Mongolian oak (*Quercus mongolica*), elm (*Ulmus glabra*), and other interspersed deciduous species.

7.1.2.3 Temperate Coniferous Forests

Temperate coniferous forest is a biome in temperate regions of the world with warm summers and cool winters with adequate rainfall. Evergreen conifers predominate in these forests; a mixture of conifers and broadleaf evergreen trees may also be found. Temperate evergreen forests are common in the coastal regions having mild winters and heavy rainfall. Temperate evergreen coniferous forests cover approximately 2.4×10^6 km^2 (Melillo et al. 1993; Landsberg and Gower 1997). Temperate conifers dominate the montane forests in North America, Europe, and China, with smaller areas in Korea, Japan, Mexico, Nicaragua, and Guatemala. In Europe, the distribution of native and introduced conifers has been expanded into areas more climatically favorable to hardwoods. Pines (*Pinus*), represented by more than 90 species, are the most widely distributed conifer and their range has increased by widespread planting in the Southern Hemisphere (Waring 2002). Other important genera include the firs (*Abies*), spruces (*Picea*), hemlocks (*Tsuga*), false-hemlocks (*Pseudotsuga*), larches (*Larix*), cypresses (*Cupressus*), cedars (*Chamaecyparis*, *Thuja*, and *Libocedrus*), and juniper (*Juniperus*). Additional conifer genera (*Araucaria*, *Fitzroya*, and *Austrocedrus*) are present in the Southern Hemisphere, where they were once more widely distributed (Axelrod et al. 1991). Temperate conifers fall within regions of >250 mm in annual precipitation that experience subfreezing conditions, down to, but not below, -45 °C, which is the limit for supercooling of water. Only boreal tree species, which include some pines and spruces, are adapted to temperatures below this limit (Waring and Running 1998). In regions where precipitation is well distributed throughout the growing season, deciduous hardwoods usually dominate over temperate conifers. Following major disturbances, however, conifers can become established and achieve temporary dominance (Landsberg and Gower 1997). In the Pacific Northwest region, where summer drought is common, the situation is reversed and many long-lived conifers replace earlier established hardwoods (Waring and Franklin 1979).

According to Waring (2002), temperate coniferous forests include representatives of the most massive forms of terrestrial life on Earth. In the coastal redwood (*Sequoia sempervirens*) forests of northern California, trees may reach heights >100 m and accumulate biomass in excess of 3000 t ha^{-1}. In the more extensive Douglas fir (*Pseudotsuga menziesii*) forests of the Pacific Northwest, the USA, and Canada, biomass in older stands frequently reaches 1,000 t ha^{-1}, with an additional 500 t ha^{-1} of standing dead and fallen trees. Aboveground biomass ranges widely, however, as does aboveground growth. Temperate coniferous forests produce an average of about 5 t ha^{-1} per year of aboveground biomass, but in climatically diverse areas production may range from <1 to >20 t ha^{-1} per year (Runyon et al. 1994; Fig. 7.7).

The continental part of the Asian temperate coniferous forest comprises the southern part of the Russian Far East as well as northeastern China, in the basin of the Amur River on the spurs of the mountain ranges Sikhote-Alin Gory, Badzhalskiy Khrebet, the Lesser Hingan Range, Changbai Shan, and Laoye Ling. The Daurian Mongolian oak area lies mostly in northeastern China and extends in the west to northeastern Mongolia. In the Russian Far East, where it stretches over the upper part of the Amur River basin called Dauriya and the basin of Ozero Khanka. The basic forest type is the Mongolian oak (*Quercus mongolica*) in pure stands or mixed with *Betula davurica*. The conifer forests in the area are dominated by *Pinus densiflora*, which forms stands on the ridges and southern slopes, occasionally affected by fires near Ozero Khanka, and *Pinus sylvestris*, forming stands on sandy soils in the upper Amur regions. The Manchurian Broad-leaved–Conifer area lies on the spurs of the Sikhote-Alin mountain range and on the southern spurs of the Lesser Hingan, the Bureinskiy and the Badzhal'skiy mountain ranges. The main vegetation type is broad-leaved Korean pine forests dominated by *Pinus koraiensis* and *Abies holophylla* and a number of broad-leaved species: *Betula costata*, *Fraxinus mandshurica*, *Kalopanax septemlobus*, *Phellodendron amurense*, *Quercus mongolica*, *Tilia amurensis*, *T. mandshurica*, and *Ulmus japonica* (Kolesnikov 1956; Krestov 1997).

7.1.2.4 Temperate Rainforests

The temperate rainforests include a range of vegetative types throughout the world (Kuppen 1918; Kuchler 1949; Franklin and Dyrness 1973; De Laubenfels 1975; Webb 1978; Alaback and Juday 1989). The rainforest climate is characterized by cool summers and wet weather year around. Rainforests are difficult to distinguish floristically or physiognomically from related forest types (Webb 1968, 1978; Cockayne 1971; De Laubenfels 1975). The conifer forests of the Pacific Coast of North America are unique and evergreen. The great stature of these forests gives a complex structure relative to many other forest types. Only the mixed *Araucaria-Nothofagus* forests of New Guinea compare in stature and structure (Enright 1982). The conifer forests of the Pacific Slope are also unique in their domination over hardwood or deciduous species. Most other parts of the world with similar climates have either a mixture of conifer and hardwood species or a domination of hardwood species with occasional conifers (Waring and Franklin 1979; Ash 1982; Wardie et al. 1983). Western hemlock occupies the majority of productive forested land in the Pacific rainforest zone. A lush understory of herbs, short and tall woody shrubs, and trees provides an optimal condition for many herbivorous mammals (Hanley et al. 1989). Mixed forests of hemlock and cedar have among the oldest trees in Southeast Alaska, but are generally less productive than pure hemlock or spruce forest types. The canopy is usually more open so that a more vigorous woody shrub understory can develop. Yellow cedar and red cedar can both exceed 1,000 years of age in this forest type. Red cedar occurs primarily in the southern portion of Southeast Alaska and at lower elevations within this region (Alaback 2010). The most diverse but least productive type of coastal rainforest occurs on poorly drained sites throughout coastal Alaska and adjacent British Columbia. In Southeast Alaska, this forest is often called the mixed conifer type because no single species appears to dominate this forest type consistently (Martin 1989). Common dominant species include mountain hemlock, shore pine, western red cedar, and yellow cedar types (Banner et al. 1990). In British Columbia, these ecosystems are called mostly mountain hemlock or western red cedar. Forests are relatively open grown and usually include a wide representation of forest forb, herb, shrub, and tree species as well as many wetland or bog species (Ver Hoer et al. 1988; Martin 1989). In the most poleward sites a distinctly open forest develops. In Alaska, this subarctic rainforest is best represented in Prince William Sound with a mixture of *Picea sitchensis*, *Tsuga heterophylla*, and *T. mertensiana* (Cooper 1942; Eck 1984; Alaback and Juday 1989; Borchers et al. 1989). Thickets of *Rubus spectabilis* and other tall woody shrubs are less common than in the spruce rainforest type. The understory is dominated by ericaceous woody shrubs and carpets of bryophytes and ferns on the forest floor. The landscape in both hemispheres is a dynamic fine-grained mosaic of bog or moorland, ice fields, and rainforest (Alaback 2010).

Temperate rainforests also occur in South Africa's Knysna-Amatole coastal forests, the Colchian rainforests of the eastern Black Sea region (Turkey and Georgia), the Caspian temperate rainforests of Iran and Azerbaijan, the mountain temperate rainforests along eastern Taiwan's Pacific Coast, southwest Japan's Taiheiyo forests, Australia's coastal New South Wales and New Zealand's North Island. Temperate rainforests are typically characterized by a rainfall of somewhere between 2,000 and 4,000 mm. These forests are highly diverse among the different types of temperate forests on the planet (Harden et al. 2006). Most tree species are evergreen conifers. Some of these trees grow taller than 70 m. They include Douglas fir (*Pseudotsuga* sp.), Sitka spruce (*Picea sitchensis*), Red cedar (*Toona ciliate*), Western hemlock (*Tsuga heterophylla*), etc. Some of these trees may be up to 500–1,000 years old and the trunk can be more than 25 m around (Wolf et al. 1995). Tasmania's rainforests contain only a few dominant tree species, including myrtle *(Nothofagus cunninghamii)*, leatherwood *(Eucryphia lucida)*, sassafras *(Atherosperma moschatum)*, deciduous beech *(Nothofagus gunnii)*, and a number of conifer species such as Huon pine *(Lagarostrobos franklinii)*, celery top pine *(Phyllocladus aspleniifolius)*, King Billy pine *(Athrotaxis selaginoides)*, pencil pine *(Athrotaxis cupressoides)*, and Diselma *(Diselma archeri)* (Fig. 7.8).

Fig. 7.8 Temperate rainforest of the Northern Pacific Coast. (Photo courtesy of Mr. Hig, Ground Truth Trekking)

7.1.2.5 Mediterranean Forests

The Mediterranean region comprises 26 countries covering an area of 8.779 million km^2. The region's forest cover amounts to 85 million ha, about 2% of the world's forest area. When the other wooded land area is added, the total area of the natural woody vegetation in the Mediterranean raises to 115 million ha. Fifteen countries with about 28% of the region land area account for about 89% of the region forest and other wooded land area (FAO 2011). The Mediterranean region is a highly complex environment due to the interactions between multiple factors such as climate, geomorphology, soil, hydrology, and land use (FAO 2013). It is estimated that around 25,000 plant species inhabit the region (Myers et al. 2000), of which about 60% are endemic (Thompson et al. 2005), making it one of the highest concentrations of endemic plants in the world (Myers et al. 2000). The region is regarded as a biodiversity hot spot because of its high endemism and because it is experiencing a widespread loss of habitat (FAO 2013). Only 5% of the original vegetation remains relatively intact in the region. According to the International Union for Conservation of Nature (IUCN) Red List, about 124 plant species found in forests in the Mediterranean region are threatened with extinction. For example, two oak species endemic to the Mediterranean (*Quercus pauciradiata* and *Q. orocantabrica*) are classed as critically endangered (Oldfield and Eastwood 2007). Region-wide recent studies (Cuttelod et al. 2009) suggest that about 18% of the species are threatened with extinction.

The characteristics of the Mediterranean region include: (i) an unusual geographical and topographical variability related to the presence of a jagged coastline and of many, relatively young mountain ranges, often quite high in elevation; (ii) a pronounced, climatic bi-seasonality with dry and hot summers and moist and cool autumns and winters; also, occasional, violent precipitation events, a large year-to-year variability of total rainfall, as well as frequent strong and dry winds that favor the spread of forest trees; (iii) a high diversity of plant and animal species, coupled with a rich variability of more or less natural vegetation types and land-use forms, giving rise to complex mosaics of patches; (iv) a long history of manipulation of trees, forests, and landscapes, as documented by the continued presence of planted forests of *Pinus pinea*, *Cupressus sempervirens*, and *Castanea sativa* in areas where these species were extended since the Greco-Roman age, and gradually have become characteristic elements of the landscape (Scarascia-Mugnozza et al. 2000).

According to FAO (2013), typical Mediterranean forests are composed of broad-leaved species (mainly oaks), both evergreen and deciduous, such as *Quercus ilex*, *Q. suber*, *Q. coccifera*, *Q. pubescens*, *Q. cerris*, *Q. pyrenaica*, *Q. toza*, *Q. calliprinos*, *Q. ithaburensis*, and others, and conifers such as *Pinus halepensis*, *P. brutia*, *P. pinea*, *P. pinaster*, and *Juniperus* species. The degradation of such forests has produced low-density woody vegetation known as the *macchia* and the *garrigue*. Where there is no water limitation (e.g., along rivers), forests of *Q. robur*, *Q. petraea*, *Fraxinus* species, *Populus alba*, and *P. nigra* can prosper. Typical of the Iberian Peninsula, the *dehesa* is characterized by pastures with scattered oaks, both evergreen and deciduous, sometimes mixed with *Pinus pinea*.

7.1.3 Boreal Forests

Boreal forests, also known as taiga, occur between 50 and 60° north latitudes and represent the largest terrestrial biome of the world. Strong seasonal variation with short, moderately warm, and moist summers and long, extremely cold, and dry winters prevails in the boreal forest region (Larsen 1980). This seasonal variation is most pronounced in the continental climates of interior Alaska and eastern Siberia, where seasonal fluctuation of mean monthly temperature is on the order of 44 °C and more than 56 °C, respectively (Rumney 1968). Pronounced interannual variation in air temperatures is coupled with this strong seasonal variation. At Fairbanks, Alaska, mean monthly January air temperature between 1971 and 1986 ranged from −35.4 °C to −7.7 °C and averaged −22.1 °C. These large seasonal and annual temperature fluctuations interact to form a distinct feature of the continental climate of the boreal forest (Larsen 1980). A more oceanic climate prevails in eastern Canada and Scandinavia and temperature fluctuations are not such extreme there. Throughout the boreal forest, annual rainfall is relatively light. In North America, winters are characteristically dry, and more than half of the annual precipitation falls in the summer. In the northwest, where high mountain ranges restrict the inland penetration of moisture-laden air, annual rainfall is less than 38 cm and, in Fort Yukon, Alaska, it is as

Fig. 7.9 Boreal forest. (Photo courtesy of Terry A. McDonald—luxborealis.com)

low as 18 cm (Rumney 1968). East of the Rocky Mountains, annual rainfall increases to 38–51 cm in central Canada and to 51–89 cm in eastern Canada. In contrast to northwestern North America, the northern European boreal forest region, west of the Ural Mountains, lacks high mountain barriers. Here, under the moderating influence of maritime air from the north Atlantic, the mean annual temperature range is relatively low (Rumney 1968; Fig. 7.9).

The boreal forest is the second largest biome in the world containing 33 % of the earth's forest cover (FAO 2001) of which approximately 25 % is natural. It is circumpolar and shares similar taxa across its range. It has approximately 20,300 identified species (Rucktuhl et al. 2008). Most of the boreal forest is still relatively intact and not accessible by road. In North America, boreal forests cover most of inland Canada and Alaska as well as parts of the extreme northern continental USA where they are known as the Northwoods. Taiga also covers most of Sweden, Finland, much of Norway, lowland and coastal areas of Iceland, much of Russia from St. Petersburg in the west to the Pacific Ocean (including much of Siberia), northern Kazakhstan, northern Mongolia, and northern Japan (on the island of Hokkaido).

Boreal forests occupy a belt up to 1,000 km wide in certain regions (Larsen 1980). Conifers occupying boreal forests include members of the genera *Picea, Abies, Pinus,* and *Larix*. In North America, white spruce (*Picea glauca*), black spruce (*Picea mariana*), jack pine (*Pinus banksiana*), balsam fir (*Abies balsamea*), and tamarack (*Larix laricina*) are the primary conifers. Hardwood tree species include members of the genera *Populus, Betula, Alnus,* and *Salix* (Graham and Jain 1998).The two major types of taiga are the closed canopy forest of the southern part, consisting of many closely spaced trees with mossy ground cover and the lichen woodland or sparse taiga, with trees that are farther-spaced and lichen ground cover common in the northernmost taiga (Sayre 1994). The forest cover is more sparse and stunted in the northernmost taiga. In Canada, Scandinavia, and Finland, the boreal forest is usually divided into three subzones: the high boreal (north boreal) or taiga zone; the middle boreal (closed forest); and the southern boreal, a closed canopy boreal forest with some scattered temperate deciduous trees among the conifers (La Roi 2011) such as maple (*Acer* sp.), elm (*Ulmus* sp.), and oak (*Quercus* sp.). The forests of the taiga are largely coniferous, dominated by larch (*Larix* sp.), spruce (*Picea* sp.), fir (*Abies* sp.), and pine (*Pinus* sp.). The woodland mix varies according to geography and climate; for example, the Eastern Canadian forests of the higher elevations are dominated by balsam fir *Abies balsamea,* while further north the Eastern Canadian Shield Taiga is dominated by black spruce *Picea mariana* and tamarack larch *Larix laricina*. Some authors separate Taiga from northern forests (boreal forests). They take the Taiga ecoregion to include most of interior Alaska and much of Canada's northern boreal forest (CEC 1997). The taiga is dominated by evergreen conifers. Dwarf birch (*Betula nana*), Labrador tea (*Ledum* spp.), willow (*Salix* spp.), bearberry (*Arctostaphylos alpina*), mosses, and sedges dominate in the mid-zone. In the south, open stands of stunted black spruce (*Picea mariana*) and jack pine (*Pinus banksiana*) are accompanied by alder (*Alnus* spp.), willow (*Salix* sp.), and tamarack (*Larix larcinia*) in fens and bogs. White spruce (*Picea glauca),* black spruce, lodgepole pine (*Pinus contorta*), quaking aspen (*Populus tremuloides*), balsam polar (*Populus balsamifera*), and paper birch (*Betula papyrifera*) grow on well-drained, warm upland sites, rivers, and streams (Geiser and Nadelhoffer 2011).

Bonan and Shugart (1989) reviewed the environmental factors and ecological processes in the boreal forests. Vegetation patterns in the boreal forest landscapes are regulated by soil moisture characteristics and air temperature (Tuhkanen 1984). At the southern edge of the North American boreal forest (Maycock and Curtis 1960) and in central Canada (Carleton and Maycock 1978), forest communities can be segregated into wet (*Picea mariana* and *Larix laricina*), mesic (*Betula papyrifera, Populus tremuloides, P. balsamifera, Picea glauca,* and *Abies balsamifera*), and dry (*Pinus banksiana*) sites. Soil moisture, along with soil temperature, is one of the primary site characteristics distinguishing vegetation patterns in interior Alaska (Van Cleve et al. 1983; Viereck 1975; Viereck et al. 1983, 1986). *Picea mariana* stands occupy the wettest sites with average seasonal forest floor and mineral soil moisture contents of 120–240 % and 60–160 %, respectively; *Picea glauca* and the successional hardwoods tend to occur on drier sites. Over much of the perennially frozen north, there is an abundant supply of soil water from seasonal snow melt and soil thawing that cannot drain owing to underlying permafrost (Lutz 1956; Wolff et al. 1977). Even in extremely arid regions, the slow thawing of permafrost soils provides sufficient moisture for tree growth

(Lutz 1956; Walter 1979). Low soil temperatures have important consequences on physical site conditions. Cold soils and permafrost impede infiltration (Ford and Bedford 1987). Combined with the slow release of water from the seasonal melting of the active soil layer, these can result in high soil moisture content throughout the growing season (Dimo 1969; Rieger et al. 1963; Viereck et al. 1983). Soil movement caused by frost heaving can result in haphazardly leaning, "drunken" trees (Zoltal 1975a, b). Up to one-third of the taiga zone in western Siberia is forested bog. With their high water content, low organic matter decomposition, and high organic matter accumulation, these bogs are important regulators of the water budget and carbon cycle (Antonovsky 1987).

Boreal forests are defined in several ways (Mayer 1984; Shugart et al. 1992), but they are dominated by coniferous trees and adapted to low temperatures and a short growing season. They have some similarities with hemiboreal and mountain forests in southern latitudes (Mayer 1984; Nikolov and Helmisaari 1992). Angelstam and Kuuluvainen (2004) argue that the transition zones to temperate forests such as the hemiboreal forest found in southern Fennoscandia, the Baltic republics, Belarus, and southwestern Russia should be included in boreal forests. Similar forest ecosystems are also found at the southern border of the boreal forest in North America and Asia (Pastor and Mladendoff 1992; Frelich 2002) as well as at higher altitudes in southern latitudes (Mayer 1984). Although the boreal forest still encompasses a large proportion of intact forests (Aksenov et al. 2002), it is also an important natural resource from which human welfare is built (Kuusela 1990; Burton et al. 2003). Therefore, the boreal forest is currently heavily impacted by different kinds of resource extraction. These include both legal and illegal logging (Ovaskainen et al. 1999; Lloyd 2000; Lopina et al. 2003), severe impacts caused by the extraction of oil and gas (Schneider 2002), pollution (Sverdrup and Stjernquist 2002), and global climate change (Watson et al. 2000).

7.1.4 Regional distribution of forests

According to Global Forest Resources Assessment 2010 (FAO 2010), the world's total forest area is estimated to be just over 4 billion ha (0.6 ha of forest per capita). Forests cover 31 % of the land area, but the forest area is very unevenly distributed. The five most forest-rich countries (the Russian Federation, Brazil, Canada, the USA, and China) account for more than half of the total forest area (53 %), while 64 countries have forest on no more than 10 % of their land area. More than 90 % of the total forest area consists of naturally regenerated forests. Ten most forest-rich countries are the Russian Federation, Brazil, Canada, the USA, China, Democratic Republic of the Congo, Australia, Indonesia, Sudan, and India. In 50 countries and areas, forests cover more than half the total land area and in 12 of these forests occupy more than 75 % of the total land area.

Table 7.1 Regional distribution of forest area in 2010 (FAO 2010)

Region/subregion	Forest area	
	1000 ha	% of total forest area
Eastern and Southern Africa	267,517	7
Northern Africa	78,814	2
Western and Central Africa	328,088	8
Total Africa	674,419	17
East Asia	254,626	6
South and Southeast Asia	294,373	7
Western and Central Asia	43,513	1
Total Asia	592,512	15
Russian Federation	809,090	20
Europe Excl. Russian Federation	195,911	5
Total Europe	1,005,001	25
Caribbean	6,933	0
Central America	19,499	0
North America	678,961	17
Total North and Central America	705,393	17
Total Oceania	191,384	5
Total South America	864,351	21
World total	4,033,060	100

The distribution of forest areas in different regions of the world is given in Table 7.1. Europe (including the Russian Federation) accounts for 25 % of the world's total forest area, followed by South America (21 %), and North and Central America (17 %). Forests and trees are planted for many purposes and make up an estimated 7 % of the total forest area, or 264 million ha. Between 2000 and 2010, the area of planted forest increased by about 5 million ha year^{-1}. Forests and trees are planted for many purposes and make up an estimated 7 % of the total forest area, or 264 million ha. Between 2000 and 2010, the area of planted forest increased by about 5 million ha year^{-1}.

7.2 General characteristics of forest soils

Under natural conditions, forest soils have developed over a long period of time under the influence of forest vegetation. Primarily, they evolved through natural succession as coupled components—soil and vegetation—of the specific forest ecosystems. Forest vegetation has profound influence on the genesis of forest soils through their canopy, biomass, litters, and roots. Forests soils, in turn, influence the dynamics of the forest vegetation. Soils and forest vegetation are mutually interdependent variables. Forest soils have some characteristics common to agricultural soils, but they also differ profoundly from them as well. Unless disturbed, forest soils retain their natural characteristics, while agricultural soils are disturbed soils and many of their characteristics are modified by soil and crop management practices. Forest soils retain many of their natural properties even long after clearing for cropping. Again, some unproductive agricultural soils have been afforested. These soils are acquiring some properties typical of forest soils.

7.2.1 Limitations of forest soils

Forest soils have some physical and chemical limitations in agricultural standard that helped them retain their vegetation and soils in the natural state. The physical limitations include: (i) extreme texture: stony, gravelly, sandy, and clayey; (ii) extreme moisture: arid, waterlogged, and submerged; and (iii) steeply sloping and erodible. The chemical limitations include: (i) acidity, (ii) alkalinity, (iii) salinity, (iv) sodicity, (v) acid sulfate conditions, and (vi) poor fertility. Stony, gravelly, excessively sandy, and stiff clay soils are not suitable for cropping. Such areas have, therefore, retained their original cover, although some have been degraded by overexploitation.

7.2.1.1 Permanent Vegetation

Forest soils develop under the influence of permanent or semi-permanent vegetations. For example, a natural forest may occupy the space for more than a millennium. On the other hand, forest plantations are usually established for 40–50 years for timber plants and 10–15 years for fast-growing energy plantations. The occupancy of a land for so long a period by a particular vegetation exerts some effects on it. The vegetation adds organic matter and its decay products into the soil; removes water and nutrients for a long period; and adds organic matter and harbors soil organisms that have tremendous modifying effects on soil fertility and productivity.

Undisturbed forests have thick, dense canopy covers, the thickness of which may exceed many meters. This canopy layer acts as an umbrella that intercepts solar radiation and rainfall. It regulates evapotranspiration and heat transmission. As a result, a microclimate develops in the ground of the forest which is entirely different from adjacent non-forest areas. For this, in the surface of the soil in the forest, the air is damp and humid. The vegetative cover keeps the soil cooler in the warm season and warmer in the cool season than the adjacent cultivated soils. The canopy may intercept almost 100% of rain reducing the beating action of rain drops. A well-developed forest canopy allows only 0–2% light to reach the ground (Thomas and Packham 2007). Significant undergrowth cannot develop at these poor light conditions.

Most roots of agricultural crops are usually limited to the furrow slice, while forest tree roots are usually deep and can penetrate some meters, if not restricted by a compacted layer or the groundwater table near the surface. Deep and extensive roots draw water and nutrients from a larger soil volume. They leave lateral and vertical channels that act as drainage and aeration routes. These channels also contribute to circulation of materials within the soil. Roots distribute organic matter and nutrients throughout the soil profile; they exude metabolites; they attract microorganisms at their vicinity; and with all these, they weather and extend soil formation to greater depths. Forest soils have more roots, deeper roots, and large roots. Therefore, forest soils have deep and well-differentiated soil profile.

7.2.2 Natural Succession

In a given place, a soil and associated vegetation develop simultaneously going through stages of birth, growth, and maturity. Thus, both soil and vegetation have a life history. The series of development is called natural succession. Natural succession on a bare rock is called lithosere and in a wetland is called hydrosere.

Let us see what happens in lithosere. We all know plants cannot grow on a bare hard rock. However, bacteria can thrive inside the cracks and obtain nutrients from the substances dissolved from rock. Bacterial colonies secrete organic substances that enhance their own growth and dissolve substances from the rock. In this way, weathering advances and at one stage lichens (an association of fungi and algae) gradually invade. As materials continue to accumulate from the wastes and dead remains of bacteria and lichens, and underlying rock mass has considerably fragmented, the first true plants can gain a hold. These are often mosses, and within deeper cracks, where more organic material has collected, there may be some grass or a small flowering herb. These plants add more organic material and slowly, as this mixes with the tiny mineral fragments from the weathered rock, soil starts to develop. Soil genesis is a slow process, but after substantial parent material has accumulated soil formation is relatively quick. At this stage, there is succession of a series of vegetation types. Herbs and shrubs come and stay for some period, after which they are replaced by pioneer trees. After soil has formed to a considerable depth, larger plants can grow in it. Their roots penetrate more deeply and leave channels after their death. These channels enhance movement of air and water. Soil materials are transformed and translocated and horizonation begins. The collection of horizons is called a soil profile. However, the horizons do not develop all at a time. As there is a succession of plants (bacteria, lichens, mosses, herbs, shrubs, pioneer trees, mature forest, and ultimately climax forest), there is also a succession of soil horizons (Fig. 7.10). At first, only the parent material, then an AC profile followed by an ABC profile, and then a mature AEBC profile develop. On the surface of many forest soils, a thick O horizon develops by the accumulation of

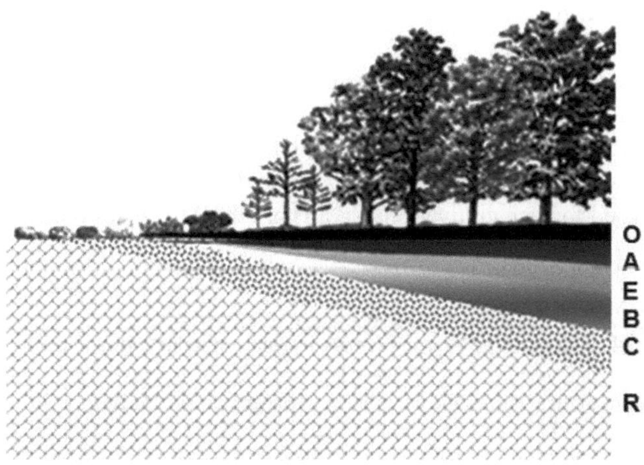

Fig. 7.10 Schematic representation of coupled evolution of forest and soil in a place

forest litters. However, for wide variations in lithology and climate, we see the development of different types of forests and associated soils (Allaby 2008). Hydrosere is a primary succession which develops in aquatic environments and results in the conversion of a water body, say a lake and its community into a land community, say a forest. In this succession, serial plant communities come and become replaced with concomitant changes in the underlying soil.

7.2.3 Forest Floor

The forest floor usually contains a thick layer of organic residues accumulated by the fall of litters or dead parts of the forest vegetation, including leaves, branches, twigs, flowers, fruits, barks, and even large logs. A plenty of mushroom and fungi proliferate on these damp litters. The forest floor is a unique and spectacular feature of the forest soil, particularly temperate. As soon as litters fall on soil surface, they are attacked by soil organisms and start decomposition. The remaining part gradually accumulates on the forest floor. The structure, composition, and functions of the forest floor are discussed in detail in Chap. 4 and are not repeated here.

7.3 Modern Soil Classification Systems

Soil classification is the orderly arrangement of soils according to some known or inferred characteristics. Soils are classified to organize information regarding their properties and behavior so that they can be used for different purposes with the maximum benefit under sustainable management. There are regional and international soil classification systems. Among international systems, Soil Taxonomy (USDA 1960, 1975) and WRB for Soil Resources (FAO 1998) are presently the most popularly used soil classification systems. Here, a brief account of the Soil Taxonomy is given to help readers to understand the features of soils in different forest biomes of the world.

7.3.1 Soil Taxonomy

Soil Classification, A Comprehensive System, popularly known as "7th Approximation" was designed by the Soil Survey Staff of the United States Department of Agriculture in 1960 (USDA 1960). It was revised in 1975 (Soil Taxonomy: A Basic System of Soil Classification for Making and Interpreting Soil Survey; USDA 1975) and in 1998 (USDA, NRCS 1998) again. Soil Taxonomy, as this system of soil classification is known now, is a hierarchical system having soil orders divided into suborders, suborders into great groups, great groups into subgroups, subgroups into families, and families into series. There are 12 orders, 64 suborders, 319 great groups, 2,484 subgroups, more than 8,000 families, and unknown number of soil series. Orders are distinguished by the presence or absence of diagnostic horizons and other diagnostic features, indicating the predominant soil-forming processes. An order includes soils whose properties suggest that they are similar in their genesis. Subdivisions of orders on the basis of genetic homogeneity are suborders. Suborders of an order are differentiated by the presence or absence of properties associated with wetness, soil moisture regimes, major parent material, and vegetation. Subdivisions of suborders are great groups. They are recognized on similar kind, arrangement, and degree of expression of horizons. Subdivisions of great groups are subgroups. Differences in texture, mineralogy, temperature, and soil depth are bases for family differentiation. Soil series is the lowest category of soil taxonomy. A soil series includes soil individuals that have similar (within an acceptable range) pedon characteristics. Readers are referred for technical terms to *Keys to Soil Taxonomy* by USDA, Natural Resources Conservation Service (NRCS 1998).

Soil Taxonomy has a systematically derived nomenclature in categories from order to subgroup. Within this hierarchy, the name of a taxon indicates the major characteristics of the soil and the higher categories to which it belongs. An example is given as follows:

- Order—Entisols (soils with little or no horizon differentiation)
- Suborder —Aquent (wet Entisols)
- Great group—Sulfaquent (wet Entisols that have sulfidic materials)
- Subgroup—typic Sulfaquent (soils that meet the central concept of Sulfaquent)

There are 12 soil orders in soil taxonomy: Alfisols, Andisols, Aridisols, Entisols, Gelisols, Histosols, Inceptisols, Mollisols, Oxisols, Spodosols, Ultisols, and Vertisols. The principal characteristics of the soil orders are summarized subsequently to understand the nature of forest soils encountered in tropical, temperate, and boreal forests. Soil orders are classified into suborders on the basis of soil moisture regimes, soil temperature regimes, nature of parent materials, accumulation of substances in profile, etc.

There are five soil moisture regimes:

Soil moisture regimes	Characteristics
Aquic	The soils are water saturated for considerable periods of the year and remain in a state of reduction as in tidal marshes.
Aridic/torric	The soil moisture control section is dry for more than half of the cumulative days per year and moist for <90 consecutive days as in arid climates.
Udic	The soil moisture control section is not dry for 90 consecutive days and dry for <45 consecutive days in the 4 months following the summer solstice as in humid climate.
Ustic	Intermediate between the aridic and udic regimes. Here plant growth is restricted for some part of the year by low moisture supply.
Xeric	The soil moisture control section is dry for 45 or more days in the 4 months following the summer solstice and moist for 45 or more days in the 4 months following the winter solstice as in Mediterranean climates.
There are nine soil temperature regimes:	
Soil temperature regimes	*Characteristics*
Cryic	If there is an organic O horizon and the soil is not water saturated, summer temperature <8 °C; if there is no O horizon and the soil is not water saturated, summer temperature< 15 °C; and if the soil is organic and mean annual temperature <6 °C.
Frigid	A soil warmer in summer than a cryic regime, but its mean annual temperature is lower than 8 °C and the difference between mean summer and mean winter soil temperatures is more than 6 °C.
Mesic	The mean annual soil temperature is 8 °C or higher but <15 °C, and the difference between mean summer and winter temperature is >6 °C.
Thermic	The mean annual soil temperature is 15°C or higher but <22 °C, and the difference between mean summer and winter temperature is >6 °C.
Hyperthermic	The mean annual soil temperature is 22°C or higher, and the difference between mean summer and winter temperature is >6 °C.
Isofrigid	The mean annual soil temperature is >8 °C, and the difference between mean summer and winter temperature is <6°C.
Isomesic	The mean annual soil temperature is 8°C or higher but <15 °C, and the difference between mean summer and winter temperature is <6 °C.
Isothermic	The mean annual soil temperature is 15°C or higher but <22 °C, and the difference between mean summer and winter temperature is <6 °C.
Isohyperthermic	The mean annual soil temperature is 22 °C or higher, and the difference between mean summer and winter temperature is <6 °C.

7.3.1.1 Alfisols

Alfisols are generally fine textured soils with high base status of the humid regions, both temperate and tropical. They have accumulation of clay in the B horizon to form argillic (a subsurface horizon with a significantly higher percentage of phyllosilicate clay than the overlying soil material), kandic (subsurface horizon that underlies a coarser textured surface horizon), or natric horizon (a subsurface horizon with a higher percentage of phyllosilicate clay and has a subhorizon with an exchangeable-sodium saturation of over 15 %), with base saturation point (BSP > 35 % in the lower part or below the argillic or kandic horizon. There is enough moisture for plant growth for 3 consecutive months during the growing seasons. They have primarily developed under deciduous forests. Alfisols comprise 9.6 % of the ice-free land of the earth. Alfisols have five suborders: Aqualfs (wet Alfisols), Cryalfs (Alfisols that show cryoturbation), Udalfs (Alfisols having udic soil moisture regime), Ustalfs (ustic soil moisture regime), and Xeralfs (xeric soil moisture regime). Nearly all Aqualfs are believed to have supported forest vegetation in the past. Aqualfs in humid regions are primarily used for rice cultivation. Cryalfs are the Alfisols of the cold regions. Cryalfs have cool short growing seasons, and most of them support boreal forests. Most Udalfs with a mesic or warmer temperature regime have or had deciduous forest vegetation and many of the frigid temperature regimes have or had mixed coniferous and deciduous trees.

7.3.1.2 Andisols

Andisols develop from andic parent materials including volcanic ash, pumice, cinders, etc., deposited during volcanic eruptions. These materials produce amorphous or poorly crystallized silicate minerals, including allophane, imogolite, ferrihydrite, etc. Andisols are young fine-textured soils which have a high content of fresh weatherable minerals and a high cation exchange capacity (CEC). They may also contain considerable organic matter. These soils are widely distributed in all geographical regions near sources of volcanoes. Andisols are generally fertile and are used for agriculture unless restricted by slope, altitude, soil moisture, temperature regimes, etc. Some Andisols are left under tundra and forests. Andisols cover approximately 0.7% of the earth's surface. There are seven suborders of Andisols: Aquands (wet), Cryands (cryoturbation), Torrands (torric soil moisture), Udants (udic soil moisture), Ustands (ustic soil moisture), Vitrands (abundance of glass), and Xerands (xeric soil moisture). Most Udants, Ustands, Vitrands, and Xerands have developed under forest vegetation. There are some Vitrands and Xerands covered with coniferous forests.

7.3.1.3 Aridisols

Typical soils of the arid region are the Aridisols. They develop mainly in the arid climates, including cold polar, cool temperate, and warm desert climates, and excluding shifting sands and deserts. They may also occur in semiarid regions due to local conditions of slope, moisture, and parent materials. Potential evapotranspiration greatly exceeds precipitation during most of the year. Little or no water percolates through the soil in most years. This hydrologic regime favors such soil-forming processes as low chemical weathering, low leaching, and high evapotranspiration. Salts (chlorides, sulfates, and carbonates) released by limited chemical weathering are not usually translocated to considerable depths but are accumulated on the surface making most soils saline and calcic or gypsic. Aridisols are mostly in xeric shrub lands with xerophytes, cactus, and thorns.

7.3.1.4 Entisols

Soils that have little or no sign of horizon differentiation are the Entisols. Most Entisols are little altered from their parent materials. Actually, they are affected to a limited extent by translocation processes. However, there is considerable darkening of the surface soil by organic matter. The presence of unweatherable parent materials, removal of soil materials by continuous erosion, continuous deposition of silts with floodwater in active floodplains, cold and dry climates, and insufficient time after rock exposure or sediment deposition are the causes of delayed soil development in Entisols. These soils are distributed over a wide geographic area and can be found in any climate and under any vegetation. Many productive forests are found on Entisols. The suborders of Entisols are Aquents (wet Entisols), Arents (disturbed Entisols), Fluvents (Entisols in floodplains), and Psamments (sandy Entisols).

7.3.1.5 Gelisols

Typical soils of the cold regions having permafrost within 2 m of the surface are the Gelisols. Freezing and thawing in the active layer influence soil formation in Gelisols. Permafrost restricts the downward movement of water, for which there are few diagnostic horizons in Gelisols. Cryoturbation results in irregular or broken horizons, organic matter accumulation on the permafrost table, oriented rock fragments, and silt caps on rock fragments. These soils are limited geographically to the high-latitude polar regions and localized areas at high mountain elevations. Gelisols have three suborders: Histels (Gelisols that have large quantities of organic matter), Turbels (cryoturbation), and Orthels (common Gelisols).

7.3.1.6 Histosols

Soils with organic soil materials and without permafrost are the Histosols. Organic soil materials consist of organic debris accumulating at the surface in which the mineral component does not significantly influence the properties of soils. Organic soil materials have either:

1. under water saturated conditions 18% organic carbon (30% organic matter) or more if the mineral fraction has 60% or more clay; or 12% organic carbon (20% organic matter) if the mineral fraction has no clay; or a proportional intermediate organic carbon for intermediate content of clay; or
2. if never saturated with water for more than a few days, 20% or more organic carbon.

Histosols typically form in settings where poor drainage inhibits the decomposition of plant and animal remains, allowing these organic materials to accumulate over time. They are commonly called bogs, moors, or peats and mucks. Peat is the name given to slightly decomposed organic material in soil, while muck is used for the rotten, highly decomposed material. Most Histosols are acidic and many are deficient in plant nutrients. Histosols occupy only 1.2% of the global ice-free land surface. Histosols have four suborders mostly distinguished on the basis of the state of organic matter and drainage: Fibrists, Folists, Hemists, and Saprists. Most Hemists support natural vegetation of widely spaced, small trees, shrubs, and grasses. Most Folists are covered with forest vegetation. Most Hemists and many Saprists have natural vegetation and are used as woodland, rangeland, or wildlife habitat.

7.3.1.7 Inceptisols

Inceptisols include soils that show the beginning of profile differentiation. They are weakly developed soils in that they have minimal horizon differentiation. They are more developed than Entisols and lack many characteristics of mature soils. Inceptisols may have many kinds of diagnostic horizons except argillic, natric, kandic, spodic (a subsurface horizon in which amorphous organic matter and aluminum accumulates), and oxic (a subsurface mineral horizon with low contents of weatherable minerals and the clay-sized material is dominated by kaolinite) horizons. The most common horizon sequence is an ochric epipedon (a light colored surface horizon of low organic carbon content) over a cambic horizon (a weakly developed subsurface mineral horizon that has few distinguishing morphological characteristics except for evidence of weathering and sometimes of gleying), with or without an underlying fragipan. Inceptisols are mainly found in humid and subhumid regions. They are often found on fairly steep slopes, young geomorphic surfaces, wet sites, and on resistant parent materials. Inceptisols occupy 9.9 % of the global ice-free land surface. Inceptisols have six suborders: Anthrepts (Inceptisols having signs of human habitation), Aquepts, Cryepts, Udepts, Ustepts, and Xerepts. Cryepts mainly support conifers or mixed conifers and hardwood forests. Udepts originally supported forests but now they are mostly cleared for cultivation. The vegetation in Xerepts is commonly coniferous forest on soils with frigid or mesic temperature regimes and shrubs, grass, and widely spaced trees on the soils with a thermic temperature regime. Other Inceptisols are mainly agricultural land.

7.3.1.8 Mollisols

Mollisols are dark-colored, base-rich, mineral soils of the grasslands. They have a mollic epipedon (a surface horizon with fine, usually granular structured, dark colored, chroma of 3 or less, value darker than 5, containing at least 1 % organic matter, and over 50 % base saturation). They may have an argillic, natric, or calcic (a horizon with $CaCO_3$ accumulation) horizon or an albic (a horizon from which clay and free iron oxides have been removed) horizon. Some have a duripan or a petrocalcic horizon. Mollisols do not have permafrost (the part of soil remaining permanently frozen), organic soil materials, and a spodic horizon. Mollisols may have any of the defined temperature regimes. Mollisols can have any soil moisture regime, but enough available moisture to support perennial grasses seems to be essential. Mollisols are used mainly for small grain in the drier regions and corn (maize) or soybeans in the warmer, humid region. Mollisols comprise 6.9 % of the ice-free land of the earth.

7.3.1.9 Oxisols

Oxisols develop under a climate characterized by small seasonal variation in soil temperature and no seasonal soil freezing, and high annual precipitation. Usually, Oxisols develop under climatic conditions where precipitation exceeds evapotranspiration for some periods of the year to facilitate the removal of soluble weathering products, and favors the residual concentration of kaolinite and sesquioxides, which are essential to form an oxic horizon. Oxisols consist mainly of quartz, kaolinite, oxides of Fe, Mn, and Al, and organic matter. Oxisols are poorly fertile weathered soils occurring on gentle slopes of geologically old surfaces in tropical and subtropical regions. Oxisols comprise 7.5 % of the global ice-free land surface. Oxisols have five suborders: Aquox, Perox (perudic soil moisture regime), Torrox, Udox, and Ustox.

7.3.1.10 Spodosols

Spodosols form in coarse-textured parent materials and have a reddish-brown spodic horizon beneath a light-colored E horizon. The particle-size class is mostly sandy, sandy skeletal, coarse loamy, loamy skeletal, or coarse silty. Spodosols are most extensive in areas of cool, humid, or perhumid climates. They may also form, however, to a limited extent, in warm, humid tropical regions, where they occur mostly in areas of quartz-rich sands with fluctuating groundwater table. Most Spodosols in cool temperate regions are covered with coniferous or, less commonly, hardwood forests. A plenty of Spodosols are found in boreal forest regions. Spodosols occupy 2.6 % of the global ice-free land surface. Spodosols have four suborders: Aquods, Cryods, Humods (large accumulation of organic carbon in the spodic horizon), and Orthods (freely drained Spodosols that have a moderate accumulation of organic carbon in the spodic horizon).

7.3.1.11 Ultisols

Ultisols are highly weathered red or yellow acidic soils, often having a pH of less than 5 and that develop in humid tropical areas under forest vegetation. They have an argillic or a kandic horizon with low base saturation, less than 35 % BSP (by summation of exchangeable bases). The low base saturation status is mainly due to formation on parent material high in silica but low in bases. In some soils, the low base status results from intense leaching of parent material, while in others, a low base status and small quantities of weatherable minerals were initial parent material characteristics. They may have any soil temperature regime and any soil moisture regime except aridic. The red and yellow colors result from the accumulation of iron oxide which is highly insoluble in water. Kaolinite, gibbsite, and aluminum-interlayered clays are common in the clay fraction. Major nutrients, such as calcium and potassium, are typically deficient in Ultisols. Ultisols occupy 8.5 % of the global ice-free land surface. Ultisols have five suborders: Aquults, Humults, Udults, Ustults, and Xerults.

7.3.1.12 Vertisols

Vertisols are clayey soils that have deep, wide cracks for a considerable time of the year and have slickensides (a shiny surface of the cracks produced in soils containing a high proportion of swelling clays) within 100 cm of the mineral soil surface. They shrink when dry and swell when moistened. They are generally sticky in the wet season and hard in the dry season. Most Vertisols have an ustic soil moisture regime and some have an aridic and a udic regime. Vertisols generally have 50–70 % clay with a relatively large proportion of fine clay in the clay fraction. The clays in Vertisols consist predominantly of 2:1 and 2:2 layer clay minerals, but some have considerable amounts of other clay minerals. The natural vegetation is predominantly grass, savanna, open forest, or desert shrub.

7.3.2 WRB for Soil Resources

A soil map of the world at the scale of 1:5,000,000 was published by FAO (1988). This map needed a legend for different soil-mapping units. This legend (FAO/UNESCO legend) which contained 28 "Major Soil Groupings" was treated as a soil classification system in different regions of the world. The FAO legend was later replaced by the WRB for Soil Resources (FAO 1994, 1998). The WRB reference groups have been adopted in 1998 by the International Union of Soil Sciences as the standard for soil correlation and nomenclature (FAO, AGL 2003). The reader is referred to *World Reference Base for Soil Resources 2006*, World Soil Resources Report 103 by FAO (2006) for recent updates. Now, there are 32 RSGs. The names and principal characteristics of the RSGs are given as follows:

7.3.2.1 Acrisols

Acrisols are soils that have higher clay content in the subsoil than in the topsoil due to clay migration forming an argic subhorizon with low activity clays and low base saturation. Acrisols are found in humid tropical, humid subtropical, and warm temperate regions of the world. The native vegetation is tropical forest. The soil taxonomy equivalents of Acrisols are the Ultisols.

7.3.2.2 Albeluvisols

Albeluvisols are soils that have an illuvial clay horizon with an irregular or broken upper boundary resulting in tonguing of bleached soil material into the illuvial horizon. Albeluvisols are developed in the continental regions of northeast Europe, northwest Asia, and southwest Canada and also in France, Belgium, Netherlands, and Germany. The soil taxonomy equivalents of Albeluvisols are the Alfisols.

7.3.2.3 Alisols

Alisols are soils that have higher clay content in the subsoil than in the topsoil due to clay migration forming an argic subsoil horizon with low base saturation and high-activity clays and higher proportion of exchangeable Al. They occur predominantly in humid tropical, humid subtropical, and warm temperate regions. Major occurrences of Alisols are found in Latin America, West Indies, West and East Africa, Southeast Asia, China, Japan, northern Australia, and southeast of the USA. The soil taxonomy equivalents of Alisols are the Ultisols.

7.3.2.4 Andosols

Andosols are soils that develop in volcanic ashes. Andosols occur in volcanic regions all over the world. Andosols are generally fertile soils and have high agricultural potential but with high P fixing problems. Andosols are Andisols in soil taxonomy.

7.3.2.5 Anthrosols

Anthrosols are soils that have been modified profoundly by human activities. Anthrosols may have plaggic horizons (resulting from manuring), irragric horizons (developed as a result of irrigation), anthraquic horizon (plough pan) overlying a hydragric horizon (resulted due to wet cultivation), and hortic horizons (deep cultivation and fertilizer application). Anthrosols fall in the orders Entisols (Anthraquents suborder) and Inceptisols (Anthraquepts suborder) of soil taxonomy.

7.3.2.6 Arenosols

Arenosols are sandy soils, including soils developed in both residual sands and recently deposited sands such as dunes in deserts and beach lands. Vast expanses of deep aeolian sands are found on the Central African plateau, Sahelian region of Africa, various parts of the Sahara, central and western Australia, and China. Some Arenosols belong to Aridisols; others are Entisols (Psamments suborder) of soil taxonomy.

7.3.2.7 Calcisols

Calcisols are soils with secondary accumulation of lime. Calcisols are common in highly calcareous parent materials and widespread in arid and semiarid environments. Most Calcisols belong to Aridisols (Calcids suborder); some may be Entisols or Inceptisols of soil taxonomy.

7.3.2.8 Cambisols

Cambisols exhibit the beginning of soil formation. A weak horizon differentiation may be noticed in the subsoil. Transformation of parent material is evident from structure formation and mostly brownish discoloration, increasing clay percentage, and/or carbonate removal. Cambisols belong to Inceptisols of soil taxonomy.

7.3.2.9 Chernozems

Chernozems are soils with a thick black surface layer that is rich in organic matter. Chernozems are distributed mainly in the middle latitude steppes (grassland plains without trees) of Eurasia and North America. Chernozems are regarded as the most productive agricultural soils. Chernozems are called Mollisols in soil taxonomy.

7.3.2.10 Cryosols

Cryosols comprise mineral soils formed in a permafrost environment. Water is present primarily in the form of ice. Cryosols are widely known as permafrost soils. These soils are classified as Gelisols in soil taxonomy.

7.3.2.11 Durisols

Durisols are very shallow to moderately deep, moderately well- to well-drained soils that contain a layer of cemented secondary silica within 100 cm of the soil surface. Extensive areas of Durisols occur in Australia, South Africa and Namibia, and USA with minor occurrences in Central and South America. They are Aridisols (Durids suborder) in soil taxonomy.

7.3.2.12 Ferralsols

Ferralsols are the classical, deeply weathered, red or yellow soils of the humid tropics. These soils have a clay assemblage dominated by low-activity clays, mainly kaolinite, and a high content of sesquioxides. They are typical soils of the humid tropics. The soil taxonomy equivalent of Ferralsols are the Oxisols which contain other RSGs as well.

7.3.2.13 Fluvisols

Fluvisols are young soils in alluvial deposits. Fluvisols occur on all continents and in all climates. Major concentrations of Fluvisols are found along rivers and lakes. Some Fluvisols have a thionic horizon (sulfidic material; acid sulfate soils). Paddy rice cultivation is widespread on tropical Fluvisols. Fluvisols belong to Entisols (Fluvents and Fluvaquent).

7.3.2.14 Gleysols

Gleysols are wetland soils that are saturated with groundwater for long enough periods to develop a characteristic gleyic color pattern. The upper part of the soil profile may be reddish, brownish, or yellowish color but essentially grayish/bluish color beneath. Some Gleysols support luxuriant swamp forests. Gleysols fall in at least four orders of soil taxonomy: Inceptisols, Entisols, Alfisols, and Mollisols.

7.3.2.15 Gypsisols

Gypsisols are soils with substantial secondary accumulation of gypsum. These soils are found in the driest parts of the arid climate zone. They were earlier called desert soils. Major occurrences of Gypsisols are in Mesopotamia, near East and Central Asian republics, Libyan and Namib deserts, southeast and central Australia, and the USA. Gypsisols belong to Aridisols (Gypsids suborder).

7.3.2.16 Histosols

Histosols are formed in organic material. They include soils developed in predominantly moss peat in boreal, arctic, and subarctic regions, moss peat, reeds/sedge peat (fen) and forest peat in temperate regions, and mangrove peat and swamp forest peat in the humid tropics. Vast majority of Histosols occurs in lowlands. Common names are peat soils, muck soils, and bog soils. Most Histosols are called in the same name in soil taxonomy; some Histosols of WRB fall into Gelisols (Histels suborder) of soil taxonomy.

7.3.2.17 Kastanozems

Kastanozems are dry grassland soils of the short-grass steppe belt. They have a similar profile to that of Chernozems but the humus-rich surface horizon is thinner and not as dark as that of the Chernozems. They show more prominent accumulation of secondary carbonates. Kastanozems belong to Mollisols (Ustolls and Xerolls suborders) of soil taxonomy.

7.3.2.18 Leptosols

Leptosols are very shallow soils over continuous rock (earlier lithosols) and soils that are extremely gravelly and/or stony. Some Leptosols on calcareous rocks belonged to Rendzinas. Leptosols are the most extensive RSG on earth. Leptosols support grasses and forests. Leptosols fall in Entisols of soil taxonomy.

7.3.2.19 Lixisols

Lixisols comprise soils that have higher clay content in the subsoil than in the topsoil due to clay migration leading to an argic subsoil horizon. Lixisols have a high base saturation and low-activity clays at certain depths. Lixisols are found in seasonally dry tropical, subtropical, and warm temperate regions on old landscapes. Lixisols belong to Alfisols of soil taxonomy.

7.3.2.20 Luvisols

Luvisols are soils that have an illuvial argic subsoil horizon with high-activity clays and a high base saturation. Luvisols extend mainly in temperate regions such as in the west and center of the Russian Federation, the USA, and Central Europe. Luvisols also belong to Alfisols of Soil Taxonomy.

7.3.2.21 Nitisols

Nitisols are soils that are deep, red, and well drained with diffuse horizon boundaries and a subsurface horizon with more than 30 % clay and moderate-to-strong angular blocky structure. These soils develop in tropical regions. Weathering is relatively advanced. Most Nitisols are found in the tropical regions. Nitisols belong to four of the orders of soil taxonomy: Inceptisols, Alfisols, Oxisols, and Ultisols.

7.3.2.22 Phaeozems

Phaeozems are soils of relatively wet grassland and forest regions in moderately continental climates. Like Chernozems and Kastanozems, they also have dark, humus-rich surface horizons but they are more extensively leached and are less rich in bases. Phaeozems belong to Mollisols (Udolls and Albolls suborders).

7.3.2.23 Planosols

Planosols are soils with a light-colored surface horizon that shows signs of periodic water stagnation and abruptly overlies a dense, slowly permeable subsoil with significantly more clay than the surface horizon. Soils of this RSG are classified into three orders of soil taxonomy: Alfisols, Mollisols, and Ultisols.

7.3.2.24 Plinthosols

Plinthosols are soils with plinthite (an Fe-rich or Mn-rich mixture of kaolinitic clay, gibbsite, etc. with quartz and other constituents forming a layer of hard nodules, or a hardpan or irregular aggregates), petroplinthite (a continuous, fractured, strongly cemented to indurated nodules), or pisoliths (discrete strongly cemented to indurated nodules). Three orders of soil taxonomy (Alfisols, Oxisols, and Ultisols) accommodate soils of this RSG.

7.3.2.25 Podzols

Podzols are soils with a typically ash-gray eluvial subsurface horizon, bleached by loss of organic matter and iron oxides, on top of a dark accumulation horizon with brown to black humus. Podzols mainly occur in humid areas in the boreal and temperate zones. Podzols are characterized by low nutrient status, low available moisture, low pH, Al toxicity, and P deficiency. These soils are classified as Spodosols in soil taxonomy.

7.3.2.26 Regosols

Regosols represent a very weakly developed mineral soil in unconsolidated materials. Regosols are extensive in eroding surfaces, in particular in arid and semiarid areas and in mountain regions. Regosols correlate with soil taxa that are marked by incipient soil formation such as Entisols in soil taxonomy.

7.3.2.27 Solonchaks

Solonchaks are soils that have a high concentration of soluble salts. They are largely confined to the arid and semiarid climate zones and to coastal regions in all climates. Solonchaks belong to Aridisols (Salids suborder) of soil taxonomy.

7.3.2.28 Solonetz

Solonetz are soils with a high proportion of adsorbed Na and/or Mg ions. Solonetz that contain free soda (Na_2CO_3) are strongly alkaline (field pH > 8.5). These soils were earlier called alkali soils and sodic soils. Soils of these RSGs fall in several orders of soil taxonomy. Most Solonetz are Aridisols; there are some Entisols and Inceptisols.

7.3.2.29 Stagnosols

Stagnosols are soils with a perched water table showing redoximorphic features caused by surface water. Stagnosols are periodically wet and mottled in the topsoil and subsoil, with or without concretions and/or bleaching. Stagnosols belong to Entisols, Inceptisols, Alfisols, Ultisols, and Mollisols of soil taxonomy.

7.3.2.30 Technosols

Technosols are soils strongly influenced by human activity and containing many artifacts. They include soils from wastes (landfills, sludge, cinders, mine spoils, and ashes), pavements with their underlying unconsolidated materials, soils with geomembranes, and constructed soils in human-made materials. In soil taxonomy, these soils are classified as Entisols.

7.3.2.31 Umbrisols

Umbrisols accommodate soils in which organic matter has accumulated within the mineral surface soil to the extent that it significantly affects the behavior and utilization of the soil. Umbrisols occur in cool, humid mountainous regions. Umbrisols in tropical mountain areas in South Asia and Oceania are under montane evergreen forest. These soils belong to Entisols and Inceptisols of soil taxonomy.

7.3.2.32 Vertisols

Vertisols are deeply and widely cracking soils with a high proportion of swelling clays. Most Vertisols occur in the semiarid tropics, with an average annual rainfall of 500–1,000 mm, but Vertisols are also found in the wet tropics. The largest Vertisol areas are on sediments that have a high content of smectitic clays. These soils are also called Vertisols in soil taxonomy.

7.4 Soils of Major Forest Types of the World

Tropical forests, temperate forests, and boreal forest biomes are situated at distinct geographical zones of the earth. Soil-forming factors, particularly climate, vegetation, and parent materials, differ significantly among these zones. Therefore, we find widely different soils in different forest types of the world. A brief account of the nature of soils in different forest types is given later. In the following discussions, characteristics of higher taxa of soils (Orders of Soil Taxonomy and RSGs of WRB) will generally be mentioned. The reader should be aware that soils of the same order may occur in different situations (for example, Alfisols occur in humid tropical and humid temperate regions), and soils of the same order have some common characteristics but may have some dissimilarities too. For example, Entisols may be deep, shallow, moist, dry, saline, acidic, alkaline, calcareous, etc. Soils are complex entities, and pedons of the same soil series may have some dissimilarities.

7.4.1 Soils of Tropical Rainforests, Evergreen Forests, and Deciduous Forests

The tropics are a region of diverse landforms ranging from lowland swamps to mountains with extensive plains. Vegetation is more diverse, including tropical rainforests, tropical evergreen forests, tropical mixed forests, tropical moist deciduous forests, tropical dry deciduous forests, tropical montane forests, tropical mangrove, and tropical swamp forests. There is high rainfall, usually greater than 2,500 mm annually, in most of the regions. In rainforests, there is no definite dry period, and in monsoon forests, there are 4–6 dry months. Parent materials in tropical forest soils have been derived from crystalline rocks, sedimentary rocks, and unconsolidated materials. As there is no clear difference between the effects of pedogenesis on igneous and metamorphic rocks of similar composition, they are considered in the same group of crystalline rocks. The minerals therein undergo intensive weathering under humid and warm conditions. On the other hand, siliceous sedimentary rocks have already passed through a weathering cycle. Therefore, they contain a few weatherable minerals. Parent materials also differ in their composition and texture. On the basis of mineralogical composition, parent materials are classified into four groups on felsic to basic axis, with a fifth class for calcareous rocks. On grain size, parent materials are coarse, medium, and fine (Young 1980). For these variations in environmental conditions and the immense interaction of these conditions for soil formation, we find development of widely diverse soils in the tropics.

Pedogenic processes in forest soils of the humid and subhumid tropics are influenced by: (1) abundant moisture, (2) high temperatures, (3) old landscapes, and (4) active biota. As the tropical regions are near the equator, they have not been affected by ice ages during the Quaternary age. Pedogenesis has, therefore, not been interrupted and in parts these soils have been weathered since the late Cretaceous, giving them ages of up to 72 million years (Marques et al. 2004) and resulting in highly weathered and deep regolith reaching depths of 80 m (Zech and Hintermaier-Erhard 2002). Excessive rainfall and high temperatures cause intensive weathering in tropical regions under forest vegetation. Hydrolysis of primary and secondary minerals in the soil and parent material releases substances into solution according to their solubility product (Ca > Na > Mg > K > Si) and are transported downward with percolating water, eventually reaching the groundwater table (Scheffer and Schachtschabel 1998). As a result of this loss of bases and silica, iron and aluminum oxides and hydroxides such as goethite (FeOOH), hematite (Fe_2O_3), and gibbsite [$Al(OH)_3$] accumulate. These substances are relatively stable and resistant to further weathering. Goethite is also a common mineral in soils of humid temperate climates and is responsible for their yellow-brown color. Hematite and gibbsite, on the other hand, are substances typical of high temperature conditions and tropical weathering. The bright red color of many tropical soils is a result of hematite formation (Zech and Hintermaier-Erhard 1997). As a result of intensive leaching and absence of free acids, the surface soil becomes mildly acid (pH 5.5–6.0). Under this condition, oxides of iron, aluminim, and manganese are precipitated as their oxides and hydroxides in the surface soil giving it a red, brown, or yellow color. Due to constant humid conditions, soils in relatively flat areas often come in contact with groundwater. In the wet subsurface horizons, iron is reduced to its soluble, mobile form (Fe^{2+}) and can undergo lateral transport by seepage water until it reaches an oxidizing environment, for example, a slope face. Here, the presence of oxygen results in the precipitation and absolute accumulation of Fe^{3+} as oxides, coating the soil minerals (kaolinite and quartz) and irreversibly hardening to form a very stable crust called petroplinthite (also laterite, ironstone or hardpan). In soils subject to frequent groundwater fluctuations, plinthite formation is a common feature. In tropical soils, iron and aluminum may be residual or brought with seepage water from higher elevations or adjacent locations. The clay fractions are dominated by low-activity clays, mostly kaolinite, and sesquioxides. The predominant soil-forming processes are laterization, latosolization, ferralitization, and desilication (Chap. 1). The high temperature and plenty of rainfall in tropical forests cause intensive weathering, dissolution of minerals, release and leaching of bases, eluviation of clay, and decomposition of organic matter. These conditions are ideal for the formation of Oxisols. High weathering and leaching also produce Ultisols and Alfisols. Many tropical forest soils are very old and impoverished, especially in regions like the Amazon basin. Vitousek and Sanford (1986) estimated that moist tropical forests have 35.3 % Oxisols, 27.7 % Ultisols, 3.6 % Alfisols, about 15 % Inceptisols, and 15 % Entisols.

Oxisols are typical soils of tropical broadleaf evergreen and deciduous forests. Oxisols have an oxic horizon—a mineral subsurface horizon of sandy loam or finer texture, with low CEC and few weatherable minerals (less than 10 %; Soil Survey Staff 1999). Most coarse fragments are coated with sesquioxides. The A horizons are moderately to strongly acidic and tend to be low in organic matter. Organic carbon contents decrease exponentially with depth in forested Oxisols (Schaetzl and Anderson 2005). Most nutrients in Oxisols ecosystems are contained in the standing vegetation and decomposing plant material. Oxisols (Soil Taxonomy) or Ferralsols (WRB) are soils that result from strong weathering, ferralitization, and desilication. Oxisols also include some Plinthosols (soils with plinthite—an Fe-rich or Mn-rich mixture of kaolinitic clay, gibbsite, etc. with quartz and other constituents forming a layer of hard nodules, or a hardpan or irregular aggregates; petroplinthite—a continuous, fractured, strongly cemented to indurated nodules; or pisoliths—discrete strongly cemented to indurated nodules) and some Nitisols (soils that are deep, red, and well drained with diffuse horizon boundaries and a subsurface horizon with more than 30 % clay and moderate-to-strong angular blocky structure) of WRB. These soils typically occur in humid tropical regions on free-draining landscapes. They develop on old land formations or over sediments that were pre-weathered from ancient regoliths (Buol 2002; Buol et al. 2003). Some Oxisols have also developed on younger materials which weather rapidly, such as basic and ultrabasic rocks and geologically old volcanic material, but these occurrences are generally of limited spatial importance (Buurman and Soepraptohardjo 1980; Beinroth 1982). Oxisols occur usually on stable topography of tropical regions where heavy rainfall and high temperature favor intensive weathering (Sanchez 1976). Pedogenesis has extremely advanced and the weathering of the mineral fraction of Oxisols releases only negligible amounts of nutrients and aluminum (Driessen et al. 2001). Moreover, P adsorption capacity is high and the total P reserve is usually small (Smeck 1985). Kaolinite clay mineral [$Al_2Si_2O_5(OH)_4$] and goethite (FeOOH), hematite (Fe_2O_3) and gibbsite [$Al(OH)_3$] are resynthesized from soluble Fe, Al, and Si in soil solution (Sombroek 1984; Buol 2002). Most Oxisols are chemically poor with an acidic reaction but most of them have well-developed physical conditions. In general, they are deep, well-drained soils which have low silt content, low bulk density, strong fine and very fine granular structure, and a high hydraulic conductivity (Sanchez 1976; Richter and Babbar 1991; Buol and Eswaran 2000; Buol 2002). Clays do not disperse in water and there is little osmotic swelling because the cation concentration at the surface of the kaolinitic clay is low. Oxisols have some capacity to accumulate organic matter through organo-mineral interactions (Dick et al. 2005; Zinn et al. 2007). Oxisols are distributed mainly in Africa (41%) and South America (57%), with 2% in Asia (http://soils.usda.gov/use/worldsoils/oxisols/117.pdf). The largest extent is in Brazil, followed by the Congo Basin of South Africa. Most Oxisols under tropical forests (usually rainforests, evergreen forests, and mixed evergreen forests) belong to the suborders Udox and Perrox.

Soils with plinthite or petroplinthite layers are common under tropical forest vegetation. Physical characteristics of these soils are usually restrictive; the soil is very compact with high bulk density and the soil structure generally weak for all horizons (Sombroek 1984). Plinthite most commonly evolves from the weathering material of basic rocks as opposed to acidic ones (Quesada et al. 2011). As formation of plinthite is associated with fluctuations in groundwater levels, these soils are, thus, often associated with lower landscape positions (Sombroek 1966; Lima et al. 2006). The development of the plinthic layer depends on the accumulation of sesquioxides through the removal of silica and bases under hydrolysis, following the discharge of weathering products. This results in a relative accumulation of weathering-resistant materials such as sesquioxides, quartz, and kaolinite (Driessen et al. 2001). These soils belong to the suborder Aquox of the order Oxisols. Some soils with plinthite may belong to the order Ultisols (Plinthaquults).

Ultisols (the prefix "ulti" from Latin ultimus, "last" denotes the ultimate stage of weathering and soil development) are strongly leached, acid forest soils of the humid tropics with relatively low native fertility. They are found typically on older, stable landscapes. Intense weathering of primary minerals has occurred, and much Ca, Mg, and K has been leached from these soils. Ultisols possess a subsurface horizon (argillic or kandic horizon) with more clay than overlying horizons and low base saturation (percentage of exchangeable Ca, Mg, K, and Na). Often, Ultisols have dense, and water-restrictive subsurface horizons (fragipans), and hardened accumulations of iron (Fe) oxides mixed with other minerals (plinthite and ironstone). Most Ultisols (> 80%) are found in the tropics (South America, Africa, China, Australia, the Pacific Islands, coastal India, and the southeastern USA). The occurrence of Ultisols suggests: 1) the degree of soil weathering has greatly advanced, 2) the land surface has been stable for a significant time, 3) the land surface has undergone long-time periodic wetting and drying, and 4) the soils are relatively infertile and have limited ability to replenish nutrients through inherent mechanisms (Shaw 2005). Ultisols are also dominated by kaolinite and gibbsite. There are two schools of thoughts regarding the genesis of gibbsite in these soils namely the direct weathering of primary A1 silicate minerals, and the other through clay mineral intermediates. It has been indicated that whether the primary A1 silicates weather directly to gibbsite or through a kaolinite intermediate, this transformation would be governed by the intensity of leaching, which is affected by a number of factors such as rainfall, temperature, parent rocks, topography, groundwater table, vegetation, and time. Hydrolysis and transformation of primary weatherable minerals occur in the parent material with some leaching of silica during the early formation of Ultisols. It results in a saprolite with little weatherable primary minerals and a dominance of high-activity clays, most likely formed by the transformation of micas. This is followed by a redistribution of clay in the soil and the formation of an argillic horizon. Alfisol formation occurs by the weathering of secondary high-activity clays. High-activity clays are unstable in environments that are depleted of silica and alkaline and alkaline earth cations. With their weathering, they release soluble aluminum and for some parent materials, iron and magnesium from the octahedral internal layers of the 2:1 clay minerals.

The idealized soil horizon sequence of well-drained Ultisols consists of a darkened surface horizon (A) with accumulation of organic matter, a lightened and bleached E horizon where materials have been removed, a subsurface B horizon where clay, iron (Fe), and aluminum (Al) oxides have concentrated, and an underlying relatively unweathered C horizon which comprises mostly of soil parent materials. The A and E horizons comprise the eluvial impoverished by leaching and erosion of soluble and colloidal soil constituents.

Most B horizons of Ultisols possess higher amounts of clay than overlying horizons and have translocated clay which is evidenced by clay films, coatings on ped surfaces, and clay bridges between sand grains. These subsurface horizons are called argillic horizons which, in combination with a low base saturation, are diagnostic of Ultisols. Kandic horizon, another diagnostic horizon, is found in some Ultisols. Clay accumulates in the subsoil as a result of the weathering of primary minerals, mainly feldspars and micas, and the intrasolum translocation of clay from the surface horizons. Primary feldspar minerals weather to form clay, and most Ultisols have very low silt contents. Where weatherable minerals are present in the surface horizons, clay skins are present in the argillic or sometimes kandic horizons, but in profiles with very low weatherable content in the surface horizons, very few or no clay skins are present in the argillic or kandic horizon. Although clay translocation is a formative process in Ultisol development, most soil clays likely move only short distances, and most of the argillic horizon clays derive in situ through mineral weathering (Shaw 2005). Ultisols have a subsurface horizon often with strong yellowish or reddish color resulting from the presence of Fe oxides. Ultisols are deep soils, and range from well-drained soils on upland ridges, rises and knolls to very poorly drained soils. The higher clay content of Ultisols gives them a nutrient retention capacity greater than Oxisols, but less than Alfisols.

Alfisols form in climatic conditions, which are thermic or warmer and mesic or cooler. Thus, Alfisols are found in humid tropical and humid temperate regions. In humid tropical and subtropical zones, Alfisols are found under broadleaf evergreen or deciduous forests. Actually, Alfisols cover a smaller area, only 4 %, of the humid tropics (NAP 1982). In West Africa, they are found in Ivory Coast, Ghana, Togo, Benin, Nigeria, and Cameroon. They are, however, the most abundant soils in Africa's subhumid and semiarid zones, covering about one third of these regions (Kampen and Burford 1980). The Alfisols are widely distributed in the subhumid and semiarid tropical regions of Africa, including large areas in western, eastern, central, and southeastern Africa. These soils are moderately leached and relatively fertile. The soil formation is characterized by eluviation of clay from A and E horizons and illuviation of clay in the B horizon. Extensive leaching often produces a light-colored E horizon below the topsoil. Alfisols are more weathered than Inceptisols, but less weathered and leached than Ultisols. The Alfisols are characterized by a subsoil horizon of increased clay content (argillic horizon) with base saturation greater than 35 % or in which the base status increases with depth. According to Aubert and Tavernier (1972), Alfisols typically occur in the transitional zone between humid and arid climates. Two main suborders occur in tropical regions: Udalfs which have only short dry periods and Ustalfs which have long periods when the soil remains dry (more than 90 cumulative days). Ustalfs represent the major Alfisols of the tropics. Alfisols of West Africa studied by Moorman et al. (1975) have well-developed argillic B horizons and modest cation retention capacity (5–10 cmolc kg^{-1} soil) in the surface horizons; also, 90 % or more of cation retention capacity is satisfied by Ca^{2+} and Mg^{2+} throughout the soil profile. Soil acidity is not a problem in these soils, nor is a shortage of Ca^{2+} or Mg^{2+}. The clay-sized fraction of tropical, as opposed to temperate, Alfisols is dominated by kaolinite and contains smaller amounts of goethite and hematite, an indication of intense weathering in the past. These profiles usually contain small amounts of micas and other weatherable minerals. Since the exchangeable cations are largely Ca^{2+} or Mg2+, these soils must have been recharged with bases in the recent past. Thus, these soils fall into the more productive high-base-status soils described by Sanchez and Boul (1975).

Sanchez (1976) and Richter and Babbar (1991) demonstrated that tropical soils are actually very diverse, encompassing all different taxa, from the lowest to the highest pedogenic levels. Many investigators including Sanchez and Buol (1975) and Quesada et al. (2011) investigated characteristics of soils in tropical rainforests of the Amazonian Basin. They used the WRB system of soil classification for description and found soils of the RSGs Acrisols, Luvisols, Lixisols, Cambisols (all belonging to the Order Inceptisols of Soil Taxonomy), Ferralsols (Oxisols), Plinthosols (Oxisols and Ultisols), Nitisols (Inceptisols, Alfisols, Oxisols, and Ultisols), and Alisols (Ultisols) in the Brazilian Amazon. There were also some Regosols (Entisols), Gleysols (Entisols and Inceptisols), Solonchacks, and Solonetz (Entisols and Inceptisols).

Much of the soil diversity in Amazonia has originated from the considerable differences in geology and geomorphology history that occur across the Basin (Quesada et al. 2011). It is seen from descriptions of Quesada et al. (2011) that Acrisols and Ferralsols occur mainly on the Brazilian and Guyana shields and in the sedimentary zone along the central and eastern portions of the Amazon River. Arenosols (Entisols), Lixisols, Nitisols, Histosols, and Podzols (Spodosols) also occur in the Brazilian and Guyana shields. Some Spodosols are found in areas of the Rio Negro basin. Plinthosols are most common in sedimentary areas near the Jurua, Purus, and Madeira rivers. Regosols and Leptosols occur along both Shields but are most common in the proximity to the Andean foothills mostly outside of the Amazonian border. Fluvisols and Gleysols are found in floodplains of the upper Amazon tributaries and along the Amazon itself. Large Gleysol patches occur in the Araguaia catchments in Brazil and in north Colombia, spreading along the Andean border in sedimentary zones in the Peruvian Amazon. There are some Cambisols and Alisols in the Andean foothills. However, Ferralsols and Acrisols account for more than 60 % of Amazonian forest soils, with Plinthosols, Gleysols, Cam-

Fig. 7.11 An Oxisol soil profile

Fig. 7.12 An Ultisol soil profile

Fig. 7.13 An Alfisol soil profile

Fig. 7.14 An Entisol soil profile

bisols, and Leptosols accounting for most of the remaining portion. Kauffman et al. (1998) recognized six major groups of soils in the humid tropical rainforests: Acrisols, Aerenosols, Cambisols, Ferralsols, Luvisols, and Podzols. In their observation, 64% soils of the humid tropics fall in the four RSGs: Acrisols, Aerenosols, Ferralsols, and Podzols. Figures. 7.11–7.18 show profile characteristics of several soil orders.

7.4.2 Soils of Tropical Montane Forests

Tropical montane forest ecotype contains soils that are cool or cold—cool in the summer, cold in the winter, and moist most of the time. Such soils generally form under coniferous forest cover and have an organic duff layer (partially decomposed leaves, etc.) underlain by either a white leached layer or a brown clay layer. If the parent material is limestone, a calcareous layer may be present. Except for the limestone-derived soils, soils in this region are usually acidic. The surface soils on steep slopes are moderately to severely eroded. Soils are often thin and rocky. Soil characteristics of several tropical montane cloud forests in southern Thailand were investigated by Jaroensutasinee et al. (2010). Automatic weather stations were installed at three study sites: Mt. Nom cloud forest, Dadfa cloud forest, and Mt. Nan Headquarters. Soil samples along the Mt. Nom elevational transect at five sites (500, 700, 900, 1,100, and 1,300 m) were collected and soil pH, soil moisture, and organic content were measured. Soil moisture and soil organic matter content increased with in-

Fig. 7.15 An Inceptisol profile

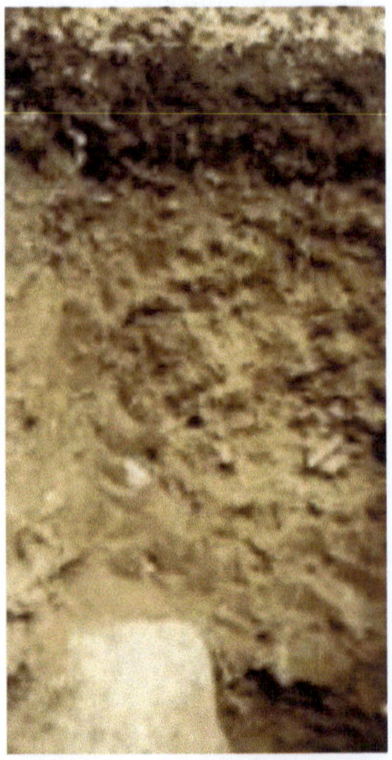

Fig. 7.16 A Spodosol profile

Fig. 7.17 A Gelisol profile

Fig. 7.18 A Histosol profile

creasing elevation. The soil pH ranged from 3.6 to 4.3 which indicated that soil at Mt. Nom is highly acidic. In a tropical montane forest in Ecuador, Wilcke et al. (2008) studied soil properties along an altitudinal transect between 1,960 and 2,450 m asl. Soil samples were collected from eight different sites and were analyzed for pH and total concentrations of Al, C, Ca, K, Mg, Mn, N, Na, P, S, Zn, polyphenols, and lignin in all soil horizons and in the mineral soil additionally the effective CEC. Soil properties changed with increasing altitude. The soils were Cambisols (Inceptisols), Planosols (Alfisols and Ultisols), and Histosols. The concentrations of Mg, Mn, N, P, and S in the O horizons and of Al, C, and all nutrients except Ca in the A horizons correlated negatively and significantly with altitude. Some alpine or subalpine areas of tropical and subtropical climatic regions, like subalpine cypress forests in Taiwan where the climatic conditions are cold and humid, provide suitable environments for podzolization. Here, Podzolic soils (true Podzols are Spodosols in Soil Taxonomy; some Podzolic soils are classified as Alfisols) are commonly found (Jien et al. 2010), as in some other perhumid tropical and subtropical subalpine regions (Hseu et al. 2004; Sauer et al. 2007; Schawe et al. 2007). These tropical or subtropical forest areas with altitudes higher than 1,900 m, high precipitation ($\geq$3,000 mm year^{-1}) and dominant coniferous vegetation are also suitable for the formation of Ultisols, Spodosols, and Podzolic soils (Chen et al. 1989, 1995; Li et al. 1998; Hseu et al. 1999; Wu and Chen 2005).

7.4.3 Soils of Tropical Freshwater Swamp Forests

Forest ecosystems of tropical swamp areas are known as tropical swamp forests. The habitat is characterized by inundation or saturation with water for most part of the year. There are four main types of swamp forests found in the tropical regions of the world: freshwater swamp forests, peat swamp forests, brackish water swamp forests (mangrove forests), and herbaceous swamps. These swamp forests are distributed throughout the tropics; however, they are found primarily in Asia and parts of the Neotropics. These forests are found in depressions or low-lying areas near rivers which are periodically inundated by floodwater. The floods may originate from rain or river water backing up in response to high tides. In depressions, floodwater comes from surrounding watersheds. Swamp forest soils are characterized by water saturation and anoxic situation, organic matter accumulation, gleying, mottling, iron/manganese segregation, oxidizing root channels and soil pore linings, and production of a reduced soil matrix. These soils are called hydric soils. A hydric soil is a soil that in its undrained condition is saturated, flooded, or ponded long enough during the growing season to develop anaerobic conditions that favor the growth and regeneration of hydrophytic vegetation (National Technical Committee for Hydric Soils 1985). This definition was revised by deleting the phrase "in its undrained condition" and any direct reference to hydrophytic vegetation (USDA/NRCS 2006). Now hydric soil is defined as a soil that is saturated, flooded, or ponded long enough during the growing season to develop anaerobic conditions in the upper part. There are mineral hydric soils and organic hydric soils.

Many mineral swamp forest soils lack any well-developed soil profiles. In Soil Taxonomy, these soils are classified as Entisols (Aquent suborder). Many soils have some horizon differentiations. These soils are Inceptisols. In WRB classification, hydric soils have been grouped into Cambisols, Fluvisols, and Gleysols. The sequence of horizons in the mineral soils may be A, E, $B_t/B_g/B_h/B_{tg}$, C_g, and C. The t subscript stands for tont meaning clay, g for gleying, and h stands for humus accumulation. In some mineral soils, an organic horizon, O, of variable thickness and differentiation, may be present. This O horizon rests on the underlying mineral soil. Subhorizons of O include Oi (fibric—little decomposed organic matter), Oe (hemic—intermediately decomposed organic matter), and Oa (sapric—highly decomposed organic matter) subhorizons. The A horizon is usually the darkest layer in the soil (moist value/chroma darker than 3.5/2). The E horizon, which is an eluvial horizon, is lighter in color (grayish to whitish) than the rest of the soil above and below. Most E horizons have a coarser texture than the A horizon. E horizons contain frequent reddish mottles and are underlain by a layer having a higher content of clay (B_t horizon) or transported organic material (B_h horizon). Hydric soils have chemically reduced anaerobic subsoil horizon, a condition conducive to gleying which produces a B_g horizon. Matrix colors of B_g horizons are usually gray, with chromas of 2 or less and values of 4 or more, usually with reddish mottles.

There are plenty of organic soils (peat soils) in swamp forests. Peat soils contain various amounts of undecomposed plant remains because of slow decomposition under anaerobic condition. Peatlands generally form in depressions, slopes, and raised bogs. Peat accumulations can be tens of meters deep (Rydin and Jeglum 2006). There are about 44.1 Mha of tropical peatlands; about 25 Mha (56 %) of tropical peatlands occur in Southeast Asia (Page et al. 2011). Organic soils contain organic soil materials in more than half of the upper 80 cm (Soil Survey Staff 2003). An organic soil material, if saturated with water, has an organic carbon content not less than 12 % (by weight) if the mineral part of the soil contains no clay, and at least 18 % organic carbon if the mineral part of the soil contains 60 % or more clay, and proportional intermediate organic carbon for intermediate clay. If not saturated with water, soil materials must contain more than 20 % (by weight) organic carbon to be organic (Soil Survey Staff 2003). Organic soil materials include peat and muck. O horizons have three subordinate horizon designations, a (sapric), e (hemic), and i (fibric), on the degree of organic matter decomposition.

7.4.4 Soils of Tropical Mangrove Forests

Mangroves develop in estuaries, deltas, and sediment-accreting open tropical coasts. They may colonize sandy shores and corals, but the common soil substrates are deposits of clay and mud. Mangrove soils are formed by the accumulation of sediment derived from coastal or riverbank erosion or eroded soils from higher areas transported down along rivers and canals. Mangroves depend on the presence of a tropical climate, fine-grained alluvium, shores free of strong wave and tidal action, brackish water or saline conditions, and a large tidal range. These factors influence the occurrence and size of mangroves, the species composition, species zonation, other structural characteristics, as well as the functions of the ecosystem. In coastal areas, tidal and wave actions have a strong influence on the zonation of plant and animal communities, sorting and ripening of sediments, biogeochemical cycling, distribution of pH and salinity of substratum, and development of soil (Fig. 7.19).

Based on the substrate, tidal range, and sedimentation, six broad classes of mangrove settings have been identified (Thom 1982; Galloway 1982):

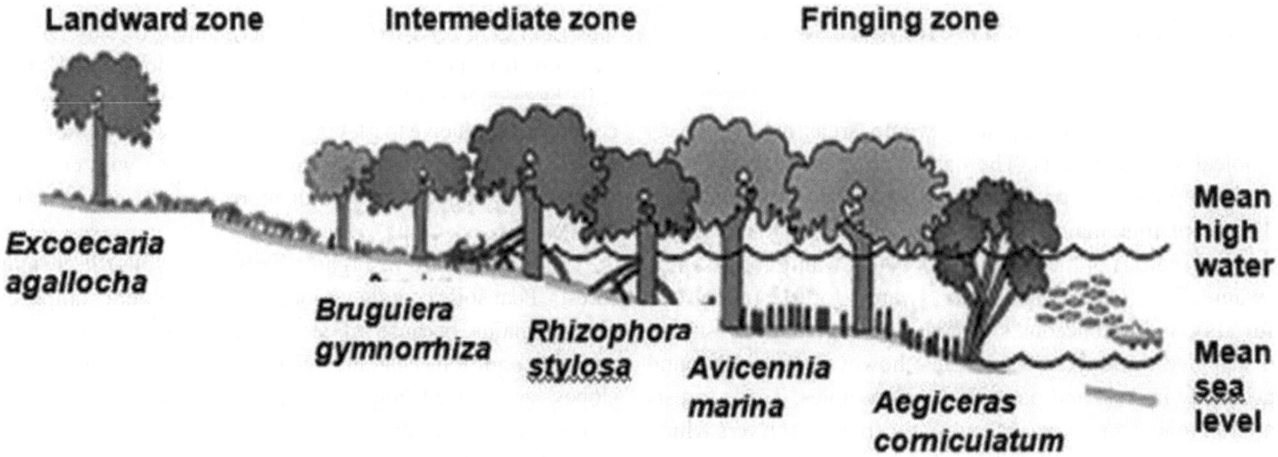

Fig. 7.19 Zonation of plant communities, physiography, and tide levels of a mangrove forest

1. Large deltaic systems occurring in low tidal range, very fine allochtonous sediments (for example, mangroves of Borneo and Sundarbans of India and Bangladesh)
2. Tidal plains where alluvial sediments are reworked by the tides, and there is the presence of large mudflats for the growth of mangroves
3. Composite plains; under the influence of both tidal and alluvial conditions (lagoons formed behind wave-built barriers where mangroves grow)
4. Fringing barriers with lagoons (high wave energy conditions with autochtonous sediments of fine sand and mud; for example, mangroves of the Philippines)
5. Drowned bedrock valleys (mangroves of Northern Vietnam and Eastern Malaysia)
6. Coral coasts (mangroves growing at the bottom of coral sand or in platform reefs; for example, mangroves of India, Indonesia, and Singapore)

Mangrove soil is commonly described as smelly, sticky, anoxic mud. The forest floor can vary from dead hard corals and boulders to quartz or carbonate sand to very fine silt and clay. There is no typical soil profile (Alongi 2009). The soils are classified mainly as Entisols (Aerenosols, Fluvisols, Gleysols, Solonchacks, and Solonetz). There are pockets where Histosols develop by the accumulation of organic debris under anaerobic conditions. Soil pH ranges from 5.5 to 8.5 and the redox potential varies from −200 to +300 mV. Soil texture and associated physicochemical properties depend on constituent trees and roots, parent rocks, geomorphology, tides, and rainfall (Alongi 2005; Ferreira et al. 2007). Soils of the Sundarbans mangroves of India were studied by Joshi and Ghose (2003) to relate the community structure and distribution of species with soil salinity and pH gradients. Soil salinity ranged from 13.0 to 31.2 ppt and decreased with increasing distance from the tidal coast. Frequency of tidal inundation seemed to affect soil salinity. Sites having regular diurnal inundation are dominated by *Acanthus ilicifolius, Avicennia alba*, and *A. marina*. There is no uniform rise or fall in pH with increasing distance from the tidal coast. *Avicennia marina* and *A. officinalis* were found in wide ranges of soil salinity. *Aegiceras corniculatum, Ceriops decandra*, and *Excoecaria agallocha* are restricted to the soil having low salinity. Wakushima et al. (1994) studied soil conditions of mangrove forests in southern Japan and found to correlate with zonal distribution of tree species. *Kandella candel* grew in soils with low salinity and low pH, *Avicennia marina, Rhizophora stylosa*, and *Sonneratia alba* grew in soils with high salinity and high pH, and *Bruguiera gymnorrhiza* grew in soils with a wide range of pH and limited range of salinity. The mangrove forests of Indonesia largely consist mainly of pure stands of *Avicennia officinalis* in the seaward zone and *Ceriops tagal* in the landward zone both of which grow on similar substrates. The soils were all weakly acidic, high in organic matter, and low in available phosphorus. They had high bulk density and moderate CEC (Sukardjo 1994). There are some acid sulfate soils and potential acid sulfate soils in mangrove areas. These soils accumulate pyrites (FeS_2) which on drainage form free sulfuric acid and give the soils a pH below 3.0. Potential acid sulfate soils may undergo alternate oxidation–reduction depending on moisture saturation. These soils are strongly acidic when dry and almost neutral when moist. These soils are classified as Entisols (suborder Sulfaquent) in Soil Taxonomy.

7.4.5 Soils of Tropical Miscellaneous Forests

Tropical dry forests occur on substrates ranging from nutrient-rich alluvial soils to nutrient-poor rock outcrops. They can occur on volcanic, limestone, or ultramorphic rocks, with soil textures ranging from sandy to clay, rocky, or organic. If soils are droughty and have low water-holding capacity, vegetation may acquire a greater xeromorphic aspect than expected. Conversely, in locations where soils store water well or where water is channelized, vegetation may acquire great stature and biomass (Murphy and Lugo 1986). These soils are Aridisols, Entisols, or Inceptisols. Aridisols are characterized by high evapotranspiration to precipitation ratio; the dry soil has relatively low leaching. Entisols are soils that lack a true soil profile; they are very young soils formed in recent alluviums or aeolian or glacial parent materials. Inceptisols are young soils that have initiated soil profile differentiation. They are found on active floodplains, tidal floodplains, or river alluviums.

7.4.6 Soils of Temperate Forests

The temperate region includes a broad range of climatic areas that have no extremes of temperature and precipitation and that have distinct summers and winters. Within this area, precipitation is distributed throughout the year in most years. The average temperature of the coldest month lies between 18 and −3 °C, and the average temperature of the warmest month is > 10 °C. A substantial part of the current temperate zone was covered for a relatively long time by continental ice sheets during the Pleistocene. The past climate has a significant influence on the sorting of parent materials and distribution of soils in this zone. Extensive deposits of glacial and fluvo-glacial materials were laid down during this time. In addition to glacial and fluvo-glacial deposits, there are extensive areas covered by wind-blown fine sands and silts, commonly known as cover sands and loess, respectively (Nortcliff 2002). Soil formation has taken place in the following broad groups of landscapes: (1) pleistocene sedimentary lowlands with glacial, fluvo-glacial, and eolian deposits and (2) uplifted and sedimentary basins, principally with Mesozoic limestones and sandstones, often with variable thickness of cover sands and loess. Soils in temperate sclerophyllous forests are frequently poor in mineral nutrients. In contrast, soils in regions of deciduous and broad-leaved evergreen forest are generally fertile. The forests that occupied the best soils in most regions, however, have been cleared almost completely to make way for agriculture. Typical temperate deciduous forest soils are mull soils, which have a high level of organic matter especially close to the surface that is well mixed with mineral matter. Variations in soil materials and fertility have a strong influence on the types of trees that will dominate the forest (Kuennecke 2008). For example, in northwestern Europe, the European beech (Fagus sylvatica) dominates deciduous forests on shallow soils that overlie chalk, while oak (Quercus) is dominant on deeper, clay soils. The richness of the ground flora beneath the trees generally increases with soil fertility. Waterlogging of soils in temperate deciduous woodlands commonly occurs in regions with higher rainfall and humidity in late winter and spring, such as the British Isles. This occurs not only because in winter rainfall is higher and evaporation lower but also because the trees, barren of foliage, transpire a minimal amount of moisture. Areas subject to waterlogging include clay-rich soils, places that have low slope angles, depressions, and spots along watercourses. These places tend to have a richer ground flora but a less luxuriant tree canopy, which consists of only a few species that are tolerant of wet soils (Encyclopedia Britannica; http://www.britannica.com/EBchecked/topic/586555/temperate-forest/70777/Environment).

High rainfall and low temperatures in the temperate forests usually cause a faster rate of plant litter accumulation than of litter decomposition resulting in thick organic layers in the soil (Bormann and Sidle 1990; Klinka et al. 1990). The unique climate of temperate rainforests leads to rapid soil morphogenesis. High precipitation results in rapid leaching and podzolization (Holdgate 1961; Bowers 1987; Alexander 1988). Heavy accumulations of organic matter, nutrient immobilization, high weathering, and leaching of bases and colloids (Ugolini and Mann 1979) lead to the development of Alfisols, Spodosols, and some Ultisols in the temperate forest regions. Soils in Southeast Alaska develop into waterlogged acidic, peat-like soils (Hennon 1986). The following soil orders of Soil Taxonomy are mainly found in the temperate forests: Alfisols, Inceptisols, Mollisols, Spodosols, and Ultisols (Lal et al. 2012). These soils have generally high inherent fertility (Martin et al. 2001; Grewer et al. 2003), soil organic matter, and favorable moisture and temperature regimes during the growing season (Lal et al. 2012). According to Binkley and Fisher (2012), soils developed under temperate rainforests commonly accumulate large amounts of organic matter and have a moderate abundance of layer lattice clay and argillic horizons. Common soils include some Spodosols (especially on coarse-textured parent materials), extensive Alfisols (Udalfs and Ustalfs), and some Inceptisols (particularly Umbrepts) and Ultisols (including Humults).

The major soil-forming processes in humid temperate climate acting on well-drained sediments are humification, weathering, eluviation and leaching, clay migration, clay destruction, ferralitization, and podzolization, while on imperfectly drained sites they consist of gleization and peat formation (Boul et al 2003). Alfisols, earlier known as brown forest soils, generally develop under broadleaf forests in parent materials rich in bases and weatherable minerals, Spodosols in coarse-textured parent materials under coniferous forests, and Histosols are found in depressions and in raised peatlands. Entisols and Inceptisols are found in recent alluviums, on shallow regoliths overlying bedrocks, highly eroded surfaces etc. Ultisols are found in deeply weathered parent materials on older landscapes. In WRB classification, the temperate forest soils are grouped into Luvisols (Alfisols), Podzols (Spodosols), Cambisols (Inceptisols), Fluvisols (Entisols), Gleysols (Entisols, Inceptisols, and Alfisols), Leptosols (Entisols), and Anthrosols (Entisols and Inceptisols).

Alfisols are mineral soils that have a pedogenic accumulation of clay in the B horizon and high base saturation and that have sufficient water during the growing season to support the growth of plants (Buol et al. 2003). Lessivage is the predominant process of Alfisols formation. The weathering of minerals in the A and E horizons yields clay minerals that are carried downward through percolating water and accumulate to form clay skins on the surface of angular and subangular blocky peds and root channels. The clay skins are not only enriched in clay but also contain more organic carbon, iron, and other elements compared to the composition of the total horizon. Alfisols develop under the temperate broadleaf deciduous forests. Broadleaf trees tend to be nutrient demanding and their leaves are rich in major nutrient elements. The litter under this forest is not as acidic as under needle-leaf trees. The autumn leaf fall provides abundant organic residues. Decomposition of the litter produces humus which gives the A horizon a dark brown color. Alfisols occur in the south of the Great Lakes, from Ohio to southern Wisconsin, and also in parts of Texas, Colorado, Montana, North Dakota, and southern California.

Spodosols are soils with subsoil accumulation of humus and sesquioxides. The chief pedogenic processes are mobilization and eluviation of aluminum and iron from O, A, and E horizons, immobilization of organometallic complexes in spodic horizon, and distributions of aluminum in the profile (Boul et al 2011). Spodosols (Podzols in WRB) are the typical soils of temperate coniferous and boreal forest. They are also the typical soils of eucalypt forests and heathlands in southern Australia. The name "Podzol" comes from Russian (под/pod = under, зола/zol = ash) for "under ash" and likely refers to the leached ash-gray E horizon exposed during plowing of a virgin soil of this type. These soils are found in areas that are wet and cold (for example, in Northern Russia). Some Spodosols are sandy and excessively drained. Others have shallow rooting zones and poor drainage due to subsoil cementation. Podzolization occurs when severe leaching leaves the upper horizon virtually depleted of all soil constituents except quartz grains. Clay minerals in the A horizon decompose by reaction with organic acids and form soluble salts. The leached material from the A horizon is deposited in the B horizon as a humus-rich horizon band, a hard layer of sesquioxides or a combination of the two. The E horizon, which is usually 4–8 cm thick, is low in Fe and Al oxides and humus. It is formed under moist, cool, and acidic conditions, especially where the parent material, such as granite or sandstone, is rich in quartz (Buol et al. 2003). It is found under a layer of organic material in the process of decomposition, which is usually 5–10 cm thick. A typical Spodosol has a distinctive profile with four master horizons: O and/or A horizons have black or dark brown color due to accumulation of organic materials; the E horizon is an ash-gray albic horizon developed as a result of weathering of nonresistant minerals and subsequent leaching and the B_{hs} or B_s horizons have dark reddish brown color resulting from illuvial accumulations of organometallic complexes. Most Spodosols have little silicate clay. The particle size class is mostly sandy, sometimes loamy. These soils are porous and permeable. Spodosols are acidic and naturally poorly fertile (Buol et al. 2003). Most Spodosols have little silicate clay. The particle-size class is mostly sandy, sandy-skeletal, coarse-loamy, loamy, loamy- skeletal, or coarse-silty.

The Ultisols are the most highly weathered and leached of the temperate zone soils. They are characterized by a thin A horizon, with a thick, strongly expressed B horizon. The soils are often very red or yellow-red. Ultisols are acidic forest soils with relatively low native fertility (Buol et al. 2003). They are found on older, stable landscapes. Because of the favorable climate regimes in which they are typically found, Ultisols often support productive forests. Most Ultisols support mixed coniferous and hardwood forest vegetation. Ultisols are usually moist but some are dry during the warm season of the year. Xerults are the typical Ultisols of the temperate region (http://sci.windwolf.org/soil/orders11.htm).

There is a considerable variety of Inceptisols in the temperate forests. Inceptisols are young soils and leaching of soluble compounds from the A horizon is at an early stage, but because they are young, all of them lack a clearly defined B horizon. The soil is often brown all the way to the C horizon. The color depends on the parent material, however, and can be distinctly reddish if this includes sandstone or red marl, a very fine clay derived from seabed or lakebed deposits. Inceptisols are well drained and fertile. Inceptisols are the predominant soils in a belt down the eastern USA from the Adirondack Mountains, through Pennsylvania and the Appalachian Mountains to as far south as northern Alabama. They also occur in the northwest, in parts of Washington. Where the climate is suitable, they support broad-leaved deciduous forest (Allaby 2008).

7.4.7 Soils of Boreal Forest

The boreal climate is characterized by long, usually very cold winters, and short, cool to mild summers. Precipitation is low and mainly in the form of snow. The evapotranspiration is also low; so there is enough available soil moisture during the growing season. The growing season is regulated by the temperature. Very cold-tolerant coniferous evergreen trees constitute the vegetation.

The soils of boreal forests, in the Northern Hemisphere extending southward from the tundra tree line and occurring across Canada, Europe, and Asia, are generally acidic and form on sandy deposits from glaciation and range in thickness from >1 m to very shallow (<1 m) underlain by bedrock (Boul et al. 2003). Because these forests have adapted to cooler climates, the decomposition of litter materials is slow, resulting in the development of a thick forest floor. Soils of boreal forests are relatively young compared to tropical forests because of glaciation. There are many soils that have little or no differentiation of soil horizons. These soils are called Entisols in Soil Taxonomy and Regosols in WRB. Sandy soils that have undergone more soil development,

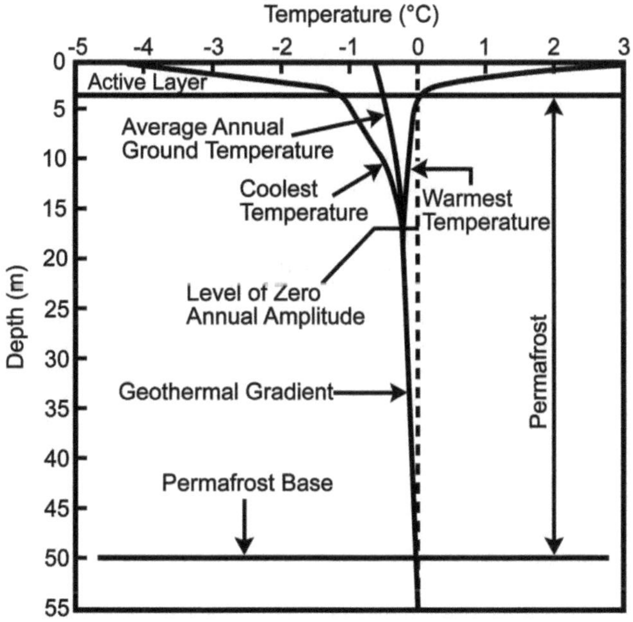

Fig. 7.20 Active layer of Gelisols

such as the development of a reddish-colored B horizon, are called Inceptisols (Cambisols) and are associated with pine forests on outwash plains. More soil development takes place in situations where plenty of carbonic acids and organic acids are produced due to incomplete decomposition of organic matter. In these soils, a thick O horizon develops on the mineral A horizon. Plenty of organic acids are produced by the decomposition of organic matter. These acids decompose the minerals and aid in leaching bases from the upper soil layer, usually the E horizon. Clay and humus are also eluviated from the A and E horizons giving them a coarser texture and bleached color. Iron, manganese, aluminum, and humus and clay may be accumulated in the B horizon. As discussed earlier, this process of soil formation is known as podzolization and the resulting soils are Spodosols in Soil Taxonomy and Podzols in earlier pedology literature and in WRB. In WRB, the principal soils of the boreal forests are Cryosols, Podzols, Histosols, Gleysols, and Leptosols. Albelluvisols and Regosols are also found extensively (Nortcliff 2002).

In the northern latitudes close to the arctic tree line, the soils are generally moist and may be either partially or totally frozen beneath the surface. When this occurs, permafrost forms a discontinuous layer (Kuennecke 2008). Above the permafrost layer, there is the active layer which becomes frozen during winter and thaws during the summer (Fig. 7.20). The soil development is recent and possibly incomplete, as indicated by the shallowness of some of these soils. These soils fall in the Soil Taxonomy order Gelisols. Gelisols are soils of the cold zone and are affected by permafrost. According to Ping (2013), permafrost includes any material—soil, rock, or water in the form of ground ice—that has temperature below the freezing point of water throughout the year for two consecutive years or more. The permafrost material that receives most attention is the ice-cemented permafrost because of the phenomena associated with ice formation, freeze–thaw cycles, and frost heave. Gelisols are defined as soils that have (1) permafrost within 100 cm of the soil surface or (2) gelic materials within 100 cm of the soil surface and permafrost within 200 cm of the soil surface (Soil Survey Staff 1999). They are classified as Cryosols in both of the WRB and the Canadian System of Soil Classification, and Cryozem in the Russian soil classification system.

Gelisols have no B horizon and have an A horizon resting on the permafrost. Most Gelisols are black or dark brown in color due to accumulation of humus. Despite the influence of glaciation in most areas where Gelisols occur, chemically they are not highly fertile because nutrients, especially calcium and potassium, are very easily leached above the permafrost. Gelisols are found mainly in Siberia, Alaska, and Canada.

Members of all three suborders of Gelisols are found in the boreal forests. Histels not only have all characteristics of Histosols but also have permafrost within 2 m below ground surface. They have 80% or more organic materials from the soil surface to a depth of 50 cm or to a glacic layer or densic, lithic, or paralithic contact, whichever is shallowest. These soils occur predominantly in subarctic and low Arctic regions of continuous or widespread permafrost. Less than one-third of the active layer or an ice layer which is at least 30 cm thick has been cryoturbated. Turbels show marked influence of cryoturbation (more than one-third of the depth of the active layer). Cryoturbation (frost churning) refers to the mixing of materials from various horizons of the soil down to the bedrock due to freezing and thawing. Cryoturbation occurs to varying degrees in most Gelisols (Ping 2013). Irregular, broken, or distorted horizon boundaries and involutions and areas with patterned ground are found. They commonly contain tongues of mineral and organic horizons, organic and mineral intrusions, and oriented rock fragments. Organic matter is accumulated on top of the permafrost and ice wedges are a common feature. Orthels show little or no cryoturbation. Orthels occur primarily within the zone of discontinuous permafrost and in alpine areas.

Study Questions

1. Define forest soil. How does it differ from agricultural soil? Discuss general characteristics of forest soils.
2. Give an account of tropical forests and soils associated with them.
3. What is a mangrove forest? How does it differ from freshwater swamp forest? Discuss characteristics of mangrove soils.
4. Distinguish between tropical and temperate climate. Describe soils of temperate forests.
5. Narrate the ecological conditions and geographical occurrence of boreal forests. Explain boreal forest soils with emphasis on cryoturbation and permafrost.

References

Aksenov D, et al. (2002) Atlas of Russia's intact forest landscapes. Global Forest Watch Russia, Moscow

Alaback PB (2010) Ecological characteristics of temperate rain forests: some implications for management of bald eagle habitat. In: Wright BA, Schempf P (eds) Bald eagles in Alaska. Hancock House, New York

Alaback PB, Juday GP (1989) Structure and composition of low elevation old-growth forests in research natural areas of Southeast Alaska. Nat Areas J 9:27–39

Alexander EB (1988) Rates of soil formation: implications for soil-loss tolerance. Soil Sci 145:37–45.

Allaby M (2008) Temperate forests. Chelsea House, New York

Alongi DM (2005) Mangrove-microbe-soil relations. In: Kristensen E, Haese RR, Kostka JE (eds) Interactions between macro- and micro-organisms in marine sediments. American Geophysical Union, Washington, DC

Alongi DM (2009) The dynamics of tropical mangrove forests. Springer, New York

Angelstam P, Kuuluvainen T (2004) Boreal forest disturbance regimes, successional dynamics and landscape structures—a European perspective. Ecol Bull 51:117–136

Antonovsky MY, Glebov FZ, Korzuhin MD (1987) A regional model of long-term wetland-forest dynamics. Work. Pap. WP-87–63. International Institute of Applied Systems Analysis, Laxenburg

Archibold OW (1995) Ecology of world vegetation. Chapman and Hall, London

Ash J (1982) The *Nothofagus* Blume Fagaceae of New Guinea. In: Gressitt JL(ed) Monogr Biol vol 42, W Junk, The Hague, p 355–380

Aubert G, Tavernier R (1972) Soil survey. In: Soils of the humid tropics. US National Research Council. Natl Acad Sci USA

Axelrod DI, Kalin-Arroyo MT, Raven PH (1991) History of temperate vegetation in the Americas. Rev Chil Hist Nat 64:413–446

Banner A, Green RN, Inselberg A, Klinka K, McLennan DS, Meidinger DV, Nuszdorfer FC, Pojar J (1990) Site classification for coastal British Columbia. Ministry of Forests, Victoria BC

Beinroth FH (1982) Some highly weathered soils of Puerto Rico 1: Morphology, formation and classification. Geoderma 27:1–73

Binkley D, Fisher RF (2012) Ecology and management of forest soils, 4th edn. Wiley, New York

Bockheim JG (1995) Permafrost distribution in the Southern Circumpolar region and its relation to the environment: a review and recommendations for further research. Permafrot Periglac 6:27–45

Bonan GB, Shugart HH (1989) Environmental factors and ecological processes in boreal forests. Ann Rev Ecol Syst 20:1–28

Borchers SL, Wattenbarger J, Ament R (1989) Forest plant associations of Montague Island, Chugach National Forest: preliminary results. In: Alexander Eb (ed) Watershed 89. March 21–23,1989, Juneau, Alaska, US Geological Agric For Serv Reg 10, Juneau, Alas

Bormann BT, Sidle RC (1990) Changes in productivity and distribution of nutrients in a chronosequence at Glacier Bay National Park, Alaska. J Ecol 78:561–578

Bowers F (1987) Effects of windthrow on soil properties and spatial variability in Southeast Alaska. PhD Thesis, University of Washington, Seattle

Brewer SW, Rejmanek M, Webb MAH, Fine PVA (2003) Relationship of phytogeography and diversity of tropical tree species with limestone topography in southern Belize. J Biogeogr 30:1669–1688

Buol SW (2002) Oxisols. In: Lal R (ed) Encyclopedia of soil science. Marcel Dekker, New York

Buol SW, Eswaran H (2000) Oxisols. Adv Agron 68:152–195

Buol SW, Southard R., Graham RC, McDaniel PA (2011) Spodosols: soils with subsoil accumulations of humus and sesquioxides. In: Soil genesis and classification, 6th edn. Wiley-Blackwell, Oxford.

Buol SW, Southard RJ, Graham RC, McDaniel PA (2003) Soil genesis and classification, 5th edn. Iowa State Press, Blackwell

Buol SW, Southard RJ, McCracken RJ, McDaniel PA (2003) Soil Genesis and Classification 5th edn. The Iowa State University Press, Ames

Burton PJ, et al. (2003) Towards sustainable management of the boreal forest. NRC Research, Ottawa

Buurman P, Soepraptohardjo M (1980) Oxisols and associated soils on ultramafic and felsic volcanic rocks in Indonesia. In: Buurman P (ed) Red soils in Indonesia. Agricultural Research Reports 889, PUDOC, Wageningen

Carleton TJ, Maycock PF (1978) Dynamics of the boreal forest south of James Bay. Can J Bot 56:1157–73

CEC (1997) Ecological regions of North America: toward a common perspective. Commission for Environmental Cooperation, Montreal

Chen ZS, Lin KC, Chang JM (1989) Soil characteristics, pedogenesis, and classification of Beichateinshan podzolic soils, Taiwan. J Chin Agric Chem Soc 27:145–155

Chen ZS, Liu JC, Chiang HC (1995) Soil properties, clay mineralogy, and genesis of some alpine forest soils in Ho-Huan Mountain area of Taiwan. J Chin Agric Chem Soc 33:1–17

Cockayne L (1971) The subtropic and subantarctic rainforests of New Zealand. In: Eyre ER (ed) World vegetation types. Columbia University Press, New York

Cooper WS (1942) Vegetation of the Prince William Sound region, Alaska; with a brief excursion into post-Pleistocene climatic history. Ecol Monogr 12:1–22

Cordova J, Valverde F, Espinosa J (1996) Phosphorus residual effect in Andisols cultivated with potatoes. Better Crops International 10(2)

Cuttelod A, Garcia N, Abdul Malak D, Temple HJ, Katariya V (2009) The Mediterranean: a biodiversity hotspot under threat. In: Vie J-C, Hilton-Taylor C, Stuart SN (eds) Wildlife in a changing world: an analysis of the 2008 IUCN Red List of Threatened Species. Gland, Switzerland, IUCN.

De Laubenfels DJ (1975) Mapping the world's vegetation. Syracuse Geogr Ser 4, Syracuse Press, Syracuse, NY

Dick DP, Goncalves CN, Dalmolin RSD, Knicker H, Klamt E, Kfgel-Knabner I, Simoes ML, Martin-Neto L (2005) Characteristics of soil organic matter of different Brazilian Ferralsols under native vegetation as a function of soil depth. Geoderma 124:319–333

Dimo VN (1969) Physical properties and elements of the heat regime in permafrost meadow-forest soils. In Permafrost Soils and Their Regime. Indian Natl Sci Doc Center, New Dehli

Dittus WPJ (1977) The ecology of a esmi-evergreen forest community in Sri Lanka. Biotropica 9(4):268–286

Doumenge C, Gilmour D, Ruiz Perez M, Blockhus J (1995) Tropical montane cloud forests. Conservation status and management issues. In Hamilton LS, Juvik JO, Scatena FN (eds): Tropical montane cloud forests. Ecological Studies 110:24–37

Driessen P, Deckers J, Spaargaren O, Nachtergaele F (2001) Lecture notes on the major soils of the world, FAO, Rome

Duivenvoorden JF (1995) Tree species composition and rain forest-environment relationships in the middle Caquetá Area, Colombia, NW Amazonia. Vegetatio 120(2):91–113

Eck KC (1984) Forest characteristics and associated deer habitat values, Prince William Sound islands. In: Meehan WR, Merrell TR, Hanley TA (eds) Fish and wildlife relationships in old-growth forests. Proc Symp, Juneau, Alaska, 12–15 April 1982. Am Inst Fish Res Biol and USDA For Serv, Pacific Northwest Res Stn, Portland, Oregon

Enright NJ (1982) The Araucaria forests of New Guinea. In: Gressitt JL (ed) Monogr.Biol, vol 42, W Junk, The Hague, p 381–399

FAO (2013) State of Mediterranean forests. FAO, Rome

FAO (1988) Soil map of the world. Revised legend, FAO–UNESCO–ISRIC. World Soil Resources Report No. 60. Rome

FAO (1992) Amenagement et conservation des forets denses en Amerique tropicale. FAO, Etude Forets 101, Rome

FAO (1994) The state of food and agriculture, 1994. http://www.fao.org/docrep/t4450e/T4450E0k.htm. FAO corporate Document Repository
FAO (1994) World reference base for soil resources. ISSS–ISRIC–FAO. Draft. Rome/Wageningen, Netherlands
FAO (1998) World Reference Base for Soil Resources, ISSS–ISRIC–FAO. World Soil Resources Report No. 84. Rome
FAO (2001) Global forest resources assessment 2000. www.fao.org. FAO forestry paper 140, FAO, Rome, Italy
FAO (2006) World reference base for soil resources 2006, a framework for international classification, correlation and communication. FAO–UNESCO–ISRIC. FAO, Rome
FAO (2007) The World's mangroves 1980–2005, FAO Forestry Paper 153. Forest Resources Division, FAO, Rome
FAO (2010) Global forest resources assessment 2000, FAO forestry paper 140. FAO, Rome
FAO (2011) State of Mediterranean Forests (SoFMF), concept paper. Arid Zone Forests and Forestry Working Paper No. 2. FAO, Rome
FAO AGL (2003) http://www.fao.org/ag/agll/default.stm. Accessed 8 July 2011
Fashing PJ, Gathua JM (2004) Spatial variability in the vegetation structure and composition of an East African rain forest. Afr J Ecol 42,189–197
Ferreira TO, Otero XL, Vidal-Torrado P, Macias F (2007) Redox processes in mangrove soils under *Rhizophora mangle* in relation to different environmental conditions. Soil Sci Soc Am J 71(2):484–491
Floyd AG (1990) Australian Rainforests in New South Wales. Surrey Beatty and Sons, Chipping Norton, NSW
Ford J, Bedford BL (1987) The hydrology of Alaskan wetlands, USA: a review. Arct Alp Res 19:209–29
Franklin JF, Dyrness CT (1973) The natural vegetation of Washington and Oregon. USDA For Serv, Pac Northwest For Range Exp Stn, Gen Tech Rep PNW-8, Portland, Oregon
Frelich LE (2002) Forest dynamics and disturbance regimes. Studies from the temperate evergreen-deciduous forest. Cambridge University Press
Galloway RW (l982) In: Clough BF (ed) Distribution and physiographic communities of Australian mangroves. Mangrove ecosystems in Australia. Australian National University Press, Canberra
Geiser LH, Nadelhoffer K (2011) Taiga. In: Pardo LH, Robin-Abbott MJ, Driscoll CT (eds) Assessment of Nitrogen deposition effects and empirical critical loads of Nitrogen for ecoregions of the United States. Gen Tech Rep NRS-80. US Department of Agriculture, Forest Service, Northern Research Station
Gentry AH (1995) Patterns of diversity and floristic composition in neotropical montane forests. In: Churchill SP, Balslev H, Forero E, Luteyn JL (eds): Biodiversity and conservation of neotropical montane forests. New York Botanical Garden, Bronx, New York
Giri C, Ochieng E, Tieszen LL, Zhu Z, Singh A, Loveland T, Masek J, Duke N (2011) Status and distribution of mangrove forests of the world using earth observation satellite data. Glob Ecol Biogeogr 20:154–159
Graham RT, Jain TB (1998) Silviculture's role in managing boreal forests. USDA Forest Service, Rocky Mountain Research Station
Guan DX, Wu JB, Zhao XS, Han SJ, Yu GR, Sun XM, Jin CJ (2006) CO_2 fluxes over an old, temperate mixed forest in northeastern China. Agricultural and Forest Meteorology 137:138–149
Hadi S, Ziegler T, Waltert M, Hodges JK (2009) Tree diversity and forest structure in northern Siberut, Mentawai islands, Indonesia. Trop Ecol 50(2):315–327
Hamilton L, Juvik JO, Scatena F (1993) Tropical montane cloud forests. Proceedings of an international symposium at San Juan, Puerto Rico, 31 May-5 June 1994. East-West Center, Honolulu
Hamilton LS (1995) Mountain cloud forest conservation and research. A synopsis. Mount Res Dev 15:259–266
Harden G, McDonald B, Williams J (2006) Rainforest trees and shrubs. Gwen Harden, Nambucca Heads, NSW
Hennon PE (1986) Pathological and ecological aspects of decline and mortality of *Chamaecyparis nootkatensis* in Southeast Alaska. PhD Thesis, Oregon State University, Corvallis
Hogarth PJ (1999) The biology of mangroves. Oxford University Press. Oxford, New York
Holdgate MW (1961) Vegetation and soils in the south Chilean islands. J Ecol 49:559–580
Hseu ZY, Chen ZS, Wu ZD (1999) Characterization of placic horizons in two subalpine Inceptisols. Soil Sci Soc Am J 63:941–947
Hseu ZY, Tsai CC, Lin CW, Chen ZS (2004) Transitional soil characteristics of Ultisols and Spodosols in the subalpine forest of Taiwan. Soil Sci. 169:457–467
http://soils.usda.gov/use/worldsoils/oxisols/117.pdf retrieved on 24.12.12
Jacobs M (1988) The tropical rain forest: A first encounter. Springer, Berlin
Jaroensutasinee M, Sangarun P, Pheera W, Jaroensutasinee K (2010) Tropical Montane cloud forest characteristics in Southern Thailand. Walailak J Sci Tech 7(2):103–113
Jien SH, Wu SP, Chen ZS, Chen TH, Chiu CY (2010) Characteristics and pedogenesis of podzolic forest soils along a toposequence near a subalpine lake in northern Taiwan. Botanical Studies 51:223–236
Joshi H, Ghose M (2003) Forest structure and species distribution along soil salinity and pH gradient in mangrove swamps of the Sundarbans. Trop Ecol 44(2):197–206
Kampen J, Burford J (1980) Production systems, soil related constraints and potentials in the semi-arid tropics with special reference to India. In: International Rice Research Institute (IRRI) (ed) Priorities for alleviating soil-related constraints to food crop production in the tropics. Los Banos, Philippines
Kauffman S, Sombroek W, Mantel S (1998) Soils of rainforest: characterization and major constraints of dominant forest soils in the humid tropics. In: Schulte A, Ruhiyat D (eds) Soils of tropical forest ecosystems. Springer, Berlin
Klinka K, Wang Q, Carter RE (1990) Relationships among humus forms, forest floor nutrient properties and understory vegetation. For Sci 36:564–581
Kolesnikov BP (1956) Kedroviye lesa Dalnego Vostoka [Korean pine forests of the [Russian] Far East]. Tr DVF AN SSSR Ser Bot Izd AN SSSR. Moscow 2:1–264 (in Russian)
Koponen P, Nygren P, Sabatier D, Rousteau A, Saur E (2004) Tree species diversity and forest structure in relation to microtopography in a tropical freshwater swamp forest in French Guiana. Plant Ecol 173(1):17–32
Krestov PV (1997) Osobennosti fitotsenoticheskogo raznoobraziya shirokolistvenno-kedrovyh lesov rossiyskogo Dalnego Vostoka [The main features of phytocenotic diversity of the broad-leaved Korean pine forests of the Russian Far East]. Komarovskiye Chteniya 46:15–42 (in Russian)
Kuchler AW (1949) A geographical system of vegetation. Geogr Rev 37:233–240
Kunnecke BH (2008) Temperate forest biomes. In Woodward SL (ed) Greenwood guides to biomes of the world. Greenwood, London
Kuppen W (1918) Klassifickation der klimate nach temperature niederschlag and jahreslauf. Petermann's Mitteilungen 64:193–203
Kuusela K (1990) The dynamics of the boreal coniferous forests. Sitra, Helsinki
La Roi GH (2011) Boreal Forests. The Canadian Encyclopedia. Historica-Dominion
LaBastille A, Pool DJ (1978) On the need for a system of cloud-forest parks in Middle America and the Caribbean. Environmental Conservation 5(3):183–190
Lal R, Lorenz K, Hüttl RF (2012) Biosphere: ecosystems and the global carbon cycle. Springer, Dordrecht
Landsberg JJ, Gower ST (1997) Applications of physiological ecology to forest management. Academic Press, San Diego

Larsen JA (1980) The Boreal ecosystem. Academic Press, New York, p 500
Letouzey R (1985) Notice de la carte phytogiographique du Cameroun au 1:500 000 M-SM: Region afromontagnarde et etage submontagnard. Institut de Ia Carte Internationale de la Végétation, Toulouse, France
Li SY, Chen ZS, Liu JC (1998) Subalpine loamy Spodosols in Taiwan: characteristics, micromorphology, and genesis. Soil Sci Soc Am J 62:710–716
Lima HN, Mello JWV, Schaefer CEGR, Ker JC, Lima AMN (2006) Mineralogia e quimica de tres solos de uma topossequencia da bacia sedimentar do Alto Solimoes, Amazonia Ocidental. Rev Bras Cienc Solo 30:59–68
Lloyd S (2000) Towards responsible Swedish timber trade? A survey of actors and origin of timber from Russia and the Baltic states. Taiga Rescue Network, Jokkmokk
Lopina O, Ptichnikov A, Voropayev A (2003) Illegal logging in northwestern Russia and export of Russian forest products to Sweden. WWF Russian Programme Office, Moscow
Lutz HF (1956) Ecological effects of forest fires in the interior of Alaska. USDA Tech Bull No 1133
Marques JJ, Schulze DG, Curi N, Mertzman SA (2004) Trace element geochemistry in Brazilian Cerrado soils. Geoderma 121:31–43
Martin JR (1989) Vegetation and environment in old growth forests of northern southeast, Alaska: a plant association classification. MS Thesis, Arizona State University, Tempe
Martin P, Nabuurs GJ, Aubinet M, et al. (2001) Carbon sinks in temperate forests, Ann Rev Energy Env 26:435–465
Maycock PF, Curtis JT (1960) The phytosociology of boreal conifer-hardwood forests of the Great Lakes region. Ecol Monogr 30:1–35
Mayer H (1984) Die Wälder Europas. Gustav Fischer, in German
Melillo JM, McGuire AD, Kicklighter DW, Moore IIIB, Vorsmarty CJ, Schloss AL (1993) Global climate change and terrestrial net primary production. Nature 363:234–240
Moorman FR, Lal R, Juo AS (1975) The soils of IITA. IITA Tech Book No. 3. Int Inst Trop Agric, Ibadan
Mumby PJ, Edwards AJ, Arias-Gonzalez JE, Lindeman KC, Blackwell PG, et al (2004) Mangroves enhance the biomass of coral reef fish communities in the Caribbean. Nature 427:533–536
Murphy PG, Lugo AE (1986) Ecology of tropical dry forest. Ann Rev Ecol Syst 17:67–88
Myers N, Mittlemeier RA, Mittlemeier CG, Da Fonseca GAB. Kent J (2000) Biodiversity hotspots for conservation priorities. Nature 403:853–858.
NAP (1982) Ecological aspects of development in the humid tropics. National Academy Press, Washington DC
National Technical Committee for Hydric Soils (1985) USDA Soil conservation service. Washington, DC
Nikolov N, Helmisaari H (1992) Silvics of the circumpolar boreal forest tree species. In: Shugart HH, Leemans R, Bonan GBA (eds) Systems analysis of the global boreal forest. Cambridge University Press
Nortcliff S (2002) Soils of cold and temperate regions. Land use, land cover and soil sciences. Encyclopedia of life support systems. UNESCO, www.eolss.net/Sample-Chapters/C19/E1–05-07–15.pd
Oldfield S, Eastwood A (2007) The Red List of oaks. Cambridge, UK, Fauna & Flora International
Ovaskainen O, Pappila M, Pötry J (1999) The Finnish forest industry in Russia. On the thorny path towards ecological and social responsibility. The Finnish Nature League Publications, Helsinki
Page SE, Rieley JO, Banks CJ (2011) Global and regional importance of the tropical peatland carbon pool. Glob Change Biol 17:798–818
Pastor J, Mladenoff DJ (1992) The southern boreal-northern hardwood forest border. In: Shugart HH, Leemans R, Bonan GB (eds) A system analysis of the global boreal forests. Cambridge University Press
Ping CL (2013) Gelisols: part I. Cryogenesis and state factors of formation. Soil Horizons. doi:10.2136/sh2013–54-3-gc
Quesada CA, Lloyd J, Anderson LO, Fyllas NM, Schwarz M, Czimczik CI (2011) Soils of Amazonia with particular reference to the rainfor sites. Biogeosciences 8:1415–1440
Ram J, Kumar A, Bhatt JP (2004) Plant diversity in six forest types of Uttaranchal, Central Himalaya, India. Curr Sci 86:638–647
Reich PB, Frelich L (2002) Temperate deciduous forests. The Earth system: biological and ecological dimensions of global environmental change, In: Munn T (ed) Encyclopedia of global environmental change, vol 2. Wiley, Chichester
Richter DD, Babbar LI (1991) Soil diversity in the tropics. Adv Ecol Res 21:315–389
Rieger S, Dement JA, Sanders D (1963) Soil survey of fairbanks area, Alaska. Series 1959, No. 25. USDA, Washington, DC
Robertson AI, Alongi DM, Boto KG (1992) Food chains and carbon fluxes. In: Robertson AI, Alongi DM (eds) Tropical mangrove ecosystems. Coastal and estuarine studies series. American Geophysical Union, Washington, DC
Ruckstuhl KE, Johnson EA, Miyanishi K (2008) Introduction. The boreal forest and global change.Phil Trans R Soc 363(1501):2243–2247
Rumney GR (1968) Climatology and the world's climates. Mac-Millan, New York:
Runyon J, Waring RH, Goward SN, Welles JM (1994) Environmental limits on net primary production and light use efficiency across the Oregon Transect Ecol Appl 4:226–237
Rydin H, Jeglum JKMurphy (2006) The biology of peatlands. Oxford University Press, New York
Saenger P (2002) Mangrove ecology, silviculture and conservation. Kluwer, Dordrecht
Sahu SC, Dhal NK, Reddy CS, Pattanaik C, Brahmam M (2007) Phytosociological study of tropical dry deciduous forest of Boudh District, Orissa, India. Res J For 1:66–72
Sanchez PA (1976) Properties and management of soils in the tropics. Wiley, New York
Sanchez PA, Buol SW (1975) Soils of the tropics and the world food crisis. Science 188:598–603
Sauer D, Sponagel H, Sommer M, Giani L, Jahn R, Stahr K (2007) Podzol: a review on its genesis, occurrence, and functions. J Plant Nutr Soil Sci 170:581–597
Sayre AP (1994) Taiga. Twenty-first century books, New York
Scarascia-Mugnozza G, Oswald H, Piussi P, Radoglou K (2000) Forests of the Mediterranean region: gaps in knowledge and research needs. For Ecol Manag 132:97–109
Schaetzl R, Anderson S (2005) *Soils*: genesis and geomorphology. Cambridge University Press, New York
Schawe M, Glatzel S, Gerold G (2007) Soil development along an altitudinal transect in a Bolivian tropical montane rainforest: Podzolization vs. hydromorphy. Catena 69:83–90
Scheffer F, Schachtschabel P (1998) Lehrbuch der Bodenkunde. Enke, Stuttgart
Schneider RR (2002) Alternative futures. Alberta's boreal forest at the crossroads. The Federation of Alberta Naturalists, Edmonton, Canada
Semwal DP, Saradhi PP, Nautiyal BP, Bhatt AB (2007) Current status, distribution and conservation of rare and endangered medicinal plants of Kedarnath wildlife sanctuary, Central Himalaya, India. Curr Sci 92:1733–1738
Semwal DP, Uniyal PL, Bhatt AB (2010) Structure, composition and dominance—diversity relations in three forest types of a part of Kedarnath wildlife sanctuary, central Himalaya, India. Not Sci Biol 2(3):128–132
Shaw JN (2005) Ultisols. In Lal R (ed) Encyclopedia of soil science. CRC, Boca Raton
Shugart HH, Leemans R, Bonan (eds) 1992. A system analysis of the global boreal forests. Cambridge University Press

Smeck NE (1985) Phosphorus dynamics in soils and landscapes. Geoderma 36:185–199

Soil Survey Staff (1999) Soil taxonomy. A basic system of soil classification for making and interpreting soil surveys, 2nd edn. Agricultural Handbook 436, USDA, Washington, DC

Soil Survey Staff (2003) Keys to *soil* taxonomy, 9th edn. United States Department of Agriculture. Natural Resources Conservation Service. *Washington, DC*

Sombroek WG (1966) A reconnaissance of the soils of the Brazilian Amazon region. Centre for Agricultural Publications and Documentation, Wageningen

Sombroek WG (1984) Soils of the Amazon region. In: Sioli H, Junk W (eds) The Amazon: limnology and landscape ecology of a mighty tropical river and its basin. Dordrecht

Spalding MD, Blasco F, Field CD (1997) World Mangrove Atlas. International Society for Mangrove Ecosystems, Okinawa

Stadtmullcr T (1987) Cloud forests in the humid tropics. A bibliographic review. United Nations University, Tokyo and CATIE, Turrialba. Costa Rica

Sukardjo S (1994) Soils of the mangrove forests of the Apar Nature Reserve, Tanah Grogot, East Kalimantan, Indonesia. Southeast Asian Studies 32(3):385–398

Suresh HS, Sukumar R (2011) Vegetative phenology of tropical montane forests in the Nilgiris, South India. J Natn Sci Foundation Sri Lanka 39(4):333–343

Sverdrup H, Stjernquist I (eds) (2002) Developing principles and models for sustainable forestry in Sweden. Kluwer

Theilade I, Schmidt L, Chhang P, McDonald JA (2011) Evergreen swamp forest in Cambodia: floristic composition, ecological characteristics, and conservation status. Nordic Journal of Botany 29:71–80

Thom BG (1982) Mangrove ecology—a geomorphological perspective. Mangrove ecosystem in Australian National Mangrove workshop. Australian National University Press, Canberra

Thomas DW, Achoundong G (I991) Montane forests of Western Africa. Proceedings of the XIII AETFAT Congress, 2–11 avril 1991, Zomba, Malawi

Thomas PA, *Packham J (2007)* Ecology of woodlands and *forests*: description, dynamics, and diversity. Cambridge University Press, New York

Thompson JD, Lavergne S, Affre L, Gaudeul M, Debussche M (2005) Ecological differentiation of Mediterranean endemic plants. Taxon 54:967–976

Tuhkanen S (1984) A circumpolar sys-tem of climatic-phytogeographical re-gions. Acta Bot Fenn 127:1–50

Ugolini FC, Mann DH (1979) Biopedological origin of peatlands in Southeast Alaska. Nature 281:366–368

USDA (1960) Soil classification, a comprehensive system. United States Department of Agriculture. Govt Printing Office, Washington DC

USDA (1975) Soil taxonomy: a basic system of soil classification for making and interpreting soil survey. United States Department of Agriculture. Govt Printing Office, Washington DC

USDA, NRCS (1998) Keys to soil Taxonomy, 8th edn. United States Department of Agriculture. Govt Printing Office, Washington DC

USDA/NRCS (2006) Field indicators of hydric soils in the United States, version 6.0. In: Hurt GW, Vasilas LM (eds) USDA, NRCS, in cooperation with the National Technical Committee for Hydric Soils

Van Cleve K, Dyrness CT, Viereck LA, Fox J, Chapin FS, Oechel W (1983) Taiga ecosystems in interior Alaska. Bioscience 33:39–44

Ver Hoer JM, Neiland BJ, Glenn-Lewin DC (1988) Vegetation gradient analysis of two sites in Southeast Alaska. Northwest Sci 62:171–180

Viereck LA (1975) Forest ecology of the Alaskan taiga. In: Proc Circumpolar Conf Northern Ecology, Ottawa: Natl Res Coun, Canada

Viereck LA, Dyrness CT, Van Cleve K, Foote MJ (1983) Vegetation, soils, and forest productivity in selected forest types in interior Alaska. Can J For Res 13:703–20

Viereck LA, Van Cleve K, Dyrness CT (1986) Forest ecosystem distribution in the taiga environment. In: Van Cleve K, Chapin FS, Flanagan PW, Viereck LA., Dyrness CT (eds) Forest ecosystems in the Alaskan taiga. Springer, New York

Vitousek PM, Sanford RL (1986) Nutrient cycling in moist tropical forest. Annu Rev Ecol Syst 17:137–167

Wakushima S, Kuraishi S, Sakurai N (1994) Soil salinity and pH in Japanese mangrove forests and growth of cultivated mangrove plants in different soil conditions. J Plant Res 107:39–46

Walter H (1979) Vegetation of the Earth. Springer-Verlag, New York

Wardie P, Buffin MJA, Dugdale J (1983) Temperate broad-leaved evergreen forests of New Zealand. In: Ovington JD (ed) Temperate broad-leaved evergreen forests. Ecosystems of the World. Elsevier, New York

Waring RH (2002) Temperate coniferous forests. The Earth system: biological and ecological dimensions of global environmental In: Munn T, Moony HA, Canadell JG(eds) Encyclopedia of global environmental change, vol 2. Wiley, Chichester

Waring RH, Franklin JF (1979) Evergreen coniferous forests of the Pacific Northwest. Science 204:1380–1386

Waring RH, Running SW (1998) Forest ecosystems: analysis at multiple scales, 2nd edn. Academic Press, SanDiego

Watson RT, et al. (2000) Land use, land-use change and forestry. Intergovernmental Panel on Climate Change, Cambridge University Press

Webb LJ (1968) Environmental relationships of the structural types of Australian rain forest vegetation. Ecol 49:296–311

Webb LJ (1978) A general classification of Australian rainforests. Aust Plants 76:349–363

Wen J (1999) Evolution of eastern Asian and eastern North American disjunct distributions in flowering plants. Annu Rev Ecol Syst 30:421–455

White F (1983) The vegetation of Africa. UNESCO, Paris

Whitmore TC (1975) Tropical rain forests of the Far East. Oxford University Press, London

Wilcke W, Oelmann Y, Schmitt A, Valarezo C, Zech W, Homeier J (2008) Soil properties and tree growth along an altitudinal transect in Ecuadorian tropical montane forest. J Plant Nutr Soil Sc 171:220–230

Wolf EC, Mitchell AP, Schoonmaker PK (1995) The rain forests of home: an atlas of people and place. Ecotrust, Portland

Wolff JO, West SD, Viereck LA (1977) Xylem pressure potential in black spruce in interior Alaska. Can J For Res 7:422–28

Wu SP, Chen ZS (2005) Characteristics and genesis of Inceptisols with placic horizons in the subalpine forest soils of Taiwan. Geoderma 125:331–341

Young A (1980) Tropical soils and soil survey (Cambridge Geographical Studies). Cambridge University Press

Zech W, Hintermaier-Erhard G (1997) Wörterbuch der Bodenkunde. Enke, Stuttgart

Zech W, Hintermaier-Erhard G (2002) Böden der Welt: Ein Bildatlas. Spektrum, Heidelberg

Zinn YL, Lal R, Bigham JM, Resck DVS (2007) Edaphic controls on soil organic carbon retention in the Brazilian Cerrado: Texture and mineralogy. Soil Sci Soc Am J 71:1204–1214

Zoltai SC (1975a) Tree ring record of soil movements on permafrost. Arct Alp Res 7:331–40

Zoltai SC (1975b) Structure of subarctic forests on hummocky permafrost terrain in northwestern Canada. Can J. For Res 5:1–9

Forest Disturbances and Soil Degradation

8

Forests have been disturbed by natural and anthropogenic agents from the day they evolved on this planet. Harvesting, fuelwood collection, windthrow, insect and disease outbreak, flooding, fires, shifting cultivation, grazing etc. are the main forest disturbance processes. Some disturbances are, however, required even for desirable forest development. At some time in the past, damages caused by natural disturbances of forests could be mended by natural healing processes, but recently natural calamities such as cyclones and floods have become drastic and frequent due to human interferences. People collect timber, fuel wood, thatching materials, canes, mushrooms, rhizomes, fruits, honey, medicines, etc. for their own consumption and for trading. Due to the ever-increasing human population and biomass needs, thrust on the forest resources has become enormous. So, forests are being degraded and deforested at an alarming rate at present with the concomitant degradation of the watersheds including their vegetation, soil, and water.

8.1 Forest Disturbances

Forest disturbances are defined as relatively discrete events that disrupt the ecosystem, community, or population structure and change the resources, substrate availability, or physical environment (White and Pickett 1985). Forest disturbances include events such as forest fire, harvesting, windthrow, insect and disease outbreak, and forest flooding. Stand-replacing disturbances, such as crown fires and windthrow, are associated with the sudden death of large cohorts of trees near one another (Pickett and White 1985, Kurz and Apps 1999). Disturbances play an important part in the lifecycle and succession dynamics of many forest systems. In boreal systems, large-scale, natural, stochastic forces tend to dominate the ecosystem dynamics, even when direct human influences are considered (Kurz et al. 1995).

Forest disturbances differ in agents and sizes, and in frequency, severity, and structure that is left after a disturbance (Turner et al. 1998). Management of forest ecosystems to sustain desired benefits requires knowledge of how forests change over time in response to natural and anthropogenic disturbances. Disturbances include both stresses and forest destruction and degradation, such as harvesting, land-use change, forest fire, and climate change and serious weather events such as hurricanes and ice storms, pollution of the air, water, and soil, etc. Some of these are natural and some are caused, in part or entirely, by humans. Landsat Thematic Mapper images covering the entire 3.9 million km^2 of forested Brazilian Amazon reveal natural change and disturbance occurring on a scale of decades to centuries. These include: 92,000 km^2 of bamboo forests undergoing synchronous mortality and regrowth; 1,500 km^2 of recently active dune fields; 900 km^2 of recent downburst blowdowns; >500 km^2 of recent forest fire scars; and an unknown area of forest mortality from flooding of high ground and alluvial forests. Fire subclimax fern savannas created by Yanoama Indians cover an additional 600 km^2. Natural and indigenous disturbances and synchronized phenologies are therefore responsible for a dynamic spectral behavior in large portions of the Amazon Basin (Nelson 1994).

Published climate-model results indicate that climate change and global warming favors increased rates of forest disturbance, as a result of weather more likely to cause forest fires (drought, wind, and natural ignition sources), convective wind storms, coastal flooding, and hurricanes. New sensitivity tests carried out with a vegetation model indicate that climate-induced increases in disturbance could, in turn, significantly alter the total biomass and compositional response of forests to future warming (Overpeck et al. 1990). Under these disturbances due to natural disasters and desperate human activities, forest decline and forest degradation have become more severe.

For thousands of years, boreal forests have adapted to large-scale natural forest disturbances such as forest fires, and wind storms which tend to disturb entire stands of trees at one time. Some boreal tree species such as jack pine and spruce actually require this kind of natural disturbance in order to regenerate and grow. Clear-cutting is one method

K. T. Osman, *Forest Soils,* DOI 10.1007/978-3-319-02541-4_8, © Springer International Publishing Switzerland 2013

Fig. 8.1 Settlement of plain landers in forested hilly areas has caused massive deforestation in Hill Tracts of Bangladesh. (Photo courtesy of Azad Kamal)

of forest harvesting which actually changes the vegetation type of a forest temporarily; within 5–10 years a new highly productive forest may develop by natural or artificial regeneration. Deforestation describes areas that are deliberately removed from the growing forestland base for uses such as roads, buildings, and agriculture. Forest insects and diseases are also disturbing agents which affect more than 4.5 million ha of forest per year. Mortality from insects and diseases is often different from mortality due to fire or weather damage. Insects and diseases tend to significantly reduce the growth rate of trees and kill individual or small groups of trees across the landscape, rather than most of the trees in a concentrated area. It should reduce the overall productivity of a forest stand.

Selective logging in tropical forests is a vast and spatially diffuse activity, involving the perforation and fracture of the canopy through large-tree thinning. In a recent study, Asner et al. (2006) used an automated, high-resolution remote-sensing technique (spectral mixture analysis) to map logged areas from 1999 to 2002 in the top five timber-producing states of Brazilian Amazonia. Their analysis of 480 Landsat enhanced thematic mapper plus (ETM+) scenes, which detected small logging gaps (0.5 ha) and associated understorey disturbance caused along skid trails and logging decks, shows that an area of 12075–19823 km^2 was selectively logged each year (Splechtna and Gratzer 2005). Studies emphasize the need of research on forest disturbance ecology, including: (1) the basic characteristics of disturbed forests; (2) the processes of natural and human disturbances; (3) the responses of forests ecosystem to the disturbances; (4) the main ecological processes or the consequential results of disturbed forests, including the change of biodiversity, soil nutrient and water cycle, eco-physiology and carbon cycle, regeneration mechanism of disturbed forests, and so on; (5) the relationships between disturbances and forest management; and (6) the principles and techniques for the management of disturbed forests (Zhu and Liu 2004).

Forest disturbance has a major role in forest succession. Most stands on the boreal forest landscape to date originated after high-severity disturbance: a combination of logging followed by burning of slash in secondary forests, or crown fire in primary forests. Disturbance severity under the current harvesting regime is very different from the historical regime. Much logging in the central part of the North American boreal forest is done during the winter. This leaves the forest floor and seedling layer intact: a moderate-severity disturbance according to our definition. Therefore, the result could be a forest with intermediate composition between the two endpoints, a mixture of jack pine, fir, birch, spruce, and cedar according to one model. According to another model, spruce, fir, birch, and cedar could occur, depending upon exactly where the discontinuity and the severity of the new harvesting regime lie along the disturbance severity axis (Frelich and Reich 1998).

There are some people who live inside forests. They are hunter-gatherers and shifting cultivators, and they are heavily dependent on forests for their livelihood on a subsistence basis. They are mainly indigenous people or people from minority ethnic groups. They are usually outside both the political and economic mainstream. There are also people who live in villages near forests; they are usually involved in agriculture outside the forest, but regularly use forest products for their own subsistence and for income generation. New populations have also migrated into forest areas, or into land near forest areas, and migration of lowland people into the forested hilly regions like Chittagong Hill Tracts of Bangladesh has some negative impacts on forests (Fig. 8.1). There are mainly two key results: (i) Forest land is often converted directly into agricultural land and (ii) pressure is put on people already living in forests. Such changes as agricultural intensification (especially reducing fallow periods in shifting cultivation) and increased forest encroachment by outsiders occur frequently. These immigrants also become a new category of forest-dependent people. Shifting cultivation, illegal logging, timber theft, forest burning, and fuel biomass harvest increase as a result.

Local rural communities obtain a range of goods and services from forests, including timber, canes, thatching materials, mushrooms, rhizomes, fruits, fuel biomass (used instead of fuelwood because they use herbs and shrubs and litters as sources of fuel too), honey, and medicines for direct household consumption, as well as trading on local and regional markets (Koziell and Saunders 2001, Kaimowitz 2003). Such human dependences on forest products for supporting their livelihood as well as meeting basic needs, and

Fig. 8.2 Harvesting machineries leave deep scars of soil disturbance. (Image courtesy of Department of Environment and Conservation)

Fig. 8.3 A section of forest that has been clear-cut. (Photo courtesy of Central Sierra Environmental Resource Centre)

the uncontrolled growth of human population have created a heavy thrust on the forest resources. The consequences of this enormous pressure on forest can be seen in severe soil erosion, landslide, flood, drought etc.

8.1.1 Forest Harvest

Timber harvest may cause changes to soil physical, chemical, and biological properties that reduce site productivity. Ground-based logging equipment such as felling machines, mechanical processors, skidders, and crawler tractors are used to harvest forests in many parts of the world. Heavy machines—tractors, harvesters, and wheeled loaders—can damage soil (Fig. 8.2). These equipments may travel over two-thirds of a site (McMahon et al. 1999). It may cause soil disturbance by displacing or mixing litter and soil, and/or compacting the soil. Removal of litter and soil alters the amount and availability of nutrients. Compaction and removal of topsoil can alter root volume by increasing resistance to root growth or reducing the ability of the soil to supply oxygen and water to plant roots. Murphy (1984) noted from a survey of 19 harvesting operations in New Zealand that "deep" disturbance covered from 5 to 25% of the harvest area and was affected by such factors as slope, type of harvesting equipment, and planning. Soil disturbance and compaction resulting from harvesting often increase seedling mortality and reduce tree growth (Skinner et al. 1989, Minore and Weatherley 1990, Wronski and Murphy 1994). In particular, the effects of soil compaction may last for decades unless remedial action is taken. Modification of the soil as a result of harvesting is of particular importance where fast-growing species or multiple-entry thinnings are utilized.

Van Rees et al. (2001) studied the effects of forest harvesting on soil compaction and surface disturbance on five harvested blocks having a range in soil properties. Soil disturbance was quantified on two 30-m transects at each grid point. Grid points were surveyed and the data used to create maps of the soil disturbance. Soil bulk density increased significantly from pre- to postharvest conditions in the two winter-harvested sites at both 10 cm and 20 cm depths, while two of the three summer-harvested sites showed significant increases at 10 cm, and one site at the 20-cm depth. Bulk densities from roadways and landings were significantly higher than preharvest levels. Bulk densities sampled from adjacent uncut stands were not significantly different than those of the preharvest bulk density levels. The United States Department of Agriculture (USDA) Forest Service uses a threshold of 15% increase in compaction for defining detrimental soil disturbance. Combining all grid points together resulted in 32% of all grid points at both depths having an increased bulk density of > 15%.

The most convenient method of forest harvesting is the clear-cutting (Fig. 8.3). Tang et al. (1997) studied the influence of forest clear-cutting patterns (location, size, shape, distribution of cut units, etc.) on the potential for landscape disturbance in Charley Creek watershed, Washington State, USA. They observed that: (1) the location of timber harvesting is more important than the geometry of harvest activity in influencing shallow landsliding; (2) forest harvest patterns strongly influence the potential for disturbance processes; and (3) a single cutting pattern will often fail to meet all landscape management goals.

Reeves (2011) used the soil disturbance-monitoring data from nine Northern Region National Forests to determine: (1) if soil disturbance monitoring data can link disturbance levels to site characteristics using monitoring data collected under disparate protocols, (2) if risk ratings can be developed to identify areas susceptible to high-impact harvest techniques using a consistent soil disturbance monitoring

Fig. 8.4 Biomass collection by nearby villagers for domestic use. (Photo courtesy of Mr. Azad Kamal)

Fig. 8.5 Harvesting plant litters from a forest plantation. (Photo courtesy of Mr. Azad Kamal)

protocol, (3) levels of detrimental soil disturbance resulting from helicopter, skyline, and ground-based harvest systems, and (4) how policies pertaining to soil and site sustainability affect management. Results indicate that a common soil-monitoring protocol is necessary to determine how site characteristics affect soil resilience to timber harvest operations. There are significant differences in soil disturbance resulting from timber harvest systems with helicopter < skyline < ground based.

Innoprise, a Malaysian company with forestry activities, and the New England Power Company, USA, have developed a technique known as "reduced impact logging" aimed at reducing damage to vegetation and soils during harvesting. It is expected to reduce damage by 50 % compared to that of conventional harvesting. The techniques employed are modifications of conventional bulldozer harvesting techniques, including pre-felling climber cutting, directional felling, skid trail design, and postharvest operations such as rehabilitation of log landings (Wan Razali and Tay 2000).

8.1.2 Fuel Biomass Collection

Villagers near forests collect stem, branches, barks, and leaves of trees, and subordinate vegetation (understorey, shrubs, and ground layer) and litter, to meet their day-to-day need and as a means of earning. But rural people no longer get enough fuel materials from dead branches and twigs of standing trees. They nowadays harvest seedlings and saplings of regenerated secondary forests, and even the herbs and shrubs left after deforestation. In the dry season, they set the bushes on fire, allow the biomass to dry in fire, and cut them for fuel. You may see many women passing through a forest road in the evening carrying sacks of minor biomass on their heads (Fig. 8.4). Commercial fuelwood collectors usually clear-cut the vegetation, harvest wood and leave small stems, branches, and leaves in the site. After clear-cutting energy plantations, the sites are reforested. The practice may temporarily damage soil properties to some extent by harvesting as well as transporting equipment and procedures. Thus, small-scale noncommercial fuelwood collection from high forests does not seem to cause large-scale damage to the forest vegetation and the forest soil. It may even go in favor of the vegetation by removing dead and diseased parts, opening the canopy for light and enhancing seed germination and regeneration. However, clear-cutting and whole-tree biomass harvest must cause some nutrient export, the amount of which depends on tree species, amount of harvested biomass, harvested tree component, etc. Because the concentration of nutrients in wood is small relative to those in other parts of trees, harvesting part of the wood only removes a relatively small nutrient store (Brown et al. 2009). Demand of fuelwood is so enormous and that of biomass has become so scarce that many of them only get fuel by sweeping forest litters from nearby plantations (Fig. 8.5). By this way, they deprive the forest soils of the benefit of natural replenishing of 5–15 t organic matter and 100–200 kg N, similar amounts of K and Ca, and10–15 kg P annually (Chap. 6). If this is continued, the forest soil is threatened to be substantially impoverished and lose the natural capacity of resilience.

8.1.3 Shifting Cultivation

Shifting cultivation is an ancient farming system in the hilly areas of the humid tropics. It has been a common agricultural practice in tropical hilly areas of Southeast Asia, the Pacific, Latin America, the Caribbean, and Africa for millennia (Lawrence and Schlesinger 2001, Eastmond and Faust 2006, Thomaz 2009). Estimates suggest that it is the basis

Fig. 8.6 A part of a hill slope left after shifting cultivation. The original cover was a tropical rainforest. (Photo courtesy of Imam Hossain)

of subsistence for at least half a billion people (Craswell et al. 1997). Shifting cultivation is also known as slash-and-burn agriculture because in this system "farmers," usually indigenous people of the mountainous regions, clear a patch of natural forest, usually on gentle slopes or on summits of hills. They slash the vegetation, allow them to dry in place and burn them, make holes in soil, and sow assorted crops in them just before the monsoon comes. Seeds germinate after the first rains, and the crop lasts in the "field" for 6–8 months. After harvest, the land is left in the hand of nature (Fig. 8.6). Shifting farmers clear a new patch of forest for cultivation in the next year. They are to return to the previous land for cropping after 15–20 years. Shifting cultivation could be considered as an early stage in the evolution of agricultural systems. The system is based on cutting and burning the vegetation in the dry season, and planting crops in the ashes early in the wet season. Declining soil productivity and increasing weed problems lead farmers to abandon fields after a few, often only 1 or 2, years of cropping. Other types of vegetation take over, and the field eventually grows into a secondary forest, before the cycle is repeated (Angelsen 1994). Recently, the rotation period has shrunk to only 2–3 years due to population pressure and reduced availability of suitable land for cropping. The consequences are lowered soil fertility, increased soil erosion losses, and reduced annual crop yields (Bruun et al. 2009, Thomaz 2009, Ziegler et al. 2009). In essence, the increasing intensity of land use for shifting cultivation has gone beyond its sustainable capacity in many tropical regions (Grogan et al. 2012).

Shifting cultivation is also known as "jhum" in India and Bangladesh, "taungya" in Myanmar, and "Sweden agriculture" in Europe. Shifting cultivation is frequently identified as the primary cause of deforestation in the tropics (Myers 1993, Riswan and Hartanti 1995). Presently, shifting cultivation is estimated to account for 70% of deforestation in Africa, 50% in Asia, and 30% in Latin America (Sodarak 1999). Over 250 million people practice shifting cultivation, which accounts for the clearance of more than 10 million hectares of tropical forest each year (ICRAF 1998). This cultivation method occupies about 30% of the world's exploitable land area and provides food for 8% of the world population (Hatch 1982). In many countries, however, shifting cultivation has been found to result in less deforestation than industrial land use such as timber extraction and oil palm as well as rubber plantations (Lawrence et al. 1998). Shifting cultivation has gradually been replaced by more intensive forms of land use in Asia and elsewhere (Spencer 1966, Watters 1971). For instance, in the mountains of Nepal, shifting cultivation had been completely replaced by intensive sedentary agriculture by the end of the nineteenth century. Such land use has been gradually replaced by agroforestry and tree-based land uses in Nagaland, India (NEPED and IIRR 1999). In the mountains of Northern Thailand and Peninsular Malaysia, and on the islands of Java and Bali in Indonesia, shifting cultivation has been largely replaced by sedentary commercial agriculture over the last few decades (Turkelboom et al. 1996, Suraswadi et al. 2000).

Earlier, shifting farmers and their advocates argued that shifting cultivation in its conventional form might be sustainable, because forests should recover by natural succession during the long fallow period. For example, Tanaka et al. (2005) suggested that the general impacts of shifting cultivation on soil properties after more than 10 years of fallow and 1 year of cultivation were limited. It has been already mentioned that overpopulation, settlement of plain landers to hills, and scarcity of available land for cropping have led to the shrinking of the rotation. Reduction of the fallow period has now created deep concerns about deforestation, loss of biodiversity, and soil degradation. It has been estimated that each year approximately 1.9–3.6 × 10^6 ha land of primary close forests, 3.4–40 × 10^6 ha land of secondary close forests, and 6.9–21.9 × 10^6 ha land of secondary open forests are being lost due to shifting cultivation (Detwiler and Hall 1988).

Shifting farmers burn the slashed debris to add nutrients of the ash to the soils and enhance fertility and crop productivity. Increases in 234 kg Ca, 55 kg Mg, and 20 kg K ha^{-1} after burning were reported by Salacedo et al. (1997). The effect of burning is, however, temporary. Rains following burning remove bases rapidly by erosion and leaching. Gafur et al. (2000) reported losses of 61 kg Ca, 13 kg Mg, 13 kg K, 0.14 kg P, 0.20 kg S, 0.05 kg Cu, 6.7 kg Fe, 6.1 kg Mn, and 0.065 kg Zn due to shifting cultivation in the Bandarban hills in Bangladesh. Thus, shifting cultivation in its distorted form leads to degradation of soil and water. Such activities affect soil physical and chemical properties, reduce nutrient stocks, and accelerate soil erosion, and sedimentation. Gafur et al. (2000) estimated soil loss of 43 Mg ha^{-1} yr^{-1} due to shifting cultivation in Bandarban, Bangladesh. There are reports of erosion loss from many other countries. Values vary due to differences in soil, slope, and climate. Alake Biswas et al. (2011) carried out an investigation on soil properties under different land use systems including shifting cultivation in Chittagong Hill Tracts, Bangladesh. They observed significantly higher nutrient levels in soils under natural forests and bushy lands than slashed, burned, and shifting cultivated sites. In this region, vast areas have been denuded by repeated slashing and burning. The effects of shifting cultivation on carbon storage and soil quality were outlined by Bruun et al. (2009). They observed that time-averaged aboveground carbon stocks declined by about 90% if the long fallow periods of traditional shifting cultivation are reduced to 4 years and by about 60% if shifting cultivation is converted to oil palm plantations. Conversion of swiddening to continuous annual cropping systems brings about substantial losses of time-averaged aboveground carbon stocks, reductions of soil C stocks and generally leads to declining soil quality. Changes of physicochemical properties of soil during different phases of shifting cultivation in Assam, India, were reported by Singh et al. (2003). Burning resulted in the significant increase in soil pH, and the acidity increased as the fallow progressed. A drastic reduction in moisture content was recorded due to burning operation. Jhuming caused appreciable change in organic matter and nutrient contents in soil. The burning of the debris after slashing the forest leads to the depletion of organic carbon and nitrogen content in soil. Exchangeable Ca, Mg, K, and available P increased just after burning and thereafter a gradual decrease was noticed in cropping phases. The nutrient status of soil was significantly higher in natural forests than in the fallow lands. Summarizing works at the International Institute of Tropical Agriculture, Lal (1987) showed very low soil erosion of a traditional farming practice resembling shifting cultivation. Studies in Thailand indicated that sediment production from shifting cultivation was limited compared to other land uses (Forsyth 1994; Ziegler et al. 2004). In Bangladesh, on the other hand, soil loss during runoff in shifting cultivation was estimated to remove 27% of the nutrients in the top 10 cm of the soil (Gafur et al. 2000), though much of the soil material lost from the fields is deposited within the watershed. Eroded soil from all the shifting fields in Chittagong Hill Tracts of Bangladesh has been estimated to carry about 4,309 tons of nitrogen along with other nutrients away. About 14,071 tons of commercial fertilizers worth of approximately US $ 1.8 million would be required annually to replace these nutrients (Gafur 2001). In Guatemala, high erosion was found to occur on shifting cultivated fields (Beach 1998). Wairiu and Lal (2003) used soil organic carbon (SOC) depletion as a proxy for soil erosion and suggested strong erosion of shifting cultivation compared to natural forest in the Solomon Islands. Some other investigators also showed SOC depletion following shifting cultivation (van Noordwijk et al. 1997, Roder et al. 1997).

8.1.4 Forest Fires

Forest fires occur mainly in the summer and autumn in the forested regions. Some fires spread along the dead leaves and branches at the bottom of trees. Some fires spread when the leafy canopy catches fire. Also, burning leaves and branches can get blown ahead of the main fire causing smaller fires to start. Forest fires are particularly destructive when there is a drought because the branches and twigs die and dry out, and plenty of fuel for the fire becomes available. A forest fire is said to be wildfire when a huge, roaring wall of open flame spreads toward all directions and thick clouds of billowing smoke rise upward from the burning fire (Fig. 8.7). A wildfire has the devastating power to wipe away thousands of hectares of forest each year. Wildfires take place all over the world. Major fires happen every few years in the forests of the USA and Indonesia, and in the bush in Australia.

Fire is a critically important disturbance agent in forest ecosystems. It can change the structure and functions of forests, including floral and faunal composition, diversity, biogeochemical cycling, etc., and alter vegetation, forest floor, soil organic matter (SOM), and nutrients (Yildiz et al. 2010). In other words, forest fires can alter soil physical, chemical, and biological properties and increase susceptibility to erosion (DeBano et al. 1998). Fires may also play a significant role in regulating ecosystem productivity and vegetation succession (Boerner et al. 2009). Many of our most valuable forest trees exist only because of past fires, which have produced conditions favoring their regeneration. At the same time, fire is one of the most destructive agents of deforestation; many hundreds of thousands of dollars are spent annually on measures aimed at protecting forests from the ravages of uncontrolled fire.

Forest fires may be prescribed or managed and wild or uncontrolled. Prescribed burning is a standard practice to reduce fuel levels, with the intention of minimizing the extent and severity of wildfires or facilitating germination and growth of desired forest species. At moderately low moisture content of soil, severity of prescribed fire is low (Walstad et al. 1990), while, wildfires occur in the presence of an abundant dry fuel load and they are very severe. Fire severity is defined in terms of: (1) the length of time fuel accumulates between fires and the amount of the accumulated fuels; (2) properties of the fuels (such as size, flammability, and moisture or mineral content); (3) how the fuels impact fire location and behavior (causing crown, surface, or ground fires); and (4) heat transfer in the soil during the combustion of aboveground fuels and surface organic layers (DeBano et al. 1998). Low- to moderate-severity fires, such as most of

Fig. 8.7 Wildfire in Florida, USA. (Image courtesy of Virginia Wildfire Information and Prevention)

those prescribed in forest management, promote renovation of the dominant vegetation through elimination of undesired species and transient increase of pH and available nutrients. However, even a low-severity fire may enhance water repellence in soil which reduces infiltration and increases erosion. On the other hand, wildfires can cause a significant loss of organic matter, deterioration of structure and porosity, and loss of nutrients through volatilization, ash entrapment in smoke columns, leaching, and erosion. Forest fires result in marked alteration of both quantity and specific composition of microbial and soil-dwelling invertebrate communities (Certini 2005).

Wildfires have devastating effects on ecosystems, and their effects have mainly concentrated on those of flora, fauna, soil, and atmosphere (Radho-Toly et al. 2001, Laughlin et al. 2004, Auld and Denham 2006, Hoinka et al. 2009). Wildfires might cause hydrological and geomorphological changes; more overland flow takes place, discharge increases, and catchment sediment yield increases (Shakesby and Doerr 2006). These changes basically depend upon fire intensity, soil type, and moisture, biomass of vegetation before and after fire and climatic conditions (Chandler et al. 1983). Often the organic matter contents of forest soils increase immediately after the occurrence of a forest fire due to incompletely burned residues left after a low-intensity burning, but this effect is largely temporary. Mangas et al. (1992) studied precipitation, runoff, and erosion in forests of Alicante, South East Spain from 1985 to 1990. They conducted a burning experiment in 1989 and observed its effects on soil at three intervals—before fire, one day after fire, and 6 months later. A significant increase was found in organic matter, total nitrogen, available phosphorus, and K^+ and Mg^{2+} contents one day after fire, but Ca^{2+} and cation-exchange capacity (CEC) showed an opposite trend. Loss of microbial species and population, reduction of invertebrates, and partial elimination of plant roots are some of the effects of a forest fire on biological properties of soil.

Fire is an important mineralizing agent, and there is a significant increase in some soil nutrients after burning. There may, however, be important losses through smoke in ecosystems subject to severe fire (Raison et al. 1985, Giovannini 1991). On the other hand, heavy rains following a high-intensity fire may cause losses of considerable amounts of organic matter, ashes, and nutrients. As to the generation of runoff and sediment yield, the effect of fire depends mainly on the intensity of the rain, the intensity and type of fire, and the slope of the land (Benito et al. 1991, Soler and Sala 1991). Soil surface may be charred due to burning and then soils may become impermeable with considerable changes in soil structure (Díaz Fierros et al. 1987, Sevink et al. 1989). On steep slopes, forest fires destroy the vegetative cover and contribute significantly to the degradation of the soil (Leite and Goncalves 2008). However, the severity of this degradation is a function of the frequency and the intensity patterns of the fires (Leita et al. 2011). Forest soils are naturally protected by all materials on or above the soil surface, such as vegetation (canopy and undergrowth), surface litter, and duff that reduce the impact of the applied forces of rain and wind (Megahan et al. 1986, McNabb and Swanson 1990). Forest fires destroy these protective agents.

A good number of systematic studies on the effects of a forest fire on soil properties have been done. Fire severity, which affects the soils most, is controlled by such environmental factors as amount, nature, and moisture of live and

Fig. 8.8 Deforestation in Sumatra, Indonesia. (Courtesy of WWF)

dead fuel, air temperature and humidity, wind speed, and topography of the site. Fire severity is indicated by both intensity and duration of fire. Although heat in moist soil is transported faster and penetrates deeper, latent heat of vaporization prevents soil temperature from exceeding 95°C until water completely vaporizes (Campbell et al. 1994); the temperature then typically rises to 200–300°C (Franklin et al. 1997). Temperature may rise at the soil surface to 500–700°C in the presence of heavy fuels (DeBano et al. 1998). It may sometimes reach values up to 850°C (DeBano 2000). Under most forest fire, soil temperatures at 5 cm below surface usually remain below 150°C, and there is little heating below 20–30 cm depth (DeBano 2000). Probably the duration of fire results in the greatest belowground damage. Fast-moving intensive fires at well-fueled sites may not transfer much heat down to more than few centimeters below the surface. Soil temperatures can remain high from a few minutes to several days, and there may be short-term and long-term soil hazards of forest fires, depending chiefly on type, severity, and frequency of fires, and post-fire climatic conditions. After heating forest soil samples in furnace at variable temperatures in the laboratory, Parlak (2011) observed that soil heating lowered clay and silt contents and, at some temperatures, partially decreased the pH and contents of lime, total nitrogen, exchangeable calcium, magnesium, sodium, and CEC. It, however, caused an increase in sand, aggregate stability, and electrical conductivity and a partial increase in exchangeable potassium, available phosphorus, and loss on ignition at some temperatures. The nutrient most vulnerable to losses due to fire is nitrogen, which can be volatilized at relatively low temperatures (200–500°C). Phosphorus can be volatilized at high burn temperatures (770°C), whereas other mineral nutrients such as calcium, magnesium, and potassium are typically converted to oxides that are relatively soluble. Water repellency results when volatilized organic compounds condense on cooler soil particles associated with steep temperature gradients below the soil surface. This results in reduction in infiltration and percolation. Reductions in the infiltration rate can be dramatic (e.g., one to two orders of magnitude), and sometimes produce very rapid surface runoff. Any mineral soil containing more than a 2% organic matter is likely to become water repellent to some degree when heated (DeBano 1981).

Soils are critical to the functioning of hydrological processes. Within a watershed, sediment and water responses to wildfire are often a function of fire severity and the occurrence of hydrologic events. For a wide range of fire severities, the impacts on hydrology and sediment loss can be minimal in the absence of precipitation. However, when a precipitation event follows a large, high-severity fire, impacts can be substantial. Increased runoff, peak flows, and sediment delivery to streams can affect fish populations and their habitat (Rinne 1997). Loss of soil from hillslopes produces several significant ecosystem impacts. Soil movement into streams, lakes, and riparian zones may degrade water quality and change the geomorphic and hydrologic characteristics of these systems and soil loss from hillslopes may alter future site productivity. The recovery of soil nutrient levels after fires can be fairly slow in some ecosystems, particularly those with limited nitrogen, and in semiarid regions where decomposition rates are slow. Klopatek (1987) determined that 35 years after a wildfire, nitrogen concentrations beneath pinyon-juniper canopies had not recovered to levels found in stands that had not burned in 300 years. In addition, soils beneath burned stands showed a twofold increase over unburned stands in the percent of total nitrogen that was changed from organic forms such as litter and humus to more mobile nitrates and ammonium. While these mobile forms of nitrogen are more accessible for plant uptake, they are also more prone to off-site movement in surface runoff and leaching (Neary et al. 1999). Changes in soil moisture, structure, and infiltration due to a forest fire can accelerate surface runoff, erosion, sediment transport, and deposition. Intense rainfall and some soil and terrain conditions can contribute to overland runoff and in-channel debris torrents. Mineralization of organic matter, interruption of root uptake, and loss of shade can further impact water quality by increasing stream temperatures and nutrient concentrations (Ice et al. 2004).

8.1.5 Deforestation

Deforestation is the removal or damage of forest vegetation to the extent that it no more supports its natural flora and fauna. In other words, deforestation is the transformation of forest land to non-forest land such as farm land, ranches, pasture, industrial complexes, and urban areas (Fig. 8.8). Forest land that includes lands under agroforestry and shifting cultivation is not a closed-canopy primary forest. Deforestation is a form of land degradation that affects the livelihoods of people in general and the rural poor in particular.

The disappearance of natural forests in developing countries is a major problem because it negatively affects the livelihoods of people dependent on forest products and services. Deforestation is the process of reducing tree canopy cover to less than 10% threshold by natural and human actions. By deforestation, a forest land is converted into a non-forest land. Almost half of the forests that originally covered the earth are already lost, and further 16 million ha disappear annually. The underlying causes of contemporary deforestation include overpopulation, poverty, lack of governance, unplanned human settlement, industrialization, and urbanization. The processes of deforestation are industrial logging, clear felling, indiscriminate use of forest materials, forest fires, shifting cultivation, encroachment, etc.

The effects of deforestation include loss of forest, loss of biodiversity, climate change, and natural disasters such as cyclones, flooding, and drought, disruption of water cycle, decline in water quality, soil erosion, and sedimentation. The rate of deforestation in the tropics is currently estimated to be about 2% of the remaining forest cover per year (Williams 1990, Myers 1991). In the neotropics, some of this deforested land is regrowing into a successional, secondary forest, some is converted to tillage agriculture (Sanchez et al. 1982), but much is being converted to pasture for cattle ranching (Fearnside 1985, 1990; Buschbacher et al. 1987). Deforestation has been extensive in North Africa, once known as "a land of continuous shade" (Thirghood 1981). At present, only 12% of total surface area of Morocco, 4% of Tunisia, and only 2% of Algeria contain forest (Salah 1994). Moroccan forests have decreased by about 5 million ha since the Roman period, and Algeria lost about 2 million ha of forests during the French occupation. More damage occurred during the War of Independence during 1954–1962. Between 1956 and 1960, 650,000 ha of forests were burned (RADP 1966). Algeria is losing about 40,000 ha forest annually, while Morocco is losing an average of 13,000 ha annually, and Tunisia 5000 ha (UNDP 1990). This change in land use can be significant at both the local and global scale in terms of: (1) land–atmosphere climate interactions (Salati and Nobre 1991); (2) hydrology and sediment transport (Meher-Homji 1991); (3) the global carbon budget (Houghton 1991); (4) atmospheric chemistry (Keller et al. 1991); (5) soil properties (Spaans et al. 1989); (6) sustainable productivity (Sinha and Swaminathan 1991); and (7) biodiversity (Ehrlich and Wilson 1991).

Forests absorb, utilize, and store great amounts of water quickly when there are heavy rains. So, trees are regulators of the hydrologic processes, including evaporation, transpiration, infiltration, and surface flow. A large part of the water that circulates in the tropical rainforest ecosystem, for instance, remains inside the plants. When forests are cleared, the climate around gets drier and the bare soil is exposed to the scorching effect of the sun and the beating action of the rains. Rainfall washes out significant amount of top soil. Due to the elevated temperature, SOM is decomposed at a faster rate. Soil aggregates are broken down by the raindrop impact. Infiltration rate is reduced and more water runs off. Moving water concentrates in narrow channels in absence of tree roots and trunks. For all these, large tracts of land become permanently impoverished due to soil erosion. The rate of increase of soil loss after forest clearing is astonishing; a study in Ivory Coast found that forested slope areas lost 0.03 Mg ha^{-1} yr^{-1}; cultivated slopes annually lost 90 Mg ha^{-1} yr^{-1} while bare slopes lost 138 Mg ha^{-1} yr^{-1} (Butler 2006).

8.1.5.1 Processes of Deforestation in Different Regions

Land clearing for agriculture is the primary driver of deforestation in boreal forests. Another important process of boreal deforestation is timber harvesting. Logging, fires, and diseases and insects are other causes. Boreal forests are not typically managed intensively for timber production, but the substantial volumes of standing timber are harvested extensively for wood products in some regions such as Scandinavia, western Russia, and parts of Canada. Virtually no regeneration of the boreal forests takes place in Eastern Canada following wildfires over the past 900 years, with a commensurate decline in forest cover. In temperate regions, forest clearing for agriculture is a historical cause of deforestation in many areas in the recent past. Clearing land for agriculture might still be a cause of deforestation in temperate developing countries. A more common cause of temperate deforestation, particularly in developed countries, is conversion of land to nonagricultural uses, notably residential and commercial development and infrastructure. Timber harvesting and natural disasters are also causes of temperate deforestation. In many areas, timber harvesting is followed by reforestation, resulting in no net deforestation over time. Agriculture, including shifting cultivation and small-scale and large-scale permanent agriculture, and wood extraction, including logging and fuelwood harvests, are the principal causes of deforestation. The anthropogenic drivers of deforestation vary among regions, and differ between rainforests and dry tropical forests. The socioeconomic conditions of tropical regions are the principal underlying causes of deforestation. Most tropical countries are very densely populated and the people are generally poor. They deforest and degrade forests to get an easy way of earning their livelihood. Shifting cultivation, livestock grazing, fuelwood collection, fodder extraction, and burning to promote grasses for fodder are common historic and continuing uses of the forest (Bhat et al. 2001, Saha 2002, Kodandapani et al. 2004, FAO 2006). These anthropogenic pressures from communities residing within and on the fringes of forest reserves continue to be enormous. In an estimate, Madhusudan (2005) reported that more than

Fig. 8.9 Remnants of a tropical rainforest. Vast hilly areas of Bangladesh have been denuded by repeated shifting cultivation, burning, fuel biomass collection, and erosion. (Photo courtesy of Mr. Azad Kamal)

5 million people have been residing within a countrywide network of 593 wildlife reserves and national parks of India. Consequently, livestock grazing, fuelwood, fodder extraction, and burning are recognized as a chronic disturbance that can have substantial impacts on the entire forest ecosystem (Tilman and Lehman 2001) including impacts on vegetation, soil and water resources, fauna, and microclimate. In some areas, rural people and forest dwellers disturb the same forest repeatedly by a sequence of processes. The land is cleared first for shifting cultivation, left for some time, and the saplings of the regeneration are cut for fuel wood. The bushes are burned and herbs and shrubs are cut for fuel. Heavy grazing follows and, as a result, vast areas of forested lands, as in Chittagong Hill Tracts of Bangladesh, have been deforested, eroded, and denuded (Fig. 8.9).

8.1.5.2 Effects of Deforestation on Forest Soils

The effects of converting lowland tropical rainforest to pasture, and of subsequent succession of pasture lands to secondary forest, were examined in the Atlantic Zone of Costa Rica by Reiners et al. (1994). Four land-use types representing disturbance–recovery sequence were sampled for changes in pedological properties, and potential nitrogen mineralization and nitrification. The four land-use types included primary forest, actively grazed pasture (10–36-year old), abandoned pasture (abandoned 4–10 year), and secondary forest (abandoned 10–20 year). Conversion and succession had obvious and significant effects on soil properties including (1) a decrease in acidity and increase in some base-exchange properties, (2) an increase in bulk density and a concomitant decrease in porosity, (3) higher concentrations of NH_4^+, (4) lower concentrations of NO_3^-, (5) lower rates of N mineralization, and (6) in some cases, lower rates of nitrification. Physical changes in density and porosity were deleterious with respect to infiltration, percolation, aeration, and, ultimately, erodibility. Burton et al. (1989) observed that forest degradation and conversion of forested land to agriculture led to losses in soil fertility during an investigation on soil properties of some land-use categories of Nepal. Soil fertility as expressed by organic carbon, total nitrogen, and CEC decreased when natural productive forest was converted into agriculture. There was also a decline in soil quality when natural forests became degraded and overutilized. Soil quality is here defined as the capacity of the soil to support plant growth without causing degradation of the soil or the environment (Carter et al. 1997, Doran and Parkin 1994). The term encompasses physical, chemical, and biological parameters such as contents of nutrients and SOM as well as pH level, pH buffering capacity, soil structure, and microbial activity. Soil quality is closely linked to soil resilience which refers to the ability of the soil to restore soil functions following disturbance—resilient soils have a high soil quality and vice versa (Lal 1997, Lal and Bruce 1999).

Forest clearance for agriculture in the Loess Plateau of Ziwuling area, China decreased biodiversity, limited naturally adapted vegetation, and simplified the ecosystem structure. The impacts of human activity on the Loess Plateau have been characterized as continuous and widespread stresses such as overgrazing and large-scale monocultures of wheat and maize on deforested land (Fu et al. 2000). Zheng (2005) reported that about 70 % of the total area of the Loess Plateau is affected by moderate to severe soil erosion. Soil erosion rate estimates range from 5000 to 25000 Mg ha^{-1} $year^{-1}$ (Fu et al. 2005). Sanchez et al. (2002) found that erosion rates had increased from 540 kg ha^{-1} $year^{-1}$ to 1110 kg ha^{-1} $year^{-1}$ when forest land was converted to pasture land on steep slopes in the Venezuelan Andes. They attributed the increase in erosion across land types to decreasing amounts of organic matter and lack of vegetative cover to intercept rainfall in the deforested areas. In the Hubbard Brook watershed experiments, Bormann et al. (1974) recorded the levels of erosion in terms of particulate matter output after clear-cutting a watershed and suppressing regeneration. In 5 years, the particulate matter output increased exponentially from 16.20 kg ha^{-1} $year^{-1}$ to 378.80 kg ha^{-1} $year^{-1}$ while the levels in a similar watershed remained relatively stable at an average of 25.42 kg ha^{-1} per year over the same period. In agreement with the bulk density levels found in the pasture, forest, and coppice in the current study, many other studies have also found bulk density to be higher in pastures, grasslands, and disturbed sites compared to natural forests (Eden et al. 1991, Hajabbasi et al. 1997, Koutika et al. 1997, Abril and Bucher 2001, Desjardins et al. 2004, Celik 2005, Abbasi et al. 2007, Zou et al. 2007, Basaran et al. 2008).

The SOC contents in different land-use classes of Armenia were reported by Rhoades (2008). He found 8.2 % organic carbon in the forest soil, 6.3 % in the regenerated coppice soil, and 5.9 % in the pasture soil. Organic matter finds its way to the soil primarily through vegetative residues such as litters, dead organisms, and roots, and a decrease in these inputs due to forest clearance can lead to a decrease in SOC and SOM (Lugo et al. 1986, Bernoux et al. 1998). Organic matter usually accumulates in the top soil, and therefore their abundance can be reduced by erosion (Dominski 1971, Murty et al. 2002, Abbasi and Rasool 2005, Zheng et al. 2005). The unplanned and unregulated clear-cutting of the coppice and the removal of all tree biomass for use as fuelwood removed organic material high in nutrients (Schwendenmann and Pendall 2006, Yimer et al. 2007, Khresat et al. 2008).

In the Neotropics, forests are often cleared by small farmers using traditional slash-and-burn agriculture (Rudel and Horowitz 1993). After a few years, the original fertility of these nutrient-poor soils decreases (Uhl 1987) and the areas are turned to pastures for cattle ranching (Myers 1989, Fearnside 1990). Secondary forests can recover in these areas (Gomez-Pompa and Vasquez-Yanes 1974) if the grasslands are not too large or if fire is not frequently used (Buschbacher 1986, Uhl 1988). Cavelier et al. (1999) studied the effects of long-term deforestation on soil physical and chemical properties in the Q. Yepes watershed in the Sierra Nevada de Santa Marta, Colombia. They suggested that deforestation since pre-Columbian times, and intensive grazing and use of fire since the arrival of the Spaniards have caused severe soil degradation. Erosion has removed approximately 50 cm of top soil in areas currently covered by grasslands. The grassland soils have lower cation concentrations, higher levels of exchangeable Al, lower pH, and lower carbon concentrations compared with soils in forest patches.

Throughout the tropics, more specifically in the Northeast Queensland wet tropical region of Australia, extensive forested ecosystems have been deforested for sugarcane, banana, and pasture, both for dairy and beef cattle production (Tucker and Murphy 1997, Tucker 2000). Clearing of woody vegetation for pasture, intensive cropping, and urban development is still continuing there (Brodie 2003). Studies have shown that deforestation and subsequent cropping decreased plant-available nutrients (Keenan et al. 1998, Lu et al. 2002), reduced microbial activity (Sahani and Behera 2001, Kiese et al. 2002), increased bulk density, soil erosion, and runoff (Rasiah and Kay 1995, Jusoff 1989), and decreased porosity, infiltration and water-holding capacity (Sahani and Behera 2001, Lu et al. 2002). Deforestation and the subsequent cropping lead to the deterioration in soil physical, chemical, and biological properties, but when some of these degraded soil are brought under well-managed pasture, significant improvements in soil properties occurred (Tisadall and Oades 1980, Rasiah and Kay 1995). In a study in Australia, Rasiah et al. (2009) observed that only insufficient improvements in nitrate-N, ammonium-N, total N, total SOC, labile C, exchangeable cations, exchangeable Al, pH, bulk density, and porosity had occurred even after abandoning a pasture for 40 years. Exchangeable Al, bulk density, and acidity increased and C, N, and CEC decreased. It indicated that the resilience of the soil in the particular tropical environment to recover from deforestation-induced land degradation was low. In 10-year-old planted forests, the recovery from degradation was not expected to be significant. The inability of the soil to recover substantially under a relatively long period of abandoned pasture indicates that extreme care should be taken in the management of the deforested lands in similar tropical environments.

Study Questions

1. What do you mean by forest disturbance? How can some disturbances be beneficial for forests? What are the reasons of intensifying forest disturbance at present? Discuss the consequences of forest disturbances.
2. Mention the demands of humankind on forests. How do forest dwellers impact forest vegetation and forest soil?
3. Distinguish between clear-cutting and deforestation. Why does some soil degradation take place during timber harvest? Suggest how harvest-related soil degradation can be minimized.
4. What do you mean by wildfire? How does it differ from prescribed fire? Narrate the consequences of wildfire on vegetation and soil.
5. What are the causes and processes of deforestation? Discuss the impacts of deforestation on soil.

References

Abbasi MK, Rasool G (2005) Effects of different land-use types on soil quality in the hilly area of Rawalakot Azad Jammu and Kashmir. Acta Agriculturae Scandinavica 55(3):221–228

Abbasi MK, Zafar M, Khan SR (2007) Influence of different land-cover types on the changes of selected soil properties in the mountain region of Rawalakot Azad Jammu and Kashmir. Nutr Cycl Agroecosyst 78(1):97–110

Abril A, Bucher EH (2001) Overgrazing and soil carbon dynamics in the western Chaco of Argentina. Appl Soil Ecol 16:243–249

Alake B, Alamgir M, Haque SMS, Osman KT (2011) Study of soils under shifting cultivation and other land use categories in Chittagong Hill Tracts, Bangladesh. J For Res 22(3):623–630

Angelsen A (1994) Shifting Cultivation and "Deforestation" A Study from Sumatra, Indonesia. Working Paper, Chr. Michelsen Institute Development Studies and Human Rights, Bergen, Norway

Asner GP, Broadbent EN, Oliveira PJC, Keller M, Knapp DE, Silva JNM (2006) Proc Natl Acad Sci 103:12947–12950

Auld TD, Denham AJ (2006) How much seed remains in the soil after a fire? Plant Ecol 187:15–24

Basaran M, Erpul G, Tercan AE, Canga MR (2008) The effects of land use changes on some soil properties in Indagi Mountain Pass—Cankiri Turkey. Env Monit Ass 136:101–119

Beach T (1998) Soil catenas, tropical deforestation, and ancient and contemporary soil erosion in the Peten, Guatemala. Phys Geogr 19:378–405

Benito G, Gutiérrez M, Sancho C (1991) Erosion patterns in rill and interill areas in badland zones of the Middle Ebro Basin, (NE-Spain). In: Sala M et al. (ed) Soil erosion studies in Spain. Geoforma ed, Logrono

Bernoux M, Cerri CC, Jener CN, de Moraes FL (1998) The use of stable carbon isotopes for estimating soil organic matter turnover rates. Geoderma 82(1–3):43–58

Bhat DM, Murali KS, Ravindranath NH (2001) Formation and recovery of secondary forests in India: a Particular reference to Western Ghats in South India. J Trop For Sci 13(4):601–620

Boerner REC, Hart S, Huang J (2009) Impacts of fire and fire surrogate treatments. Ecological Applications 19(2):338–358

Bormann FH, Likens GE, Siccama TG, Pierce RS, Eaton JS (1974) The export of nutrients and recovery of stable conditions following deforestation at Hubbard Brook. Ecol Monogr 44 (3):255–277

Brodie J (2003) The effects of land use on water quality in Australian Northeast coastal catchments. Australian centre for tropical freshwater research. James Cook University, Townsville.

Brown G, Tolsma A, Murphy S, Miehs A, McNabb E, York A (2009) Ecological impacts of firewood collection—a literature review to inform firewood management on public land in Victoria. Victorian Government Department of Sustainability and Environment, Melbourne

Bruun TB, de Neergaard A, Lawrence D, Ziegler AD (2009) Environmental consequences of the demise in swidden cultivation in Southeast Asia: carbon storage and soil quality. Humn Ecol 37:375–388

Burton S, Shah PB, Schreier H (1989) Soil degradation from converting forest land into agriculture in the Chitawan district of Nepal. Mt Res Dev 9(4):393–404

Buschbacher RJ (1986) Tropical deforestation and pasture development. Bioscience 36:22–28

Buschbacher RJ, Uhl C, Serrao EAS (1987) Large-scale development in eastern Amazonia. In: Jordan CF (ed) Amazonian rain forests. Ecosystem disturbance and recovery. Springer-Verlag, New York

Butler R (2006) A place out of time: tropical rainforests and the perils they Face. http://www.mongabay.com/. (Accessed on 12 Feb, 2012)

Campbell GS, Jungbauer JD Jr, Bidlake WR, Hungerford RD (1994) Predicting the effect of temperature on soil thermal conductivity. Soil Sci 158:307–313

Carter MR, Gregorich EG, Anderson DW, Doran JW, Janzen HH, Pierce FJ (1997) Concepts of soil quality and their significance. In Gregorich EG, Carter MR (eds) Soil quality for crop production and ecosystem health. Elsevier, Amsterdam

Cavelier J, Mitchell T, Aide M, DupuyJM, Eusse AM, Santos C (1999) Long-term effects of deforestation on soil properties and vegetation in a tropical lowland forest in Colombia. Ecotropicos 12(2):57–68

Celik I (2005) Land-use effects on organic matter and physical properties of soil in a southern Mediterranean highland of Turkey. Soil Till Res 83:270–277

Certini G (2005) Effects of fire on properties of forest soils: a review. Oecologia 143:1–10

Chandler C, Cheney P, Thomas L, Williams D (1983) Fire in foresty Vol 1. Forest fire behaviour and effects. John Wiley & Sons, New York

Craswell ET, Sajjapongse A, Howlett DJB, Dowling AJ (1997) Agroforestry in the management of sloping lands in Asia and the Pacific. Agrofor Syst 38:121–137

DeBano LF (1981) Water repellant soils: a state-of-the-art. Gen Tech Rep PSW-46. US Department of Agriculture, Forest Service, Pacific Southwest Forest and Range Experiment Station, Berkeley, CA

DeBano LF (2000) The role of fire and soil heating on water repellence in wildland environments: a review. J Hydrol 231:195–206

DeBano LF, Neary DG, Ffolliott PF (1998) Fire effects on ecosystems. Wiley, New York

Desjardins T, Barros E, Sarrazin M, Girardin C, Mariotti A (2004) Effects of forest conversion to pasture on soil carbon content and dynamics in Brazilian Amazonia. Agric Ecosyst Env 103:365–373

Detwiler RP, Hall CAS (1988) Tropical forest and global carbon cycle. Science 239:42–47

Diaz-Fierros F, Benito E, Perez R (1987) Evaluation of the USLE for prediction of erosion in burnt forest areas in Galicia (N.W. Spain). Catena 14:189–199

Dominski AS (1971) Nitrogen transformations in a northern hardwood podzol on cutover and forested sites. Dissertation, Yale University, New Haven, Connecticut, USA

Doran JW, Parkin TB (1994) Defining and assessing soil quality. In: Doran JW (ed.) Defining Soil Quality for a Sustainable Environment. Soil Science Society of America Special Publication No. 35. Soil Science Society of America and American Society of Agronomy, Madison, WI, pp. 3 – 21

Eastmond A, Faust B (2006) Farmers, fires, and forests: a green alternative to shifting cultivation for conservation of the Maya forest. Landsc Urban Plan 74:267–284

Eden MJ, Furley PA, McGregor DFM, Milliken W, Ratter JA (1991) Effect of forest clearance and burning on soil properties in northern Roraima, Brazil. For Ecol Manage 38:283–290

Ehrlich PR, Wilson EO (1991) Biodiversity studies: science and policy. Science 253:758–762

FAO (2006) Global forest resources assessment 2005. Progress towards sustainable forest management. FAO Forestry Paper 147. Food and Agriculture Organization of the United Nations: Rome

Fearnside PM (1985) Predominant land uses in Brazilian Amazonia. In: Anderson AB (ed) Alternatives to deforestation: steps toward sustainable use of the Amazon rain forest. Columbia University Press, New York

Fearnside PM (1990) Predominant land uses in Brazilian Amazonia. In: Anderson AB (ed) Alternatives to deforestation: steps toward sustainable use of the Amazon rain forest. Columbia University Press, New York, USA

Forsyth TJ (1994) The use of Cesium-137 measurements of soil erosion and farmer's perceptions to indicate land degradation amongst shifting cultivators in northern Thailand. Mountain Res Dev 14:229–244

Franklin JF, Berg DR, Thornburgh DA, TappeinerJC (1997) Alternative silvicultural approaches to timber harvesting: variable retention harvest systems. In: Kohm KA, Franklin JF (eds) Creating a forestry for the 21st century. Island Press, Washington, DC

Frelich LE, Reich PB (1998) Disturbance severity and threshold responses in the boreal forest. Conservation Ecology [online] 2(2):7

Fu BJ, Chen LD, Ma KM (2000) The relationship between land use and soil conditions in the hilly area of Loess Plateau in Northern Shanxi. Catena 39:69–77

Gafur A (2001) Effects of shifting cultivation on soil properties, erosion, nutrient depletion, and hydrological responses in small watershed of the Chittagong Hill Tracts of Bangladesh. PhD Dissertation, The Royal Veterinary and Agricultural University, Copenhagen

Gafur A, Borggaard OK, Jensen JR, Petersen L (2000) Changes in soil nutrient content under shifting cultivation in the Chittagong Hill Tracts of Bangladesh. Geografisk Tidsskrift, Danish J Geogr 100:37–46

Giovannini G (1991) The effects of fire on soil quality. Proceedings of the conference on soil erosion and degradation as a consequence of forest fires. September 1991, Barcelona and Valencia

Gomez-Pompa A, Vasquez-Yanes C (1974) Studies on the secondary succession of tropical lowlands: the life cycle of secondary species. In: van Dobben WH, Lowe-Mcconell (eds) Proceedings of the first international congress of ecology, Junk, The Hague

Grogan P, Lalnunmawia F, Tripathi SK (2012) Shifting cultivation in steeply sloped regions: a review of management options and research priorities for Mizoram state, Northeast India. Agroforest Syst 84:163–177

Hajabbasi MA, Jalalian A, Karimzadeh HR (1997) Deforestation effects on soil physical and chemical properties, Lordegan, Iran. Plant Soil 190:301–308

Hatch T (1982) Shifting cultivation in Saravak (a review), Technical paper no 8, soil division research branch. Department of Agriculture, Saravak

Hoinka KP, Carvalho A, Miranda AI (2009) Regional-scale weather patterns and wildland fires in central Portugal. Int J Wildland Fire 18:36–49

Houghton RA (1991) Tropical deforestation and atmospheric carbon cycle. Climate Change 19:99–118

Ice GG, Neary DG, Adams PW (2004) Effects of wildfire on soils and watershed processes.J Forest 102: 16–20

ICRAF (1998) The alternatives to Slash-and-Burn Programme. http://198.93.235.8/training/icraft/sys_wide/asb1.htn

Jusoff K (1989) Physical soil-properties associated with recreational use of forested reserve area in Malaysia. Environ Conserv 16:339–342

Kaimowitz D (2003) Forest law enforcement and rural livelihoods. Int Forest Rev 5(3):199–210

Keenan RJ, Hambleton A, Robson K, Webb M (1998) Growth response of three rainforest cabinet timber to fertilizer application in plantations in North Queensland. In: Schulte A, Ruhiyat D (eds) Soils of tropical forest ecosystems: characteristics, ecology and management. Springer, Berlin

Keller M, Jacobs DJ, Wofsy SC, Harris RC (1991) Effects of tropical deforestation on global and regional atmospheric chemistry. Climate change 19:139–158

Khresat S, Al-Bakri J, Al-Tahhan R (2008) Impacts of land use/cover change on soil properties in the Mediterranean region of northwestern Jordan. Land Degrad Dev 19(4):397–407

Kiese K, Papen H, Zumbusch E, Butterbach-Bahl L (2002) Nitrification activity in tropical rainforest soils of the coastal lowlands and Atherton Tablelands, Queensland, Australia. J Plant Nutr 165, 682–685

Klopatek JM (1987) Nitrogen mineralization and nitrification in mineral soils of pinyon ± juniper ecosystems. Soil Sci Soc Am J 51:453–457

Kodandapani N, Cochrane MA, Sukumar R (2004) Conservation threat of increasing fire frequencies in the Western Ghats, India. Conserv Biol 18(6):1553–1561

Koutika LS, Bartoli F, Andreux F, Cerri CC, Burtin G, Chone T, Philippy R (1997) Organic matter dynamics and aggregation in soils under rain forest and pastures of increasing age in the eastern Amazon Basin. Geoderma 76:87–112

Koziell I, Saunders J (eds) (2001) Living off biodiversity: exploring livelihoods and biodiversity. IIED, London

Kurz WA, Apps MJ (1999) A 70-year retrospective analysis of carbon fluxes in the Canadian forest sector. Ecol Appl 9:526–547

Kurz WA, Apps MJ, Beukema SJ, Lekstrum T (1995) 20th century carbon budget of Canadian forests. Tellus 47B:170–177

Lal R (1987) Need for approaches to and consequences of land clearing and development in the tropics. In: IBSRAM (ed) Tropical land clearing for sustainable agriculture. Proceedings of an IBSRAM inaugural workshop held in Jakarta and Bukittingi, Indonesia 27 Aug–3 Sept, 1985. IBSRAM, Bangkok

Lal R (1997) Degradation and resilience of soils. Philosophical transactions of the Royal Society of London Series B—Biological Sciences 352:869–889

Lal R, Bruce JP (1999) The potential of world cropland soils to sequester C and mitigate the greenhouse effect. Environ Sci Policy 2:177–185

Laughlin DC, Bakker JD, Stoddard MT, Daniels ML, Springer JD, Gildar CN, Green AM, Covington W (2004) Toward references conditions: Wildfire effects on flora in an old-growth ponderosa pine forest. Forest Ecol Manag 119:137–142

Lawrence D, Peart DR, Leighton M (1998) The impact of shifting cultivation on a rainforest landscape in West Kalimantan: spatial and temporal dynamics. Landscape Ecol 13:135–148

Lawrence D, Schlesinger WH (2001) Changes in soil phosphorus during 200 years of shifting cultivation in Indonesia. Ecology 82:2769–2780

Leite FF, Gonçalves AB, Vieira A, da Vinhac L (2011) The frequency of fires and land degradation—Cabeço da Vessada do Monte, Serra da Cabreira. In: Gonçalves AB, Vieira A (eds) Fire effects on soil properties. Proceedings of the 3rd international meeting of fire effects on soil properties. 15–19 March 2011, University of Minho, Guimaraes, Portugal

Leite FF, Gonçalves BAJ (2008) Alterações no ciclo de fogo na serra da Cabreira. Actas XI Colóquio Ibérico de Geografia, Universidad de Alcalá de Henares, Espanha

Lu D, Moran E, Mausel P (2002) Linking Amazonian secondary succession forest growth to soil properties. Land Degrad Dev 13:331–343

Lugo AE, Sanchez MJ, Brown S (1986) Land use and organic carbon content of some subtropical soils. Plant and Soil 96:185–196

Madhusudan MD (2005) The global village: linkages between international coffee markets and grazing by livestock in a south Indian wildlife reserve. Conserv Biol 19(2):411–420

Mangas VJ, Sánchez JR, Ortiz C (1992) Effects of a fire on runoff and erosion on Mediterranean forest soils in SE-Spain. Pirineos 140:37–51

McMahon S, Simcock R, Dando J, Ross C (1999) A fresh look at operational soil compaction. New Zeal J Forest 44(3): 33–37

McNabb DH, Swanson FJ (1990) Chap. 14. Effects of fire on soil erosion. In: Walstad JD, et al. (ed) Natural and prescribed fire in Pacific Northwest forests. Oregon State University Press, Corvallis

Megahan WF, Seyedbagheri KA, Mosko TL, Ketcheson GL (1986) Construction phase sediment budget for forest roads on granitic slopes in Idaho. In: Hadley RF (ed) Drainage basin sediment delivery, proceedings. Albuquerque, IAHS Publication 159. Wajlingford, Oxon

Meher-Homji VM (1991) Probable impact of deforestation on hydrological processes. Clim Change 19:163–173

Minore D, Weatherley HG (1990) Effects of site preparation on Douglas-fir seedling growth and survival. West J Appl For 5(2):49–51

Murphy G (1984) A survey of soil disturbance caused by harvesting machinery in New Zealand plantation forests. Bulletin 69, Forest Research Institute, New Zealand

Murty D, Kirschbaum MUF, McMurtrie RE, McGilvary H (2002) Does conversion of forest to agricultural land change soil carbon and nitrogen? A review of the literature. Glob Change Biol 8:105–123

Myers N (1989) Deforestation rates in tropical forest and their climatic implications. Friends of the Earth, London

Myers N (1991) Tropical forests: present status and future outlook. Climatic Change 19:3–32

Myers N (1993) Tropical forests: the main deforestation fronts. Env Conserv 20:9–16

Nagaland Environmental Protection and Economic Development (NEPED), International Institute of Rural Reconstruction (IIRR) (1999) Building upon traditional agriculture in Nagaland, India. Nagaland, India and Silang, Cavite

Neary DG, Klopatek CC, DeBano LF, Ffolliott PF (1999) Fire effects on belowground sustainability: a review and synthesis. For Ecol Manag 122:51–71

Nelson BW (1994) Natural forest disturbance and change in the Brazilian Amazon. Remote Sens Rev 10(1–3):105–125

Overpeck JT, Rind D, Goldberg R (1990) Climate-induced changes in forest disturbance and vegetation. Nature 343:51–53

Parlak M (2011) Effect of heating on some physical, chemical and mineralogical aspects of forest soil. Bartın Orman Fakültesi Dergisi 19:143–152

Radho-Toly S, Majer JD, Yates C (2001) Impact of fire on leaf nutrients, arthropod fauna and herbivory of native and exotics eucalypts in Kings Park, Petrh, Western Australia. Australia Ecol 26:500–506

RADP (Republique Algerienne Democratique et Populaire). (1966) MARA (Ministere de I'Agriculturee t de la ReformeA graire) La Statistique A gricole. No 1. Algiers

Raison RJ, Khana PK, Woods PV (1985) Transfer of elements to the atmosphere during low-intensity prescribed fires in three Australian subalpine eucalypt forests. Can J For Res 15(1):657–664

Rasiah V, Florentine SK, Williams BL, Westbrooke ME (2009) The impact of deforestation and pasture abandonment on soil properties in the wet tropics of Australia. Geoderma ISS 120:35–45

Rasiah V, Kay BD (1995) Runoff and soil loss as influenced by selected stability parameters and cropping and tillage practices. Geoderma 68:321–329

Reeves D (2011) Soil disturbance monitoring of timber harvest operations in the USDA Forest Service Northern Region. MS Thesis, University of Idaho

Reiners WA, Bouwman AF, Paesons WFJ, Keller M (1994) Tropical rain forest conversion to pasture: changes in vegetation and soil properties. Ecol Appl 4(2):363–377

Rhoades JL (2008) Impacts of deforestation and land cover change on mountain soils in Hrazdan, Armenia. MS (For) thesis, Michigan Techological University, USA

Rinne JN (1997) Short†term effects of wildfire on fishes and aquatic macroinvertebrates in the southwestern United States. N Am J Fish Manage 16:653–658

Riswan R, Hartanti L (1995) Human impacts on tropical forest dynamics. Vegetatio 121:41–52

Roder W, Phengchanh S, Maniphone S (1997) Dynamics of soil and vegetation during crop and fallow period in slash-and-burn fields of northern Laos. Geoderma 76:131–144

Rudel KT, Horowitz B (1993) Tropical deforestation. Small farmers and land use clearing in the Ecuatorian Amazon. Columbia University Press, New York

Saha S (2002) Anthropogenic fire regime in a tropical deciduous forest of central India. Current Science 82:101–104

Sahani U, Behera N (2001) Impact of deforestation on soil physicochemical characteristics, microbial biomass and microbial activity of tropical soil. Land Degrad Dev 12:93–105

Salacedo IH, Tiessen H, Sampaio EVSB (1997) Nutrient availability in soil samples from shifting cultivation sites in the semi-arid Caatinga of NE Brazil. Agric Ecosyst Environ 65:177–186

Salah EZ (1994) The consequences of rapid deforestation: a North African example. Ambio 23(2):136–140

Salati E, Nobre CA (1991) Possible climatic impacts of tropical deforestation. Climatic Change 19:177–196

Sanchez LA, Ataroff M, Lopez R (2002) Soil erosion under different vegetation covers in the Venezuelan Andes. Environmentalist 22(2):161–172

Sanchez PA, Bandy DE, Villachica JH, Nicholaides JJ (1982) Amazon basin soils: management for continuous crop production. Science 216:821–827

Schwendenmann L, Pendall E (2006) Effects of forest conversion into grassland on soil aggregate structure and carbon storage in Panama: evidence from soil carbon fractionation and stable isotopes. Plant Soil 288:217–232

Sevink J, Imenson AC, Verstraten JM (1989) Humus form development and hillslope runoff, and the effects of fire management, under Mediterranean forest in NE-Spain. Catena 16:461–475

Shakesby RA, Doerr SH (2006) Wildfire as a hydrological and geomorphological agent. Earth-Sci Rev 74:269–307

Singh J, Bora IP, Baruah A, Hussain M (2003) Effect of shifting cultivation on nutrient status of soil in Silonijan (Karbi-Anglong) Assam. Indian For 129(11):1329–1338

Sinha SK, Swaminathan MS (1991) Deforestation, climate change and sustainable nutrition security: a case study of India. Clim Change 19:201–209

Skinner MF, Murphy G, Robertson ED, Firth JG (1989) Deleterious effects of soil disturbance on soil properties and the subsequent early growth of second-rotation radiata pine. In: Dyck WJ, Mees CA (eds) Research strategies for longterm site productivity. Proceedings, IEA/BE A3 Workshop, Seattle, WA, August 1988. IEA/BE A3 Report No. 8. Forest Research Institute, New Zealand

Sodarak H (1999) Shifting cultivation practices by Hmong, Khamu and Lao ethnic categories in the nam Nane watershed, Nane district, Luang Prabang province, Lao PDR. MS thesis, Swedish University of Agricultural Sciences, Department of Rural Development Studies, Uppsala

Soler M, Sala M (1991) Runoff and erosion after clearcutting and burning. In: Conference on Soil Erosion and degradation as a consequence of forest fires. September 1991, Excursion Guide-Book: 24–26, Barcelona-Valencia

Spaans EJ, Baltissen AGAM, Bouma J, Miedema R, Lansu ALE, Schoonderbeek D, Wielemaker WG (1989) Changes in physical properties of young and old volcanic surface soils in Costa Rica after clearing of tropical rain forest. Hydrol Processes 3:383–392

Spencer JE (1966) Shifting cultivation in southeastern Asia. University of California Press, Berkeley

Splechtna BE, Gratzer G (2005) Natural disturbances in Central European forests: approaches and preliminary results from Rothwald, Austria. For Snow Landsc Res 79(1/2):57–67

Suraswadi P, Thomas DE, Pragtong K, Preechapanya P, Weyerhauser H (2000) Changing land use mosaics of (former) shifting cultivators in watersheds of north Thailand. In: Sanchez P (ed) Alternatives to slash-and-burn. American Society of Agronomy, Madison

Tanaka S, Kendawang JJ, Yoshida N, Shibata K, Jee A, Tanaka K, Ninomiya I, Sakurai K (2005) Effects of shifting cultivation on soil ecosystems in Sarawak, Malaysia—IV. Chemical properties of the soils and runoff water at Niah and Bakam experimental sites. Soil Sci Plant Nutr 51:525–533

Tang SM, Franklin JF, Montgomery DR (1997) Forest harvest patterns and landscape disturbance processes. Landscape Ecol 12(6):349–363

Thirghood JV (1981) Man and the African Forest. Academic Press, London

Thomaz EL (2009) The influence of traditional steep land agricultural practices on runoff and soil loss. Agric Ecosyst Environ 130:23–30

Tilman D, Lehman C (2001) Human-caused environmental change: impacts on plant diversity and evolution. Proceedings of the national academy of sciences of the United States of America 98(10):5433–5440

Tisadall JM, Oades JM (1980) The management of ryegrass to stabilize aggregates in a red-brown earth. Aust J Soil Res 18:415–422

Tucker NIJ (2000) Linkage restoration: interpreting fragmentation theory for the design of a rainforest linkage in the humid Wet Tropics of northeast Queensland. Eco Manag Rest 1:35–39

Tucker NIJ, Murphy TM (1997) The effect of ecological rehabilitation of vegetation recruitment: some observations from the wet tropics of north Queensland. J For Ecol Manag 99:133–152

Turkelboom F, Van KK, Ongprasert S, Sutigoolabud P, Pelletier J (1996) The changing landscape of the Northern Thai hills: adaptive strategies to increasing land pressure. Chiang Mai University, Thailand

Turner MG, Baker WL, Peterson CJ, Peet PK (1998) Factors influencing succession: lessons from large, infrequent natural disturbances. Ecosystems 1:511–523

Uhl C (1987) Factors controlling succession following slash-and-burn agriculture in Amazonia. J Ecol 75:377–407

Uhl C (1988) Restoration of degraded lands in the Amazon Basin. In: Wilson EO (ed) Biodiversity. National Academy of Science, Washington DC

UNDP (1990) Human Development Report. Oxford University Press

van Noordwijk M, Cerri C, Woomer PL, Nugroho K, Bernoux M (1997) Soil carbon dynamics in the humid tropical forest zone. Geoderma 79:187–225

Van Rees KCJ, Pennock DJ, Block R (2001) Impacts of forest harvesting on soil compaction and disturbance in North-Central Saskatchewan. Final report submitted to The Prince Albert Model Forest Association

Wairiu M, Lal R (2003) Soil organic carbon in relation to cultivation and topsoil removal on sloping lands of Kolombangara, Solomon Islands. Soil Tillage Res 70:19–27

Walstad JD, Radosevich SR, Sandberg DV (1990) Natural and prescribed fire in the pacific Northwest forest. Oregon State University Press, Corvallis

Wan Razali WM, Tay J (2000) Forestry carbon emission offset and carbon sinks project: examples of opportunity in carbon sequestration and its implications from Kyoto Protocol. International workshop on the response of tropical forest ecosystems to long term cyclic climate change. Science and Technology Agency, Japan

Watters RF (1971) Shifting cultivation in Latin America. Food and Agricultural Organization, Rome

White PS, Pickett STA (1985) Natural disturbance and patch dynamics: an introduction. In: Pickett STA, White PS (eds) The ecology of natural disturbance and patch dynamics. Orlando FL, Academic Press

Williams M (1990) Forests. In: Turner BL II, Clark WC, Kates RW, Richards JF, Mathews JT, Meyer WB (eds) The earth as transformed by human action. Cambridge University Press, England

Wronski EB, Murphy G (1994) Responses of forest crops to soil compaction. In: Soane BD, van Ouwerkerk C (ed) Soil compaction in crop production. Elsevier Science

Yildiz O, Esen D, Sarginci M, Toprak B (2010) Effects of forest fire on soil nutrients in Turkish pine (Pinus brutia, Ten) Ecosystems. J Env Biol 31:11–13

Yimer F, Ledin S, Abdelkair A (2007) Changes in soil carbon and total nitrogen contents in three adjacent land use types in the Bale Mountains, south-eastern highlands of Ethiopia. For Ecol Manage 242:337–342

Zheng F, He X, Gao X, Zhang C, Tang K (2005) Effects of erosion patterns on nutrient loss following deforestation on the Loess Plateau of China. Agriculture Ecosyst Env 108:85–97

Zheng FL (2005) Effects of accelerated soil erosion on soil nutrient loss after deforestation on the loess plateau. Pedosphere 15:707–715

Zhu J, Liu Z (2004) A review on disturbance ecology of forest. Ying Yong Sheng Tai Xue Bao 15(10):1703–10

Ziegler AD, Giambelluca TW, Sutherland RA, Nullet MA, Yarnasarn S, Pinthong J, Preechapanya P, Jaiaree S (2004) Toward understanding the cumulative impacts of roads in upland agricultural watersheds of northern Thailand. Agric Ecosyst Environ 104:145–158

Ziegler AD, Bruun TB, Guardiola-Claramonte M, Giambelluca TW, Lawrence D, Lam NT (2009) Environmental consequences of the demise in Swidden cultivation in montane mainland Southeast Asia: hydrology and geomorphology. Hum Ecol 37:361–373

Zou C, Wang K, Wang T, Xu W (2007) Overgrazing and soil dynamics in eastern Inner Mongolia of China. Ecological Research 22:135–142

Forest Soil and Climate Change

9

Climate change affects forests, and forests affect climate change; they are linked together. Changes in global climate cause stresses on forests through elevated temperatures, altered precipitation patterns, and more frequent and extreme weather events. On the other hand, forests and the wood they produce trap and store carbon dioxide (CO_2), playing a major role in mitigating climate change. Actually, forests play multiple and vital roles in climate change: (i) they are a great reservoir of stabilized carbon, and their carbon sequestration potential can considerably be improved by appropriate management; (ii) they have the potential to absorb about one-tenth of global carbon emissions projected for the first half of this century into their biomass, soils, and products, and store them; (iii) they react sensitively to a changing climate, and when managed sustainably, they produce wood fuels as a benign alternative to fossil fuels; and (iv) they currently contribute about one-sixth of global carbon emissions when cleared, overused, or degraded. Forest soils constitute a large stock of carbon. By soil respiration, decomposition, and burning, a substantial amount of CO_2 would be produced from the soil. Thus, forest soils may act as both a source and a sink of carbon dioxide. At present, however, they are carbon sinks on a global scale. Carbon sequestration in forest soils would decrease the rate of enrichment of atmospheric concentration of CO_2. Increase in C stock of forest soils can be achieved through forest management, including forest conservation, fire management, afforestation, species selection, fertilizer use, and soil amendments. Forest harvesting and deforestation may decrease C stock, at least temporarily.

9.1 Climate Change and Forests

Healthy forests can play a key role in climate change and climate change mitigation, the toughest environmental challenge of the twenty-first century. Through a loss of the climate change mitigation potential of forests, many important ecosystem benefits are also lost. Forests and climate change are inherently linked. As greenhouse gases (GHGs) build up in the atmosphere, they trap heat. This increased heat leads to changes in climate patterns that affect everything on the planet. Carbon dioxide produced by such human actions as burning fossil fuel, deforestation, mismanagement of land and biota for agriculture, etc. is one of the biggest contributors to climate change. By absorbing carbon dioxide and converting it into clean oxygen, trees decrease the concentration of carbon dioxide, which they release, and sequester carbon, which they store. Therefore, healthy forests are the most efficient, inexpensive, and natural system to combat climate change. Forests are important carbon pools that continuously exchange CO_2 with the atmosphere. Forests have several carbon pools: vegetation, dead wood and litter, soil organic matter, and the humus. At the global level, 19% of the carbon in the earth's biosphere is stored in plants, and 81% in the soil. In all forests, tropical, temperate, and boreal together, approximately 31% of the carbon is stored in the biomass and 69% in the soil. In tropical forests, approximately 50% of the carbon is stored in the biomass and 50% in the soil (IPCC 2000).

Forests have a role in climate change through the sequestration or emission of carbon dioxide, and other important GHGs. Biological growth can increase forest stocks, and deforestation can increase carbon emissions. Carbon is captured in tree biomass and in forest soils (Sedjo and Sohngen 2012). Forests account for approximately 30% of terrestrial land cover (FAO 2006) and store about 45% of the carbon in terrestrial ecosystems (Anderegg et al. 2012). An estimate shows that global forest vegetation stores 283 Gt of carbon in its biomass, 38 Gt in dead wood, and 317 Gt in soils (top 30 cm) and litter (however, there may be some variations among different estimates). The total carbon content of forest ecosystems has been estimated at 638 Gt for 2005, which is more than the amount of carbon in the entire atmosphere. This standing carbon is combined with a gross terrestrial uptake of carbon, which was estimated at 2.4 Gt a year, a good deal of which is sequestration by forests. Forests also have a potentially significant role to play in climate change adaptation planning through maintaining ecosystem services

K. T. Osman, *Forest Soils*, DOI 10.1007/978-3-319-02541-4_9, © Springer International Publishing Switzerland 2013

and providing livelihood options (FAO 2005). There is more carbon stock in forests than in other terrestrial ecosystems. Price et al. (2012) suggest that the nature and condition of forests can play a critical role in soil carbon sequestration and storage processes. However, there are concerns about the possible impacts of losing boreal and temperate forests, although existing data show little, if any, net deforestation in these forest biomes. However, deforestation is substantial in tropical forests and it is continuing, and it releases large amounts of CO_2. Actually, forests are carbon sinks in their natural state; they store more carbon than they release (Gorte and Sheikh 2010).

Peterman and Bachelet (2012) indicated that there may be changes in vegetation cover in response to climate change. Changes in structure and composition of some higher latitude forests are already evident. However, forests can either enhance or mitigate climate change on regional scales, but they have the potential to play a significant role in mitigating the pressures of global environmental change (Hoogeveen 2007). Again, forests are biologically diverse systems that are being increasingly threatened due to deforestation, fragmentation, climate change, and other stressors linked to human activities. Climate change may impact forest biodiversity and the ability of forests to provide soil and water protection. Forest ecosystems particularly vulnerable to the impacts of climate change include mangroves, boreal forests, tropical forests, cloud forests, and dry forests. Dry forest soils are particularly susceptible to wind and water erosion. According to the Millennium Ecosystem Assessment, drylands occupy 41 % of the earth's land area and are home to more than 2 billion people. Intensive human intervention, for example, fire, grazing, agriculture, and firewood collection, has adversely transformed many dry forests. Those dry forest systems that have not been completely destroyed are generally impoverished and fragmented. Original vegetation types have been converted to drier, less productive, and less resistant forest types, exposing large numbers of people to the threat of desertification and associated disastrous ecological, social, and economic impacts. On the other hand, primary forests and forests that are adequately managed for diversity and multiple benefits are more resilient to disturbances. These ecosystems maintain healthy, stable soils, provide natural habitats for forest biodiversity, and provide a more stable store of carbon. According to the Intergovernmental Panel on Climate Change, the conservation and restoration of forests can considerably reduce emissions of GHGs (FAO 2005; Millenium Ecosystem Assessment 2005; IPCC 2007).

Forests in peat swamps of tropical montane and lowland areas play a significant role in the global carbon cycle as they store a considerable amount of carbon in their soils (Page et al. 2004; Jaenicke et al. 2008). Peat soils form from decayed woody plant debris decomposing in high-precipitation and -temperature conditions in swamp forests at low elevations in river valleys. Peat swamps dry as a result of logging or draining for agricultural use, but, when these soils dry, they are extremely flammable (Langner et al. 2007; Langner and Siegert 2009). Clearing and burning of peat swamp forest in southeast Asia could contribute to 3 % of total global human emissions (Ballhorn et al. 2009). According to scientific estimates, forest fires in Indonesia in 1998 burned some 8 million ha of land and released between 0.48 and 2.57 Gt of CO_2 into the atmosphere, which is between 13 and 40 % of the mean annual global carbon emissions from fossil fuels. The exact amount remains uncertain (Page et al. 2002).

There is an indication that the Amazon forests will convert to grasslands by the end of the twenty-first century (Cox et al. 2000, 2004) causing dramatic changes in soil and hydrologic conditions. The HadCM3 general circulation model projects that regional warming and drying will also lead to large-scale forest dieback (Betts et al. 2008). Deforestation in the Amazon up to 2001 had reduced the original forest area from 6.2 to 5.4 million km^2. Amazon forests may further be reduced to 3.2 million km^2 by 2050 (Soares-Filho et al. 2006). It is alarming because Amazon forests currently contain 90–140 billion tonnes of carbon, which is about 9–14 decades of the current anthropogenic carbon emissions (Canadell et al. 2007). In the temperate region, the importance of many tree species is changing rapidly. Some tree species are experiencing dieback in response to precipitation and temperature changes, while others are seeing shifts in species dominance. In a regression tree analysis of 80 common trees species under 5 future climate scenarios for 2,100 counties in the eastern USA, Iverson and Prasad (2001) project that average species richness may remain the same or even increase with climate change, but there are likely to be dramatic changes in forest type in this region. As the climate warms, tree species are expected to migrate to higher altitudes and higher latitudes into areas previously characterized by low temperatures. The compositions of high-elevation forests are changing rapidly. Climate projections show the greatest and earliest warming trends at higher latitudes (45–65° N), especially in continental interiors (Kattenberg et al. 1996; Serreze et al. 2000). At the thermal ecotone, small changes in temperature can have large consequences (Serreze et al. 2000). Climate change may affect the ecosystem productivity and allocation of aboveground versus belowground biomass (Joyce and Birdsey 2000). It has been predicted that the standing biomass (aboveground C stock) in minimally disturbed northern forests would increase and the soil organic carbon (SOC) stock would decrease (Morgan et al. 2001). However, the response of forested ecosystems to the projected increase in temperature is complex. Both climate and disturbance interact to influence latitudinal patterns of vegetation and soil C storage (McGuire et al. 2002). The changes in surface hydrology and soil temperature affect forest and tundra vegetation. In Alaska, the tree line is shifting northward into the tundra (Peterman and Bachelet 2012).

Deforestation enhances atmospheric GHG emissions through the combustion of forest biomass and decomposition of plant and soil carbon. Carbon emissions from deforestation and forest degradation have been estimated to account for about 20 % of global anthropogenic CO_2 emissions (Denman et al. 2007; Gibbs and Herold 2007; Schrope 2009). van der Werf et al. (2009) suggested that carbon emissions from both deforestation and fossil fuel combustion in 2008 should be around 12 %. Baccini et al. (2012) reported that deforestation contributes 6–17 % of global anthropogenic CO_2 emissions to the atmosphere. Large uncertainties in emission estimates arise from inadequate data on the carbon density of forests and the regional rates of deforestation. Deforestation is the long-term reduction of tree canopy cover to below 10–30 % (Penman et al. 2003). Deforestation is generally followed by conversion of forestland to cropland or pasture. Forest degradation is partial deforestation. Deforestation in the boreal forest through wildfire, insect or disease infestation, or timber harvest may result in the emission of some of the carbon contained in the trees and soils. As almost five-sixths of the carbon of boreal forest is stored in the soil, soil disturbance is the most important factor for carbon release from these forests. The loss of forest cover from deforestation may accelerate the oxidation of carbon near the soil surface and cause increased emissions. The albedo effect is another concern of boreal deforestation. When boreal forests are cleared, there is a change in the albedo due to changed color of canopy. In addition, snow accumulates in clearings, and reflects more sunlight than snow under the trees. Thus, boreal forest clearings are cooler than the surrounding forest which may slow down decomposition. Temperate forests have lower carbon stock than tropical or boreal forests; therefore, deforestation of temperate forests generally has less severe consequences for climate. However, insect outbreaks and fires in Canadian temperate forests have transformed these forests, since 2000, from a carbon sink to a projected carbon source (Kurz et al. 2008). Timber harvesting and logging could also affect carbon emissions from temperate forests directly. Tropical deforestation is responsible for the largest share of CO_2 released to the atmosphere from land-use changes (IPCC 2007). Clearing tropical forests could release an additional 87–130 Gt C as CO_2 to the atmosphere by 2100. Tropical deforestation, including both the permanent conversion of forests to croplands and pastures and the temporary or partial removal of forests for shifting cultivation and selective logging, is estimated to have released 1–2 Pg C $year^{-1}$ (15–35 % of annual fossil fuel emissions) during the 1990s. The magnitude of emissions depends on the rates of deforestation, the biomass of the forests deforested, and other reductions in biomass that result from forest use. If emissions of methane, nitrous oxide, and other chemically reactive gases are considered in addition to carbon dioxide, annual emissions during the 1990s as a result of deforestation and subsequent uses of the land accounted for about 25 % of the total anthropogenic emissions of GHGs (Houghton 2005). Tropical deforestation can quickly deplete the moderate levels of soil carbon. Additionally, shifting cultivation may cause soil carbon depletion. The soil carbon impacts of tropical deforestation are particularly important for Indonesia, because of its relatively extensive peatlands. Commercial logging in the tropics also affects climate through the relatively large release of carbon, compared to logging in temperate and boreal forests (Gorte and Sheikh 2010). Total GHG emissions from deforestation and degradation that can be attributed to the Dutch economy amounts to 12–25 million tonnes $year^{-1}$ for the period up to 2005. These emissions represent 5–10 % of total Dutch domestic emissions (Croezen and van Valkengoed 2009).

Afforestation and reforestation of degraded land may enhance soil C stock. Akala and Lal (2001) suggested that reforestation of abandoned and marginal agricultural lands can increase SOC stock. Schauvlieghe and Lust (1999) observed total C stock of 128 Mg ha^{-1} under pasture, 173 Mg ha^{-1} under 29-year-old forest, and 232 Mg ha^{-1} under 69-year-old stand in Northern Belgium. The total C stock was 117 Mg ha^{-1} under 27-year-old pin oak stand and 227 Mg ha^{-1} under 69-year-old oak beech stand. Johnston et al. (1996) reported an average increase in SOC stock at the rate of 0.8 Mg C ha^{-1} $year^{-1}$ in the mineral soil over a 40-year period due to afforestation of degraded agricultural soils in east-central Minnesota. Establishing bioenergy plantation crops is another option for enhancing SOC stock and offsetting CO_2 emission. All woody crops sequester considerable organic matter belowground as stumps and roots. McPherson et al. (1997) studied the impact of urban forests on air quality in Chicago. They observed that each year Chicago forests sequester 0.32 Tg C, and contain stock of 5.6 Tg C equivalent to 14–18 mg C ha^{-1}. Large (> 77 cm dbh) and medium-sized (31–46 cm dbh) trees store an average of 3,186 and 399 kg of C and annually sequester 93 and 19 kg of C, respectively.

9.2 Climate Change and Soil

Global temperatures are apprehended to increase by 1.1–6.4 °C during the twenty-first century (Brevik 2012). This increase may have positive and negative impacts on land and biota at regional scales, but the global effect is threatened to be negative. Precipitation patterns will also be altered by climate change. As soils are intricately linked to the atmospheric–climate system through the carbon, nitrogen, and hydrologic cycles, altered climate will also affect soil processes and properties, and, at the same time, the soils themselves will have an effect on climate. Climate change will impact soil organic matter dynamics, and associated soil properties, soil water, and soil erosion. Some soils, particularly of high-latitude regions with currently permanently frozen soils, may become net sources of atmospheric carbon as temperature rises (Brevik 2012).

GHG emissions have alarmingly increased in the past due to land-use changes, cultivation, and burning fossil fuels. Concentrations of CO_2, CH_4, and N_2O in the atmosphere have increased by 37, 156, and 19%, respectively, over 1,750 values. Global temperature has risen by 0.75 °C since 1850 and the projected temperature rise in the twenty-first century ranges from 1.8 to 4 °C. Sea level has risen in many places, arctic ice is melting, and permafrosts are in danger of thawing. Extreme events such as floods and droughts have become more frequent. Climate change has tremendous bearing on land and biota, including soil and forest. Soil carbon is the largest terrestrial pool of carbon (Batjes 1996, p. 2,200). A large stock of carbon (1.5×10^{18} g C) is stabilized in the form of soil organic matter, particularly in peatlands, wetlands, and permafrost. Price et al. (2012) state that carbon that remains in soil does so because it is stabilized by its own intrinsic chemical properties, through production of secondary microbial compounds, by physical separation from microbial breakdown, by molecular interactions with metals or other biomolecules, or by freezing, inundation from flooding, or carbonization. If carbon stored belowground were transferred to the atmosphere by a warming-induced acceleration of its decomposition, a positive feedback to climate change would occur. Melting of permafrost soils with predicted warming trends within the circumpolar region could result in the release of 30–60 Pg C by the year 2040. Such predictions, however, are limited by uncertainty in both climatic changes and soil response to these changes (DeLuca and Boisvenue 2012). Conversely, if increases in plant-derived carbon inputs to soils exceed the increases in decomposition, the feedback would be negative (Davidson and Janssens 2006). Therefore, soil can be both a source and a sink of carbon. In other words, soils can aggravate the climate change situation or mitigate climate change, depending on the way they are managed and manipulated. Responses may differ between upland mineral soils and lowland organic soils. Upland mineral soils that have good drainage and aeration favor decomposition, resulting in relatively low carbon densities. In contrast, wetlands and peatlands have anaerobic conditions that slow down organic matter decomposition, and favor accumulation of deep layers of organic matter on top of mineral layers. In permafrost soils, drainage is often poor, and organic matter may become buried in deep soil layers through cryoturbation (Bockheim and Tarnocai 1998). Thus, wetlands, peatlands, and permafrost soils generally contain higher carbon densities than upland mineral soils, and together they make up enormous stocks of carbon globally. These soils are at a high risk of human disturbance leading to enhanced GHG emissions. The environmental constraints that can temporarily or indefinitely affect apparent temperature sensitivities of decomposition include the following:

1. Physical protection of organic matter against decomposition by binding humus particles to the interior of aggregates (Oades 1988; Six et al. 2002), where microorganisms and their enzymes may have limited access, or by low water solubility or occurrence of enzymes in hydrophobic domains of humified organic matter (Spacini et al. 2002). Clay content exerts a powerful control on the size of the older soil carbon pool in Amazonian soils (Telles et al. 2003).
2. Chemical protection of organic matter by adsorbing on mineral surfaces through covalent or electrostatical bonds (Oades 1988).
3. Drought which reduces the thickness of soil water films, thus inhibiting the diffusion of extracellular enzymes and soluble organic-C substrates and lowering substrate availability at reaction microsites.
4. Flooding which slows oxygen diffusion to decomposition reaction sites, often allowing only anaerobic decomposition, which includes fewer and generally slower degradative enzymatic pathways. The importance of anoxic conditions for the preservation of organic matter in boreal peatlands is without doubt (Fierer and Schimel 2002; Borken et al. 2003; Jarvis et al. 2007).
5. Freezing for which the diffusion of substrates and extracellular enzymes within the soil is extremely slow where the extracellular soil water is frozen.

The decomposition of organic matter in a highly weathered mature tropical forest soil with high clay content may, on average, have low apparent temperature sensitivity because of chemical protection of a large fraction of soil C on mineral surfaces. In contrast, the apparent temperature sensitivity may, on average, be higher in a recently tilled temperate prairie soil because of lower proportions of substrates under environmental constraints to decomposition (Davidson and Janssens 2006). Microbial decomposition can also, itself, facilitate stabilization through the production of secondary microbial products and the majority of carbon that is most resistant to decay has microbial—and not plant—signatures (Grandy and Neff 2008). Carbon sequestration in forest soils has a potential to decrease the rate of enrichment of atmospheric concentration of CO_2. Increase in C stock of forest soils can be achieved through forest management including site preparation, fire management, afforestation, species management or selection, use of fertilizers, and soil amendments. Forest harvesting may decrease C stock, at least temporarily (Lal 2005).

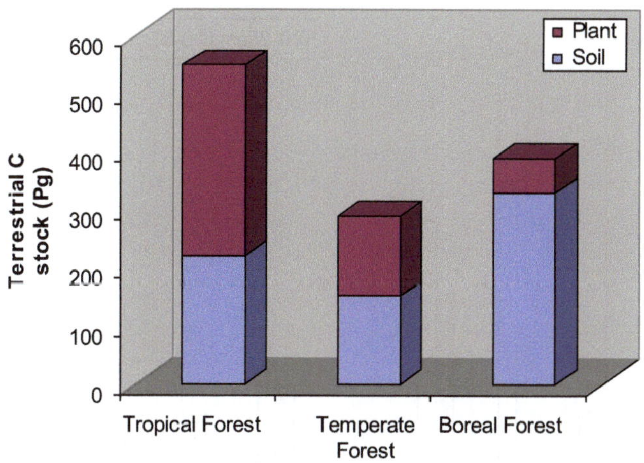

Fig. 9.1 Carbon stocks in different forest types of the world. (Based on data of Vogt et al. 1998; Eswaran et al. 1995; Goodale et al. 2002; Guo and Gifford 2002)

9.3 Carbon Stock in Forest Soils

Soil carbon is one of the key variables for estimating terrestrial C dynamics, since it is a major determinant of the amount of CO_2 released to the atmosphere through the decomposition of organic matter in soils. Improved knowledge of the amount and spatial distribution of the C stock in soils is crucial to estimating changes in the terrestrial C dynamics (Bhatti et al. 2002). Soil C storage is determined by the balance of C input from plant primary production and C release through decomposition. Disturbance, land-use history, climate, soil texture, topography, and hydrology are the primary variables that influence both production and decomposition processes and, therefore, the total amount of soil C stocks (Ju and Chen 2005).

Dixon et al. (1994) stated that forests contain about 1,240 Pg of carbon in their vegetation and soils. Of the C stock in forest biomes, 37 % is in low-latitude forests, 14 % in mid-latitude forests, and 49 % in high-latitude forests. Tropical and boreal forests represent the largest stores. Typical plant C density ranges from 40 to 60 Mg C ha^{-1} in boreal forests, 60 to 130 Mg C ha^{-1} in temperate forests, and 120 to 194 Mg C ha^{-1} in tropical forests. An undisturbed tropical rainforest contains as high C density as 250 Mg C ha^{-1}. Forest soils contain more than two-thirds of the terrestrial C in forest ecosystems. Soil C stock contributes 85 % of the terrestrial C stock in the boreal forest, 60 % in temperate forests, and 50 % in tropical rainforest. Boreal forests contain large quantities of soil C (Rapalee et al. 1998). In the boreal forest ecosystem of West Alberta, Canada, Banfield et al. (2002) reported that the regional average aboveground C stock ranged from 43 to 50 Mg C ha^{-1} compared with the soil organic carbon stock of 83–156 Mg C ha^{-1}. Kolchugina and Vinson (1993) estimated that the forest biome of the former Soviet Union contained 319 Pg C in soil and 128 Pg in the phytomass. Soils of the northern latitudes (arctic and boreal biomes) may also contain a large proportion of labile C (10–40 %) that could be easily decomposed (Neff and Hooper 2002). However, soils of tropical rainforests contain low SOC density. Powers and Schlesinger (2002) reported that Costa Rican tropical forest soils contain only 50–140 Mg C ha^{-1}. Soil organic carbon content in forest soils may range from 0 % in very young soils to as much as 50 % (w/w) in some organic or wetland soils (Trettin and Jurgensen 2003). However, most soils contain 0.3–11.5 % in the surface 20 cm of mineral soil (Perry 1994). Soil organic carbon concentration in mineral soils is also lower in tropical forests and higher in montane and boreal forests (Jones 1989). According to Price et al. (2012), the soil carbon stocks of tropical, temperate, and boreal forests may be taken to be around 220, 150, and 330 Pg, respectively (Fig. 9.1)

Tarnocai et al. (2009) developed the Northern Circumpolar Soil Carbon Database to determine carbon pools in soils of the northern circumpolar permafrost region. In the northern permafrost region, organic soils (peatlands) and cryoturbated permafrost-affected mineral soils have the highest mean soil organic carbon contents (32.2–69.6 kg m^{-2}). They reported a new estimate of the carbon pools in soils of the northern permafrost region, including deeper layers and pools not accounted for in previous analyses. Carbon pools were estimated to be 191.29 Pg for the 0–30 cm depth, 495.80 Pg for the 0–100 cm depth, and 1,024.00 Pg for the 0–300 cm depth. However, many of these soils do not bear forests. Price et al. (2012) gave values of carbon stock in soils of different orders of soil taxonomy, without distinguishing forest and non-forest soils (Fig. 9.2).

9.4 Temperature and Soil Organic Carbon

The decomposition of soil organic carbon may depend on temperature, and this dependence may be described by the Q_{10} value, which is defined by the difference in respiration rates over a 10 °C interval. It is extremely variable, with a range from 1 (no effect of temperature on respiration) to 5 (five times higher respiration rate with a 10 °C rise in temperature) under different combinations of soil moisture and soil temperature (Reichstein et al. 2003). The nature of organic compounds in soils and environmental constraints, such as limiting soil moisture or nutrients, complicate efforts to fully explain the temperature sensitivity of microbial respiration (Davidson and Janssens 2006). In an analysis of sources of uncertainty in the soil carbon model SWIM, Post et al. (2008) identified the carbon mineralization rate, carbon use efficiency, Q_{10}, soil bulk density, and initial carbon content as the most critically sensitive parameters. Improved models are needed to differentiate the direct effects of drying, wetting, and carbon substrate supply to soil microbes

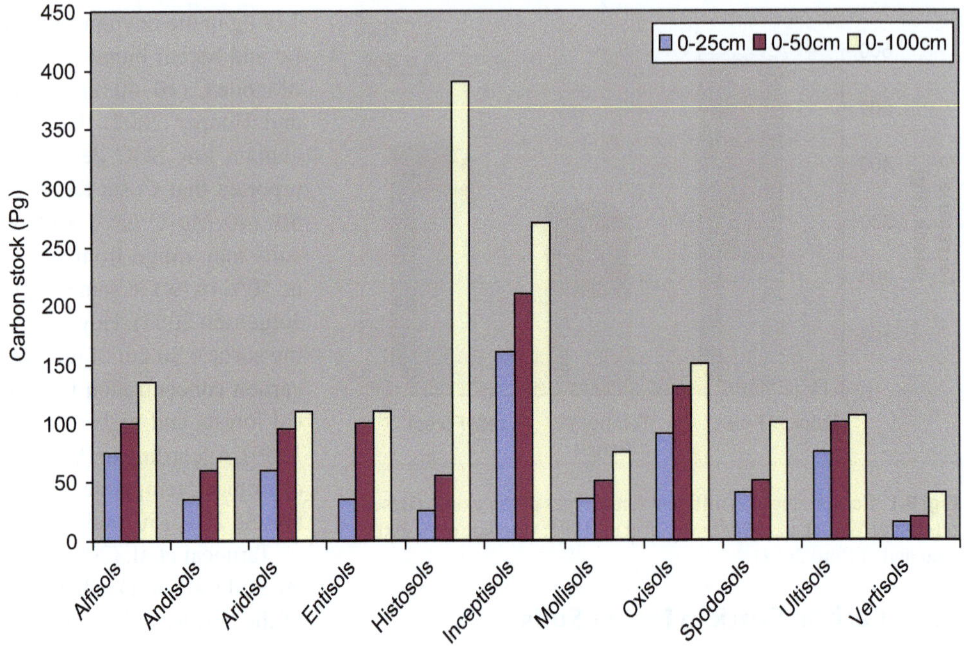

Fig. 9.2 Soil carbon stock in different soil orders

from the indirect effects of soil water content and temperature on diffusion of carbon substrates to the microbial population (Davidson et al. 2006). Models incorporating realistic spatial relationships and mechanistic workings give the most accurate results (Hanson et al. 2004). Liski et al. (2005) suggested that all applications may not be suited to process models due to the extensive inputs required. Yet, leaving out the process may yield projections for soil carbon stocks that differ to when biological response is explicitly modeled. For example, Allison et al. (2010) showed that microbial biomass decreases with warming, which negates the loss of soil carbon with temperature increase as projected with conventional models, because of the reduction in microbes that actually carry out decomposition.

9.5 GHG Emissions from Forest Soils

Forest soils may emit several GHGs, including CO_2, N_2O, and CH_4. Reactive nitrogen compounds (e.g., NH_3 and NO_x) from agriculture, transport, and industry lead to increased emissions of the GHG nitrous oxide (N_2O) from forests in Europe. These compounds are mainly of anthropogenic origin. After their deposition to forests via air (nitrogen deposition), they are partly converted into nitrous oxide (N_2O). Following carbon dioxide and methane, nitrous oxide is the third most important GHG. Increased deposition of reactive nitrogen to forests not only results in climate-damaging nitrous oxide emissions from forest soils but also causes loss of plant and animal diversity and enhanced nitrate emissions into water. Nitrous oxide (N_2O) emission from the soil surface is an important pathway for loss of soil nitrogen and a source of atmospheric pollution (Hahn et al. 2000). It contributes to climate change and participates in reactions that destroy stratospheric ozone as well (Cicerone 1987). Soil moisture, soil temperature, pH, organic matter, O_2 status, and the capacity of soils to supply nitrifiable N affect N_2O release from soil. A key factor controlling denitrification in forest soils is aeration (Sexstone et al. 1985). Nitrification requires aerobic conditions, whereas denitrification is favored by low O_2 concentrations. However, the development of aerobic and anaerobic microsites within the same soil aggregate permits both nitrification and denitrification to occur simultaneously (Davidson et al. 1986). Factors affecting carbon availability such as freezing and thawing or wetting and drying should also be important regulators of denitrification in forest soils (Groffman and Tiedje 1988).

Ullah et al. (2009) presented soil fluxes of CH_4, N_2O, and CO_2 from 12 sites located in four major boreal forest types, black spruce (*Picea marina*), jack pine (*Pinus banksiana*), aspen (*Populus* spp.) and alder (*Alnus* spp.) in Quebec. Well-drained black spruce, jack pine, and aspen forest soils were net sinks of CH_4 (-0.33 mg m^{-2} day^{-1}), while alder-dominated wetland soils were sources of CH4 (0.45 mg m^{-2} day^{-1}). The cut-over alder wetland soil produced 131 times more CH_4 than the undisturbed wetland soil. Soil moisture and temperature mainly regulated CH_4 fluxes. CO_2 emissions from these soils ranged from 0.20 to 2.7 g m^{-2} day^{-1} and were mainly controlled by soil temperature. Even small proportions of hydromorphic soils can significantly influence the net fluxes of the GHGs CO2, N_2O, and CH_4 in terrestrial ecosystems. From the great variety of hydromorphic soils, mainly histosols, have been investigated for GHG emissions. Fewer studies are available on hydric mineral soils. Flessa et al. (2006) reported CH4 emissions from some mineral hydromorphic soils. Strong CH_4 emissions occurred from all soils at a water table of −5 cm. The groundwater table was much more important for the emission of GHGs from hydromorphic soils than the soil type. The depth of the groundwater table seems to be an appropriate factor to stratify groundwater-influenced soils with respect to GHG emissions. For all soils, the total emission of CO2 equivalents was the highest at a water table depth of 20 cm.

Several independent studies have demonstrated that N_2O is emitted in significant quantities from forest ecosystems (Ambus and Christensen 1995; Bowden et al. 1990; Schmidt et al. 1989; Struwe and Kjøller 1994; Papen and Butterbach-Bahl 1999). In Europe, forest ecosystems cover about 1.9 million km^2 about 28 % of the total land area and may thus have significant importance for European-scale N_2O budgets. However, the importance of temperate forests for the atmospheric N_2O budget still remains highly uncertain with a 20-fold range in the emission estimates (IPCC 2001). Ambus et al. (2006) reported nitrous oxide emissions from some forest soils of Europe. They observed that N_2O emission varied significantly among the different forest soils, and was inversely correlated to the soil C:N ratio. The N_2O emissions were significantly higher from the deciduous soils (13 ng N_2O–N cm^{-3} day^{-1}) than from the coniferous soils (4 ng N_2O–N cm^{-3} day^{-1}). Nitrate was the dominant substrate for N_2O with an average contribution of 62 %. The average contribution of ammonium (NH_4^+) to N_2O averaged 34 %. Earlier studies suggested that tropical forests are larger sources for N_2O than temperate and boreal forests, but recent studies have indicated that temperate beech (*Fagus Sylvatica*) forests can have N_2O fluxes similar to those observed in tropical forests (Papen and Butterbach-Bahl 1999). Beech forests with high annual N_2O fluxes have a seasonal emission pattern with high N_2O fluxes in summer and low N_2O fluxes in winter (Brumme and Borken 2009).

Pihlatie et al. (2010) measured fluxes of the GHGs CO_2, CH_4, and N_2O during a 2-month campaign at a drained peatland forest in Finland by the eddy covariance technique, and automatic and manual chambers. They observed that the forest was a net CO_2 sink during the 2 months and the fluxes of CO_2 dominated the GHG exchange. The peat soil was a small sink of atmospheric CH_4 and a small source of N_2O. Both CH_4 oxidation and N_2O production took place in the topsoil, whereas CH_4 was produced in the deeper layers of the peat, which were unfrozen throughout the measurement period. The CO_2 peak followed the increase in soil temperature, whereas the N_2O peak occurred with a delay after the thawing of the litter layer.

9.6 Forestry Activities to Adapt to and Mitigate Climate Change

New forest plantation, degraded forest rehabilitation, and existing forest enrichment contribute to adaptation to and mitigation of climate change because these actions increase the rate and quantity of carbon sequestration in biomass. Agroforestry and the planting of multiple-use trees (fruit trees, rubber wood, etc.) also contribute to this objective.

9.6.1 Planting Forests

Tree planting projects improve carbon sequestration, inasmuch as carbon storage in durable products such as boards, plywood, furniture, railway slippers, etc., complements the permanent stock in standing trees. The life span of these products is limited, but 20–100 years is still significant; it can help avoid concentration peaks of CO_2 in the atmosphere.

9.6.2 Protecting Existing Forests

The carbon reservoir in the forest biomass and soils is very large. Destabilizing the forests, deforestation, and forest degradation would enhance the release of CO_2 to the atmosphere, so that forests can be converted to a source from a sink of GHGs, which emphasizes the importance of conserving natural forest.

9.6.3 Improved Forestry Techniques to Reduce Emissions

GHG emissions can be reduced by applying appropriate forestry measures. For example, forest-harvesting operations can cause serious damage to the soil and the forest stand when carried out inappropriately. Reduced impact logging and harvest plans may involve a set of techniques, such as preplanning skidding trails; optimizing landings; directional felling; and employing appropriate skidders, which reduce damage to soils, harvested trees, and the remaining stand. Timber processing also generates a considerable quantity of waste wood, which could either be reduced or be used as a raw material for production or as fuel.

9.6.4 Replacing High-energy Materials with Climate-friendly Substitutes

Using lumber instead of materials requiring large amounts of energy during production helps fight the greenhouse effect, e.g., in replacing concrete or steel constructions with wood as frames, beams, etc. Using 1 m^3 of lumber in buildings sequesters 1 t of CO_2 for an average period of 20 years, and reduces net emissions by 0.3 t of CO_2 if concrete is replaced, and 1.2 t of CO_2 if steel is substituted.

9.6.5 Fossil Fuel Substitution Through Renewable Wood Energy

Producing wood for energy purposes mitigates climate change by combining sink action with emission reduction. Substituting fossil fuels, such as coal, natural gas, or oil, by fuelwood for domestic use, electricity production, or industrial use, e.g., in iron smelters, reduces CO_2 emissions-because wood is renewable. The expected sequestration of carbon through the growth of trees after sustainable harvest compensates for the CO_2 emitted by combustion.

Study Questions

1. Why is climate changing? Why are we alarmed by climate change? How can forests mitigate climate change?
2. How can forest soils act as both a source and a sink of GHGs? Discuss forest soils' stock of organic carbon. Why is some soil organic carbon highly stabilized?
3. Explain carbon sequestration. How do forests sequester carbon? How do soils sequester carbon? Discuss the relative contribution of tropical, temperate, and boreal forests to the forest carbon stock.
4. Hydromorphic soils are great sinks of carbon. They can be converted into great sources of carbon, too—how?
5. What will be the fate of soil carbon reservoir in a warm climate? Discuss the contribution of forestry practices to climate change.

References

Akala VA, Lal R (2001) Soil organic carbon pools and sequestration rates in reclaimed mine soils in Ohio. J Environ Qual 30:2098–2104
Allison SD, Wallenstein MD, Bradford MA (2010) Soilcarbon response to warming dependent on microbial physiology. Nat Geosci 3:336–340
Ambus P, Christensen S (1995) Spatial and seasonal nitrous oxide and methane fluxes in Danish forest-, grassland-, and agroecosystems. J Environ Qual 24:993–1001
Ambus P, Zechmeister-Boltenstern S, Butterbach-Bahl K (2006) Sources of nitrous oxide emitted from European forest soils. Biogeosciences 3:135–145
Anderegg WR, Berry JA, Smith DD, Sperry JS, Anderegg LD, Field CB (2012) Proc Natl Acad Sci USA 109:233–237
Baccini A, Goetz SJ, Walker WS, Laporte NT, Sun M, Sulla-Menashe D, Hackler J, Beck PSA, et al. (2012) Estimated carbon dioxide emissions from tropical deforestation improved by carbon-density maps. Nature Clim Change 2:182–185
Ballhorn U, Siegert F, Mason M, Limin S (2009) Derivation of burn scar depths and estimation of carbon emissions with LIDAR in Indonesian peatlands. Proc Natl Acad Sci USA 106:21213–21218
Banfield GE, Bhatti JS, Jiang H, Apps MJ, Karjalainen T (2002) Variability in regional scale estimates of carbon stocks in boreal forest ecosystems: results from west-central Alberta. Forest Ecol Manag 169:15–27
Batjes NH (1996) Total carbon and nitrogen in the soils of the world. Soil Science 47 (2):151–163
Betts R, Sanderson M, Woodward S (2008) Effects of large-scale Amazon forest degradation on climate and air quality through fluxes of carbon dioxide, water, energy, mineral dust and isoprene. Philos Trans R Soc London Ser B 363:1873–1880
Bhatti JS, Apps MJ, Jiang H (2002) Influence of nutrients, disturbances and site conditions on carbon stocks along a boreal forest transect in Central Canada. Plant Soil 242:1–14
Bockheim JG, Tarnocai C (1998) Recognition of cryoturbation for classifying permafrost-affected soils. Geoderma 81:281–293
Borken W, Davidson EA, Savage K, Gaudinski J, Trumbore SE (2003) Drying and wetting effects on carbon dioxide release from organic horizons. Soil Sci Soc Am J 67:1888–1896
Bowden RD, Steudler PA, Melillo JM (1990) Annual nitrous oxide fluxes from temperate forest in the Northeastern United States. J Geophys Res 95:13997–14005
Brevik EC (2012) Soils and climate change: gas fluxes and soil processes. Soil Sci Soc Am J 53(4). doi: 10.2136/sh12-04-0012
Brumme R, Borken W (2009) N_2O emission from temperate beech forest soils. Ecol Stud 208:353–367
Canadell JG, Quere CL, Raupach MR, Field CB, Buitenhuis ET, Ciais P, Conway TJ, Gillett NP, Houghon RA, Marland G (2007) Contributions to accelerating atmospheric CO_2 growth from economic activity, carbon intensity, and efficiency of natural sinks. Proc Natl Acad Sci 104:18866–18870
Cicerone RJ (1987) Changes in atmospheric ozone. Science 237:35–42
Cox PM, Betts RA, Collins M, Harris PP, Huntingford C, Jones CD (2004) Amazonian forest dieback under climate-carbon cycle projections for the 21st century. Theor Appl Climatol 78:137–156
Cox PM, Betts RA, Jones CD, Spall SA, Totterdell IJ (2000) Acceleration of global warming due to carbon-cycle feedbacks in a coupled climate model. Nature 408:184–187
Croezen HJ, van Valkengoed MPJ (2009) GHG emissions due to deforestation. CE Delft, Delft
Davidson E, Swank WT, Perry T (1986) Distinguishing between nitrification and denitrification as sources of gaseous nitrogen production in soil. Appl Environ Microbiol 52(6):1280–1286
Davidson EA, Janssens IA (2006) Temperature sensitivity of soil carbon decomposition and feedbacks to climate change. Nature 440:165–173
Davidson EA, Janssens IA, Lou Y (2006) On the variability of respiration in terrestrial ecosystems: moving beyond Q_{10}. Glob Change Biol 12:154–164
DeLuca TH, Boisvenue C (2012) Boreal forest soil carbon: distribution, function and modeling. Forestry (2012) doi: 10.1093/forestry/cps003
Denman KL, Brasseur G, Chidthaisong A, Ciais Ph, Cox P, Dickinson RE, Hauglustaine D, Heinze C, Holland E, Jacob D, Lohmann U, Ramachandran S, da Silva Dias DL, Wofsy SC, Zhang X (2007) Couplings between changes in the climate system and biogeochemistry, Chapter 7. In: Climate change 2007: The physical science basis, The IPCC fourth assessment report, Intergovernmental Panel on Climate Change, Cambridge University Press, Cambridge
Dixon RK, Brown S, Houghton RA, Solomon AM, Trexler MC, Wisniewski J (1994) Carbon pools and fluxes of global forest ecosystems. Science 263:185–190
Eswaran H, Van den Berg E, Reich P, Kimble J (1995) Global soil carbon resources In: Lal R, Kimble JM, Eswaran H, Stewart BA (eds) Global change and pedogenic carbonate. CRC, Boca Raton

FAO (2005) Global forest resources assessment 2005: progress towards sustainable forest management. FAO Forestry Paper 147, Food and Agriculture Organization of the United Nations, Rome

FAO (2006) Global forest resource assessment 2005, Main Report, 2006, Forestry Paper 147, FAO, Rome

Fierer N, Schimel JP (2002) Effects of drying-rewetting frequency on soil carbon and nitrogen transformations. Soil Biol Biochem 34:777–787

Flessa H, Jungkunst H, Fiedler S (2006) Greenhouse gas emissions from hydromorphic forest soils: effect of soil type and ground water table. Geophys Res Abs 8, 07120

Gibbs HK, Herold M (2007) Tropical deforestation and carbon emissions: Introduction to special issue. Environmental Research Lab

Goodale CL, Apps MJ, Birdsey RA, Field CB, Heath LS, Houghton RA, Jenkins JC, Kohlmaier GH, Kurz W, Liu S, Nabuurs GJ, Nilsson S, Shvidenko AZ (2002) Forest carbon sinks in the Northern Hemisphere. Ecol Appl 12:891–899

Gorte RW, Sheikh PA (2010) Deforestation and climate change. CRC Report for Congress

Grandy AS, Neff JC (2008) Molecular C dynamics downstream: the biochemical decomposition sequence and its impact on soil organic matter structure and function. Sci Total Environ 404:297–307

Groffman PM, Tiedje JM (1988) Denitrification hysteresis during wetting and drying cycles in soil Soil Sci Soc Am J 152:1626–1629

Guo LB, Gifford RM (2002) Soil carbon stocks and land use change: a meta analysis. Glob Change Biol 8:345–360

Hahn M, Gartner K, Zechmeister-Boltenstern S (2000) Greenhouse gas emissions (N_2O, CO_2 and CH_4) from three forest soils near Vienna (Austria) with different water and nitrogen regimes. Die Bodenkultur 51(2):115–125

Hanson PJ, Amthor JS, Wullschleger SD, Wilson KB, Grant RF, Hartley A, Hui D, Hunt ER Jr, Johnson DW, Kimball JS, King AW, Luo Y, McNulty SG, Sun G, Thornton PE, Wang S, Williams M, Baldocchi DD, Cushman RM (2004) Oak forest carbon and water simulations: model intercomparisons and evaluations against independent data. Ecol Monogr 74:443–489

Hoogeveen H (2007) Climate change 2:36–46

Houghton RA (2005) Tropical deforestation as a source of greenhouse gas emissions. In: Moutinho P, Schwartzman S (eds) Tropical deforestation and climate change. Amazon Institute for the Environment (IPAM) and Environmental Defense Fund, Washington

IPCC (2000) Special report on land use, land-use change, and forestry. In: Watson RT, Noble IR, Bolin B, Ravindranath NH, Verardo DJ, Dokken DJ (eds) Intergovernmental Panel on Climate Change (IPCC), Cambridge University Press, Cambridge

IPCC (2001) Climate change 2001: The scientific basis. Contribution of working group I to the third assessment report of the Intergovernmental Panel on Climate Change. Houghton J, Ding T, Griggs Y, Noguer DJ, van der Linden M, Dai PJ, Maskell XK, Johnson CA (eds). Cambridge University Press, Cambridge, UK

IPCC (2007) Contribution of working groups I, II and III to the fourth assessment report of the intergovernmental panel on climate change. Pachauri RK, Reisinger A (eds) Core writing team, IPCC, Geneva, Switzerland

Iverson LR, Prasad AM (2001) Potential changes in tree species richness and forest community types following climate change. Ecosystems 4:186–199

Jaenicke J, Riley JO, Mott C, Kimman P, Siegert F (2008) Determination *of* the amount *of* carbon stored in Indonesian peatlands. Geoderma 147:151–158

Jarvis P, Rey A, Petsikos C, Wingate L, Rayment M, Pereira J, Banza J, David J, Miglietta F, Borghetti M, Manca G, Valentini R (2007) Drying and wetting of Mediterranean soils stimulates decomposition and carbon dioxide emission: the "Birch effect". Tree Physiol 27:929–940

Johnston MH, Homan PS, Engstrom JK, Grigal DF (1996) Changes in ecosystem carbon storage over 40 years on an old field/forest landscape in east-central Minnesota. Forest Ecol Manag 83:17–26

Jones JA (1989) Environmental influences on soil chemistry in central semi-arid Tanzania. Soil Sci Soc Am J 53: 1748–1758.

Joyce LA, Birdsey R (2000) The impact of climate change on America's forests: a technical document supporting the 2000 USDA Forest Service RPA Assessment. USDA Forest Service No. RMRS-GTR 59, Washington, DC

Ju W, Chen JM (2005) Distribution of soil carbon stocks in Canada's forests and wetlands simulated based on drainage class, topography and remotely sensed vegetation parameters. Hydrol Process 19:77–94

Kattenberg A, Giorgi AF, Grassl H, Meehl GA, Mitchell JFB, Stouffer RJ, Tokioka T, Weaver AJ, Wigley TML (1996) The science of climate change: contribution of working group I to the second assessment of the IPCC, Cambridge University Press, UK

Kolchugina TP, Vinson TS (1993) Equilibrium analysis of carbon pools and fluxes of forest biomes in the former Soviet Union. Can J Forest Res 23:81–88

Kurz WA, Stinson G, Rampley GJ, Dymond CC, Neilson ET (2008) Risk of natural disturbances makes future contribution of Canada's forests to the global carbon cycle highly uncertain. Proceedings of the National Academy of Sciences 105:1551–1555

Lal R (2005) Forest soils and carbon sequestration. For Ecol Manag 220:242–258

Langner A, Miettinen J, Siegert F (2007) Land cover change 2002–2005 in Borneo and the role of fire derived from MODIS imagery. Global Change Biol 13:2329–2340

Langner A, Siegert F (2009) Spatiotemporal fire occurrence in Borneo over a period of 10 years. Global Change Biol 15:48–62

Levine E (ed) Soils and global change. CRC, Boca Raton

Liski J, Palosuo T, Peltoniemi M, Sievanen R (2005) Carbon and decomposition model yasso for forest soils. Ecol Model 189:168–182

McGuire AD, Apps WC, Beringer M, Clein J et al. (2002) Environmental variation, vegetation distribution, carbon dynamics and water/energy exchange at high latitudes. J Veg Sci 13:301–314

McPherson EG, Nowak D, Heisler G, Grimmand S, Souch C, Grant R, Rowntree R (1997) Quantifying urban forest structure, function and value: the Chicago Urban forest climate project. Urban Ecosyst 1:49–61

Millennium Ecosystem Assessment (2005) Summary for decision makers. In Ecosystems and human well-being: synthesis. Island Press, Washington DC

Morgan MG, Pitelka LF, Shevliakova E (2001) Elicitation of expert judgments of climate change impacts on forest ecosystems. Climatic Change 49:279–307

Neff JC, Hooper DU (2002) Vegetation and climate controls on potential CO_2, DOC and DON production in northern latitude soils. Global Change Biol 8:872–884

Nepstad DC, Stickler CM, Soares-Filho B, Merry F (2008) Interactions among Amazon land use, forests and climate: prospects for a near-term forest tipping point. Phil Trans R Soc B 363 (1498):1737–1746

Oades JM (1988) The retention of organic matter in soils. Biogeochemistry 5:35–70

Page SE, Siegert F, Rieley JO, Boehm HDV, Jaya A, Limin S (2002) The amount of carbon released from peat and forest fires in Indonesia during 1997. Nature 420:61–65

Page SE, Wuest RAJ, Weiss D, Riley JO, Shotyk W, Limin SH (2004) J Quat Sci 19:625–635

Papen H, Butterbach-Bahl K (1999) A 3-year continuous record of nitrogen trace gas fluxes from untreated and limed soil of a N saturated spruce and beech forest ecosystem in Germany 1. N_2O emissions, J Geophys Res-Atmos 104:18487–18503

Penman J, Gytarsky M, Hiraishi T, Krug T, Kruger D, Pipatti R, Buendia L, Miwa K, Ngara T, Tanabe K, Wagner F (2003) Good practice guidance for land use, land-use change and forestry. IPCC national greenhouse gas inventories programme, Intergovernmental Panel on Climate Change

Perry DA (1994) Forest ecosystems. The John Hopkins University, London

Peterman W, Bachelet D (2012) Climate change and forest dynamics: a soils perspective. In: Hester RE, Harrison RM (eds) Issues in environmental science and technology No. 35, Soils and Food Security. The Royal Society of Chemistry, www.rsc.org

Pihlatie MK, Kiese R, Bruggemann N, Butterbach-Bahl K, Kieloaho AJ, Laurila T, Lohila A, Mammarella I, Minkkinen K (2010) Greenhouse gas fluxes in a drained peatland forest during spring frost-thaw event. Biogeosciences 7:1715–1727

Post J, Hattermann FF, Krysanova V, Suckow F (2008) Parameter and input data uncertainty estimation for the assessment of long-term soil organic carbon dynamics. Environ Modell Softw 23:125–138

Powers, J.S., Schlesinger, W.H., 2002. Relationships among soil carbon distributions and biophysical factors at nested spatial scales in rainforests of northeastern Costa Rica. Geoderma 109:165–190

Price SP, Bradford MA, Ashton MS (2012) Characterizing organic carbon stocks and flows in forest soils. In: Ashton MS et al. (eds) Managing forest carbon in a changing climate. Springer Science + Business Media BV

Rapalee G, Trumbore SE, Davidson EA, Harden JW, Veldhuis H (1998) Soil carbon stocks and their rates of accumulation and loss in a boreal forest landscape. Global Biogeochem Cycles 12:687–701

Reichstein M, Rey A, Freibauer A, Tenhunen J, Valentini R, Banza J, Casals P, Cheng Y, Grunzweig JM, Irvine J, Joffre R, Law BE, Loustau D, Miglietta F, Oechel W, Ourcival JM, Pereira JS, Peressotti A, Ponti F, Qi Y, Rambal S, Rayment M, Romanya J, Rossi F, Tedeschi V, Tirone G, Xu M, Yakir D (2003) Modeling temporal and large-scale spatial variability of soil respiration from soil water availability, temperature and vegetation productivity indices. Glob Biogeochem Cycles 17:15–11

Schauvlieghe M, Lust N (1999) Carbon accumulation and allocation after afforestation of a pasture with pin oak (Quercus palustris) and ash (Fraxinus excelsior). Silva Gandevensis 64:72–81

Schmidt J, Seiler W, Conrad R (1989) Emission of nitrous oxide from temperate forest soils into the atmosphere. J Atmos Chem 6:95–115

Schrope M(2009) When money grows on trees. Nat Rep Climate Change 3:101–103

Sedjo R, Sohngen B (2012) Carbon sequestration in forests and soils. Annu Rev Resour Econ 4:127–144

Serreze MC, Walsh JE, Chapin FS, Osterkamp T, Dyurgerov M, Romanovsky V, Oechel WC, Morison J, Zhang T, Barry RG (2000) Observational evidence of recent change in the northern high-latitude environment. Clim Chang 46:159–207

Sexstone AJ, Parkin TB, Tiedje JM (1985) Temporal response of soil denitrification rates to rainfall and irrigation. Soil Sci Soc Am J 49:99–103

Six J, Callewaert P, Lenders S, De Gryze S, Morris SJ, Gregorich EG, Paul EA, Paustian K (2002) Measuring and understanding carbon storage in afforested soils by physical fractionation. Soil Sci Soc Am J 66:1981–1987

Soares-Filho BS, Nepstad DC, Curran LM, Cerqueira GC, Garcia RA, Ramos CA, Voll E, McDonald A, LeFebvre P, Schlesinger P (2006) Modelling conservation in the Amazon basin. Nature 440:520–523

Spacini R. et al (2002) Increased soil organic carbon sequestration though hydrophobic protection by humic substances. Soil Biol Biochem 34:1839–1851

Struwe S, Kjøller A (1994) Potential for N_2O production from beech (*Fagus sylvaticus*) forest soils with varying pH. Soil Biol Biochem 26:1003–1009

Tarnocai C, Canadell JG, Schuur EAG, Kuhry P, Mazhitova G, Zimov S (2009) Soil organic carbon pools in the northern circumpolar permafrost region. Global Biogeochem Cycles 23, GB2023, doi:10.1029/2008GB003327, 2009

Telles ECC, de Camargo PB, Martinelli LA, Trumbore SE, da Costa ES, Santos J, Higuchi N, Oliveira RC Jr (2003) Influence of soil texture on carbon dynamics and storage potential in tropical forest soils of Amazonia. Glob Biogeochem Cycles 17:X1-X12

Trettin CC, Jurgensen MF (2003) Carbon cycling in wetland forest soils. In: Kimble JM, Heath LS, Birdsey RA, Lal R (eds) The potential of U.S. forest soils to sequester carbon and mitigate the greenhouse effect. CRC, Boca Raton, pp 311–331

Ullah S, Frasier R, Pelletier L, Moore TR (2009) Greenhouse gas fluxes from boreal forest soils during the snow-free period in Quebec, Canada. Can J For Res 39:666–680

van der Werf GR, Morton DC, DeFries RS, Olivier JGJ, Kasibhatla PS, Jackson RB, Collatz GJ, Randerson JT (2009) CO_2 emissions from forest loss. Nat Geosci 2: 737–738

Vogt KA, Vogt DJ, Bloomfi eld J (1998) Analysis of some direct and indirect methods for estimating root biomass and production of forests at an ecosystem level. Plant Soil 200:71–89

10 Forest Soil Management and Silvicultural Treatments

Natural forests are managed with the objectives of conservation of biodiversity and ecosystem health so that they sustain ecosystem functions and services. Forest plantations are managed for enhancing and sustaining site quality and productivity and ensure economic returns from higher yields of wood, timber, fuelwood, pulp, etc. Site quality and productivity are influenced by soil and environmental and management factors. The range of management practices includes seedling development, site preparation, planting, tending, weeding, fertilization, irrigation, and harvesting. These procedures may be manual (particularly in developing countries) and mechanical (particularly in developed countries). These practices have both positive and negative impacts on growth of trees, soil, and environment. For example, suppressing competition through weeding by herbicides, and promoting growth of trees by sewage sludge and wastewater application can deteriorate water quality unless carefully managed.

10.1 Objectives of Forest Soil Management

The ultimate objectives of forest soil management are to improve forest soil quality and productivity and sustain ecosystem functions and services. There are some short-term objectives as well.

The short-term objectives of forest soil management are as follows:

a) To make the soil suitable for planting or sowing.
b) To improve survival and growth of seedlings.
c) To reduce competition.
d) To improve nutrient cycling.
e) To improve forest tree nutrition and biomass production.
f) To increase water and nutrient use efficiency.
g) To prevent damages to soil during harvesting.
h) To mend up rutting, soil displacement and other harvesting damages.
i) To conserve forest soil.
j) To rehabilitate forestland.

The long-term objectives are as follows:

a) To meet the people's basic needs for fuelwood, fodder, timber, and other forest products, and to contribute to food production through an effective interaction between forestry and farming practices.
b) To protect the land against degradation by soil erosion, floods, landslides, desertification, and other effects of ecological imbalance.
c) To conserve the ecosystems and genetic resources.
d) To protect environment by primary production and carbon sequestration.
e) To improve water yield and quality in watersheds.

10.2 Soil Management-Related Terminologies

There are several soil management-related terms in literature that often may create confusion among the readers. An explanation of such terms as land capability, soil quality, soil productivity, site quality, and site productivity is given below.

10.2.1 Land Capability

Land capability is the ability of land to accept a type and intensity of land use permanently, or for a specified period under a certain management without long-term degradation (Houghton and Charman 1986). It denotes the potential of a land for use in agriculture, horticulture, forestry, and other uses based on the degree of limitation imposed by its biophysical properties. The United States Department of Agriculture (USDA) Soil Conservation Service devised the land capability classification system during the late 1930s and early 1940s (Helms 1977). Some form or adaptation of land capability classification is used throughout the world (Olson 1974; FAO 1999). Land capability classification is a system of grouping soils primarily on the basis of their capability to produce common cultivated crops and pasture plants without

K. T. Osman, *Forest Soils*, DOI 10.1007/978-3-319-02541-4_10, © Springer International Publishing Switzerland 2013

deteriorating over a long period of time. There are eight land capability classes; forests can be grown on all land capability classes but classes VI, VII, and VIII are mainly used for forests, pastures, and rangelands. Characteristics of land capability classes are given as follows:

Land capability class I Class I soils have few limitations and are suited to wide range of plants; they can be safely used for cultivated crops, pasture, range, woodland, and wildlife. Soils are nearly level with low erosion hazard. Soils are deep, fertile, and responsive to fertilizers.

Land capability class II Soils of this class have some limitations reducing choice of crops and may require moderate conservation practices. These soils require careful management and conservation to prevent deterioration under cultivation.

Land capability class III Soils of this class have severe limitations that reduce the choice of plants or require special conservation practices, or both. These soils have more restrictions than those in class II and when used for cultivated crops, the conservation practices are usually more difficult to apply and maintain. They may be used for cultivated crops, pasture, woodland, range, or wildlife food and cover.

Land capability class IV Soils in this class have severe limitations that restrict the choice of plants or require very careful management, or both. The restrictions in use for soils in this class are greater than those in class III and the choice of plants is more limited. These soils may be used for cultivated crops, pasture, woodland, range, or wildlife food and cover.

Land capability class V Soils in this class have little or no erosion hazard but have other limitations that are impractical to remove, thereby limiting their use largely to pasture, range, woodland, or wildlife food and cover. Soil limitations restrict the kinds of plants that can be grown.

Land capability class VI Soils in this class have severe limitations that make them generally unsuited to cultivation and limit their use largely to pasture or range, woodland, or wildlife food or cover. Physical conditions of soils in this class are such that it is practical to apply range or pasture improvements. Limitations include steep slope, severe erosion hazard, stoniness, shallow rooting, etc.

Land capability class VII Soils in this class have very severe limitations that make them unsuited to cultivation and that restrict their use largely to grazing, woodland, or wildlife. Limitations include very steep slopes, erosion, shallow soil, stoniness, wet soil, salts, unfavorable climate, etc.

Land capability class VIII Soils and landforms in this class have limitations that preclude their use for commercial plant production and restrict their use to recreation, wildlife, or water supply or to aesthetic purposes. Limitations include erosion, stoniness, wet soil, low moisture, salts, severe climate, etc.

10.2.2 Soil Quality

The soil quality concept evolved in the early 1990s (Doran and Safely 1997; Wienhold et al. 2004), and the first official application of the term was approved by the Soil Science Society of America Ad Hoc Committee on Soil Quality. It was discussed elaborately by Karlen et al. (1997). Soil quality has been defined as "the capacity of a reference soil to function, within natural or managed ecosystem boundaries, to sustain plant and animal productivity, maintain or enhance water and air quality, and support human health and habitation." Subsequently, the two terms soil quality and soil health are used interchangeably (Karlen et al. 2001), although it is important to distinguish that, soil quality is related to soil function (Karlen et al. 2003; Letey et al. 2003), whereas soil health presents the soil as a finite nonrenewable and dynamic living resource (Doran and Zeiss 2000). The quality of soil is rather dynamic and can affect the sustainability and productivity of land use. It is the end product of soil degrading or conserving processes and is controlled by chemical, physical, and biological components of a soil and their interactions (Papendick and Parr 1992). Eswaran et al. (1999) divided world soils into nine land quality (in this classification, land and soil are used synonymously) classes. They defined land quality, soil resilience, and soil performance as:

Land quality: The ability of the land to perform its function of sustainable agriculture production and enable it to respond to sustainable land management.

Soil resilience: The ability of the land to revert to a near-original production level after it is degraded, as by mismanagement. Land with low resilience is permanently damaged by degradation.

Soil performance: The ability of the land to produce (as measured by yield of grain, or biomass) under moderate levels of inputs in the form of conservation technology, fertilizers, pest, and disease control. The definitions of the land quality classes according to Eswaran et al. (1999) are shown as follows:

Combining three soil performance classes with three soil resilience classes, there were nine land quality classes.

Land quality class performance and resilience	
Class I	high performance, high resilience
Class II	high performance, medium resilience
Class III	medium performance, high resilience
Class IV	high performance, low resilience
Class V	medium performance, medium resilience
Class VI	low performance, high resilience
Class VII	low performance, low resilience
Class VIII	low performance, medium resilience
Class IX low performance, low resilience	

Soil quality class I: Class I land is the prime land. The soils are highly productive with few limitations. Moisture and temperature conditions are ideal for annual crops. Soil management consists largely of sensible conservation practices to minimize erosion, maintain appropriate fertilization, and maximize use of best available plant materials. Risk for sustainable grain crop production is generally <20 %.

Soil quality classes II and III: The soils in these classes are good and have few problems for sustainable production. Care must be taken to reduce degradation, particularly for class II soils. The lower resilience characteristics of class II soils make them more risky, particularly for low-input grain crop production. However, their productivity is generally very high and consequently, response to management is high. Conservation tillage is essential, buffer strips are generally required, and fertilizer use must be carefully managed. Due to the relatively good terrain conditions, the land is suitable for national parks and biodiversity zones. Risk for sustainable grain crop production is generally 20–40 %, but risks can be reduced with good conservation practices.

Soil quality classes IV, V, and VI: All three classes require important inputs of soil conservation and management. The lack of plant nutrients is a major constraint and so a good fertilizer use plan must be adopted. Soil degradation must be continuously monitored. Productivity (in agricultural standard) is not high, and the lands can be reserved for national parks or as biodiversity zones. In the semiarid areas, they can be managed for range. Risk for sustainable grain crop production is 40–60 %.

Soil quality class VII: These soils are definitely not suitable for low-input grain crop production; their low resilience makes them easily prone to degradation. They should be retained under natural forests or range and some localized areas can be used for recreational purposes. Biodiversity management is crucial in these areas. Risk for sustainable grain crop production is 60–80 %.

Soil quality classes VIII and IX: These are soils belonging to very fragile ecosystems. They should be retained under their natural state. Some areas may be used for recreational purposes but under very controlled conditions. In class IX, which is largely confined to the Boreal area, timber harvesting must be done very carefully with considerable attention to ecosystem damage. Risk for sustainable grain crop production is >80 %.

The concept of soil quality includes assessment of soil properties and processes as they relate to the ability of soil to function effectively as a component of a healthy ecosystem. Soil quality, like site quality or forest productivity, is a value-based concept related to the objectives of ecosystem management, and hence will be management and ecosystem dependent (Schoenholtz et al. 2000).

Soil quality classes IV–IX can be used for forestry. Classes VIII and IX should better be left under natural vegetation without any disturbance.

10.2.3 Soil Productivity

Soil productivity is the ability of soil to give optimum yields under standard set of management practices (Eswaran et al. 1997). Yield is the amount of biomass production. Yield may be biological (total biomass) or economic (grains, vegetables, wood, fiber, or other products such as resins, sugars, oils, etc.). A poorly fertile soil may be made productive by fertilizer addition, and a dry soil may be made productive by irrigation. A waterlogged soil may be naturally productive for a swamp forest tree such as *Barringtonia acutangula* but not for an upland tree species such as *Shorea robusta*. Draining the soil may make it productive for *Shorea robusta*, too. The total productivity of a forest is difficult to measure, and not all parts of a forest vegetation are harvested at a time. Thus, primary productivity, biomass productivity, timber productivity, and litter productivity are usually viewed separately. Potential of a soil for producing timber is given particular emphasis in forestry. Since timber productivity depends not only on soil, but also on species, topography, microclimate, and management, a more meaningful term is site productivity, which is actually a measure of site quality. Site and soil characteristics, combined with management and natural events, affect survival, health, density, and growth. Good sites with suitable soils are capable of supporting more species of trees, higher densities of trees, and larger, faster growing trees as compared to poor sites. Good sites tend to be more flexible in terms of management options.

10.2.4 Site Quality and Site index

Site quality is based on the volume produced and site index is based on the height attained by a forest crop within a given time. Site-quality measures forest soil productivity in terms of the amount of the maximum timber that the land can produce in a definite time. Site quality determines the kind and magnitude of problems and opportunities that a forester has in managing a stand. No valid silvicultural decision can be made without reference to site quality and other site conditions. Site quality is the sum of environmental factors: soil depth, soil texture, profile characteristics, mineral composition, steepness of slope, aspect, microclimate, species, and management (Daniel et al. 1979). Site quality can be improved by forest and soil management; but it can also be rapidly lowered by mismanagement, fertility depletion, and erosion. Site quality can vary with tree species and the time element chosen. Thus, if the period is held constant at 100 years, then a particular area could have a different site quality depending on the tree species it supported. There are fast- and slow-growing species, and a particular species does not grow at the same rate throughout its life span. Site quality is determined by growing fully stocked stand of the desired species on a site for a designated period. Estimating site quality for forestland has long been a primary concern of American foresters, and extensive volume of literature has accumulated regarding methods for estimating site quality in the USA (Carmean 1975). Canadian foresters have also long been concerned with site-quality estimation and with methods of forestland classification and mapping (Rennie 1962; Rowe 1962, 1971; Burger 1972).

Site index is the height to which a tree will grow in forests in 100 years for most old growth Western species, or in 50 or 25 years for Eastern and Southern species and young-growth stands managed on shorter rotations. For fast-growing energy plantations, 10 years may be considered. Site index has also been expressed as the height of a dominant tree at a particular diameter (McLintock and Bickford 1957; Pflugbeil 1960). Site index is the most widely accepted method for estimating site quality in the USA (Jones 1969; Husch et al. 1972; Daubenmire 1976; Carmean 1975). The site index method is easy to use and correlates well with timber yields. It is a widely accepted means of evaluating site productivity in forests or woodlands. It is also valuable in estimating the potential of an area for tree planting for a number of different purposes (Geyer and Lynch 1987). Height and age measurements from free-growing, uninjured, dominant, or dominant and codominant trees are used with a set of height–age site index curves to estimate total tree height at a specified index age. If it is stated that an area has a site index of 70 at 50 years for loblolly pine, then we expect loblolly seedlings planted now on that area to be 70 feet tall in 50 years (Geyer and Lynch 1987). The site index method is accurate and simple to use (1) when suitable trees are available for measuring site index and (2) when accurate site index curves are available. Such site curves are needed for all important and valuable timber tree species. Several investigators report significant differences in site index between habitat types for several tree species (Stanek 1966; Roe 1967; Hoffman 1976; Mathiasen et al. 1986). Because height growth patterns may vary from area to area, we may have to construct separate site curves to fit certain localities, or certain soil types or climatic conditions (Carmean 1977). Curves or tables have been published for many commercial tree species in the USA (Geyer and Lynch 1990).

The site index curves are constructed predominantly on the performance of even-aged monocultures. However, most of the world's forests are dominated by mixed species stands (Gerlarch et al. 2002). Obtaining site quality of natural forests of complex structures is difficult, considering the great floristic diversity of the tropical rain forests. In inventories, the number of tree species > 10 cm diameter is typically found to be 80–300+ per hectare (Lieberman et al. 1985; Foster and Hubbell 1990; Gentry 1992; Valencia et al. 1994). How difficult would it be to measure growth of all these tree species and develop realistic growth models in the context of hard accessibility in such forests (Clark and Clark 1999). However, there are some studies for the assessment of site quality in mixed stands. For example, Ma Jianlu (1997) evaluated site quality of six Korean mixed forests including Mongolian oak type, Yezo spruce type, Korean spruce type, Amur linden type, Ribbed birch type, and Manchurian ash type on the south slope of Xiaoaingian Mountains. These are, however, low-diversity montane forests.

Although the most direct measure of site quality is stand volume or yield (Gevorkiantz and Scholz 1944), site index or an age–height relationship is more easily obtained and, for years, has been the standard criterion in site evaluation and development of predictive models (Carmean 1965, 1967, 1975). Basal area increment has also been used as the basis for estimating growth and yield in some models (Pienaar and Turnbull 1973; Hall 1982; Pienaar et al. 1985; Dolph 1988; Thomas and Parresol 1989). Site index curves become asymptotic after a given age; stand basal area also becomes asymptotic (Fralish 1994).

10.2.5 Soil-site Evaluation

Soil-site evaluation techniques can be based on estimates of environmental factors (Geyer and Lynch 1987). This approach is used to estimate the growth of a tree species in areas where it is not present. For example, Geyer et al. (1980) developed a mathematical equation utilizing soil and topographic characteristics in native woodlands in northeastern Kansas to estimate black walnut growth potential. It is expressed as:

$$Y = 42.39 + 3.739X_1 + 0.874X_2 + 4.773X_3 + 2.197X_4,$$

where Y is the site index in feet at age 50 years; X_1 is the soil origin, either alluvial or residual (+1 or −1); X_2 is the A horizon thickness in inches or depth to a significant textural or structural change; X_3 is aspect defined as cosine (Azimuth $-45°)+1$; and X_4 is slope, either less than 10% or 10% and greater (+1 or −1). Determining the four independent variables with this formula allows a mathematical evaluation of the site index. There is another more practical approach for the evaluation of site productivity, based on soil characteristics, which requires estimation of several components of the site related to physical condition of the soil, available moisture, nutrient availability, and aeration. Baker and Broadfoot (1979) give evaluation formulas for several tree species in the Mississippi bottomland hardwood area. However, Carmean (1996) observed that site index for jack pine, black spruce, and trembling aspen was found to be poorly related to soil types described in the Northwestern Ontario Forest Ecosystem Classification (NWOFEC). Statistical analyses showed that average site indices for most soil types and groupings of soil types were not significantly different from each other. Site index curves and an edaphic discriminant rule for Scots pine stands in the High Ebro Basin (northern Spain) were presented to estimate site index under different silvicultural situations by Bravo and Montero (2001).

10.3 Forest Management

Sustainable forest management is a key concept that underpins modern forestry practice by recognizing the need to balance the social, ecological, and economic outputs from forests, as outlined in Europe in the principles agreed upon through the Ministerial Conference on the Protection of Forests in Europe (MCPFE 2003). The situation is complex, however. For example, European forests cover a wide range of climatic zones and forest types, ranging from the spruce–pine forests of boreal Scandinavia to the mixed oak and pine forests of Mediterranean Europe (European Environment Agency (EEA 2006)). In addition, there are extensive plantation forests of conifers in Atlantic Europe and broadleaved plantations in Hungary and other Central European countries. In each of these forest types, a range of silvicultural operations can be applied, from intensive systems based on clear felling and artificial regeneration to the fostering of irregular stand structures based on natural regeneration. Each coherent set of silvicultural operations applied to a given forest forms a silvicultural system that may be defined as "the process by which the crops constituting a forest are tended, removed, and replaced by new crops, resulting in the production of stands of distinctive form" (Matthews 1989).

10.3.1 Management of Natural Forests

Forests provide various benefits desired by the people, including clean air, clean water, environmental cooling, open space, opportunity for outdoor recreation, wood and other forest products, and economic activity and employment associated with forest products near villages, towns, and cities (Carey 2006). So, forests must be sustainably managed. A central goal of modern sustainable forest management is maintaining the biodiversity and resilience of forest landscapes in conjunction with ensuring the availability of a wide variety of ecosystem services (Lindenmayer and Franklin 2002). In many countries, safeguarding biodiversity has been set as one of the leading principles and goals in forest policy. Attaining the goals of biodiversity conservation and the resilience of managed ecosystems depends on how well the natural ecosystem structures and processes at various scales can be maintained (Timo Kuuluvainen 2009). The prolonged overexploitation of forest resources culminated in severe resource-management problems. Forests were indiscriminately logged, burned, and grazed without concern for regeneration or watershed protection (Sharitz et al. 1992). However, natural resource management must integrate commercial development and use of forest resources with the maintenance of ecological values. However, the objectives of forest management include improvement of aesthetic quality, forage for livestock and wildlife, recreational use, quantity and quality of water, and reducing fire and flood hazards, invasive exotic plant species, etc. (Poff et al. 2010). These multidimensional and often conflicting objectives are hard to achieve; and concerted efforts are needed for it (Gregory et al. 2006). Climber cutting, suppression of epiphytes, enrichment planting (Fig. 10.1), prescribed burning, rehabilitation of degraded forests, and harvesting are some of the processes of natural forest management. These processes are also adopted for management of man-made forests or forest plantations.

Fig. 10.1 Enrichment planting in a degraded garjan (*Dipterocarpus turbinatus*) forest in Dulahazara, Bangladesh. (Photo courtesy of Md Imam Hossain)

Table 10.1 Areas of forest plantations and abundance of species groups in different regions of the world (FAO 2001)

Region	Total area	Planting rate of new plantation	Plantation area (000 ha) by species groups							
	000 ha	000 ha per year	*Acacia* species	*Eucalyptus* species	*Hevea brasiliensis*	*Tectona grandis*	Other broadleaf	*Pinus* species	Other conifer	Unspecified species
Africa	8,036	194	345	1,799	573	207	902	1,648	578	1,985
Asia	115,847	3,500	7,964	10,994	9,058	5,409	31,556	15,532	19,968	15,365
Europe	32,015	5	–	–	–	–	15	–	–	32,000
North/Central America	17,533	234	–	198	52	76	383	15,440	88	1,297
Oceania	3,201	50	8	33	20	7	101	73	10	2,948
South America	10,455	509	-	4,836	183	18	599	4,699	98	23
World total	187,086	4,493	8,317	17,860	9,885	5,716	33,556	37,391	20,743	53,618

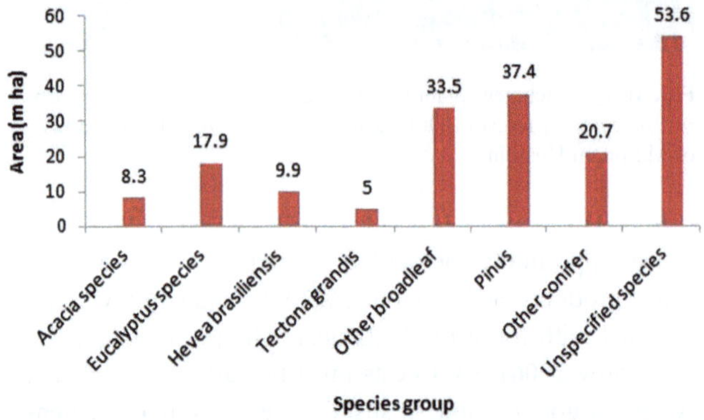

Fig. 10.2 Area of coverage of different tree species groups

10.4 Forest Plantations

Forest plantations are also called man-made forests or artificial forests. They are developed by planting seedlings or sowing seeds of forest tree species either in a place where there was a forest earlier (reforestation) or in former agricultural or otherwise used lands (afforestation). For convenience of management, forest plantations are usually composed of even-aged, uniformly spaced monocultures. They are intensively managed systems like orchards. There are some mixed plantations too, which are composed of two or more species at varying proportions. Many old plantations do not look different from natural forests. Examples of such forests are teak forests of Java, New Forest of UK, and Black Forest of Germany (Evans 2009).

In an estimate in 2000, there were 187 million hectares of forest plantations in the world consisting of 40% broadleaf species, 31% conifers, and 29% unspecified tree species. Asia accounted for 62%, Europe 17.1%, North and Central America 9.4%, South America 5.6%, Africa 4.3%, and Oceania 1.6%. The forest plantation area has increased at the rate of 4.5 million hectares between 1995 and 2000, Asia and South America contributing 89% (FAO 2001). Table 10.1 gives the distribution of global plantation areas with abundance of species groups. The different regions have preferences of tree species. Figure 10.2 shows the areas under different tree species groups. Asia has about 64 million hectares (57%) of broadleaf species, including *Acacia* sp., *Eucalyptus* sp., *Hevea brasiliensis*, *Tectona grandis*, and others, and about 35 million hectares (30%) of conifers. On the other hand, most of the plantation tree species of North and Central America (more than 15 million hectares) are pines. In Europe, there are 32 million ha of unspecified tree species. In Africa, broadleaves make up 47%, mainly with *Eucalyptus*, *Hevea*, *Acacia*, and *Tectona*. Readers should be aware, however, that wide discrepancies exist on data of world plantation areas among different sources. The main reasons for discrepancies between sources are related to differences in the definition and classification of forest plantations (ITTO 2009).

By 2050, the total area of plantations is expected to exceed 200 million hectares, and owing to their generally higher productivity than natural forests, forest plantations are expected to be the main source of wood products globally. Successful forest plantations require species well matched with sites, improvement of genetic stock, control of competing vegetation, and attention to thinning and related operations that enhance tree quality and stand growth (Evans 2009).

10.4.1 Major Plantation Tree Species

A wide variety of tree species are used in forest plantations around the world. The most common tree species are pines, acacias, eucalypts, teak (Pandey 1995), and poplar. Teak and pine have, probably, the longest history of plantation. *Pinus radiata* is the main species used in Australia and New Zealand (FAO Corporate Document Repository;

http://www.fao.org/docrep/w7715e/w7715e05.htm#plantation%20species).

The following species are becoming important recently: Robinia pseudoacacia, Gmelina arborea, Melia azaderach, Casuarina equisetifolia, Calliandra spp., Paraserianthes falcataria, Dalbergia sissoo, Swietenia macrophylla, and

Leucaena leucocephala. More than 1 million ha are planted with P. elliottii, P. taeda, and P. caribaea (Pan 1991) and poplars (Wang 1991). The two acacias, A. mangium and A. auriculiformis (Hong 1991), Eucalyptus camaldulensis, and Tectona grandis (Bhumibhamon 1992) are of major importance in India (Pandey 1995), Malaysia, Indonesia, and Thailand (Haines 1994). Pinus caribaea and Araucaria are also popular in Malaysia (Ahmad and Ang 1993; cf. FAO Corporate Document Repository; http://www.fao.org/docrep/w7715e/w7715e05.htm#plantation%20species).

10.4.1.1 Acacia auriculiformis A. Cunn. Ex. Benth

Acacia auriculiformis occurs naturally in the savannas of Papua New Guinea and Irian Jaya, the islands of the Torres Strait, and Northern Australia (Turnbull 1986). The tree has been extensively planted in moist habitats throughout the tropics and in many subtropical areas. *A. auriculiformis* is a fast-growing, medium-sized tree that may reach 30 m in height. It tolerates a wide range of soil types from light sands to heavy clays, including laterite soils, Vertisols on marl, and eroded soils (Banerjee 1973; Wiersum and Ramlan 1982). It has been grown on uranium mine soils with a pH of 3.0 and on alkaline dune sands with pH 9 (National Research Council 1980). Hu et al. (1983) observed that seedlings of the species grow equally well within the pH range of 4.3–8 in pots. A. auriculiformis can tolerate prolonged waterlogging (Turnbull 1986). This species can grow on poorly fertile soils deficient in nitrogen (National Research Council 1980; Chakraborty and Chakraborty 1989). The species can be used for rehabilitating degraded soils (Banerjee 1973).

10.4.1.2 Acacia mangium Wild

Acacia mangium (mangium) originates from the humid tropical forests of northeastern Australia, Papua New Guinea, and the Molucca Islands of eastern Indonesia (Krisnawati et al. 2011a). *A. mangium* is one of the most widely used fast-growing tree species in plantation forestry programs throughout Asia and the Pacific. Its desirable properties include rapid growth, good wood quality, and tolerance of a wide range of soils and environments (National Research Council 1983). *A. mangium* has been successful in reclamation of soils mined for bauxite, gold, cooper, charcoal, iron, and tin especially in tropical Asia, Australia, and Brazil (Ang and Ho 2002; Ferrari and Wall 2004; Maiti 2006). It can be planted on hard compact soil, savannah areas, dry ridge tops and slopes of hills, moist foothills, and infertile dry soils (Chaw and Mitlohner 2011).

10.4.1.3 Dalbergia sissoo Roxb

Dalbergia sissoo (Indian Rosewood or sisu/sheesham) is native to the Indian Subcontinent and Southern Iran. It is a medium to large deciduous tree with a light crown which can grow up to a maximum of 25 m in height and 2–3 m in diameter. It has been planted both in its natural zone of India and Pakistan and widely outside its natural range. It has been introduced in many countries, including Nepal, Sri Lanka, Indonesia, Mauritius, Ghana, Kenya, Sudan, Nigeria, Zimbabwe, Cyprus, Iraq, the USA, the Gaza strip, South Africa, and Tanzania. *D. sissoo* occurs in open deciduous forest, on alluvial soils which are periodically inundated and along rivers up to 1,500 m (Louppe et al. 2008).

10.4.1.4 Eucalyptus camaldulensis Dehnh

Eucalyptus camaldulensis (red river gum) grows on riverine sites, whether of permanent or seasonal water (Brooker et al. 2002). It is most extensive on grey heavy clay soils along riverbanks and on floodplains subject to frequent or periodic flooding (Costermans 1989). It is also found along channels of sandy watercourses and creeks (Boland et al. 1984). It can also occur in the higher reaches of creeks in major valleys of hilly countries (Cunningham et al. 1981). It is commonly found on brown and red clays (Dalton 1990). E. camaldulensis is a commonly planted tree throughout much of southern Australia and numerous other countries (Eldridge et al. 1993). It grows successfully on a wide range of recharge sites. It is fast growing, readily established, and is mainly cultivated for remediation of degraded sites, particularly mine sites and those affected by salinity (Marcar and Crawford 2004).

10.4.1.5 Gmelina arborea Roxb

Gmelina arborea (gamar) is a fast-growing deciduous tree, occurring naturally throughout greater parts of India at altitudes up to 1,500 m. It also occurs naturally in Myanmar, Thailand, Laos, Cambodia, Vietnam, and in southern provinces of China, and has been planted extensively in Sierra Leone, Nigeria, and Malaysia. *G. arborea* is among the tropical timber species most often planted for pulpwood (Ogbonnaya and Kinako 1993), because of its exceptional productivity with rapid and sustained growth throughout an 8—12-year rotation period (Palmer 1973; Greaves 1981). The productivity of *G. arborea* appears to be closely related to soil fertility. On fertile soils, for example, yields can exceed 30 $m^3\ ha^{-1}\ year^{-1}$, whereas on less fertile soils yields are in the range of 7–18 $m^3\ ha^{-1}\ year^{-1}$ (Palmer 1973).

10.4.1.6 Leucaena leucocephala (Lam.) De Wit

Leucaena leucocephala (ipil-ipil) occurs naturally in the Yucatan Peninsula and the Isthmus of Tehuantepec in southern Mexico and is widely distributed (naturalized) throughout the tropics. It was probably introduced into the Philippines in the 16th century as a feed for ruminant livestock. Subsequently, it spread throughout Asia-Pacific region. In its native range, it grows on shallow limestone soils, coastal sands and seasonally dry, self-mulching vertisol soils of pH 7.0–8.5. In exotic locations, it requires well-drained soils with pH above 5.0 where aluminum saturation is very low. It is intolerant of soils with low pH, low P, low Ca, high aluminum saturation, high salinity, and waterlogging (Fact Sheet Tropical Forages; http://www.tropicalforages.info/key/Forages/Media/Html/Leucaena_leucocephala.htm).

10.4.1.7 Melia azaderach L

Melia azedarach (white cedar or chinaberry tree) is a deciduous tree in the mahogany family, Meliaceae, that is native to Pakistan, India, Bangladesh, Indochina, Southeast Asia, and Australia. It is relatively drought and frost tolerant. Because of its ornamental value, it is often planted as a street tree. M. azedarach is a fast-growing tree which can be grown on wide range of soils, including sandy loam to heavy clay, slightly acidic to slightly alkaline, and shallow to deep well-drained soils (Doran and Turnbull 1997).

10.4.1.8 Paraserianthes falcataria (L.) Nielsen

Paraserianthes falcataria (white albizia or batay) is native to the Bismarck Archipelago including the Admirality Islands, the Moluccas, New Guinea, and the Solomon Islands. It is widely planted throughout the tropics. It is one of the most important pioneer multipurpose tree species in Indonesia. It is preferred for industrial forest plantations in Indonesia because of its very fast growth, its ability to grow on a variety of soils, its favorable silvicultural characteristics, and its acceptable quality of wood for the panel and plywood industries. It plays an important role in both commercial and traditional farming systems in several sites in Indonesia (Krisnawati et al. 2011c).

10.4.1.9 Swietenia macrophylla King

Swietenia macrophylla (big-leaf mahogany) is one of three species that yields genuine mahogany timber, the others being *S. mahagoni* and *S. humilis*. Its natural range stretches from Mexico at 23 N of the equator down the Central American Atlantic coastal strip into South America, continuing in a broad southeasterly arc from Venezuela through Amazonian regions to points as far south as 18 S in Bolivia. Countries where mahogany naturally occurs include Mexico, Belize, Guatemala, Honduras, Nicaragua, Costa Rica, Panama, Venezuela, Colombia, Ecuador, Peru, Brazil, and Bolivia (Lamb 1966; Cornelius et al. 2004). In 2002, *S. macrophylla* was listed in Appendix II (species that may face extinction if trade is not controlled) of the Convention on International Trade in Endangered Species of Wild Fauna and Flora (CITES) (Grogan and Barreto 2005). The largest plantations of *S. macrophylla* have been reported in South and Southeast Asia and the Pacific region. *S. macrophylla* is widely used for avenue planting in some Asian countries including Indonesia, India, and Sri Lanka (Krisnawati et al 2011b). According to Mayhew and Newton (1998), the earliest recorded introduction of *S. macrophylla* into any country is to Indonesia in 1870 with seeds from India. *S. macrophylla* also has great potential for reforestation and afforestation, particularly for improving soil. In the Philippines, this tree species is recommended for revegetation of scrubland and denuded areas (Soerianegara and Lemmens 1993).

10.4.1.10 Tectona grandis L

Tectona grandis (Teak) is one of the most well-known and valued timbers of the world. It has attractive color and grain, durability, lightness with strength, ease of seasoning without splitting and cracking, ease of working and carving, resistance to termite, fungus, weathering, etc. Teak is found naturally in India, Myanmar, Thailand, and Laos (Kaosa-ard 1983); it is an old introduction to Indonesia (Kaosa-ard 1981). Due to its high timber qualities, market demand, and ease of domestication and cultivation, teak plantations have been widely established throughout the tropics from the 1850s (FAO 1956, 1957). Teak has been successfully established as an exotic plant in many countries, e.g., Sri Lanka, Bangladesh, and China in Asia; Ghana, Nigeria, Ivory Coast, Senegal, Togo, and Benin in West Africa; Sudan and Tanzania in East Africa; Trinidad, Puerto Rico, and Panama in Central America; and Brazil and Ecuador in South America (FAO 1957; Keogh 1987).

Teak has been grown on a wide range of climatic and edaphic conditions. It has been cultivated in a range of rainfall and temperature of 500–3,500 mm and 2–48 °C (minimum and maximum range), respectively (Kaosa-ard 1981). Teak grows best on deep, well-drained alluvial soils derived from limestone, schist, gneiss, and shale (and some volcanic rocks, such as basalt). However, it performs very poorly on dry sandy soil, shallow soil, acidic soil ($pH < 6.0$), laterite or peatbog, and on compacted or waterlogged soil (Kaosa-ard 1981). Teak soil is relatively fertile with high calcium, phosphorus, potassium, nitrogen, and organic matter contents (Bhatia 1954; Seth and Yadav 1958; Kaosa-ard 1981). Ranasinghe (2012) concluded that both physical and chemical parameters as well as the effective soil depth for tree growth had shown significant impact on the growth of Teak.

10.4.1.11 Araucaria cunninghamii Aiton ex D.Don

Araucaria cunninghamii (hoop pine) is found in the dry rainforests of New South Wales and Queensland and in New Guinea. It is a large, unbuttressed, symmetrical tree of 50–70 m height; with clean straight cylindrical bole up to 30 m or more. Most natural stands in Australia and Papua New Guinea have been depleted by logging. It is now mainly found on timber plantations. Stands are found on loam, clay loam, or peat soils derived from breccias, agglomerates, coralline or limestone formations, lacustrine deposits, or old volcanic deposits. It is a nitrogen-sensitive rainforest tree species and can be grown without fertilizers as an underplanting to established stands of pines, on soils which will not support open plantings of pure hoop pine without fertilizing with N at the rate of 112 kg/ha year^{-1} (Bevege and Richards 1970).

10.4.1.12 Casuarina equisetifolia L

Casuarina equisetifolia (she-oak or Australian pine; it is not a pine at all) is native to the range extending from Burma and Vietnam throughout Malaysia east to French Polynesia, New Caledonia, and Vanuatu, and south to Australia (north of Northern Territory, north and east Queensland, and northeastern New South Wales) (Boland et al. 2006). There are plantations in Madagascar, Southern USA, and West Africa. It is an invasive species in Florida. Despite an ongoing statewide ban on cultivation, Australian pine is now widely distributed in central and south Florida. It is also an established exotic species in California, Arizona, and Hawaii (Snyder 1992). Worldwide, *C. equisetifolia* has become broadly established and Duke (1983) notes that it is one of the most common trees on the upland margins of beaches in frost-free zones.

10.4.1.13 Populus deltoides W. Bartram ex Marshall

Populus deltoides (cottonwood) is native to North America, throughout the eastern, central, and southwestern USA, the southernmost part of eastern Canada, and northeastern Mexico. It is a large tree growing to 20–40 m height and with a trunk up to 1.8 m diameter. It needs bare soil and full sun for successful germination and establishment; in natural conditions, it usually grows near rivers, with mud banks left after floods providing ideal conditions for seedling germination; human soil cultivation has allowed it to increase its range away from such habitats (Eckenwalder 1977).

10.4.1.14 Pinus caribaea Morelet

Pinus caribaea (Caribbean pine) is a hard pine native to Central America, Cuba, the Bahamas, and the Turks and Caicos Islands. It is a medium-sized evergreen tree with a straight, cylindrical trunk and a rounded to pyramid-shaped crown. It inhabits tropical and subtropical coniferous forests, including lowland savannas and montane forests. Two varieties are mainly planted, var. *caribarea* and var. *hondurensis*. Growth increment and quality of yields vary considerably between the two varieties of *P. caribaea*. In general, var. *hondurensis* is faster growing but of poorer quality (Martin 1971; Lamb 1973; Rollet 1980; Lamprecht 1990). P. *caribaea* has been widely planted outside of its natural range, throughout the tropics in the Americas, Asia, Africa, and Australia on a wide variety of soils (Salazar and Joker 2000; World Agroforestry Centre 2011).

10.4.1.15 Pinus elliottii Engelm

Pinus elliottii (slash pine) is a native to the southeastern USA, from southern South Carolina west to southeastern Louisiana, and south to the Florida Keys (Moore et al. 2008). It is fast-growing short-lived (200 years) tree, and prefers humid climates and moist soils. It is widely cultivated in Australia. It has been extensively planted in forestry plantations in sub-tropical regions for its good quality timber, and is also commonly grown as a garden ornamental and street tree. Slash pine grows best on pond or stream margins where soil moisture is ample, but not excessive, and drainage is poor (Lohrey and Kossuth 1990). Established stands grow well on flooded sites, but flooding restricts seedling establishment (Hebb and Clewell 1976). Soils include Spodosols, Ultisols, and Entisols. Slash pine's native range was probably more restricted by frequent fire than by soil types or soil moisture. With fire suppression, slash pine has spread to drier sites (Hebb et al. 1976; Abrahamson and Hartnett 1990).

10.4.1.16 Pinus radiata D. Don

Pinus radiata (radiata pine or Monterey pine) naturally occurs on the Californian coast (35–37 N) and further south on Guadeloupe Island. For about a century, it has been grown as large plantations in the temperate Southern Hemisphere countries of Australia, New Zealand, South Africa, and Chile (Lamprecht 1990). It has been reported to grow on a range of soils, but the best growth occurs on deep sandy loams derived from marine sediments and with a thick duff layer (McDonald and Laacke 1990). It has adapted in Southern Hemisphere plantations to a wide range of soils from recently formed sand dunes and volcanic soils to older leached red earths and podzols (Turner and Lambert 1986). In Spain, the species has been planted on soils with pH ranging from <4.0 to neutral (Romanya and Vallejo 2004; Afif-Khouri et al. 2010). Many studies have shown that soil groups and parent materials are useful indicators of radiata pine productivity (Turner and Lambert 1986; Gerding and Schlatter 1995; Toro 2004; Romanya and Vallejo 2004; Ross et al. 2009; Watt et al. 2010).

10.4.1.17 Pinus taeda L

Pinus taeda (loblolly pine) is native to the Southeastern USA, from central Texas east to Florida, and north to Delaware and Southern New Jersey. *P. taeda* plantations occur in 11.7 million ha in the southern USA alone (Burns and Honkala 1990). Loblolly is extensively planted in other parts of the world for timber and pulpwood because it can be grown quickly on a variety of sites. It occurs naturally in wet soil conditions and their best growth is found on moderately acid soils in areas of poor surface drainage which consists of a thick, medium-textured surface layer, and fine-textured subsoil. They do not grow well, however, on very wet or waterlogged sites, or in areas with shallow or eroded soils. It is mostly found on Ultisol soils, although small areas of Entisols, Spodosols, and Alfisols are scattered throughout its range. This tree is planted on Histosols following artificial drainage and site preparation.

(http://www.ncsu.edu/project/dendrology/index/plantae/vascular/seedplants/gymnosperms/conifers/pine/pinus/australes/loblolly/habitat.html)

10.4.1.18 Robinia pseudoacacia L

Robinia pseudoacacia (black locust) is native to the USA, including Ohio, Alabama, Georgia, South Carolina, Missouri, Arkansas, and Oklahoma. Its outlying native populations existed in southern Indiana and Illinois, Kentucky, Alabama, and Georgia. It inhabits a broad range of forest types and conditions within its native range, though it develops and grows best in the cove or mixed-mesophytic forests of the central and southern Appalachian region (Huntly 1990). It occurs in a variety of soil types within its native range. Some investigators associate black locust with limestone (Bryant et al. 1980; Bryant 1989), but it is also associated with sandstone (Abrams et al 1998), chert (Read 1952), and mica–gneiss substrates (Mansberg and Wentworth 1984). Soils tend to be loams (Greller 1988) or sandy loams (Oswald et al. 1996).

10.4.1.19 Miscellaneous Hardwoods

In some areas of the USA, plantations of sweetgum (*Liquidambar styraciflua*), sycamore (*Platanus occidentalis*), green ash (*Fraxinus pennsylvanica*), and various oaks (*Quercus spp.*) have been developed. Floodplain hardwood species are planted in the states of Mississippi, Arkansas, and Louisiana. Current plans for restoration on public and private land suggest that as many as 200,000 ha could be restored over the next decade to create wildlife habitat and improve or protect surface water quality (Stanturf et al. 2001; Stanturf and Madsen 2002).

10.4.2 Plantation Development

Development of forest plantations needs a series of activities, including seed collection, preparation of seedbed and containers in the nurseries, seedling development, and planting out. These activities require skilled labor, money, and equipment. Nowadays, nursery business has become popular with the increasing demand of forest and horticultural planting. One can find a large stock of seedlings in a big nursery even in a remote village area of a tropical country. In Bangladesh, there are "tree fairs" (fairs where tree seedlings are sold) in major cities during the planting season, i.e., June–July encompassing the International Environment Day on 5th June.

10.4.2.1 Forest tree Seeds

Provenance

Provenance denotes the place or area or region where the mother trees of the seeds for plantation development were growing. In cases where seeds are collected from an exotic plantation (Jones and Burley 1973), there has been some inconsistency in usage; some authors would define provenance as the place in which the immediate parents were growing as exotics, others would confine its use to the place where the original progenitors were growing in natural forest. Provenances of the same species may differ in growth and wood quality.

Seed Collection

There is little problem for an experienced seed collector to collect seeds from tree species that bear ripe seed in adequate quantities at all times. Some species bear seed throughout the year but only a little at any one time, e.g., *Pinus merkusii* in Indonesia (Keiding 1973), and this makes seed collection slow and expensive. For most tree species, the seeding season is narrow and the collector must collect as much of mature seeds as possible within the short time. Large indehiscent or fleshy fruits may be collected from the ground, but even in these cases collection must be done quickly to avoid losses of seeds due to the destruction by animals, fungi, or premature germination. Seeds are usually collected from accessible and selected plantations or seed orchards. Collection from multi-structured inaccessible natural forests is difficult. It is a general practice to collect seeds when they are mature, because they have a higher germination capacity and a greater longevity in storage than immature seeds. An alternative method is to collect fruits prior to ripening and to store them in relatively cool, well-ventilated conditions which permit after-ripening of the seeds within the fruit. Seeds can be collected from standing trees, bottom of trees, forest floor, and soil below the canopy (FAO; A Guide to Forest Tree Seed Handling; http://www.fao.org/docrep/006/ad232e/AD232E06.htm, retrieved on 16.6.2013).

Woessner and McNabb (1979) found that collection of green or yellow fruits of *Gmelina arborea* from the forest floor gave the best results. They could be stored temporarily in sacks during transit from the field to the fruit-processing depot, without serious loss of viability. Mohammad and Ibrahim (1980) obtained similar results in Malaysia where green and yellow fruits collected from the ground gave more than 90 % germination, but brown fruits, on the other hand, produced only 53 %. Seeds of some hard-coated species may remain viable on the forest floor for years, especially in temperate conditions. In Hungary, seed of *Robinia pseudoacacia* is collected from the forest floor under 30-year-old stands in the Pusztavacs forest district (Keresztesi 1979). A special machine screens the top 10 cm of soil and yields about 770 kg of seed per hectare, which is the equivalent of the yield of approximately 10 seed years. After collection, seeds and fruits need several stages of processing and storage. For detail information, the reader is referred to Food and Agriculture Organization (FAO) (A Guide to Forest Tree Seed Handling; http://www.fao.org/docrep/006/ad232e/AD232E06.htm, retrieved on 16.6.20013).

Seed Storage

Seeds are stored for preservation from the time of collection to the time of sowing (Holmes and Buszewicz 1958). Seeds may need to be stored for variable periods depending on viability and season of sowing: up to 1 year when both seed production and afforestation are regular annual events, but it is necessary to await the best season for sowing; 1–5 years or more when a species bears an abundant seed crop at intervals of several years and enough seed must be collected in a good year to cover annual afforestation needs; long-term storage for purposes of conserving genetic resources. In general, the goal of seed storage is to conserve seeds in a way that maintains their viability and vigor for the longest possible time from harvest to sowing (Hilli et al. 2003; Rajjou and Debeaujon 2008). Detailed storage processes and facilities can be found in a number of literatures (Holmes and Buszewicz 1958; Magini 1962; Stein et al. 1974; Wang 1974; Barner 1975).

Seed Germination

Seeds of some tree species germinate immediately after maturation and before falling from the tree. The example of such a tree species is *Bruguiera gymnorrhiza*, a mangrove plant. This is known as viviparous germination. Germinated seeds are detached from the trees, fall on the mud below, and get anchored. On the other hand, seeds of some species may remain dormant but alive for many years, capable of germinating if an event occurs to break the dormancy state. Between these two extremes, there are many seeds capable of germination under favorable environmental conditions after seed shed. These conditions are (1) adequate moisture, (2) favorable temperatures, (3) adequate gas exchange, and, for some species, (4) light (Krugman et al. 1974). Different species require length of time to complete germination.

Seed Dormancy

Seed dormancy is a condition in a viable seed which prevents it from germinating even if provided with suitable environmental conditions. A viable seed is defined as one which can germinate under favorable conditions, provided any dormancy that may be present is removed (Roberts 1972). The causes of dormancy may be physical, such as hard and thick seed coat impermeable to moisture; chemical, such as inhibitors in pericarp or seed coat; and physiological, such as immature embryo. For example, some studies (Asakawa 1964; Zhang et al. 1981) indicated that Korean pine (*Pinus koraiensis*) seeds possess very deep dormancy caused by both a thick seed coat and an immature embryo, which slow down a series of physiological and biochemical activities (Zhang et al. 1983; Tan et al. 1984). Dormancy is a mechanism of seed protection. However, dormancy poses a serious constraint on efficient nursery management (Bonner et al. 1974). For germinating such seeds, several types of pre-treatments are employed. Recently, Pawłowski (2010) employed the proteomic approach for dormancy breaking of tree seeds generating new data which allow better understanding of the mechanism of deep physiological dormancy in tree seeds like beech (*Fagus sylvatica* L.), Norway maple (*Acer platanoides* L.), and sycamore (*Acer pseudoplatanus* L.).

Seed Pre-treatments

Hard seed coats that prevent the imbibition of water and exchange of gases necessary for seed germination occur most frequently in tree species adapted to alternating dry and wet seasons, including several leguminous genera such as *Acacia, Prosopis, Ceratonia, Robinia, Albizzia*, and *Cassia*. In some species, e.g., *Tectona grandis* and *Pterocarpus angolensis*, the pericarp or fruit forms the hard layer. Pre-treatments to remove physical seed coat dormancy are designed to soften, puncture, wear away, or split the seed coat in order to render it permeable, without damaging the embryo and endosperm within. It is done by some physical and biological methods, dry heating, and soaking in water or chemical solutions. These processes are known as scarification (Bonner 1984). The presence of some chemicals located in the seed coat may inhibit germination of the embryo. This chemical dormancy can usually be overcome by some form of liquid treatment which leaches out the chemicals. Scarification can occur under natural condition through heating of the soil or by alternating temperatures (Gonçalves et al. 2011). This permits the entrance of water to the interior of the seed. This process can also occur through the action of acids during seed digestion by dispersing animals and the action of soil microorganisms (Eira et al. 1993).

Physical Methods

Cutting, drilling, or filing in the seed coat before sowing is the most widely used scarification of forest tree seeds (Goor and Barney 1976). These methods are suitable for seeds of *Afzelia, Albizzia, Intsia* and *Sindora* (Seeber and Agpaoa 1976), *Acacia, Prosopis*, and *Enterolobium*. Chipping of bristly pericarp of *Pterocarpus angolensis* at one edge by knife was found helpful (Laurie 1974). The hard seed coat of *Eusideroxylon* is cracked by a hammer. In *Calophyllum*, complete removal of the seed coat has produced better germination than nicking (Seeber and Agpaoa 1976). Abrasion by sandpaper may also be used to reduce seed coat thickness (Nisa and Qadir 1969). Mechanical scarification is more suitable than the manual method for treatment of large quantities of seeds. Seeds may be tumbled or churned in a concrete mixer with sharp gravel or sand, or in a special drum lined with abrasive material such as sandpaper, cement, or crushed glass or incorporating rotating abrasive disks (Kemp 1975; Goor and Barney 1976). Mechanical scarification is used successfully to overcome seed coat dormancy in *Acacia catechu, A. nilotica subsp. indica, Albizzia falcataria, A. lebbek, Cassia fistula, C. javanica, C. nodosa, Delonix regia, Dichrostachys cinerea, Santalum album, Terminalia arjuna*, and *T. tomentosa* (FAO; A Guide to Forest Tree Seed Handling; http://www.fao.org/docrep/006/ad232e/AD232E06.htm, retrieved on 16.6.20013).

Soaking in Water

For breaking dormancy, forest tree seeds are sometimes soaked in water or other liquids for variable periods. These treatments cause softening hard seed coats and leaching out chemical inhibitors. Seeds having a little resistance to germination may respond well to soaking for 24 h in water at ambient temperature (Kemp 1975). This may be the result of a more rapid imbibition than can be achieved in a moistened nursery bed. It is recommended for some species after manual, mechanical, or acid scarification (Seeber and Agpaoa; Elamin 1975). Matias et al. (1973) found that 48 h of soaking of *Pinus caribaea* seeds in water at room temperature gave more uniform germination than in the untreated seed. Soaking in water for periods varying from 2 to 48 h accelerates germination in *Acacia mearnsii, A. melanoxylon, A. nilotica subsp. kraussiana, Adenanthera microsperma, Albizzia amara, A. procera, Grevillea robusta*, and *Trewia nudiflora*. Alternate wetting and drying of seed is done in hot climates, e.g., for *Tectona* fruits. In Thailand, after scarification, alternate soaking and drying was applied, four times soaking and three times drying, each for 30–45 min (Bryndum 1966). Sometimes, alternate soaking, drying, and baking is repeated a number of times, usually between five and ten cycles, until signs of germination appear (Laurie 1974). Alternate soaking and drying removes chemical inhibitors from *Tectona grandis* and *Terminalia ivorensis* fruits and improves seed germination (Brookman-Amissah 1973). Hot water treatment has given good results with a number of leguminous seeds. The seeds are usually placed into boiling water which is immediately removed from the heat source and left to cool gradually; the seeds remain in the water for about 12 h (Kemp 1975). Hot water treatment is relatively easy and safe to apply, as well as effective with some species. It is not well adapted to large lots because of the difficulty in handling and sowing the swollen seeds (Heit 1967b).

Acid Treatment

Concentrated sulfuric acid is sometimes used for breaking forest tree seed dormancy. For some species, it is more effective than hot water treatment. Seeds stored for a long period may need a longer time in the acid than fresh seeds do. Fresh seeds could be severely damaged by the same length of treatment with acid (Kemp 1975). Great care is needed in the handling of sulfuric acid and this method is not suitable for use by unskilled workers. Dry stored seeds are immersed in the acid for the required period, making sure that all are covered. Treatments are carried out at temperatures between 18 °C and 27 °C (Heit 1967a). Lower temperatures require longer soaking times than do higher temperatures. Seeds are removed from the acid after the required time, washed immediately and thoroughly in cool running water for 5–10 min to remove traces of acid.

10.4.2.2 Seedling Development

Forest Nursery

A nursery is an essential component of plantation forestry. Here, the seedlings or the planting materials are developed from seeds, root or shoot cuttings, suckers, stolons, and any other plant material with necessary silvicultural treatments. A forest nursery can be large or small, temporary or permanent, but it accommodates varieties of tree seedlings for planting in private or public lands. Nursery managers obtain seeds of popular plantation tree species from forest villagers, seed orchards, and seed dealers, screen them for healthy seeds, and sow them in nursery beds, polybags, or trays and take care of them. Planting stocks are also developed by grafting, budding, layering, etc. Although there are some common aspects of seedling tending, different tree species may need different treatments. An area of 0.25 ha is sufficient to raise a small but viable nursery where 1,25,000 seedlings can be raised. The size of the nursery may be increased if need be to increase capacity and demand for planting material. The nursery should be on a gently sloping land to ensure proper drainage. Initially, the seedlings are produced in nursery beds and are pricked out in polypots later. The shape of the nursery should be rectangular for convenience of constructing nursery beds or polybag beds. Plastic-covered shade houses or green houses can be used for seedling production. In these systems, seed trays are placed where seeds are sown for quick germination under controlled humidity and temperature. After germination, when seedlings produce 2–4 leaves, they are picked out and transplanted into polybag or into other types of containers. For vegetative propagation, shoot or root cuttings are placed on tray soil and covered with plastic sheets for some days to keep the soil and cuttings moist. Alternatively, vegetative propagation can be carried out in specially made propagators.

Nursery Bed and Polybag Beds

Seeds are initially sown in nursery beds. After germination and growth of seedlings for some days, they are pricked in polybags. Nursery beds are rectangular in shape and for convenience of taking care, they should have the size of 10 m x 1 m. The thickness of the nursery beds should be 15 cm. Nursery bed is prepared with a forest top soil rich in organic matter. The soil should contain neither too much sand nor too much clay. It should be medium textured (sandy loam to sandy clay loam) to accommodate balanced water and air and to remain well drained. Warkentin (1984) suggested that the desirable physical properties of forest nursery soil are those that provide the optimum environmental conditions for root growth and function. Solid grains (sand), about half the soil volume, provide the framework for stable soil pores and an anchor for plant roots. The large pores, about half the pore space, allow for necessary gas exchange; the small pores store water for plant use.

Containers and Container Media

Seeds may be sown in different types of containers, including earthen or plastic pots and polybags. For large scale planting, seedlings are generally raised in polybags. They are cheap, lightweight, easy to handle, and convenient for distant transport (Thang 1994). Ornamental plants are usually raised in earthen or plastic pots. Various sizes of transparent 200 gauge polybags are used, including 5x10 cm, 7.5x18 cm, 10x20 cm, etc., depending on the size of the ultimate seed lings and the duration of stay in the nursery. Black polythene pots are now being used to raise seedlings as it is slower to degrade under the effects of sunlight and also prevents the growth of algae within the pots. Growth of *Leucaena leucocephala* seedlings grown in polybag, nursery bed, and root trainers was investigated with an aim to select suitable container for quality seedlings for large-scale plantation programs. Seed germination and seedling growth were assessed for shoot length, root length, collar diameter, fresh weight, dry matter production, leaf number, and number of nodules of the seedlings. Germination percentage was found better in nursery bed. Seedlings raised in polybags of 23 × 15 cm size revealed best performance with respect to germination and other growth parameters (Ferdousee et al. 2010). Napoles et al. (1990) studied the effects of four bag sizes (19.6 × 30, 19.6 × 18, 12.6 × 22, and 12.6 × 20 cm) on the development of robusta coffee seedlings in Cuba. Plant growth characters were measured at the age of 5 months. Best results were obtained with bag sizes of 19.6 × 30 and 12.6 × 22 cm. The use of smaller bags reduced the production costs.

Forest top soil, sand plus soil (1:1), peat, soil–peat mix, soil–compost mix, vermiculite, etc. can be used as container media. Inorganic fertilizers are not usually added. However, Wightman et al. (2001) grew seedlings of three economically important and ecologically different native hardwoods, *Cordia alliodora*, *Hyeronima alchorneoides*, and *Calophyllum brasiliense* in root trainers (a book-type container), paper pots, and plastic bags filled with either soil, soil with fertilizer, or compost substrates. In the nursery, species responded primarily to substrate type. *Cordia* grew better in bags of soil with Nitrogen-Phosphorous-Potassium (NPK) fertilizer and compost than in unamended soil, probably responding to higher nitrogen availability. Despite large treatment differences at planting, there were no significant differences in plant size after 1year in the field between book containers and bags. Plants produced in compost were also bigger after 1 year's field growth. Plants produced with soil or in paper pots had higher mortality. Thankamani et al. (1996) evaluated 24 container media prepared from different combinations of sand, soil, farmyard manure, leaf litter, coir dust, vermicompost, and coir dust compost for growth of clove seedlings in the nursery. Plant height, number of branches, tap root length, and root dry weight were recorded after 1 year. The best results were obtained with mixture of soil and vermicompost (1:1). The time to transplanting could also be shortened to 1 year by adopting this medium in the clove nursery.

Fig. 10.3 Women sowing seeds in polybags in a nursery. (Photo courtesy of Mr. Tarik Kabir)

Sowing of Seed

Seeds may be sown either directly into the pots or first in seedbeds or seed trays and the seedlings are later picked out into the pots. Generally, seeds are sown directly into the pots only with fairly large seeds having a good germination percentage. Seeds *Acacia catechu, Artocarpus lakcoof cha, Bassia butyracea, Albizea* spp., *Pinus,* etc. are sown in this way. Seeds may be hand-placed, broadcasted, line sown, or pit sown. Lines are usually 30–45 cm apart; seed to seed distance may be 3–5 cm. The optimum time for sowing seeds on most sites is during the pre-monsoon rains as soon as the soil has sufficient moisture. Figure 10.3 shows seed sowing in a forest nursery.

Bare-root Plants

In this system, seedlings are not grown in separate containers but seeds are sown and a large number of seedlings are grown in a bed. It is best suited to species with medium to large size seed, e.g., *Pinus* spp. These seedlings are much easier to transport than plants in containers.

Large Ball-rooted Seedling

These seedlings are grown in nursery bed up to 75–150 cm for 12 months or longer, and lifted with a ball of soil around their roots 10–15 cm in diameter. The soil is usually wrapped in grass, sacking etc. to keep it moist during transports. This is traditional method in Europe, practically for raising ornamental trees.

Stump (Roots and Shoots Cuttings)

Stumps are prepared by digging up plants from nursery beds, cutting off the stems a few centimeters above ground level, and shortening and trimming the roots. With this method, it

Fig. 10.4 Seedlings of Swietenia mahagoni in a forest nursery. (Photo courtesy of Mr. Tarik Kabir, Forest Dept. Cox's Bazar)

is much easier to transport the plants than container-raised plants and less subject to damage than bare-root plants. Teak was earlier grown from stumps.

Figure 10.4 shows seedlings in a forest nursery ready for planting.

10.4.2.3 Plantation Establishment and Management

Plantation establishment and management practices include site preparation, planting, weeding, fertilizer application, irrigation, thinning, and protection. The following is a brief account of all these practices.

I. **Site preparation**

USDA, Natural Resources Conservation Service (NRCS)—practice code 490 (NRCS 2012) (www.nrcs.usda.gov/Internet/FSE_ DOCUMENTS/nrcs143_026116.pdf)—states that forest site preparation aims at encouraging natural regeneration of desirable trees and shrubs or providing optimum site conditions for planting or direct seeding of desirable woody species. A purpose of site preparation is to prepare the land for establishing a stand of desirable woody vegetation by controlling undesirable vegetation, removing slash and debris, or altering site conditions. In plantation establishment, site preparation is key to the efficiency of subsequent silvicultural treatments in the stand. Efficient and effective site preparation is, therefore, paramount to the success of the forest plantations.

USDA-NRCS drafted the following criteria for site preparation for planting trees (NRCS 2012):

A. **General criteria**

1. Protection of existing desirable vegetation.
2. Management of slash and debris so that they do not harbor harmful levels of pests, hinder equipment operations, or create undue fire hazard.
3. Minimizing erosion and/or runoff caused by site preparation by other conservation practices.
4. Choosing cost-effective methods and protecting culture resources, springs, seeps, wetlands, and other unique areas.
5. Evaluating impacts on wildlife habitat as part of the planning process.

B. **Mechanical site preparation criteria**

When removal of competing vegetation is a necessity or where a low ridge is desired to prevent loss of young trees, this method can be followed to expose mineral soil for establishing desired species. The criteria for mechanical soil preparation include:

1. A heavy disk, bulldozer, KG Blade, heavy plow, or other suitable implements can be used to prepare the site.
2. To minimize erosion or runoff, the following criteria shall be implemented:
3. Not more than 50% of the soil surface will be disturbed over the area treated.
4. On sloping areas of 6% or more, all operations will be done on the contour as far as practicable.
5. Additional erosion control practices will be installed as identified in an approved conservation plan.
6. Soil compaction and displacement will be minimized.
7. Ridges prepared for planting on wet sites will be no less than 36 inches (90 cm) wide and no more than 12 inches (30 cm) high.
8. This method is not suitable for steep and stony soils.

C. **Chemical site preparation criteria**

The following criteria will be used when site preparation with herbicide is selected:

1. Herbicides labeled for only forestry use will be used.
2. The amount, timing, and method of application will be according to the label recommendations of the manufacturer.
3. Care will be exercised to prevent damage to wildlife and fish populations, agricultural crops, and desired vegetation in the vicinity due to herbicide drift, overspray or runoff.
4. Herbicides will be applied under the direction of a qualified individual. Restricted use herbicides must be applied by a licensed applicator.

D. **Prescribed burning site criteria**

1. A burning permit should be obtained from an appropriate authority.
2. Burning will not be done on organic soils.
3. All burning must be done according to a burning plan developed by proper authority.

There are several methods of mechanical site preparation, including shearing or KG blading, drum chopping, lopping, disking, trenching, burning, bedding, draining, etc. Shearing or KG blading is used to remove large numbers of stems more than 4 inches (10 cm) in diameter. KG blades are bulldozer blades with a sharpened lower edge, angled or V-shaped and have a "stinger" for splitting large trees and stumps. Drum chopping can be effective where brush competition is dense but of small diameter (<4 inches). A rolling drum chopper with offset blades is used to uproot, chop, and compact brush. Lopping involves using hand tools or chain saws to fell residual stems and leaving the felled stems where they fall. Lopping is cost-effective where scattered large diameter residuals are present. Disking with heavy disks may be all that is required to prepare a site for natural seeding or direct seeding.

Disk trenching is a relatively light soil preparation method. The main idea in disk trenching is to expose mineral soil and to reduce competition between the crop trees and other vegetation (Uotila et al. 2010). Mounding is a more intensive method and it has become more common especially in connection with spruce planting (Saksa et al. 2005). In addition to reduction of competition, intensive preparation accelerates nutrient release and enhances soil temperature more than disk trenching does (Örlander et al. 1990; Sutton 1993; Nordborg et al. 2006). On spot-mounded sites, the share of exposed mineral soil of the total area is less than that in disk trenching (Luoranen et al. 2007).

Various mechanical site preparation techniques for treatment of soils and vegetation are important tools to prevent low seedling survival and growth. Mechanical site preparation often results in improved seedling survival and growth. However, if not intensive, methods with much soil disturbance are used, it is a rather ineffective tool for controlling competing vegetation (Magnus et al. 2012). Wallertz and Malmqvist (2013) conducted an experiment on a fresh clearcut at the Asa experimental forest (57°10′ N). The effects of five different site preparation treatments were investigated: control, patch, mound, invert, and mix. In each treatment, 40 seedlings of Norway spruce and 40 of Douglas fir were planted in each of four blocks. Site preparation had little or no effect on the survival and growth of Norway spruce: Only a few seedlings died during the first 2 years. For Douglas fir, however, all site preparation treatments increased survival compared with the control, where mortality was high. Pearson et al. (1995) described the effects of site preparation on tree growth and the associated understory woody and herbaceous plant succession. During the first 3 years after treatment, the mechanical site preparation method (shear–windrow–burn) reduced woody plant heights more than the underplant-release method; however, these height differences disappeared by the 6th year of posttreatment. Different site preparation techniques have been used for many years in Turkey. Lately, machine (a rake-equipped bulldozer) clearing has become the sole technique used for this purpose (Yildiz et al. 2010). Machine site preparation, however, may remove substantial amounts of topsoil. Thus, this method may result in significant ecosystem nutrient losses, which eventually may reduce seedling growth (Yildiz et al. 2007).

In developing countries all site preparation procedures are manual, including clear-felling (in case of reforestation), and slashing and burning (in both reforestation and afforestation). Damage to soil is minimum there; still there is considerable erosion and fertility depletion due to clear-felling followed by artificial regeneration (Southgate and Whitaker 1992). Burning and forest fire effects on soil have been discussed in Chap. 8.

II. **Planting**

Earlier planting operations were largely manual. Seedlings were planted in notches and holes made with a spade, shovel, or boring with an auger. There are single notch and double notch planting methods for bare-root seedlings. In single notch system, a single slot is made by pushing the spade to the required depth and by rocking backward and forward, and the roots are inserted. The spade is withdrawn and the slot is closed with the ball of the foot. In double notch system, two slots are made with the purpose of creating space to allow the roots to be distributed evenly. Once the seedling has been positioned in the slot, the spade is removed and the soil is firmed with the ball of the foot. Larger holes can be made with spades and shovels for seedlings in comparatively large bags or containers. Small basins (double the size of ball) are prepared for placement of seedlings with root ball. Auger hole method is being used for tree planting since the 1980s (Abrol and Sandhu 1980). Auger of 15–20 cm, corroborating with the polybag diameter, is pushed into the desired depth of soil. It is convenient for vertical draining, piercing the compact layers, and placing fertilizers and manures into the auger holes, and it causes minimum soil disturbance. However, it is hard to make auger holes in the dry season or on compacted soils (Jha and Sen Sharma 2009). Motorized augers can be used in reforestation and are becoming popular for restoration projects. Augers work best in deep soils without too many large rocks or roots and are the best planting tool to use for larger, taller stock types (Dumroese and Haase 2010). One concern of auger hole digging particularly in wet conditions with clay subsoil is compaction at the sides of the holes (Williamson and Coston 1990). In these systems, seedlings are so placed that the collar remains at the ground level. Seedlings are planted at lines. Plant spacing is specific for species or silvicultural systems. In Indian sub-continent, most tree species were earlier spaced 6x 6 ft (1.8x1.8 m). Such close growing makes the stem tall and straight, and fewer branching from the initial stage of development, particularly of teak plantations (Troup 1921). Spacings of 1.8x1.8 m, 3.7x 3.7 m, and 4.6x 4.6 m have been used for loblolly and slash pines. Studies have shown that there is difference in height growth, which is higher in less density, at different ages among different spacings (Smith and Strub 1991).

Planting machines are being used in both America and Europe for some years. Initially, their use in the UK was mainly restricted to the sandy soils of East Anglia. Now, interest in mechanical planting machines has widened to other soil types too (Forestry Commission 1977). Planting machines offer a means of planting a woodland quickly using a small team of operators. While the quality of machine planting may be lower than standard manual planting, it is consistent and can achieve better results than operational manual planting. Manually controlled machines can achieve a higher planting quality than semiautomatic planters because the operator can influence the positioning of the plant in the furrow (Forestry Commission 1977). Dumroese and Haase (2010) suggest that container plants are ideal for machine planting because of their compact root system and uniformity. Steadily increasing labor costs and difficulty in finding skilled planters have motivated many reforestation and restoration specialists to look at machine planting (Hallonborg 1997). Luoranen et al. (2011) evaluated the effects of planting machines (Bracke and Ecoplanter) on the quality and field performance 1 and 3 years after planting of Norway spruce (*Picea abies* (L.) Karst.) seedlings in central Finland. Both machine types planted, on an average, 1,800 seedlings per hectare, and after 3 years approximately 1,600 (Bracke) and 1,200 (Ecoplanter) survived. This study suggests that planting with a Bracke machine can achieve better regeneration rates than those observed in privately owned Finnish forests.

10.4.2.4 The Taungya System of Plantation Establishment

The taungya (Burmese taung = hill; ya = cultivation) method can be defined as a reforestation method in which tree saplings and agricultural crops are simultaneously planted at the initial stage of reforestation (Combe 1982; Evans 1982). The taungya system was evolved in Burma (Myanmar) during the nineteenth century (1856) and it had spread to other parts of the world (Blanford 1958). In this system, peasants from areas within or surrounding the forest and sometimes plantation workers were allowed to cultivate food crops within the forest during plantation establishment and early stages of growth. In addition to growing food crops, their responsibility was to plant and tend tree seedlings according to silvicultural prescriptions. They did the entire job with simple hand tools. They used the land for cropping for 3–5 years until the canopy closure. They moved to a new place, where they slashed, burned, and developed a new forest plantation there. They were given no wages (Chamshama et al. 1992). Intercropping maize with young *Cupressus lusitanica* or *Pinus patula* in taungya system in Kenya did not affect maize growth, nutrient uptake, and yield during the first 2 years indicating the suitability of the taungya system for establishment of plantation forests. The taungya system was effective in improving the survival, growth, and nutrient uptake and utilization by newly established forest plantations by significantly reducing weed competition (Imo 2008).

III. **Firebreaks**

The establishment of firebreaks is necessary to protect plantations in the early stages. Plantations are divided into blocks of about 25 ha for establishing firebreaks. However, different planting sites can have different block sizes. The block should not be too small because it would need more area, and it should not be too big because it cannot prevent the fire effectively. These blocks are separated from each other by firebreaks which can be trails, clear strips of land, creaks, and green vegetation strips. The width of the strips could be about 10 m.

IV. **Weeding**

Unwanted vegetation or weeds compete with young seedlings or plants for light, water, space, and nutrients. A thick ground cover may also harbor harmful animal pests, which can destroy the plants. However, the intensity of the weeding operation should be chosen so that establishment of tree crop can be achieved at a reasonable cost and at the same time, weeding operations do not create environmental concern. Until a few years ago, weed control in forestry was generally done by hand (Lyle 1981). Manual weeding is still the weed control practice in forestry of developing countries. Manual weeding can be heavy or light. Heavy weeding is done by removing the ground flora after the monsoon. Light weeding is done by removing vegetation from a circular band in the base of the saplings.

At present, chemical weed control is gaining ground. It involves matching of chemicals to combinations of weeds and trees to favor the trees, and to do this within the limits of the law, soil stability, and environmental safety (Lyle 1981). Some inorganic salts such as sodium arsenate and ammonium sulfate were used as herbicides for vegetation management as early as 1915 (Sutton 1958). Use of herbicides in forestry became popular after 1945 with the use of 2,4,5-trichlorophenoxyacetic acid (2,4,5-T) on woody vegetation. Within a few years after the war, its use was widely accepted (Malik and Vanden Born 1986). A few years later, a large number of organic herbicides such as chloro-striazines, amitrole, and chlorpropham were produced. In 1954, about 27,500 ha of brush were treated in Canada (Suggit 1956), and this increased to 34,000 ha in 1955 (Suggit 1957). In 2001, herbicides were used on 74,46 ha in the National Forest System (including rangeland) of the USA (Shepard et al. 2004).

Anon (1975) states that a given application of herbicide is merely part of a program of vegetation management. It is always wise to work out a plan for managing ground cover for the life of the plantation, and to prescribe for weed control with the most suitable tools currently available. Such a plan generally considers: (1) Total weed control is usually necessary for first-year establishment of nursery seedlings; (2) weeds resistant to the first year's applications will make up a large proportion of the subsequent years' weed problems; (3) complete weed control will result in surface soil movement in the third and subsequent years of management unless stabilization measures are taken; and (4) in the third and subsequent years, animal populations, especially gophers, will be attracted to plantations with residual weed growth. Atrazine (2-chloro-4-ethyllamino-6-isopropylamino-1,3,5-triazine) is one of the most extensively used herbicides worldwide. It has been intensively studied (Means and Wijayaratne

1982; Davis-Carter and Burgoa 1993; Sprague et al. 2000). It has moderate water solubility, and is recommended for either pre-emergence or early post-emergence control of grasses and weeds. It is relatively non-toxic to humans and animals, non-corrosive to spray equipment, and non-flammable. It has been used for herbaceous weed control in young pine plantations alone and in combination with other herbicides. It is readily sorbed in soil particles and is rather resistant to decomposition in soils, and has relatively low solubility resulting in low mobility and high persistence (Pinho et al. 2007). Broadleaf weeds are generally treated with 2,4-dichlorophenoxyacetic acid (2,4-D). About 27,000 lbs of 2,4-D and 5,900 lbs of triclopyr were used by the USA in 2002 (USDA Forest Service 2002). Usually, ground/aerial broadcast, high/low volume foliar, and backpack sprayers control weeds by applying a mixture of herbicides (Zhang et al. 2011). Due to their high efficiency and low mammalian toxicity, pyridine and phenoxy herbicides are essential for many agricultural uses. Triclopyr (Garlon 4; 3,5,6-trichloro-2-pyridinyloxyacetic acid) and 2,4-D, two broad-spectrum broadleaf herbicides which act as auxin agonists in woody plants, and in annual and perennial weeds, are formulated as esters which are hydrolyzed in human body to acids and rapidly excreted with urine (Zhang et al. 2011). Glyphosate is also used for chemical weed control in forest plantations (Lamb 1975). Indaziflam is a new herbicide which potentially provides pre- and post-emergence control of broadleaf and grass weeds. It was first registered in 2010 for use on turf by the US Environmental Protection Agency (EPA). Indaziflam represents a new chemistry from the alkylazine chemical class (Kaapro and Hall 2012). Indaziflam has the potential to control a broad spectrum of weeds and provides long-lasting efficacy at low application rates.

According to Malik and Vanden Born (1986), there was a setback in the use of phenoxy herbicides in North American forestry in the early 1970s because of concern over their suspected environmental hazards. The Rebuttable Presumption Against Registration (RPAR) process initiated by the EPA in the USA in 1978 restricted the use of 2,4,5-T and 2,4,5-TP, and the EPA banned all uses of these herbicides in October 1983. A Nova Scotia court case in 1983 and later the ruling of a British Columbia court against the use of herbicides in the Skeena River area have shaken the confidence of the general public in the safe use of herbicides. The biggest public concern about use herbicides in forests is the potential for toxicity to non-targeted organisms, primarily humans, pets, livestock, and wildlife (Guynn et al. 2004; Tatum 2004). Information, so far gathered, suggests that at recommended rates and normal use scenarios, herbicides used in forest management operations pose little, if any, acute toxicity hazard to wildlife species. They are not mutagenic or oncogenic, and are rapidly eliminated from animal systems once ingested/absorbed (McComb and Hurst 1987; Miller and Miller 2004; Tatum 2004).

V. **Thinning**

If seedlings are planted initially at 2 x 2 m spacing, then one hectare of land would accommodate about 2,000–2,500 trees. This is a very high density. This density is used initially so that the saplings become straight and have fewer branching. Oosterbaan et al. (2009) suggested that the ultimate aim of plantations was to produce high quality, knot-free, and large boles which give good quality timber. In case of broadleaf tree species, a mature stand retains only 350–500 trees at the harvest time. The excess trees are periodically removed from the stand by thinning. Natural thinning occurs when stand density is high. The lowest branches of the surviving trees die and fall off to a certain extent, depending on the tree species. Stem diameter increases rather slowly because of the competition between the trees. The first thinning is done when young trees have completely closed the canopy, offered competition among themselves, and the stems have attained a mean diameter at breast height of > 10 cm. During this stage, a great part of height growth takes place and there is a high competition between the young trees (Oliver and Larson 1996). In Indian silviculture, for broadleaf hardwood species, every alternate tree (spacing 1.8 x 1.8 m) is removed by cutting at 1.30 m height at 5 years of age so that the canopy opens and the trees can subsequently grow in girth. This is called mechanical thinning. Another mechanical thinning is done after next 5 years (Champion et al. 1965). Mechanical thinning is also called pre-commercial thinning which is an age-old practice in forestry (Ligne 2004). Pre-commercial thinning is defined as the removal of trees not for immediate financial return but to reduce stocking to concentrate growth on the more desirable trees (Helms 1998), and cutting in an immature crop or stand to improve crop spacing and to accelerate the diameter increment of favored trees and/or form the trees that remain (Anon. 2000). Afterwards, faulty, diseased, malformed, and competing trees are removed and this is called selective thinning. Ligne (2004) studied an alternative method of thinning which could reduce competition for light and other factors. It was a process of beheading in which the apex of the tree is cut and removed. Cutting the apex of a tree breaks its apical dominance and, thus, enhances compensatory growth (Maschinski & Whitham 1989; Rinne 1994).

Thinning and associated managements in forest plantations can influence the functions and structure of forest ecosystems as well as the status and dynamics of soil nutrients and organic matter that affect the forest ecosystem and related environment. Litter accumulation and decomposition rate are also affected by different thinning treatments. Thinning treatments affect concentrations of K^+, Fe^{3+}, Al^{3+}, and anions in soil and biomass (Chiang et al. 2010). Contents of N, P, K, Ca, and Mg in *Pinus sylvestris* needles, branches, and stems before and after thinning were studied in 37- and 32-year-old plantations by Blanco et al. (2006a). Thinning significantly decreased all nutrient pools but no significant differences were found between thinning intensities. Many researchers (Huebschmann et al. 1999; Caldentey et al. 2001) have reported that stand thinning decreases litterfall production with concomitant reductions in nutrient inputs to the soil from biomass (Chertov et al. 1999; Blanco et al. 2006a, b). Agren and Knecht (2001) found in *Pinus sylvestris* and *Pinus contorta* forests that this may be a temporary effect, because litterfall values returned to normal in the following years.

VI. **Fertilizer application in forest plantations**

Miller (1981) proposed three guideline concepts for fertilization in pines which can be applicable to other forest types as well.

(1). Fertilizers are generally of benefit to the trees, not the site, and measureable permanent site improvement is only likely if the amount of nutrient applied is large in relation to the soil capital.

(2). In the absence of any site improvement, fertilizer response is best explained, and costed, by using the simple analogy of an acceleration through time, leading to a reduction in rotation length.

(3). Consideration of the variations with tree age in the patterns of nutrient uptake, cycling, and immobilization, and in the capture and retention of atmospheric nutrients, leads to the suggestion of three distinct nutritional stages. Stage I: Prior to canopy closure, tree growth is very dependent on soil nutrient concentrations and response to a number of nutrients can be expected. Stage II: After thinning on low nitrogen capital sites, immobilization of nitrogen in biomass and humus as trees age can lead to the progressive development of deficiency. Stage III: A deficiency that may eventually disappear as tree demands fall with age.

It was widely accepted in the 1930s that nitrogen (N) limits primary production in temperate forests (Hesselman 1937; Mitchell and Chandler 1939). Nitrogen was found to be the primary limiting nutrient for tree growth on mineral soils in most boreal and temperate forests (Tamm 1991; Vitousek et al. 1997). Co-limitation by N and phosphorus (P) has also been suggested, and on some soils, calcium (Ca) and base cations are in short supply (Matthew and Vadeboncoeur 2010). On the other hand, several investigators (Vitousek 1984; Vitousek and Sanford 1986; Townsend et al. 2007) observed that tropical forests on old and highly weathered soils are generally rich in nitrogen but poor in plant available P and possibly other rock-derived nutrients. To overcome these limitations, fertilizers are used in intensively managed forest plantations, otherwise considered as tree crops. Many investigators demonstrated the benefits of forest fertilization (Henry 1986; Edmonds and Cole 1977; Allen 1994; Allen and Lein 1998). Forest fertilization has become a basic component of most intensively managed plantation forests around the globe (Binkley et al. 1995). Fertilization programs are well developed for forests in Australia, New Zealand, South Africa, and several forest regions in Europe and North America (Cole and Gessel 1965; Bengtson 1979; Miller 1983; Schutz 1976; Turner 1983). Many programs are directed to ameliorating nutrient deficiencies which adversely affect forest plantation establishment, while others are for the purpose of enhancing growth rates in natural and man-made forests (Chappell et al. 1991). Forest fertilization is a more common practice now; area under fertilized pine plantations increased 30-fold in 1998 over that in 1988 in nine southern states of USA (NCSFNC 1998). Forest fertilization is a profitable investment, generally yielding real internal rates of return exceeding 10% (Laakkonen et al. 1983). Apart from the increment in wood volume, another important revenue factor associated with forest fertilization is the effect on tree dimensions. An increased stem diameter translates into increased timber proportions as well as an increased timber price. Furthermore, a greater stem diameter reduces logging costs (Brunberg 2007).

The cost of fertilizer is a major item of expenditure, normally accounting for up to two-thirds of total fertilization costs. Composite fertilizers, containing, for example, phosphorus (P), potassium (K) and magnesium (Mg), are considerably more expensive than nitrogen-only fertilizers. Therefore, in order to keep the fertilization budget as low as possible, it is vital to avoid standard nutrient addition regimes if they do not result in increased production in a particular forest stand (Jacobson and Pettersson 2010). The most widespread practice is fertilization with nitrogen, or nitrogen plus phosphorus. Nitrogen is commonly applied as urea (especially in North America), ammonium nitrate (particularly in Europe), or diammonium phosphate (particularly in the southeastern USA). The application of 150 kg N ha^{-1} is a normal commercial dose in Sweden for middle-aged and older stands (stands in which the canopy has closed, and after the first thinning). It results in an average relative growth increase of about 30–55% over a period of 7–11 years (Pettersson 1994). Mergen and Voight (1959) studied the effects of three formulations of fertilizer applied at five rates to two age classes of slash pine growing in Florida. The fertilizers caused increases in the nutrient contents of needles and branches of the younger trees and increased basal area growth as much as 32% in the older trees. According to Miller and Fight (1979), the optimum fertilizer dosage for established stands of Douglas fir is between 150 and 300 pounds of nitrogen per acre (168–336 kg N ha^{-1}). Lesser amounts may fail to provide measurable effects. Larger amounts may provide slightly more volume and longer duration of response, but in some locations can lead to increased mortality (Lee 1974) or snow or winter breakage brought on by sudden stimulation of foliar growth (Miller and Pienaar 1973). The Pacific Northwest forest industry and agencies have long used forest fertilization to enhance the yield of second-growth coastal Douglas fir. Fertilizer can be applied by hand over snow without problems. It melts through the snow cover (up to 45 cm) rapidly and dissolves into the soil. Applications over snow covered tree canopies are not recommended (Walker and Gessel 1991).

Fertilizing effects of organic wastes have also been tested on forest plantations. Da Silva et al. (2011) suggested that sewage sludge applications might reduce the amounts of mineral fertilizers needed to sustain the productivity on infertile tropical soils, but sewage sludge must be applied with care to avoid soil and water pollution. Several studies have assessed the effectiveness of applying organic waste residues to forest areas and the effects on tree growth (Henry and Cole 1997). Early investigations in the USA have shown positive effects of sewage sludge applications on the development of conifer plantations (Harrison et al. 1994; Henry et al. 1993). Sewage sludge releases nutrients slowly and continuously into soil solutions during decomposition, which is an advantage in comparison with mineral fertilizations (Guedes and Poggiani 2003). The slow release of nutrients makes it possible to restore soil nutrient stocks throughout the development cycle of forest plantations (Garciano et al. 2008). Previous research suggests that the application of sewage sludge might significantly improve the economic performance of forest plantations due to increases in wood production (Kimberly et al. 2004). Convincing growth responses have been reported following application of municipal sewage sludge to forests of coastal Douglas fir (*Pseudotsuga menziesii*) (Cole et al. 1984), white spruce (*Picea glauca*) (Gagnon 1973), loblolly pine (*Pinus taeda*) (McKee et al. 1986), Corsican pine (*Pinus nigra* var, *maritima*) (Moffat et al. 1991), and Sitka spruce (*Picea sitchensis*) (Bayes et al. 1987). To avoid high rates of nitrate leaching from sludge-amended forest, sewage sludge had been mixed with wood ashes, sawdust or pulp, and paper sludges to raise the C/N ratio of the sludge (McDonald et al. 1994). Mixtures of these wastes in appropriate proportion have been effective at reducing nitrate losses and promoting tree growth (Henry 1986; Vogt et al. 1980). On the assumption that heavy metals in sewage sludge can accumulate in edible mushrooms on the forest floor, Benbrahim et al. (2006) conducted a study in six sites of maritime pine forests in the South-West of France. Municipal sludge (dry matter) was applied at a rate of 6 t ha^{-1}. Edible mushrooms were collected 2 years after sludge application. The concentration of trace metals increased in the soils, and Cd, Pb, and Hg concentrations slightly increased in mushrooms, but sludge application did not affect As, Cu, Se, and Zn concentrations in carpophores. This effect was dependent on sludge type and sites.

Miller and Miller (1988) suggested that there was increasing concern over the level of nitrogen inputs to forests particularly in N-saturated ecosystems in which the trees do not show positive response to the addition of further nitrogen, nitrogen addition leads to growth disturbances, or added nitrogen leads to increased losses of nitrate in stream water. Stream water quality in forested watershed is better than that in areas under other major land use systems (USEPA 1995). Total nitrogen and phosphorus concentrations in water draining agricultural areas in the USA are about nine times greater than those in forested streams (Omernik 1977). The concentration of nitrate-N, which is particularly important for water quality, average about 0.23 mg N l^{-1} for very large forested watersheds in the USA, compared with 3.2 mg N l^{-1} for streams in large agricultural watersheds (Omernik 1976). However, some forest practices may significantly alter water quality (Binkley et al. 1999). Binkley and Brown (1993) suggested that forest fertilization may increase concentrations of nutrients in stream water. However, many studies concluded that forest fertilization poses little or no risk to water quality parameters (Tamm et al. 1974; Fredriksen et al. 1975; Norris et al. 1991; Bisson et al. 1992; Shepard 1994; Burnham and Allen 1997).

VII. **Irrigation in forest plantations**

Irrigated forest plantations can be established for the commercial production of fuelwood, posts, construction lumber, and fodder. The use of irrigation practices also allows the use of more exacting fast-growing tree and shrub species. Four methods of irrigation can be adopted: surface flooding, border check, basin irrigation, and furrow irrigation. Reclaimed or treated wastewater can be an important source of irrigation water for forestry and agroforestry facilitating wood and fuelwood production and creating windbreaks and shelterbelts for farmers and smallholders (Armitage 1985; FAO 1989; Braatz and Kandiah 1996). Treated sewage effluent is often reused for irrigation in fast-growing forest plantations in arid and semiarid regions. Treated effluent water, however, contains high levels of salts, toxic ions, heavy metals, and organic residues (Hassanli et al. 2008). Accumulation of these pollutants in soil poses a threat to the environment (Abbott and El-Quosy 1996). Some chemical substances, particularly nitrogen can be leached limiting the sustainability of effluent-irrigated plantations. Soils can build up slowly soluble P compounds which can be washed into streams by runoff and overland flow (Myers et al. 1999). However, irrigation of tree crops has become an effective means of effluent disposal on land with minimal health risks and recycling effluent nutrients (Stewart et al. 1990). Many examples of effluent-irrigated forest plantations are available in the world (Stewart and Flinn 1984; Cromer et al. 1984). The effects of effluent irrigation on soil properties were studied by Cromer (1980) and Stewart et al. (1990). They reported that effluent significantly increased pH, phosphorus, sodium, calcium, and magnesium of the surface soil. According to them, the overall soil chemical properties were not adversely affected by effluent irrigation. The nutrient enrichment occurred mainly in the upper 0.35 m soil. Cromer et al. (1984) concluded that irrigation with wastewater substantially increased tree growth. Hassanli and Javan (2005) reported that trickle irrigation system can be employed for treated municipal effluent application in plantations without serious concern of adverse effect on the environment. High sodium concentration in irrigation water can replace exchangeable Ca^{2+} and Mg^{2+} from clay surfaces causing their dispersion, permeability reduction, and poor aeration. Toxic elements in reclaimed water are absorbed by trees and remain locked into biomass (Myers et al. 1999).

VIII. **Forest harvesting**

Forest harvesting is the extraction of trees from the forests at the end of the rotation age or after the trees have attained a desired size. The trees are usually clear-felled (i.e., all the trees are cut), but sometimes, particularly in natural forests, selective felling (only the larger trees are cut) is done. Harvesting can be done manually or mechanically. In both systems and means, some soil disturbance such as displacement and compaction must occur. This has been discussed in some detail in connection with soil disturbance in Chap. 8.

10.5 Effects of Forest Plantations on Soil

Forest plantations were introduced to tropical regions to supply fuelwood, charcoal, fodder, sticks, and building materials. They were also planted to restore degraded lands, to control soil erosion, or to serve as buffer zones around roads and areas of natural forests (Evans 1999; Hartemink 2003; Jagger and Pender 2003; Mishra et al. 2003). Today, afforestation is considered an option to mitigate climate change through carbon sequestration in tree biomass and soils (IPCC 2000; Paul et al. 2002; Turner et al. 2005). Forest plantations are generally believed to sustain a low diversity of wildlife; they are high consumers of water and nutrients and increase soil acidification (Cannell 1999; Jagger and Pender 2003). Soil properties under forest plantations in comparison to natural forests, pastures, natural savannas, and croplands have been investigated all over the world (Lilienfein et al. 2000; Zdenko 2002; Hartemink, 2003; Jobbágy and Jackson 2003; Mishra et al. 2003; Turner et al. 2005). Results of these investigations have mostly indicated that forest plantations gradually improve the physical conditions (structure, density, porosity) and fertility (supply of nutrients including N, P, K, Ca, and Mg) of degraded soils. Nsabimana et al. (2008) investigated the effects of forest plantations made up of monodominant stands of 17 different native and exotic tree species as well as a stand with mixed native tree species in Rawanda. Their results revealed that the plantation species composition influenced the soil chemical properties. Soil pH, cation-exchange capacity (CEC), total C and N, and available P differed significantly between plantation stands of different species. Mixed native tree species increased the levels of soil organic C, total N, CEC, and base saturation. They recommended that afforestation of abandoned and less productive lands should utilize fast-growing tree species in combination with agroforestry and native species in order to maintain or improve soil chemical properties. The effect of different tree species on soil pH is most significant in the first few centimeters of the topsoil. Augustoa et al. (2002) observed that topsoil pH under *Picea abies* and *Pinus sylvestris* was significantly lower than under *Fagus sylvatica*, *Quercus petraea*, or *Quercus robur. Abies alba* and *Pseudotsuga menziesii* appeared to be intermediate. All these trees exerted an acidifying effect on soil. The order of decreasing soil-acidifying ability was (*Picea abies*; *Picea sitchensis*; *Pinus sylvestris* > (*Abies alba*; *Pseudotsuga menziesii*) > (*Betula pendula*; *Fagus sylvatica*; *Quercus petraea*; *Quercus robur*) > (*Acer platanoides*; *Carpinus betulus*; *Fraxinus excelsior*; *Tilia cordata*). The base content of litter might have a bearing on the species' ability to affect soil pH. The processes involved in creating the observed variation in pH among species are likely to be threefold: (1) interspecific differences in the production of organic acids from decomposing litter that change the relative quantities of exchangeable base (Ca, Mg) and acid (Al, Fe) cations in soils, (2) differences in cation uptake and allocation to biomass pools with differing turnover times, and (3) variation in the mineral content of the parent material. Higher organic acid production and lower soil pH have been noted in sites dominated by species whose litter is relatively recalcitrant to the decomposition process. Calcium uptake and/or allocation to biomass pools with differing turnover times may explain the high soil pH and the large quantity of exchangeable Ca beneath sugar maple. Sugar maple trees are characterized by deep crowns and relatively high leaf litter Ca concentrations (Finzi et al. 1998). Singwane and Malinga (2012) studied soils under pine (*Pinus elliottii*) and Eucalyptus (*Eucalyptus grandis*) plantations in Swaziland. The results revealed that forest plantations have positive impacts on soil organic matter content.

Aweto and Dikinya (2003) observed that growing *Combretum apiculatum* and *Peltophorum africanum* in a semiarid savanna rangeland in Botswana lowered the bulk density of the soil. They attributed this change in soil to the addition of litter materials that increased organic matter content and aggregation. They also observed improvement in the levels of organic matter, exchangeable potassium, calcium, magnesium and CEC of soils by *Combretum apiculatum* and *Peltophorum africanum*. *Acacia nilotica* trees have enriched the organic matter and nutrient levels in the soil under their canopies in central India (Pandey et al. 2000). A similar effect has been reported by Hailu et al. (2000) for the soil under the canopy of *Milletia ferruginea* in Ethiopia. The content of nutrients in litter differs greatly with the type of tree species and so does the rate of litter decomposition and release

Fig. 10.5 Absence of undergrowth, and erosion and root exposure in a teak plantation. (Photo courtesy of Dr. Animesh Biswas)

of nutrients. Foliage of hardwood species usually has higher concentrations of N, K, Ca, and Mg than coniferous species. Thus, hardwood litters can be richer in nutrients than coniferous species (Phillips and Marion 2006). Enrichment of surface soil with nutrients may also be due to absorption of nutrients by tree roots from the subsoil and recycling them to the topsoil. Roos and Allsopp (1997) reported that growing *Acacia sieberana* in South African Savanna improved organic matter content, total nitrogen, mineralizable nitrogen, resin extractable phosphorus, total phosphorus, and mineralizable phosphorus of soil. Development of forest floor is also influenced by the tree species. For instance, the litter weight under *Picea abies* could be up to twice that of hardwood species like *Fagus sylvatica*. The decomposition rate of litter depends on characteristics, which are tree species dependent such as hardness, morphology, lignin/N ratio, foliage longevity, or the content of hydrosoluble components. Numerous studies have provided evidence that canopy composition has an impact on nitrogen mineralization and nitrification.

Again, some investigators (Davis and Lang 1991; Chen et al. 2003) reported that there may be some short-term beneficial effects associated with the change in land use from grassland to forest such as increased phosphorus availability. Most afforestation studies have involved paired site comparisons of soils under forest 15–20 years after establishment with adjacent deforested or degraded sites (Chen et al. 2000; Chen et al. 2003). Positive changes in soil organic matter and nutrient availability have been observed (Chirino et al. 2010). However, some plantation tree species such as teak (*Tectona grandis*) and Gamar (*Gmelina arborea*) and some coniferous tree species in tropical and temperate environment are blamed to have suppressed undergrowth vegetation and favor erosion (Fig. 10.5). Under teak, Bell (1973) found soil erosion 2.5–9 times higher than under natural forest. Soil deterioration by successive rotations of Cunninghamia lanceolata monocultures was reported from subtropical China.

Study Questions

1. What is seed dormancy? Give example of dormancy in some forest tree seeds. How can dormancy be overcome?
2. Write down nursery practices for raising seedlings? Mention different types of planting materials with examples. Can container size or container media affect health of seedlings?
3. Discuss methods of site preparation and planting seedlings. What are the spacings in used in forest plantations? How do they affect growth of trees?
4. What do you mean by thinning? Distinguish between mechanical thinning and selective thinning. Give an account of chemical weed control in forestry and its impact on environment.
5. Discuss fertilization and irrigation in forest plantations with emphasis on effects of organic wastes and wastewater on environment.

References

Abbott CL, El-Quosy DED (1996) A procedure to assess the impact of drainwater reuse. HR Wallingford Report No OD 134

Abrahamson WG, Hartnett DC (1990) Pine flatwoods and dry prairies. In: Myers RL, Ewel JJ (eds) Ecosystems of Florida. University of Central Florida, Orlando

Abrams MD, Ruffner CM, DeMeo TE (1998) Dendroecology and species co-existence in an old-growth Quercus–Acer–Tilia talus slope forest in the central Appalachians, USA. For Ecol Manag 106(1):9–18

Abrol IP, Sandhu SS (1980) Growing trees in alkali soils. Indian Farming 30(6):19–20

Afif-Khouri E, Cámara Obregón MA, Oliveira-Prendes JA, Gorgoso-Varela JJ, Canga-Líbano E (2010) Relationship among soil parameters, tree nutrition and site index of *Pinus radiata* D. Don in Asturias, NW Spain. Forest Systems 19(1):77–88

Agren GI, Knecht M (2001) Simulation of soil carbon and nutrient development under *Pinus sylvestris* and *Pinus contorta*. For Ecol Manage 141:117–129

Ahmad and Ang (1993) cf FAO Corporate Document Repository. http://www.fao.org/docrep/w7715e/w7715e05.htm#plantation%20species

Allen HL (1994) Enhancing southern pine productivity with fertilization. The Consultant (Summer):12–17

Allen HL, Lein S (1998) Effects of site preparation, early fertilization, and weed control on 14-year old loblolly pine. Proc South Weed Sci Soc 51:104–110

Ang LH, Ho WM (2002) Afforestation of tin tailings in Malaysia. Proceedings of 12th ISCO Conference, Beijing, China

Anon (1975) Chemical weed control in christmas tree plantations. A pacific northwest cooperative extension publication PNW 148, Washington State University and the University of Idaho

Anon (2000) Dictionary of forestry. Presses de lÚniversité Laval, Canada

Armitage FB (1985) Irrigated forestry in arid and semi-arid lands. International Development Research Centre, Ottawa, Canada

Asakawa S (1964) A brief report on the respiration of *Pinus koraiensis* seeds. J Jap Forest. Soc 46(10):336–338

Augustoa L, Ranger J, Binkley D, Rothe A (2002) Impact of several common tree species of European temperate forests on soil fertility. Ann For Sci 59(3):233–253

Aweto AO, Dikinya O (2003) The beneficial effects of two tree species on soil properties in a semi-arid savanna rangeland in Botswana. Land Contamination and Reclamation 11(3):339–344

Baker JB, Broadfoot WM (1979) A practical field method of site evaluation for commercially important southern hardwoods. Gen Tech Rep SO-26. USDA For Ser, South For Exp Sta, New Orleans, Louisiana

Banerjee AK (1973) Plantations of *Acacia auriculiformis* (Benth.) A. Cunn. in West Bengal. Indian Forester 99(9):533–540

Barner H (1975) The storage of tree seeds. In: Report on FAO/DANIDA training course on forest seed collection and handling, vol. 2. FAO, Rome

Bayes CD, Davis JM, Taylor CMA (1987) Sewage sludge as a forest fertilizer: Experience to date. J Inst Water Pollut Control 86:158–171

Bell TIW (1973) Erosion in Trinidad teak plantation. Comm For Rev 52:233–235

Benbrahim M, Denaix L, Thomas AL, Balet J, Carnus JM (2006) Metal concentrations in edible mushrooms following municipal sludge application on forest land. Environ Pollut 144(3):847–854

Bengtson GW (1979) Forest fertilization in the United States: progress and outlook. J For 78:222–228

Bevege DI, Richards BN (1970) Nitrogen in the growth of Araucaria Cunninghamii Ait. Underplanted in Pinus stands. Ecology 51:134–142

Bhatia KK (1954) Calcium a factor in the ecology of *Teak* (Tectona Grandis, Linn.). J Indian Bot Soc 33:27–34

Bhumibhamon S (1992) Problems and Strategies of *Eucalyptus camaldulensis* Plantation in Thailand. In Horng FW, Yang JC (eds) Proceedings of the symposium on current development of fast-growing plantation silviculture and management in South-East Asian region. Taiwan Forestry research Institute, Taipei

Binkley D, Brown TC (1993) Forest practices as nonpoint sources of pollution in North America. Water Res Bull 29:729–740

Binkley D, Burnham H, Lee Allen H (1999) Water quality impacts of forest fertilization with nitrogen and phosphorus. For Ecol Manag 121:191–213

Binkley D, Carter R, Allen HL (1995) Nitrogen fertilization practices in forestry. In: Bacon PE (ed) Nitrogen fertilisation in the environment. Marcel Dekker, New York, pp 421–441

Bisson PA, Ice GG, Perrin CJ, Bilby R (1992) Effects of forest fertilization on water quality and aquatic resources in the Douglas-fir region. In: Forest fertilization: sustaining and improving nutrition and growth of western forests. University of Washington, Seattle

Blanco J, Bosco IJ, Castillo F (2006a) Effects of thinning on nutrient content pools in two Pinus sylvestris forests in the western Pyrenees. Scand J Forest Res 21(2):143–150

Blanco JA, Imbert BJ, Castillo FJ (2006b) Influence of site characteristics and thinning intensity on litterfall production in two *Pinus sylvestris* L. forests in the western Pyrenees. For Ecol Manag 237:342–352

Blanford HR (1958) Highlights of one hundred years of forestry in Burma. Empire Forestry Review 37(1):33–42

Boland DJ, Brooker MIH, Chippendale GM, Hall N, Hyland BPM, Johnston RD, Kleinig DA, Turner JD (1984) Forest trees of Australia. Nelson and CSIRO, Melbourne

Boland DJ, Brooker MIH, Chippendale GM, McDonald MW (2006) Forest trees of Australia, 5th edn. CSIRO Publishing, Collingwood

Bonner FT (1984) Glossary of seed germination terms for tree seed workers. USDA Forest Service. Gen Tech Rep SO-49, Southern Forest Experiment Station

Bonner FT, McLEmore BF, Barnet JP (1974) Presowing treatment of seed to speed germination. In: Seeds of woody plants in the United States, Agriculture Handbook No. 450. For Service, USDA, Washington DC

Braatz S, Kandiah A (1996) The use of municipal waste water for forest and tree irrigation. Unasylva 47(185):45–51

Bravo F, Montero G (2001) Site index estimation in Scots pine (Pinus sylvestris; L.) stands in the High Ebro Basin (northern Spain) using soil attributes. Forestry 74(4): 395–406

Brooker MIH, Connors JR, Slee AV, Duffy S (2002) EUCLID: eucalypts of southern Australia (CD Rom), CSIRO Publishing, Collingwood

Brookman-Amissah J (1973) Seed problems as they affect forestry practice in Ghana. In: Seed processing proc symposium IUFRO Wkg group on seed problems, Bergen. vol. II, Paper 6

Brunberg T (2007) Basic data for productivity norms for extra large single-grip harvesters in final felling. The Forestry Research Institute of Sweden, Redogörelse 2. (In Swedish with English summary)

Bryant WS (1989) Red cedar (Juniperus virginiana L.) communities in the Kentucky River Gorge area of the Bluegrass Region of Kentucky. In: Rink G, Budelsky CA (eds) Proceedings, 7th central hardwood conference, March 5–8, 1989, Carbondale, IL Gen Tech Rep NC-132. US Department of Agriculture, Forest Service, North Central Forest Experiment Station St. Paul

Bryant WS, Wharton ME, Martin WH, Varner JB (1980) The blue ash-oak savanna–woodland, a remnant of pre-settlement vegetation in the Inner Bluegrass of Kentucky. Castanea 45(3):149–165

Bryndum K (1966) The germination of teak. Natural History Bulletin. Siam Society 21:75–86

Burger D (1972) Forest site classification in Canada. Mitt Ver Forstl Standortskunde Forstpflanzenztichtung 21:4–36

Burnham HM, Allen HL (1997) Forest fertilization and water quality: status, recommendations, and research needs. In: Issues concerning the use of fertilizers in southern pine forests.NCSFNC Research Note #15, North Carolina State University, Raleigh

Burns RM, Honkala BH (1990) Silvics of North America conifers. Agriculture handbook, vol 1. 654. USDA Forest Service, Washington, DC

Caldentey J, Ibarra M, Hernandez J (2001) Litter fluxes and decomposition in *Northofagus pumilio* stands in the region of Magallanes, Chile. For Ecol Manag 148:145–157

Cannell MGR (1999) Growing trees to sequester carbon in the UK: answers to some common questions. Forestry 72(3):237–247

Carey AB (2006) Active and passive forest management for multiple values. Northwestern Naturalist 87(1):18–30

Carmean WH (1965) Black oak site quality in relation to soil and topography in southeastern Ohio. Proceedings of the Soil Science Society of America 29:308–312

Carmean WH (1967) Soil survey refinements for predicting site quality in southeastern Ohio. Proceedings of the Soil Science Society of America 31:805–810

Carmean WH (1975) Forest site quality evaluation in the United States. Adv Agron 27:209–269

Carmean WH (1977) Site classification for Northern forest species. Gen Tech Rep, US Department of Agriculture, Forest Service, Northeastern Forest Experiment Station

Carmean WH (1996) Forest site quality estimation using forest ecosystem classification in Northwestern Ontario. Environmental Monitoring and Assessment 39:493–508

Chakraborty RN, Chakraborty D (1989) Changes in soil properties under Acacia auriculiformis plantations in Tripua. Indian Forester 115(4):272–273

Champion HG, Seth SK, Khattak GM (1965) Manual of silviculture for Pakistan. Pakistan Forestry Institute, Peshawar

Chamshama SAO, Monela GC, Sekiete KEA, Persson A (1992) Suitability of the taungya system at North Kilimanjaro forest plantation, Tanzania. Agroforest Syst 17(1):1–11

Chappell HN, Colel DW, Gesselt SP, Walker RB (1991) Forest fertilization research and practice in the Pacific Northwest. Fert Res 271:129–140

Chaw CS, Mitlöhner (2011) *Acacia mangium* Willd: ecology and silviculture in Vietnam. Centre for International Forestry Research, Bogor, Indonesia

Chen CR, Condron LM, Davis MR, Sherlock RR (2000) Effects of afforestation on phosphorus dynamics

Chen CR, Condron LM, Davis MR, Sherlock RR (2003) Seasonal changes in soil phosphorus and associated microbial properties under adjacent grassland and forest in New Zealand. For Ecol Manag 177:539–557

Chertov OG, Komarov AS, Tsiplianovsky AM (1999) A combined simulation model of Scots pine, Norway spruce and silver birch ecosystems in the European boreal zone. For Ecol Manage 116:189–206

Chiang PN, Yu JC, Wang YN, Tzou YM (2010) Effect of nutrient cycle by different thinning practice in temperate forest. 19th world congress of soil science, soil solutions for a changing world, 1–6 August 2010, Brisbane, Australia

Chirino I, Condron L, McLenaghen R, Davis M (2010) Effects of plantation forest species on soil properties. 19th world congress of soil science, soil solutions for a changing world. 1–6 August 2010, Brisbane, Australia

Clark DA, Clark DB (1999) Assessing the growth of tropical rain forest trees: Issues for forest management. Ecological Applications 9(3):981–997

Cole DW, Gessel SP (1965) Movement of elements through a forest soil as influenced by tree removal and fertilizer additions. In Youngberg CT (ed) Forest-Soil Relationships in North America. Proc Second North Amer Forest Soils Conf, Corvallis

Cole DW, Rinehart ML, Briggs DG, Henry CL, Mecifi F (1984) Response of Douglas-fir to sludge application: volume growth and specific gravity. TAPPI Research and Development Conference

Combe J (1982) Agroforestry techniques in tropical countries:Potentials and limitations. Agrofor syst 1:13–27

Cornelius J, Wightman K, Grogan J, Ward S (2004) *Swietenia* (American mahogany). In: Burley J (ed) Elsevier encyclopedia of forest sciences. Elsevier, Amsterdam

Costermans LF (1989) Native trees and shrubs of south-eastern Australia, Weldon, Sydney

Cromer RN, Tompkins D, Barr NJ, Hopmans P (1984) Irrigation of Monterey pine with wastewatereffect on soil chemistry and groundwater composition. J Env Qual 13:539–542

Cunningham GM, Mulham WE, Milthorpe PE, Leigh JH (1981) Plants of western new South Wales, Soil Conservation Service of New South Wales, Sydney

Da Silva PHM, Poggiani F, Laclau JP (2011) Applying sewage sludge to Eucalyptus grandis plantations: effects on biomass production and nutrient cycling through litterfall. Applied and environmental soil science, vol 2011. Article ID 710614, 11 pages, doi:10.1155/2011/710614

Dalton K (1990) Managing our river red gums. Soil conservation service of New South Wales, Sydney

Daniel TW, Helms JA, Baker FS (1979) Principles of silviculture. McGraw Hill Series in forest resources. MCGraw Hill, New York

Daubenmire (1976) The use of vegetation in assessing the productivity of forest lands. Bot Review 42:115–140

Davis MR, Lang LH (1991) Increased nutrient availability in top soils under conifers in the South Island high country. NZ J For Sci 21:165–179

Davis-Carter JG, Burgoa B (1993) Atrazine runoff and leaching losses from soil in tilted beds as influenced by three rates of lagoon effluent. J Environ Sci Health B 28:1–18

Dolph KL (1988) Prediction of periodic basal area increment for young-growth mixed conifers in the Sierra Ne- vada. Research Paper PSW- 1 90. USDA Forest Service, Pacific Southwest Forest and Range Experiment Station, Berkeley, California

Doran JC, Turnbull JW (eds) (1997) Australian Trees and Shrubs: species for land rehabilitation and farm planting in the tropics. Australian Centre for International Agricultural Research, Canberra

Doran JW, Safley M (1997) Defining and assessing soil health and sustainable productivity. In: Pankhurst C, Doube BM, Gupta VVSR (eds) Biological indicators of soil health CAB international. Wallingford, Oxon, UK

Doran JW, Zeiss MR (2000) Soil health and sustainability: managing the biotic component of soil quality. Appl Soil Ecol 15:3–11

Duke JA (1983) Casuarina equisetifolia JR and G Forst. Center for New Crops and Plant Products, Purdue University

Dumroese RK, Haase DL (2010) Container tree nursery manual, vol 7: Seedling Processing, Storage, and Outplanting. USDA

Eckenwalder JE (1977) North American cottonwoods (Populus, Salicaceae) of sections Abaso and Aigeiros. J Arnold Arbor 58(3):193–208

Edmonds RL, Cole DW (1977) Use of De-watered sludge as an amendment for forest growth. Center for ecosystem studies, vol. 2. College of Forest Resources, University of Washington

Eira MTS, Freitas RWA, Mello CMC (1993) Breaking seed dormancy of Enterolobium contortisiliquum (VELL) Morong-Leguminosae. Journal of Seeds 15(1):177–182

Elamin HM (1975) Germination and development of the Sudan acacias. Sudan Sylva III 20:23–33

Eldridge K, Davidson J, Harwood C, Wyk Gv (1993) Eucalypt domestication and Breeding. Clarendon, Oxford

Eswaran H, Beinroth FH, Reich P (1999) Global land resources and population supporting capacity. Am J Alternative Agric 14:129–136

Eswaran, H, Almaraz R, van den Berg E, Reich P (1997). An assessment of soil resources of Africa in relation to soil productivity. Geoderma

European Environment Agency (EEA) (2006) European forest types. Categories and types for sustainable forest management reporting and policy. Technical report No. 9. EEA, Copenhagen, Denmark

Evans J (1982) Plantation forestry in the tropics. Clarendon, Oxford

Evans J (1999) Sustainability of forest plantations: the evidence. Report commissioned by DFID. London, UK

Evans J (2009) Forest plantations. In Evans J (ed) Planted forests: Uses, Impacts and Sustainability. FAO, Rome

FAO (1956) Country reports on *teak*. First session, *Teak* Sub-Commission, 9–18 February, Bangkok

FAO (1957) Report on teak growing under exotic conditions. FAO/TSC-57/3, FAO, Rome

FAO (1989) Irrigated forest plantations. In: FAO arid zone forestry: a guide for field technicians. Chapter VII FAO Conservation Guide 20. Rome

FAO (1999) AEZWIN an interactive multiple-criteria analysis tool for land resources appraisal (with CD-ROM included). World Soil Resources Report—Series 87. http://www.fao.org/icatalog/inter-e.htm

FAO (2001) Global forest resources assessment 2000 (FRA 2000). Main report, FAO Forestry Paper 139, FAO, Rome

Ferdousee N, Jabbar F, Hossain MK, Hoque ATMR (2010) Comparative growth performance of *Leucaena leucocephala* seedlings raised in nursery bed, polybag and root trainers. Proc. of international conference on environmental aspects of Bangladesh (ICEAB10), Japan, Sep 2010

Ferrari AE, Wall LG (2004) Utilization de arboles fijadores de nitrogeno para la revegetacion de suelos degrades. Revista de la Facultad de Agronomia 105(92):63–87

Finzi A, Van Breemen N, Canham C (1998) Canopy tree soil interactions within temperate forest: species effects on pH and cations. Ecol Appl 8:447–454

Forestry Commission (1977) E(E) Work study branch report No 42, initial trials of the quickwood planting machine in Thetford forest E(E). Forestry commission technical Development Branch, Ae Village, Dumfries

Foster RB, Hubbell SP (1990) Estructura de la vegetaci6n y composici6n de especies de un lote de cincuenta hectareas en la isla de Barro Colorado. In: Leigh Jr EG, Rand AS, Windsor DM (eds) Ecologfa de un bosque tropical: ciclos estacionales y cambios a largo plazo. Smithsonian Tropical Research Institute, Balboa, Panama

Fralish JS (1994) The effect of site environment on forest productivity in the Illinois Shanee Hills. Ecological Applications 4(1):134–143

Fredriksen RL, Moore DG, Norris LA (1975) The impact of timber harvest, fertilization, and herbicide treatment on stream water quality in western Oregon and Washington. In: Bernier B, Winget CH (eds) Management forest soils and forest. Land University of Laval, Quebec

Gagnon JD (1973) Environmental aspects of sewage derived fertilizers. Proc Forest Fertilization Symposium, USDA For Serv Gen Tech Rep NE-3

Garciano C, Goya JF, Arturi M, Pérez C, Frangi JL (2008) Fertilization in a fourth rotation Eucalyptus grandis plantation with minimal management. J Sustain Forest 26(2):155–169

Gentry AH (1992) Tropical forest biodiversity: distributional patterns and their conservational significance. Oikos 63:19–28

Gerding V, Schlatter JE (1995) Variables y factores del sitio de importancia para la productividad de *Pinus radiata* D. Don en Chile. Bosque 16(2):39–56

Gerlach JP, Gilmore DW, Puettmann KJ, Zasada JC (2002) Mixed-species forest ecosystems in the Great Lakes region: a bibliography. Staff paper series number 155, Department of Forest Resources, College of Natural Resources and Minnesota Agricultural Experiment Station, University of Minnesota, St Paul, Minnesota

Gevorkiantz SR, Scholz HF (1944) Determining site quality in understocked oak stands. J Forest 42:808–811

Geyer WA, Lynch KD (1987) Use of site index as a forestry management tool. Transactions of the Kansas Academy of Science 90(1–2):46–51

Geyer WA, Lynch KD (1990) Green ash growth curves. Transactions of the Kansas Academy of Sciences 93(1–2):12–16

Geyer WA, Marquard RD, Barber JF (1980) Black walnut site quality in relation to soil and topographic characteristics in northeastern Kansas. J Soil Water Conserv 35:135–137

Gonçalves EP, de Jesus Soares FS, Silva SS, de Souza Tavares D, Viana JS, Cardoso BCC (2011) Dormancy Breaking in *Ormosia arborea* Seeds. Int J Agron 2011:5

Goor AY, Barney CW (1976) Forest tree planting in arid zones, 2nd edn. Ronald Press, New York

Greaves A (1981) Gmelina arhrea. For Abstr 42:1231–258

Gregory R, Ohlson D, Arvai J (2006) Deconstructing adaptive management: Criteria for applications to environmental management. Ecol Appl 16(6):2411–2425

Greller AM (1988) Deciduous forest. In: Barbour MG, Billings WD (eds) North American terrestrial vegetation. Cambridge University Press, Cambridge

Grogan J, Barreto P (2005) Big-leaf *Swietenia macrophylla* on CITES Appendix II: big challenge, big opportunity. Conserv Biol 19(3):973–976

Guedes MC, Poggiani F (2003) Variação dos teores de nutrientes foliares em eucalipto fertilizado com biossólido. Scientia Forestalis 63:188–201

Guynn DC, Glfynn ST, Wigly TB, Miller DA (2004) Herbicides and forest biodiversity-what do we know and where do we go from here? Wildlife Society Bulletin 32:1085–1092

Hailu T, Negash L, Olsson O (2000) *Millettia ferruginea* from southern Ethiopia: impacts on soil fertility and growth of maize. Agrofor Syst 48:9–24

Haines (1994) Biotechnology in Forest Tree Improvement. FAO Forestry Paper 118. Food and Agriculture Organization of the United Nations, Rome

Hall FC (1982) Growth basal area: a field method for appraising forest site potential for stockability. Canadian J For Res 13:70–77

Hallonborg U (1997) Aspects of mechanized tree planting. SLU. Acta Universitatis Agriculturae Sueciae Silvestria 29

Harrison R, Xue D, Henry C, Cole DW (1994) Long-term effects of heavy applications of biosolids on organic matter and nutrient content of a coarse-textured forest soil. For Ecol Manag 66(1–3):165–177

Hartemink AE (2003) Soil fertility decline in the tropics with case studies on plantations. CABI, London, UK

Hassanli AH, Javan M, Saadat Y (2008) Reuse of municipal effluent with drip irrigation and evaluation of the effect on soil properties in a semi-arid area. Environ Monit Assess 144:151–158

Hassanli AM, Javan M (2005) Evaluation of municipal eflluent quality as an approperate alternative for green space development and mitigating the environment effects. Tehran University Journal of Environment, No.38 (in Persian).

Hebb EA, Clewell AF (1976) A remnant stand of old-growth slash pine in the Florida panhandle. Bull Torrey Bot Club 103(1):1–9

Heit CE (1967a) Propagation from seed. Part 6: Hardseededness—a critical factor. Am Nurseryman 125(12):10–12

Heit CE (1967b) Propagation from seed. Part 10: Storage methods for conifer seeds. Am Nurseryman 126(8):14–16

Helms D (1977) Land capability classification: the U.S. experience. In Yaalon DH, Berkowicz S (ed) History of soil science: international perspectives. Advances in GeoEcology 29:159–175

Helms JA (1998) The dictionary of forestry. Society of American Foresters, Berkely, USA

Henry CL (1986) Growth response, mortality and foliar nitrogen concentrations of four tree species treated with pulp and paper and municipal sludges. In: Cole DW, Henry CL, Nutter WL (eds) The forest alternatve for treatment and utilization of municipal and industrial wastewater and sludge. University of Washington Press, Seattle

Henry CL, Cole DW (1997) Use of biosolids in the forest: technology, economics and regulations. Biomass Bioenerg 13(4–5):269–277

Henry CL, Cole DW, Hinckley TM, Harrison RB (1993) The use of municipal and pulp and paper sludges to increase production in forestry. J Sustain Forest 1(3):41–55

Hesselman H (1937) Om humustäckets beroende av beståndets ålder och sammansättning i den nordiska granskogen av blåbärsrik Vacciniumtyp och dess inverkan på skogens föryngring och tillväxt. Meddelanden från Statens Skogsförsöksanstalt 30: 529–716 (In Swedish with German summary)

Hilli A, Tillman-Sutela E, Kauppi A (2003) Germination of pretreated Scots pine seeds after long-term storage. Can J For Res 33:47–53.

Hoffman LJ (1976) Height growth of Douglas-fir in relation to habitat types in northern Idaho. Un- published thesis, University of Idaho, Moscow

Holmes GD, Buszewicz G (1958) The storage of seed of temperate forest tree species. For Abstr 19:313–322

Hong J (1991) Advances in research on tree improvement and breeding in China. In: Development of forestry science technology in China. Scientific and Technological Information Center, Ministry of Forestry, Beijing

Houghton PD, Charman PEV (1986) Glossary of terms used in soil conservation. Soil Conservation Service of NSW, Australia

Hu PW, Cheng WE, Shen TA (1983) Growth of seedling of four leguminous tree species in relation to soil pH in a pot test. Nitrogen Fixing Tree Research Reports 1:24–25

Huebschmann MM, Lynch TB, Wittwer RF (1999) Needle litterfall prediction models for even-aged natural short-leaf pine (*Pinus echinata* Mill) stands. For Ecol Manag 117:179–186

Huntley JC (1990) Robinia pseudoacacia L. black locust. In: Burns RM, Honkala BH (eds) Silvics of North America. Volume 2. Hardwoods. Agric Handb 654. US Department of Agriculture, Forest Service, Washington, DC

Husch B, Miller C, Beers TW (1972) Forest Mensuration. Ronald, New York

Imo M (2008) Interactions amongst trees and crops in taungya systems of western Kenya. Agrofor Syst. DOI 10.1007/s10457–008-9164-z

IPCC (2000) Special report on land use, land-use change, and forestry. In: Watson RT, Noble IR, Bolin B, Ravindranath NH, Verardo DJ, Dokken DJ (eds) Intergovernmental panel on climate change (IPCC), Cambridge University Press, Cambridge, UK

ITTO (2009) Encouraging industrial forest plantations in the tropics. International Timber Tree Organization Technical Series 33

Jacobson S, Pettersson F (2010) An assessment of different fertilization regimes in three boreal coniferous stands. Silva Fennica 44(5):815–827

Jagger P, Pender J (2003) The role of trees for sustainable management of less-favored lands: the case of *Eucalyptus* in Ethiopia. Forest Policy and Economics 5(1):83–95

Jha LK, Sen Sharma PK (2009) Agroforestry: Indian Perspective. APH Publishing Corporation, New Delhi

Jobbágy EG, *Jackson* RB *(2003)* Patterns and mechanisms of soil acidification in the conversion of grasslands to *forests*. Biogeochem 64:205–229

Jones JR (1969) Review and comparison of site evaluation methods. USDA Forest Service, Res Pap RM-75

Jones N, Burley J (1973) Seed certification, provenance nomenclature and genetic history in forestry. Silvae Genetica 22:3

Kaapro J, Hall J (2012) Indaziflam—A new herbicide for pre-emergent control of weeds in turf, forestry, industrial vegetation and ornamentals. Pak J Weed Sci Res 18:267–270

Kaosa-ard A (1981) Teak, Its natural distribution and related factors. Nat His Bull Siam Soc 29:55–74

Kaosa-ard A (1983) Teak in ASEAN: a survey report. Aasean Canada Forest Tree Seed Centre

Karlen DL, Andrews SS, Doran JW (2001) Soil quality: current concepts and applications. Adv Agron 74:1–40

Karlen DL, Doran JW, Weinhold BJ, Andrews SS (2003) Soil quality: human kind's foundation for survival. J Soil Water Conserv 58:171–179

Karlen DL, Mausbach MJ, Doran JW, Cline RG, Harris RF, Schuman GE (1997) Soil quality: a concept, definition, and framework for evaluation. Soil Sci Soc Am J 61:4–10

Keiding H (1973) Report on a journey to Sumatra, Thailand and India for the Danish/FAO Forest Tree Seed Centre, Jan. – March 1970. In Forest Genetic Resources Information No. 1. FAO, Rome

Kemp RH (1975) Seed pretreatment and principles of nursery handling. In Report on FAO/DANIDA training course on forest seed collection and handling, vol. II. FAO, Rome

Keogh RM (1987) The care and management of teak plantations. A practical guide for foresters in the Caribbean, Central America, Venezuela and Colombia. Universidad Nacional de Costa Rica and Gorta

Keresztesi B (1979) Mini-monograph on *Robinia pseudoacacia*. In : Documents presented to technical consultation on fast-growing broadleaved trees for Mediterranean and Temperate Zones, vol. 1. Lisbon Oct. 1979. FO: FGB-79, FAO, Rome

Kimberley MO, Wang H, Wilks PJ, Fisher CR, Magesan GN (2004) Economic analysis of growth response from a pine plantation forest applied with biosolids. For Ecol Manag 189(1–3):345–351

Krisnawati H, Kallio M, Kanninen M (2011a) *Acacia mangium* Willd. Ecology, silviculture and productivity. Center for International Forestry Research, Bogor, Indonesia

Krisnawati H, Kallio M, Kanninen M (2011b) *Swietenia macrophylla* king: ecology, silviculture and productivity. CIFOR, Bogor, Indonesia

Krisnawati H, Varis E, Kallio M, Kanninen M (2011c) *Paraserianthes falcataria* (L.) Nielsen: ecology, silviculture and productivity. CIFOR, Bogor, Indonesia

Kruggman SL, Stein WI, Schmitt DM (1974) Seed biology. In: Seeds of woody plants in the United States. Agriculture Handbook No. 450. For Service. USDA, Washington, DC

Laakkonen O, Keipi K, Lipas E (1983) Profitability of nitrogen fertilization in mature forests on mineral soils. Folia Forestalia 577:1–20

Lamb AFA (1973) *Pinus caribaea*. Fast growing timber trees of the lowland tropics, vol. I. No. 6. CFI, Oxford

Lamb D (1975) Weed control in tropical forest plantations with glyphosate. PANS 21(2):177

Lamb FB (1966) Mahogany of Tropical America: its ecology and management. University of Michigan, Ann Arbor, MI

Lamprecht H (1990) Silviculture in the tropics: tropical forest ecosystems and their tree species-possibilities and methods for their long-term utilization. GTZ, Eschborn. Germany

Laurie MV (1974) Tree planting practices in African Savannas. FAO For Dev Paper No. 19 FAO, Rome

Lee YJ (1974) Four-year basal area growth response of a 25-year-old Douglasfir stand to thinning and urea fertilization. Can J For Res 4:568–571

Letey J, Sojka RE, Upchurch DR, Cassel DK, Olson KR, Payne W A, Petrie SE, Price GH, Reginato RJ, Scott HD, Smethurst PJ, Triplett GB (2003) Deficiencies in the soil quality concept and its application. Journal of Soil and Water Conservation 58:180–187

Lieberman D, Lieberman M, Peralta R, Hartshorn GS (1985) Mortality patterns and stand turnover rates in a wet tropical forest in Costa Rica. J Ecol 73:915–924

Ligne D (2004) New technical and alternative silvicultural approaches to pre-commercial thinning. Doctoral Thesis, Swedish University of Agricultural Sciences, Umeå

Lilienfein J, Wilcke W, Ayarza MA, Vilela L, Lima SdC, Zech W (2000) Soil acidification in *Pinus caribaea* forests on Brazilian savanna Oxisols. For Ecol Manag 128:145–157

Lindenmayer DB, Franklin JF (2002) Conserving forest biodiversity – A comprehensive multiscaled approach. Island Press, Washington, DC

Lohrey RE, Kossuth SV (1990) Pinus elliottii Engelm. slash pine. In: Burns RM, Honkala BH (eds) Silvics of North America. vol 1. Conifers. Agric Handb 654. U.S. Department of Agriculture, Forest Service, Washington, DC

Louppe D, Oteng -Amoako AA, Brink M (2008) Timbers 1, vol. 1–7. IITO

Luoranen J, Rikala R, Smolander H (2011) Machine planting of Norway spruce by Bracke and Ecoplanter: an evaluation of soil preparation, planting method and seedling performance. Silva Fennica 45(3):341–357

Luoranen J, Saksa T, Finér L, Tamminen P (2007) Metsämaan muokkausopas. [Guide for silvicultural soil preparation.] Metsäntutkimuslaitos, Suonenjoen toimintayksikkö. (In Finnish)

Lyle GR (1981) Chemical weed control in forestry. South Afr Forest J 117 (1):61–64

Ma Jianlu (1997) Site Quality Evaluation for Mixed, Uneven-aged, Natural Forest Dominated by Korean Pine in Northeast China. J Forest Res 8(2):70–73

Magini E (1962) Forest seed handling, equipment and procedures: II Seed treatments, storage, testing and transport. Unasylva 16(1):20–35

Magnus L, Daniel D, Rafael N, Douglass J (2012) Mechanical site preparation for forest restoration. New Forests 43(5–6):825–848

Maiti SK (2006) Properties of mine soil and its effect on bioaccumulation of metals in tree species: case study from a large opencast coal mining project. Int J Surf Min Reclam Environ 20(2):96–110

Malik N, Vanden Born WH (1986) Use of herbicides in forest management. Can For Serv, North For Cent, Edmonton, Alberta. Inf Rep NOR-X-282

Mansberg L, Wentworth TR (1984) Vegetation and soils of a serpentine barren in western North Carolina. Bulletin of the Torrey Botanical Club 111(3):273–286

Marcar NE, Crawford DF (2004) Trees for Saline Landscapes. RIRDC Publication Number 03/108, Canberra

Martin B (1971) Premiers travaux d'amélioration génétique des arbres foretières en Republique Populaire du Congo. Bois et Forêts des Tropiques 140:13–31

Maschinski J, Whitham TG (1989) The continuum of plant responses to herbivory: the influence of plant association, nutrient availability, and timing. Am Nat 134:1–19

Mathiasen RL, Blake EA, Edminster CB (1986) Estimates of site potential for Douglas-fir based on site index for several southwestern habitat types. Great Basin Nat 46(2):277–280

Matias AR, Betancourt A, Zayas A, Pena A, Rivero R (1973) Forest seed in Cuba. In Seed processing proc. symposium IUFRO Wkg group on seed problems, Bergen. vol. II, Paper 31

Matthew A, Vadeboncoeur (2010) Meta-analysis of Fertilization Experiments Indicates Multiple Limiting Nutrients in Northeastern Deciduous Forests. Can J For Res 40(9):1766–1780

Matthews JD(1989) Silvicultural systems. Clarendon, Oxford, UK

Mayhew JE, Newton AC (1998) The silviculture of *S. macrophylla*. CABI, New York

McComb W, Hurst G (1987) Herbicides and wildlife in southern forests. In: Dickson JG, Maughan OE (eds) Managing southern forests for wildlife and fish, a proceedings. USDA Forest Service, General Technical Report SO-65

McDonald MA, Hawkins BJ, PrescottCE, Kimmins JP (1994) Growth and foliar nutrition of western red cedar fertilized with sewage sludge, pulp sludge, wood ash on northern Vancouver Island. Can J For Res 24:297–301

McDonald PM, Laacke RJ (1990) *Pinus radiata* D.Don Monterey Pine. In Burns RM, Honkala BH (eds) The silvics of North America. vol 1 Conifers. Agricultural handbook No. 654. US Forest Service, Washington, DC

McKee WH, McLeod KW, Davis CE, McKelvin MR, Thomas HA (1986) Growth response of loblolly pine to municipal and industrial sewage sludge applied at four ages on Upper Coastal Plain sites. In: Cole DW, Henry CL, Nutter WL (eds) The forest alternative for treatment and utilization of municipal and industrial wastewater and sludge. University of Washington Press, Seattle

McLintock TF, Bickford CA(1957) A proposed site index for red spruce in the Northeast. US Forest Ser Northeastern Exp Sta Pap 93

Means JC, Wijayaratne R (1982) Role of natural colloids in the transport of hydrophobic pollutants. Science 25:968–970

Mergen F, Voight GK (1959) Effects of fertilizer applications on two generations of slash pine. Soil Sci Soc Am J 24(5):407–409

Miiller G (1983) Forest fertilization in Sweden. In: Ballard R and Gessel SP (eds) Proc IUFRO Sympos on

Miller HG (1981) Forest fertilization: some guideline concepts. Forestry 54 (2):157–167

Miller HG, Miller JD (1988) Response to heavy nitrogen applications in fertilizer experiments in British forests. Environ Pollut 54(3–4):219–231

Miller K, Witt J (1990) Impacts of forestry herbicides on wildlife. In: Coleman S, Neary D (eds) United States Department of Agriculture Forest Service, General Technical Report SE-70

Miller KV, Miller JH (2004) Forestry herbicide influences on biodiversity and wildlife habitat in southern forests. Wildlife Society Bulletin 32:1049–1060

Miller RE, Fight RG (1979) Fertilizing Douglas-fir forests. USDA For Serv Gen Tech Rep PNW 83. Pacific Northwest Forest and Range Experiment Station. USDA Forest Service

Miller RE, Pienaar LV (1973) Seven-year response of 35 year-old Douglas-fir to nitrogen fertilizer. USDA For Serv

Ministerial Conference on the Protection of Forests in Europe (MCPFE) (2003) Improved pan-European indicators for sustainable forest management as adopted by the MCPFE Expert Level Meeting 7–8 October 2002, Vienna, Austria. http://www.mcpfe.org/files/u1/publications/pdf/improved_indicators.pdf

Mishra A, Sharma SD, Khan GH (2003) Improvement in physical and chemical properties of sodic soil by 3, 6, and 9 years old plantation of *Eucalyptus tereticornis* Biorejuvenation of sodic soil. For Ecol Manag 184:115–124

Mitchell HL, Chandler RF (1939) The nitrogen nutrition and growth of certain deciduous trees of northeastern United States. With a discussion of the principles and practice of leaf analysis as applied to forest trees. Black Rock Forest Bulletin 11:1–94

Moffat AJ, Matthews RW, Hall JE (1991) The effects of sewage sludge on growth and foliar and soil chemistry in pole stage Corsican pine at Ringwood Forest, Dorset, UK. Can J For Res 21:902–909

Mohammad A Bin, Ibrahim Z Bin (1980) Grading of *Gmelina arborea* (Yemane) fruits by colour. Malayan Forester 43(3):

Moore G, Kershner B, Craig T, Daniel M, Nelson G, Spellenberg R, Thieret JW, Purinton T, Block A (2008) National wildlife federation field guide to trees of North America. Sterling, New York

Myers RG, Thien SJ, Pierzynski GM (1999) Using an ion sink to extract microbial phosphorus from soil. Soil Sci Soc Am J 63:1229–1237

Napoles S, Lacerra JA, Tabares G, Cordero R, Arena R (1990) Effects of bag size and number of seedlings/bag on the development of robusta coffee seedlings. Ciencia y Tecnica en la Agricultura. Cafe y Cacao 12(1–2):51–58

National Research Council (1980) Firewood crops: shrub and tree species for energy plantations. BOSTID Rep 27. National Research Council 1980, Washington, DC

National Research Council (1983) Mangium and other Acacias for the humid tropics. National Academy Press, Washington, DC

NCSFNC (North Carolina State Forest Nutrition Cooperative) (1998) Effects of site preparation and early fertilization and weed control on 14 Year old loblolly pine growth. NCSFNC Report No. 36. Department of Forestry, North Carolina State University

Nisa SH, Qadir SA (1969) Seed germination of common cultivated trees, shrubs and some wild grasses. Pakistan J For 19(2):195–220

Nordborg F, Nilsson U, Gemmel P, Örlander G (2006) Carbon and nitrogen stocks in soil, trees and field vegetation in conifer plantations 10 years after deep soil cultivation and patch scarification. Scand J For Res 21(5):356–363

Norris LA, Lorz HW, Gregory SV (1991) Forest chemicals. In: Meehan W (ed) Influences of forest and rangeland management on salmonid fishes and their habitat. American Fisheries Society Special Publication #19, Bethesda, Maryland

NRCS (2012) Tree/shrub site preparation Code 490. Natural resources conservation service practice standard. USDA-NRCS Draft 12.1.2012

Nsabimana D, Klemedtson L, Kaplin BA, Wallin G (2008) Soil carbon and nutrient accumulation under forest plantations in southern Rwanda. Afr J Env Sci Tech 2(6):142–149

Ogbonnaya CI, Kinako PDS (1993) Growth and mineral nutrition of Gmelina arborea Roxb. seedlings fertilized with four sources of nitrogen on a latosolic soil. Tree Physiology 12:29 l-299

Oliver CD, Larson BC (1996) Forest stand dynamics. Wiley, New York

Olson GW (1974) Interpretative land classification in english-speaking countries (a working paper on some of the systems in current use and some of their adaptations). In: Edouard Saouma (ed) Approaches to land classification. Soils Bulletin 22. FAO. Rome, Italy

Omernik JM (1976) The influence of landuse on stream nutrient levels. US Environmental Protection Agency, EPA Pub. 600/3–76-014, Seattle, WA

Omernik JM (1977) Nonpoint source-stream nutrient level relationships: a nationwide study. US Environmental Protection Agency Report EPA-600/3–77-1056. Corvallis, OR

Oosterbaan A, Hochbichler E, Nicolescu VN, Spiecker H (2009) Silvicultural principles, goals and measures in growing valuable broadleaved tree species. Die Bodenkultur 60(3):45–51

Örlander G, Gemmel P, Hunt J (1990) Site preparation: Swedish overview. FRDA Report 105

Oswald BP, Zhang L, Green TH, Ward KE (1996) Differences in species composition and stand characteristics of mixed upland hardwood forests of North Alabama as reflected by the presence of eastern red cedar (Juniperus virginiana L.). J Sustain Forest 3(4):75–100

Palmer ER (1973) *Gmelina arhorea* as a potential source of hardwood pulp. Tropic Sci 18:243–260

Pan Z (1991) Achievements and an outlook on introduction of forests trees in China. In: Development of forestry science technology in

China. Scientific and Technological Information Center, Ministry of Forestry, Beijing

Pandey CB, Singh AK, Sharma DK (2000) Soil properties under *Acacia nilotica* trees in a traditional agroforestry syatem in Central India. Agrofor Syst 49:53–61

Pandey D (1995) Forest Resources Assessment 1990. Tropical Forest Plantation Resources. FAO Forestry Paper 128. Food and Agriculture Organization of the United Nations, Rome

Papendick RI, Parr J (1992) Soil quality—the key to a sustainable agriculture. Am J Altern Agric 7:2–3

Paul KI, Polglase PJ, Nyakuengama JG, Khanna PK (2002) Change in soil carbon following afforestation. For Ecol Manag 168:241–257

Pawlowski TA (2010) Proteomic approach to analyze dormancy breaking of tree seeds. Plant Mol Biol 73(1–2):15–25

Pearson HA, Wolters GL, Thill RE, Martin Jr A, Baldwin VC (1995) Plant response to soil, site preparation and initial pine planting density. J Range Managt 48(6):511–516

Pettersson F (1994) Predictive functions for calculating the total response in growth to nitrogen fertilization, duration and distribution over time. The Forestry Research Institute of Sweden, Report 4

Pflugbeil E (1960) Proposed site indices for Englemann spruce on the college forest of the Utah State University. MS Thesis, Utah State University, USA

Phillips JD, Marion DA (2006) Biomechanical effects of trees on soil and regolith: beyond treethrow. Ann Assoc Am Geogr 96(2):233–247

Pienaar LV, Shiver BD, Grider GE (1985) Predicting basal area growth in thinned slash pine plantations. For Sci 31:731–741

Pienaar LV, Turnbull KL (1973) The Chapman- Richards generalization of Von Bertalanffy's growth model for basal area growth and yield in even-aged stands. For Sci 19:2–22

Pinho AP, Matos AT, Morris LA, Costa LM (2007) Atrazine and picloram adsorption in organic horizon forest samples under laboratory conditions. Planta daninha 25(1)

Poff B, Tecle A, NearyDG, Geils B (2010) Compromise programming in forest management. J Ariz-Nev Aca Sci 42(1):44–60

Rajjou L, Debeaujon I (2008) Seed longevity: Survival and maintenance of high germination ability of dry seeds. Comptes Rendues Biologies 331:796–805

Ranasinghe H (2012) Comparison of some physical and chemical properties of soils under different aged Teak plantations in Kadjuwatte with other sites having healthy growth of Teak in Mahaweli System. Proceedings of international forestry and environment symposium, Sri Lanka. Published by Department of Forestry and Environmental Science, University of Sri Jayewardenepura

Read RA (1952) Tree species occurrence as influenced by geology and soil on an Ozark north slope. Ecology 33(2):239–246

Rennie KJ (1962) Methods of assessing forest site capacity. Seventh Intl Soc Soil Sci Comm IV and V Trans:3–18

Rinne P (1994) Hormonal regulation of bud burst and coppice shoot growth in *Betula* Acta Universitatis Ouluensis, Series A. Scientiae Rerum Naturalium 0(255):1–40

Roberts EH (1972) Viability of seeds. Chapman and Hall, London

Roe AL (1967) Productivity indicators in western larch forests. USDA Forest Service, Res Note INT-59

Rollet B (1980) Jari: Succès ou échec? Un exemple de développement agro-sylvo-pastoral et industriel en Amazonie brésilienne, Bois et Forêts des Tropiques, n°192, Juillet-Août

Romanyà J, Vallejo VR (2004) Productivity of *Pinus radiata* plantations in Spain in response to climate and soil. For Ecol Manag 195:177–189

Roos PC, Allsopp N (1997) Soil nutrient ecology associated with *Acacia sieberana* at different tree densities in a South African savanna. Afr J Range Forage Sci 14(2):39–44

Ross CW, Watt MS, Parfitt RL, Simcock R, Dando J, Coker G, Clinton PC, Davis MR (2009) Soil quality relationships with tree growth in exotic forests in New Zealand. For Ecol Manag 258:2326–2334

Rowe JS (1962) Soil, site and land classification. For Chron 38:420–432

Rowe JS (1971) Why classify forest land? For Chron 47:144–148

Saksa T, Heiskanen J, Miina J, Tuomola J, Kolström T (2005) Multilevel modelling of height growth in young Norway spruce plantations in southern Finland. Silva Fennica 39(1):143–153

Salazar R, Joker D (2000) *Seed Leaflet No. 40*: Pinus caribaea; *Morelet*. Danida Forest Seed Centre, Denmark

Schoenholtz SH, Van Miegroet H, Burger JA (2000) A review of chemical and physical properties as indicators of forest soil quality: challenges and opportunities. For Ecol Manag 138:335–356

Schutz CJ (1976) A review of fertilizer research on some of the more important conifers and eucalypts planted in subtropical and tropical countries, with special reference to South Africa. Bulletin No 53, Department of Forestry, RSA

Seeber G, Agpaoa A (1976) Forest Tree Seeds. In Manual of Reforestation and Erosion Control for the Philippines. German Agency for Technical Co-operation, Eschborn

Seth SK, Yadav JSP (1958) Teak Soils. Proceedings of the All India Teak Study Tour and Symposium: 121–137 Forest Research Institute, Dehra Dun

Sharitz RR, Boring LR, Van Lear DH, Pinder III JE (1992) Integrating Ecological Concepts with Management of Southern Forests. Ecol Appl 2(3):226–237

Shepard JP (1994) Effects of forest management on surface water quality in wetland forests. Wetlands 14:18–26

Shepard JP, Creighton J, Duzan H (2004) Forestry Herbicides in the United States: An Overview. Wildlife Society Bulletin 32(4):1020–1027

Singwane SS, Malinga P (2012) Impacts of pine and eucalyptus forest plantations on soil organic matter content in Swaziland—case of Shiselwen. J Sustain Dev Afr 14(1):137–151

Smith WD, Strub MR (1991) Initial spacing: how many trees to plant. In: Durya ML, Dougherty PM (eds) Forest Regeneration Manual. Kluwer

Snyder SA (1992) Casuarina spp. In: Fire effects information system. USDA, Forest Service, Rocky Mountain Research Station, Fire Sciences Laboratory

Soerianegara I, Lemmens RHMJ (1993) Plant resources of South-east Asia 5(1): timber trees: major commercial timbers. Pudoc Scientific, Wageningen, Netherlands

Southgate D, Whitaker M (1992) Promoting resource degradation in Latin America: tropical deforestation, soil erosion, and coastal ecosystem disturbance in Ecuador. Econ Dev Cultural Change 40:787–807

Sprague LA, Herman JS, Hornberger GM, Mills AL (2000) Atrazine adsorption and colloid-facilitated transport through the unsaturated zone. J Env Qual 29:1632–1641

Stanek W (1966) Relative quality of the major forest associations of the southern British Columbia interior for growth of lodgepole pine, Engelmann spruce, Douglas-fir, and alpine fir. For Chron 42:306–313

Stanturf J A, Madsen P (2002) Restoration concepts for temperate and boreal forests of North America and Western Europe. Plant Biosystems 136:143–158

Stanturf JA, Schoenholtz SH, Schweitzer CJ, and Shepard JP (2001) Indicators of restoration success: Myths in bottomland hardwood forests. Resto Ecol 9(2):189–200

Stein WI, Slabaugh PE, Plummer AP (1974) Harvesting, processing and storage of fruits and seeds. In: Seeds of woody plants in the United States. Agriculture Handbook No. 450. For Service, USDA, Washington, DC

Stewart HTL, Flinn DW (1984) Establishment and early growth of trees irrigated withwastewater at four sites in Victoria, Australia. For Ecol Manag 8:243–256

Stewart HTL, Hopmans P, Flinn DW, Hillman TJ (1990) Nutrient accumulation in trees and soil following irrigation with municipal effluent in Australia. Environ Poll 63:155–177

Suggit JW (1956) Brush control acreage sprayed in Canada-1954. Pages 103–106 in 9th Meet East Sect, Natl Weed Comm (Can), Truro, Nova Scotia
Suggit JW (1957) Brush control acreage sprayed in Canada-1955. Proc. 10th Meet. East. Sect., Natl. Weed Comm (Can), Quebec, Quebec, pp 43–45
Sutton RF (1958) Chemical herbicides and their uses in the silviculture of forests of eastern Canada. Can Dep For, Ottawa, Ontario. Tech. Note 68
Sutton RF (1993) Mounding site preparation: a review of European and North-American experience. New Forests 7:151–192
Tamm CO (1991) Nitrogen in terrestrial ecosystems. Ecological Studies 81. Springe, Berlin
Tamm CO, Holmen H, Popovic B, Wiklander G (1974) Leaching of plant nutrients from soils as a consequence of forestry operations. Ambio 3:211–221
Tan ZY, Dong YD, Fang YR, Cao RF (1984) The relations of seed dormancy with abscisic acid and seed coat in Korean pine. Sc\en\\aSm\ ca, China. 27(2):178–186
Tatum V (2004) Toxicity, transport, and fate of forest herbicides. Wildlife Society Bulletin 32:1042–1048
Thang HC (1994) Management Practices of *Acacia mangium* Plantations in Peninsular Malaysia. In: Mohd WRW, Ibrahim S, Appanah S, Rashid MFA (eds) Proceedings of the symposium on harvesting and silviculture for sustainable forestry in the tropics. Forest Research Institute Malaysia, Kepong
Thankamani CK, Sivaraman K, Kandiannan K (1996) Response of clove (Syzygium aromaticum (L.) Merr. & Perry) seedlings and black pepper (Piper nigrum L.) cuttings to propagating media under nursery conditions. J Spices and Aromatic Crops. 5(2):99–104
Thomas CE, Parresol BR (1989) Comparing basal area growth rates in repeated inventories: Simpson's paradox in forestry. Forest Science 35:1029–1037
Timo Kuuluvainen (2009) Forest management and biodiversity conservation based on natural ecosystem dynamics in Northern Europe: The Complexity Challenge. Ambio 38(6):309–315
Toro J (2004) Alternativas silvicolas para aumentar la rentabilidad de las plantaciones forestales. Bosque 25(2):101–113
Townsend AR, Cleveland CC, Asner GP, Bustamante MM (2007) Controls over foliar N:P ratios in tropical rain forests. Ecology 88:107–118
Troup RS (1921) The *silviculture of Indian trees*. Oxford University Press, Oxford
Turnbull JW (1986) Multipurpose Australian trees and shrubs: lesser known species for fuelwood and agroforestry. AGIAR Monograph 1. Center for International Agricultural Research, Canberra
Turner J (1983) A review of forest fertilization programs in Australia. USDA Forest Service general technical report PNW - United States, Pacific Northwest Forest and Range Experiment Station, U.S. Department of Agriculture. Forest Service. Portland, Oregon
Turner J, Lambert MJ (1986) Nutrition and nutritional requirements of *Pinus radiata*. Annu Rev Ecol Syst 17:325–350
Turner J, Lambert MJ, Johnson DW (2005) Experience with patterns of soil carbon resulting from forest plantation establishment in eastern Australia. For Ecol Manag 220:259–269
Uotila K, Rantala J, Saksa T, Harstela P (2010) Effect of soil preparation method on economic result of Norway spruce regeneration chain. Silva Fennica 44(3):511–524
USDA Forest Service (2002) National report of pesticide use o n national forest system lands in fiscal year 2002. http://www.fs.fed.us/foresthealth/pesticidedocs/2002pesticideusereport. Accessed Mar 2011
USEPA (1995) National water quality inventory, 1994. Report to Congress. EPA841-R-95-005. Office of Water, USEPA, Washington, DC
Valencia R, Balslev H, Paz G, Mino C (1994) High tree alpha-diversity in Amazonian Ecuador. Biodivers Conserv 3:21–28
Vitousek PM (1984) Litterfall, nutrient cycling, and nutrient limitation in tropical forests. Ecology 65:285–298
Vitousek PM, Aber JD, Howarth RW, Likens GE, Matson PA, Schindler DW, Schlesinger WH, Tilman DG (1997) Human alteration of the global nitrogen cycle: sources and consequences. Ecol Appl 7:737–750
Vitousek PM, Sanford RL Jr (1986) Nutrient cycling in moist tropical forest. Ann Rev Ecol Syst 17:137–167
Vogt KA, Edmonds RL, Vogt DJ (1980) Nitrate leaching in soils after sludge application. In: Bledsoe CS (ed) Municipal sludge application to Pacific Northwest forest lands. College of Forest Resources Contrib 41, University of Washington, Seattle
Walker RB, Gessel SP (1991) Mineral deficiencies of coastal northwest conifers. Institute of Forest Resources Contribution No. 70. University of Washington
Wallertz K, Malmqvist C (2013) The effect of mechanical site preparation methods on the establishment of Norway spruce (Picea abies (L.) Karst.) and Douglas fir (*Pseudotsuga menziesii*; (Mirb.) Franco) in southern Sweden. Forestry 86 (1):71–78
Wang BSP (1974) Tree seed storage. Department of Environment, Can For Service Pub No 1335, Ottawa
Wang H (1991) A review of introduction, improvement, silviculture and utilization of eucalypts in China. In: Development of forestry science technology in China. Scientific and Technological Information Center, Ministry of Forestry, Beijing
Warkentin BP (1984) Physical Properties of forest nursery soils: Relation to seedling growth. In: Durya ML, Landis TD (eds) Forest nursery manual. Martinus Nihhoff/Dr W. Junk, The Hague
Watt M, Palmer DJ, Kimberley MO, Hock BK, Payn TW, Lowe DJ (2010) Development of models to predict *Pinus radiata* productivity throughout New Zealand. Canadian J For Res 40:488–499
Wienhold BJ, Andrews SS, Karlen DL (2004) Soil quality: a review of the science and experiences in the USA. Environ Geochem Hlth 26:89–95
Wiersum KF, Ramlan A (1982) Cultivation of *Acacia auriculiformis* on Jaya, Indonesia. Comm Fo Rev 61(2):135–144
Wightman KE, Shear T, Goldfarb B, Haggar J (2001) Nursery and field establishment techniques to improve seedling growth of three Costa Rican hardwoods. New Forests 22:75–96
Williamson JG, Coston DC (1990) Planting method and irrigation rate influence vegetative and reproductive growth of peach planted at high density. J Am Soc Hort Sci 115(2):207–212
Woessner RA, McNabb KL (1979) Large scale production of *Gmelina arborea* Roxb. seed—a case study. Common For Review 58(2)
World Agroforestry Centre (2011) Agroforestry tree database— Pinus caribaea (March 2011). http://www.worldagroforestrycentre.org/sea/Products/AFDbases/af/asp/SpeciesInfo.asp?SpID=1299
Yildiz O, Esen D, Karaoz OM, Sarginci M, Toprak B, Soysal Y (2010) Effects of different site preparation methods on soil carbon and nutrient removal from Eastern beech regeneration sites in Turkey's Black Sea region. Applied Soil Ecology 45 (2010):49–55
Yildiz O, Sarginci M, Esen, D, Cromack Jr K (2007) Effects of vegetation control on nutrient removal and Fagus orientalis, Lipsky regeneration in the Western Black Sea region of Turkey. For Ecol Manage 240:186–194
Zdenko R (2002) Handbook of plant growth: pH as the master variable. Marcel Dekker Incorporated New York
Zhang LC, Guo WM, Chen YS (1981) Experimental investigation on the physiology of after-ripening of Korean pine seed. J Northeast Forestry University, China. No.4:43–50
Zhang LC, Guo WM, Chen YS (1983) Studies on the relationship between peroxidase activity and germination of *Pinus koraiensis* seeds and on metabolic inhibitors in the seeds. Acta Botanical Sinica 25(1):53–61
Zhang X, Acevedo S, Chao Y, Chen Z, Dinoff T, Driver J, Ross J, Williams R, Krieger R (2011) Concurrent 2,4-D and triclopyr biomonitoring of backpack applicators, mixer/loader and field supervisor in forestry. J Environ Sci Health 46:281–293

Index

MIX
Papier aus verantwortungsvollen Quellen
Paper from responsible sources
FSC® C105338

If you have any concerns about our products,
you can contact us on
ProductSafety@springernature.com

In case Publisher is established outside the EU,
the EU authorized representative is:
Springer Nature Customer Service Center GmbH
Europaplatz 3, 69115 Heidelberg, Germany

Printed by Libri Plureos GmbH
in Hamburg, Germany